ANALYSIS AND DESIGN OF ELASTIC BEAMS

To Samantha Jane

CONTENTS

PREFACE

This book treats the analysis and design of beams, with a particular emphasis on computational approaches for thin-walled beams. The underlying formulations are based on the assumption of linear elasticity. Extension, bending, and torsion are discussed. Beams with arbitrary cross sections, loading, and boundary conditions are covered, as well as the determination of displacements, natural frequencies, buckling loads, and the normal and shear stresses due to bending, torsion, direct shear, and restrained warping. The Wiley website (http://www.wiley.com/go/pilkey) provides information on the availability of computer programs that perform the calculations for the formulations of this book.

Most of this book deals with computational methods for finding beam cross-sectional properties and stresses. The computational solutions apply to solid and thin-walled open and closed sections. Some traditional analytical formulas for thin-walled beams are developed here. A systematic and thorough treatment of analytical thin-walled beam theory for both open and closed sections is on the author's website.

The technology essential for the study of a structural system that is modeled by beam elements is provided here. The cross-sectional properties of the individual beams can be computed using the methodology provided in this book. Then, a general-purpose analysis computer program can be applied to the entire structure to compute the forces and moments in the individual members. Finally, the methodology developed here can be used to find the normal and shear stresses on the members' cross sections.

Historically, shear stress-related cross-sectional properties have been difficult to obtain analytically. These properties include the torsional constant, shear deformation coefficients, the warping constant, and the shear stresses themselves. The formulations of this book overcome the problems encountered in the calculation of these properties. Computational techniques permit these properties to be obtained efficiently and accurately. The finite element formulations apply to cross sections with arbitrary shapes, including solid or thin-walled configurations. Thin-walled cross sections can be open or closed. For thin-walled cross sections, it is possible to prepare a computer program of analytical formulas to calculate several of the cross-sectional properties efficiently and with acceptable accuracy.

Shape optimization of beam cross sections is discussed. The cross-sectional shape can be optimized for objectives such as minium weight or an upper bound on the stress level. Essential to the optimization is the proper calculation of the sensitivity of various cross-sectional properties with respect to the design parameters. Standard optimization algorithms, which are readily available in existing software, can be utilized to perform the computations necessary to achieve an optimal design.

For cross-sectional shape optimization, B-splines, in particular, NURBS can be conveniently used to describe the shape. Using NURBS eases the task of adjusting the shape during the optimization process. Such B-spline characteristics as knots and weights are defined.

Computer programs are available to implement the formulations in this book. See the Wiley website. These include programs that can find the internal net forces and moments along a solid or thin-walled beam with an arbitrary cross-sectional shape with any boundary conditions and any applied loads. Another program can be used to find the cross-sectional properties of bars with arbitrary cross-sectional shapes, as well as the cross-sectional stresses. One version of this program is intended to be used as an "engine" in more comprehensive analysis or optimal design software. That is, the program can be integrated into the reader's software in order to perform cross-sectional analyses or cross-sectional shape optimization for beams. The input to this program utilizes NURBS, which helps facilitate the interaction with design packages.

The book begins with an introduction to the theory of linear elasticity and of pure bending of a beam. In Chapter 2 we discuss the development of stiffness and mass matrices for a beam element, including matrices based on differential equations and variational principles. Both exact and approximate matrices are derived, the latter utilizing polynomial trial functions. Static, dynamic, and stability analyses of structural systems are set forth in Chapter 3. Initially, the element structural matrices are assembled to form global matrices. The finite element method is introduced in Chapter 4 and applied to find simple non–shear-related cross-sectional properties of beams. In Chapter 5 we present Saint-Venant torsion, with special attention being paid to the accurate calculation of the torsional constant. Shear stresses generated by shear forces on beams are considered in Chapter 6; these require the relatively difficult calculation of shear deformation coefficients for the cross section. In Chapter 7 we present torsional stress calculations when constrained warping is present. Principal stresses and yield theories are discussed in Chapter 8. In Chapters 9 and 10 we introduce definitions and formulations necessary to enable cross-sectional shape optimization. In Chapter 9 we introduce the concept of B-splines and in Chapter 10 provide formulas for sensitivities of the cross-sectional properties. In the two appendixes we describe some of the computer programs that have been prepared to accompany the book.

The work of Dr. Levent Kitis, my former student and now colleague, has been crucial in the development of this book. Support for some of the research related to the computational implementation of thin-walled cross-sectional properties and stresses was provided by Ford Motor Company, with the guidance of Mark Zebrowski and Victor Borowski. As indicated by the frequent citations to his papers in this book, a major contributor to the theory here has been Dr. Uwe Schramm, who spent several years as a senior researcher at the University of Virginia. Some of the structural matrices in the book were derived by another University of Virginia researcher, Dr. Weise Kang. Most of the text and figures were skillfully crafted by Wei Wei Ding, Timothy Preston, Check Kam, Adam Ziemba, and Rod Shirbacheh. Annie Frazer performed the calculations for several example problems. Instrumental in the preparation of this book has been the help of B. F. Pilkey.

WALTER PILKEY

CHAPTER 1

BEAMS IN BENDING

This book deals with the extension, bending, and torsion of bars, especially thin-walled members. Although computational approaches for the analysis and design of bars are emphasized, traditional analytical solutions are included.

We begin with a study of the bending of beams, starting with a brief review of some of the fundamental concepts of the theory of linear elasticity. The theory of beams in bending is then treated from a strength-of-materials point of view. Both topics are treated more thoroughly in Pilkey and Wunderlich (1994). Atanackovic and Guran (2000), Boresi and Chong (1987), Gould (1994), Love (1944), and Sokolnikoff (1956) contain a full account of the theory of elasticity. References such as these should be consulted for the derivation of theory-of-elasticity relationships that are not derived in this chapter. Gere (2001), Oden and Ripperger (1981), Rivello (1969), and Uugural and Fenster (1981) may be consulted for a detailed development of beam theory.

1.1 REVIEW OF LINEAR ELASTICITY

The equations of elasticity for a three-dimensional body contain 15 unknown functions: six stresses, six strains, and three displacements. These functions satisfy three equations of equilibrium, six strain–displacement relations, and six stress–strain equations.

1.1.1 Kinematical Strain–Displacement Equations

The displacement vector **u** at a point in a solid has the three components $u_x(x, y, z)$, $u_y(x, y, z)$, $u_z(x, y, z)$ which are mutually orthogonal in a Cartesian coordinate system and are taken to be positive in the direction of the positive coordinate axes. In vector notation,

$$\mathbf{u} = \begin{bmatrix} u_x \\ u_y \\ u_z \end{bmatrix} = \begin{bmatrix} u_x & u_y & u_z \end{bmatrix}^{\mathrm{T}} \tag{1.1}$$

Designate the normal strains by ϵ_x, ϵ_y, and ϵ_z and the shear strains are $\gamma_{xy}, \gamma_{xz}, \gamma_{yz}$. The shear strains are symmetric (i.e., $\gamma_{ij} = \gamma_{ji}$). In matrix notation

$$\boldsymbol{\epsilon} = \begin{bmatrix} \epsilon_x \\ \epsilon_y \\ \epsilon_z \\ \gamma_{xy} \\ \gamma_{xz} \\ \gamma_{yz} \end{bmatrix} = \begin{bmatrix} \epsilon_x & \epsilon_y & \epsilon_z & \gamma_{xy} & \gamma_{xz} & \gamma_{yz} \end{bmatrix}^{\mathrm{T}} = \begin{bmatrix} \epsilon_x & \epsilon_y & \epsilon_z & 2\epsilon_{xy} & 2\epsilon_{xz} & 2\epsilon_{yz} \end{bmatrix}^{\mathrm{T}} \tag{1.2}$$

As indicated, $\gamma_{ik} = 2\epsilon_{ik}$, where γ_{ik} is sometimes called the engineering shear strain and ϵ_{ik} the theory of elasticity shear strain.

The linearized strain–displacement relations, which form the *Cauchy strain tensor*, are

$$\begin{aligned} \epsilon_x &= \frac{\partial u_x}{\partial x} & \epsilon_y &= \frac{\partial u_y}{\partial y} & \epsilon_z &= \frac{\partial u_z}{\partial z} \\ \gamma_{xy} &= \frac{\partial u_y}{\partial x} + \frac{\partial u_x}{\partial y} & \gamma_{xz} &= \frac{\partial u_z}{\partial x} + \frac{\partial u_x}{\partial z} & \gamma_{yz} &= \frac{\partial u_z}{\partial y} + \frac{\partial u_y}{\partial z} \end{aligned} \tag{1.3}$$

In matrix form Eq. (1.3) can be written as

$$\begin{bmatrix} \epsilon_x \\ \epsilon_y \\ \epsilon_z \\ \gamma_{xy} \\ \gamma_{xz} \\ \gamma_{yz} \end{bmatrix} = \begin{bmatrix} \partial_x & 0 & 0 \\ 0 & \partial_y & 0 \\ 0 & 0 & \partial_z \\ \partial_y & \partial_x & 0 \\ \partial_z & 0 & \partial_x \\ 0 & \partial_z & \partial_y \end{bmatrix} \begin{bmatrix} u_x \\ u_y \\ u_z \end{bmatrix} \tag{1.4}$$

or

$$\boldsymbol{\epsilon} = \mathbf{D} \quad \mathbf{u}$$

with the differential operator matrix

$$\mathbf{D} = \begin{bmatrix} \partial_x & 0 & 0 \\ 0 & \partial_y & 0 \\ 0 & 0 & \partial_z \\ \partial_y & \partial_x & 0 \\ \partial_z & 0 & \partial_x \\ 0 & \partial_z & \partial_y \end{bmatrix} \tag{1.5}$$

Six strain components are required to characterize the state of strain at a point and are derived from the three displacement functions $u_x,\ u_y,\ u_z$. The displacement field must be continuous and single valued, because it is being assumed that the body remains continuous after deformations have taken place. The six strain–displacement equations will not possess a single-valued solution for the three displacements if the strains are arbitrarily prescribed. Thus, the calculated displacements could possess tears, cracks, gaps, or overlaps, none of which should occur in practice. It appears as though the strains should not be independent and that they should be required to satisfy special conditions. To find relationships between the strains, differentiate the expression for the shear strain γ_{xy} with respect to x and y,

$$\frac{\partial^2 \gamma_{xy}}{\partial x\, \partial y} = \frac{\partial^2}{\partial x\, \partial y}\frac{\partial u_x}{\partial y} + \frac{\partial^2}{\partial x\, \partial y}\frac{\partial u_y}{\partial x} \tag{1.6}$$

According to the calculus, a single-valued continuous function f satisfies the condition

$$\frac{\partial^2 f}{\partial x\, \partial y} = \frac{\partial^2 f}{\partial y\, \partial x} \tag{1.7}$$

With the assistance of Eq. (1.7), Eq. (1.6) may be rewritten, using the strain–displacement relations, as

$$\frac{\partial^2 \gamma_{xy}}{\partial x\, \partial y} = \frac{\partial^2 \epsilon_x}{\partial y^2} + \frac{\partial^2 \epsilon_y}{\partial x^2} \tag{1.8}$$

showing that the three strain components γ_{xy}, ϵ_x, ϵ_y are not independent functions. Similar considerations that eliminate the displacements from the strain–displacement relations lead to five additional relations among the strains. These six relationships,

$$\begin{aligned}
2\frac{\partial^2 \epsilon_x}{\partial y\, \partial z} &= \frac{\partial}{\partial x}\left(-\frac{\partial \gamma_{yz}}{\partial x} + \frac{\partial \gamma_{xz}}{\partial y} + \frac{\partial \gamma_{xy}}{\partial z}\right)\\
2\frac{\partial^2 \epsilon_y}{\partial x\, \partial z} &= \frac{\partial}{\partial y}\left(-\frac{\partial \gamma_{xz}}{\partial y} + \frac{\partial \gamma_{xy}}{\partial z} + \frac{\partial \gamma_{yz}}{\partial x}\right)\\
2\frac{\partial^2 \epsilon_z}{\partial x\, \partial y} &= \frac{\partial}{\partial z}\left(-\frac{\partial \gamma_{xy}}{\partial z} + \frac{\partial \gamma_{yz}}{\partial x} + \frac{\partial \gamma_{xz}}{\partial y}\right)\\
\frac{\partial^2 \gamma_{xy}}{\partial x\, \partial y} &= \frac{\partial^2 \epsilon_x}{\partial y^2} + \frac{\partial^2 \epsilon_y}{\partial x^2}\\
\frac{\partial^2 \gamma_{xz}}{\partial x\, \partial z} &= \frac{\partial^2 \epsilon_x}{\partial z^2} + \frac{\partial^2 \epsilon_z}{\partial x^2}\\
\frac{\partial^2 \gamma_{yz}}{\partial y\, \partial z} &= \frac{\partial^2 \epsilon_z}{\partial y^2} + \frac{\partial^2 \epsilon_y}{\partial z^2}
\end{aligned} \tag{1.9}$$

are known as the strain *compatibility conditions* or *integrability conditions*. Although there are six conditions, only three are independent.

1.1.2 Material Law

The kinematical conditions of Section 1.1.1 are independent of the material of which the body is made. The material is introduced to the formulation through a *material law*, which is a relationship between the stresses $\boldsymbol{\sigma}$ and strains $\boldsymbol{\epsilon}$. Other names are the *constitutive relations* or the *stress–strain equations*.

Figure 1.1 shows the stress components that define the state of stress in a three-dimensional continuum. The quantities σ_x, σ_y, and σ_z designate stress components normal to a coordinate plane and τ_{xy}, τ_{xz}, τ_{yz}, τ_{yx}, τ_{zx}, and τ_{zy} are the shear stress components. In the case of a normal stress, the single subscript indicates that the stress acts on a plane normal to the axis in the subscript direction. For the shear stresses, the first letter of the double subscript denotes that the plane on which the stress acts is normal to the axis in the subscript direction. The second subscript letter designates the coordinate direction in which the stress acts. As a result of the need to satisfy an equilibrium condition of moments, the shear stress components must be symmetric that is,

$$\tau_{xy} = \tau_{yx} \qquad \tau_{xz} = \tau_{zx} \qquad \tau_{yz} = \tau_{zy} \tag{1.10}$$

Then the state of stress at a point is characterized by six components. In matrix form,

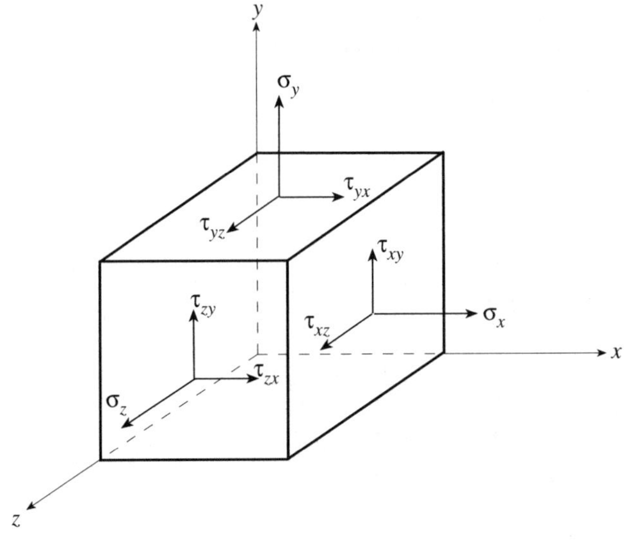

Figure 1.1 Notation for the components of the Cartesian stress tensor.

$$\boldsymbol{\sigma} = \begin{bmatrix} \sigma_x \\ \sigma_y \\ \sigma_z \\ \tau_{xy} \\ \tau_{xz} \\ \tau_{yz} \end{bmatrix} = \begin{bmatrix} \sigma_x & \sigma_y & \sigma_z & \tau_{xy} & \tau_{xz} & \tau_{yz} \end{bmatrix}^{\mathrm{T}} \tag{1.11}$$

For a solid element as shown in Fig. 1.1, a face with its outward normal along the positive direction of a coordinate axis is defined to be a positive face. A face with its normal in the negative coordinate direction is defined as a negative face. Stress (strain) components on a positive face are positive when acting along a positive coordinate direction. The components shown in Fig. 1.1 are positive. Components on a negative face acting in the negative coordinate direction are defined to be positive.

An *isotropic* material has the same material properties in all directions. If the properties differ in various directions, such as with wood, the material is said to be *anisotropic*. A material is *homogeneous* if it has the same properties at every point. Wood is an example of a homogeneous material that can be anisotropic. A body formed of steel and aluminum portions is an example of a material that is inhomogeneous, but each portion is isotropic.

The stress–strain equations for linearly elastic isotropic materials are

$$\begin{aligned} \epsilon_x &= \frac{\sigma_x}{E} - \frac{\nu}{E}(\sigma_y + \sigma_z) \\ \epsilon_y &= \frac{\sigma_y}{E} - \frac{\nu}{E}(\sigma_x + \sigma_z) \\ \epsilon_z &= \frac{\sigma_z}{E} - \frac{\nu}{E}(\sigma_x + \sigma_y) \\ \gamma_{xy} &= \frac{\tau_{xy}}{G} \\ \gamma_{xz} &= \frac{\tau_{xz}}{G} \\ \gamma_{yz} &= \frac{\tau_{yz}}{G} \end{aligned} \tag{1.12}$$

where E is the *elastic* or *Young's modulus*, ν is *Poisson's ratio*, and G is the *shear modulus*. Only two of these three material properties are independent. The shear modulus is given in terms of E and ν as

$$G = \frac{E}{2(1+\nu)} \tag{1.13}$$

$$
\begin{bmatrix} \epsilon_x \\ \epsilon_y \\ \epsilon_z \\ \cdots \\ \gamma_{xy} \\ \gamma_{xz} \\ \gamma_{yz} \end{bmatrix} = \frac{1}{E} \left[\begin{array}{ccc|ccc} 1 & -\nu & -\nu & & & \\ -\nu & 1 & -\nu & & \mathbf{0} & \\ -\nu & -\nu & 1 & & & \\ \cdots & \cdots & \cdots & \cdots & \cdots & \cdots \\ & & & 2(1+\nu) & 0 & 0 \\ & \mathbf{0} & & 0 & 2(1+\nu) & 0 \\ & & & 0 & 0 & 2(1+\nu) \end{array} \right] \begin{bmatrix} \sigma_x \\ \sigma_y \\ \sigma_z \\ \cdots \\ \tau_{xy} \\ \tau_{xz} \\ \tau_{yz} \end{bmatrix}
$$

$$
\boldsymbol{\epsilon} = \mathbf{E}^{-1} \boldsymbol{\sigma} \tag{1.14}
$$

Stresses may be written as a function of the strains by inverting the six relationships of Eq. (1.12) that express strains in terms of stresses. The result is

$$
\begin{aligned}
\sigma_x &= \lambda e + 2G\epsilon_x \\
\sigma_y &= \lambda e + 2G\epsilon_y \\
\sigma_z &= \lambda e + 2G\epsilon_z \\
\tau_{xy} = G\gamma_{xy} \qquad & \tau_{xz} = G\gamma_{xz} \qquad \tau_{yz} = G\gamma_{yz}
\end{aligned} \tag{1.15}
$$

where e is the change in volume per unit volume, also called the *dilatation*,

$$
e = \epsilon_x + \epsilon_y + \epsilon_z \tag{1.16}
$$

and λ is *Lamé's constant*,

$$
\lambda = \frac{\nu E}{(1+\nu)(1-2\nu)} \tag{1.17}
$$

The matrix form appears as

$$
\begin{bmatrix} \sigma_x \\ \sigma_y \\ \sigma_z \\ \cdots \\ \tau_{xy} \\ \tau_{xz} \\ \tau_{yz} \end{bmatrix} = \frac{E}{(1+\nu)(1-2\nu)} \left[\begin{array}{ccc|ccc} 1-\nu & \nu & \nu & & & \\ \nu & 1-\nu & \nu & & \mathbf{0} & \\ \nu & \nu & 1-\nu & & & \\ \cdots & \cdots & \cdots & & & \\ & & & \frac{1-2\nu}{2} & 0 & 0 \\ & \mathbf{0} & & 0 & \frac{1-2\nu}{2} & 0 \\ & & & 0 & 0 & \frac{(1-2\nu)}{2} \end{array} \right] \begin{bmatrix} \epsilon_x \\ \epsilon_y \\ \epsilon_z \\ \cdots \\ \gamma_{xy} \\ \gamma_{xz} \\ \gamma_{yz} \end{bmatrix}
$$

$$
\boldsymbol{\sigma} = \mathbf{E} \boldsymbol{\epsilon} \tag{1.18}
$$

For uniaxial tension, with the normal stress in the x direction given a constant positive value σ_0, and all other stresses set equal to zero,

$$\sigma_x = \sigma_0 > 0 \qquad \sigma_y = \sigma_z = \tau_{xy} = \tau_{yz} = \tau_{xz} = 0 \tag{1.19a}$$

The normal strains are given by Hooke's law as

$$\epsilon_x = \frac{\sigma_0}{E} \qquad \epsilon_y = \epsilon_z = -\frac{\nu\sigma_0}{E} \tag{1.19b}$$

and the shear strains are all zero. Under this loading condition, the material undergoes extension in the axial direction x and contraction in the transverse directions y and z. This shows that the material constants ν and E are both positive:

$$E > 0 \qquad \nu > 0 \tag{1.20}$$

In hydrostatic compression p, the material is subjected to identical compressive stresses in all three coordinate directions:

$$\sigma_x = \sigma_y = \sigma_z = -p \qquad p > 0 \tag{1.21}$$

while all shear stresses are zero. The dilatation under this loading condition is

$$e = -\frac{3p}{3\lambda + 2G} = -\frac{3p(1 - 2\nu)}{E} \tag{1.22}$$

Since the volume change in hydrostatic compression is negative, this expression for e implies that Poisson's ratio must be less than $\frac{1}{2}$:

$$\nu < \tfrac{1}{2} \tag{1.23}$$

and the following properties of the elastic constants are established:

$$E > 0 \qquad G > 0 \qquad \lambda > 0 \qquad 0 < \nu < \tfrac{1}{2} \tag{1.24}$$

Materials for which $\nu \approx 0$ and $\nu \approx \frac{1}{2}$ are very compressible or very incompressible, respectively. Cork is an example of a very compressible material, whereas rubber is very incompressible.

1.1.3 Equations of Equilibrium

Equilibrium at a point in a solid is characterized by a relationship between internal (volume or body) forces $\overline{p}_{Vx}, \overline{p}_{Vy}, \overline{p}_{Vz}$, such as those generated by gravity or acceleration, and differential equations involving stress. Prescribed forces are designated with a bar placed over a letter. These *equilibrium* or *static* relations appear as

$$\frac{\partial \sigma_x}{\partial x} + \frac{\partial \tau_{xy}}{\partial y} + \frac{\partial \tau_{xz}}{\partial z} + \overline{p}_{Vx} = 0$$

$$\frac{\partial \tau_{xy}}{\partial x} + \frac{\partial \sigma_y}{\partial y} + \frac{\partial \tau_{yz}}{\partial z} + \overline{p}_{Vy} = 0 \tag{1.25}$$

$$\frac{\partial \tau_{xz}}{\partial x} + \frac{\partial \tau_{yz}}{\partial y} + \frac{\partial \sigma_z}{\partial z} + \overline{p}_{Vz} = 0$$

where $\overline{p}_{Vx}, \overline{p}_{Vy}, \overline{p}_{Vz}$ are the body forces per unit volume. In matrix form,

$$\underbrace{\begin{bmatrix} \partial_x & 0 & 0 & \vdots & \partial_y & \partial_z & 0 \\ 0 & \partial_y & 0 & \vdots & \partial_x & 0 & \partial_z \\ 0 & 0 & \partial_z & \vdots & 0 & \partial_x & \partial_y \end{bmatrix}}_{\mathbf{D}^{\mathrm{T}}} \underbrace{\begin{bmatrix} \sigma_x \\ \sigma_y \\ \sigma_z \\ \cdots \\ \tau_{xy} \\ \tau_{xz} \\ \tau_{yz} \end{bmatrix}}_{\boldsymbol{\sigma}} + \underbrace{\begin{bmatrix} \overline{p}_{Vx} \\ \overline{p}_{Vy} \\ \overline{p}_{Vz} \end{bmatrix}}_{\overline{\mathbf{p}}_V} = \underbrace{\begin{bmatrix} 0 \\ 0 \\ 0 \end{bmatrix}}_{\mathbf{0}} \tag{1.26}$$

where the matrix of differential operators $\mathbf{D}^{\mathrm{T}}$ is the transpose of the $\mathbf{D}$ of Eq. (1.5). These relationships are derived in books dealing with the theory of elasticity and, also, in many basic strength-of-materials textbooks.

1.1.4 Surface Forces and Boundary Conditions

The forces applied to a surface (i.e., the boundary) of a body must be in equilibrium with the stress components on the surface. Let S_p denote the part of the surface of the body on which forces are prescribed, and let displacements be specified on the remaining surface S_u. The surface conditions on S_p are

$$\begin{aligned} p_x &= \sigma_x n_x + \tau_{xy} n_y + \tau_{xz} n_z \\ p_y &= \tau_{xy} n_x + \sigma_y n_y + \tau_{yz} n_z \\ p_z &= \tau_{xz} n_x + \tau_{yz} n_y + \sigma_z n_z \end{aligned} \tag{1.27}$$

where n_x, n_y, n_z are the components of the unit vector $\mathbf{n}$ normal to the surface and p_x, p_y, p_z are the surface forces per unit area.

In matrix form,

$$\underbrace{\begin{bmatrix} p_x \\ p_y \\ p_z \end{bmatrix}}_{\mathbf{p}} = \underbrace{\begin{bmatrix} n_x & 0 & 0 & \vdots & n_y & n_z & 0 \\ 0 & n_y & 0 & \vdots & n_x & 0 & n_z \\ 0 & 0 & n_z & \vdots & 0 & n_x & n_y \end{bmatrix}}_{\mathbf{N}^{\mathrm{T}}} \underbrace{\begin{bmatrix} \sigma_x \\ \sigma_y \\ \sigma_z \\ \cdots \\ \tau_{xy} \\ \tau_{xz} \\ \tau_{yz} \end{bmatrix}}_{\boldsymbol{\sigma}} \tag{1.28}$$

Note that $\mathbf{N}^{\mathrm{T}}$ is similar in form to $\mathbf{D}^{\mathrm{T}}$ of Eq. (1.26) in that the components of $\mathbf{N}^{\mathrm{T}}$ correspond to the derivatives of $\mathbf{D}^{\mathrm{T}}$. The relations of Eq. (1.27) are referred to as *Cauchy's formula*.

Surface forces (per unit area) $\mathbf{p}$ applied externally are called prescribed surface *tractions* $\overline{\mathbf{p}}$. Equilibrium demands that the resultant stress be equal to the applied surface tractions $\overline{\mathbf{p}}$ on S_p:

$$\mathbf{p} = \overline{\mathbf{p}} \qquad \text{on } S_p \tag{1.29}$$

These are the *static (force, stress*, or *mechanical) boundary conditions.* Continuity requires that on the surface S_u, the displacements $\mathbf{u}$ be equal to the specified displacements $\overline{\mathbf{u}}$:

$$\mathbf{u} = \overline{\mathbf{u}} \qquad \text{on } S_u \tag{1.30}$$

These are the *displacement* (*kinematic*) *boundary conditions.*

Unit Vectors on a Boundary Curve It is helpful to identify several useful relationships between vectors on a boundary curve. Consider a boundary curve lying in the yz plane as shown in Fig. 1.2a. The vector $\mathbf{n}$ is the unit outward normal $\mathbf{n} = n_y\mathbf{j} + n_z\mathbf{k}$ and $\mathbf{t}$ is the unit tangent vector $\mathbf{t} = t_y\mathbf{j} + t_z\mathbf{k}$, where $\mathbf{j}$ and $\mathbf{k}$ are unit vectors along the y and z axes. The quantity s, the coordinate along the arc of the boundary, is chosen to increase in the counterclockwise sense. As shown in Fig. 1.2a, the unit tangent vector $\mathbf{t}$ is directed along increasing s. Since $\mathbf{n}$ and $\mathbf{t}$ are unit vectors, $n_y^2 + n_z^2 = 1$ and $t_y^2 + t_z^2 = 1$. The components of $\mathbf{n}$ are its direction cosines, that is, from Fig. 1.2b,

$$n_y = \cos\theta_y \quad \text{and} \quad n_z = \cos\theta_z \tag{1.31}$$

since, for example, $\cos\theta_y = n_y/\sqrt{n_y^2 + n_z^2} = n_y$.

From Fig. 1.2c it can be observed that

$$\begin{aligned} \cos\varphi &= n_y \qquad & \sin\varphi &= n_z \\ \sin\varphi &= -t_y \qquad & \cos\varphi &= t_z \end{aligned} \tag{1.32}$$

As a consequence,

$$n_y = t_z \qquad n_z = -t_y \tag{1.33}$$

and the unit outward normal is defined in terms of the components t_y and t_z of the unit tangent as

$$\mathbf{n} = t_z\mathbf{j} - t_y\mathbf{k} = \mathbf{t} \times \mathbf{i} \tag{1.34}$$

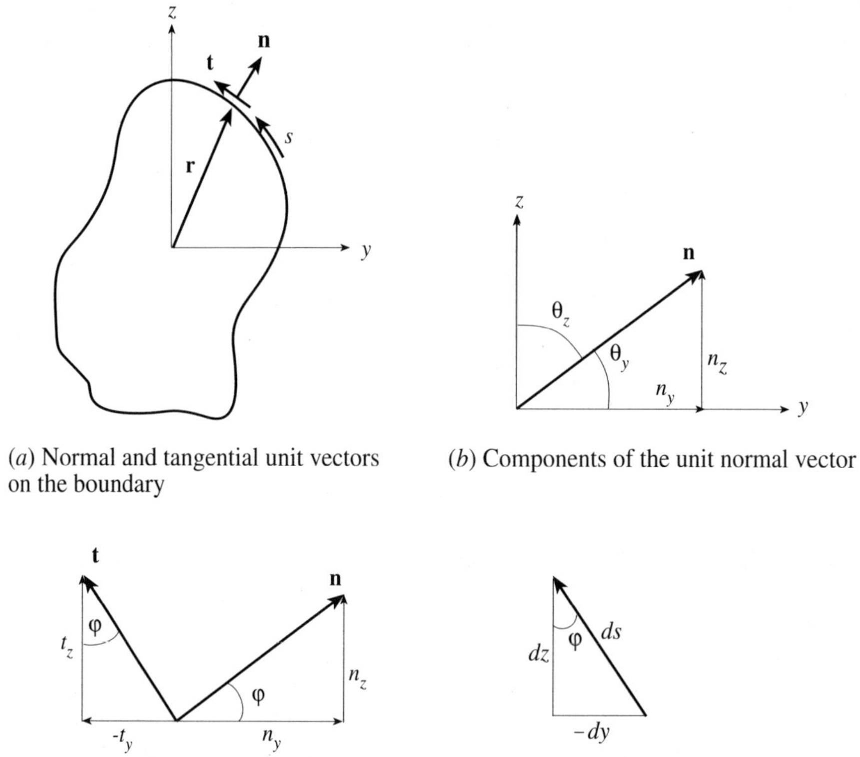

(*a*) Normal and tangential unit vectors on the boundary

(*b*) Components of the unit normal vector

(*c*) Unit normal and tangential vectors

(*d*) Differential components

Figure 1.2 Geometry of the unit normal and tangential vectors.

From Fig. 1.2d it is apparent that

$$\sin\varphi = -\frac{dy}{ds} \quad \text{and} \quad \cos\varphi = \frac{dz}{ds} \tag{1.35}$$

Thus,

$$n_y = t_z = \frac{dz}{ds} \qquad n_z = -t_y = -\frac{dy}{ds} \tag{1.36}$$

The vector $\mathbf{r}$ to any point on the boundary is

$$\mathbf{r} = y\mathbf{j} + z\mathbf{k}$$

Then

$$d\mathbf{r} = dy\,\mathbf{j} + dz\,\mathbf{k} = \frac{d\mathbf{r}}{ds}\,ds = \left(\frac{dy}{ds}\mathbf{j} + \frac{dz}{ds}\mathbf{k}\right)\,ds = \mathbf{t}\,ds \tag{1.37}$$

1.1.5 Other Forms of the Governing Differential Equations

The general problem of the theory of elasticity is to calculate the stresses, strains, and displacements throughout a solid. The kinematic equations $\boldsymbol{\epsilon} = \mathbf{Du}$ (Eq. 1.4) are written in terms of six strains and three displacements, while the static equations $\mathbf{D}^T\boldsymbol{\sigma} + \overline{\mathbf{p}}_V = \mathbf{0}$ (Eq. 1.26) are expressed as functions of the six stress components. The constitutive equations $\boldsymbol{\sigma} = \mathbf{E}\boldsymbol{\epsilon}$ (Eq. 1.18) are relations between the stresses and strains. The boundary conditions of Eqs. (1.29) and (1.30) need to be satisfied by the solution for the 15 unknowns.

In terms of achieving solutions, it is useful to derive alternative forms of the governing equations. The elasticity problem can be formulated in terms of the displacement functions u_x, u_y, u_z. The stress–strain equations allow the equilibrium equations to be written in terms of the strains. When the strains are replaced in the resulting equations by the expressions given by the strain–displacement relations, the equilibrium equations become a set of partial differential equations for the displacements. Thus, substitute $\boldsymbol{\epsilon} = \mathbf{Du}$ into $\boldsymbol{\sigma} = \mathbf{E}\boldsymbol{\epsilon}$ to give the stress–displacement relations $\boldsymbol{\sigma} = \mathbf{EDu}$. The conditions of equilibrium become

$$\mathbf{D}^T\boldsymbol{\sigma} + \overline{\mathbf{p}}_V = \mathbf{D}^T\mathbf{EDu} + \overline{\mathbf{p}}_V = 0 \tag{1.38}$$

or, in scalar form,

$$\begin{aligned}
(\lambda + G)\frac{\partial e}{\partial x} + G\nabla^2 u_x + \overline{p}_{Vx} &= 0 \\
(\lambda + G)\frac{\partial e}{\partial y} + G\nabla^2 u_y + \overline{p}_{Vy} &= 0 \\
(\lambda + G)\frac{\partial e}{\partial z} + G\nabla^2 u_z + \overline{p}_{Vz} &= 0
\end{aligned} \tag{1.39}$$

where ∇^2 is the Laplacian operator

$$\nabla^2 = \frac{\partial^2}{\partial x^2} + \frac{\partial^2}{\partial y^2} + \frac{\partial^2}{\partial z^2} \tag{1.40}$$

The dilatation e is a function of displacements

$$e = \frac{\partial u_x}{\partial x} + \frac{\partial u_y}{\partial y} + \frac{\partial u_z}{\partial z} = \boldsymbol{\nabla} \cdot \mathbf{u} \tag{1.41}$$

where $\mathbf{u}$ is the displacement vector, whose components along the x, y, z axes are u_x, u_y, u_z, and $\boldsymbol{\nabla}$ is the gradient operator. The displacement vector is expressed as $\mathbf{u} = u_x\mathbf{i} + u_y\mathbf{j} + u_z\mathbf{k}$, where $\mathbf{i}$, $\mathbf{j}$, $\mathbf{k}$ are the unit base vectors along the coordinates x, y, z, respectively. The gradient operator appears as

$$\boldsymbol{\nabla} = \mathbf{i}\frac{\partial}{\partial x} + \mathbf{j}\frac{\partial}{\partial y} + \mathbf{k}\frac{\partial}{\partial z} \tag{1.42}$$

To complete the displacement formulation, the surface conditions on S_p must also be written in terms of the displacements. This is done by first writing these surface conditions of Eq. (1.27) in terms of strains using the material laws, and then expressing the strains in terms of the displacements, using the strain–displacement relations. The resulting conditions are

$$\begin{aligned}
\lambda e n_x + G\mathbf{n}\cdot\boldsymbol{\nabla} u_x + G\mathbf{n}\cdot\frac{\partial \mathbf{u}}{\partial x} &= p_x \\
\lambda e n_y + G\mathbf{n}\cdot\boldsymbol{\nabla} u_y + G\mathbf{n}\cdot\frac{\partial \mathbf{u}}{\partial y} &= p_y \\
\lambda e n_z + G\mathbf{n}\cdot\boldsymbol{\nabla} u_z + G\mathbf{n}\cdot\frac{\partial \mathbf{u}}{\partial z} &= p_z
\end{aligned} \tag{1.43}$$

where $\mathbf{n} = n_x\mathbf{i} + n_y\mathbf{j} + n_z\mathbf{k}$. If boundary conditions exist for both S_p and S_u, the boundary value problem is called *mixed*. The equations of equilibrium written in terms of the displacements together with boundary conditions on S_p and S_u constitute the displacement formulation of the elasticity problem. In this formulation, the displacement functions are found first. The strain–displacement relations then give the strains, and the material laws give the stresses.

1.2 BENDING STRESSES IN A BEAM IN PURE BENDING

A beam is said to be in *pure bending* if the force–couple equivalent of the stresses over any cross section is a couple $\mathbf{M}$ in the plane of the section

$$\mathbf{M} = M_y\mathbf{j} + M_z\mathbf{k} \tag{1.44}$$

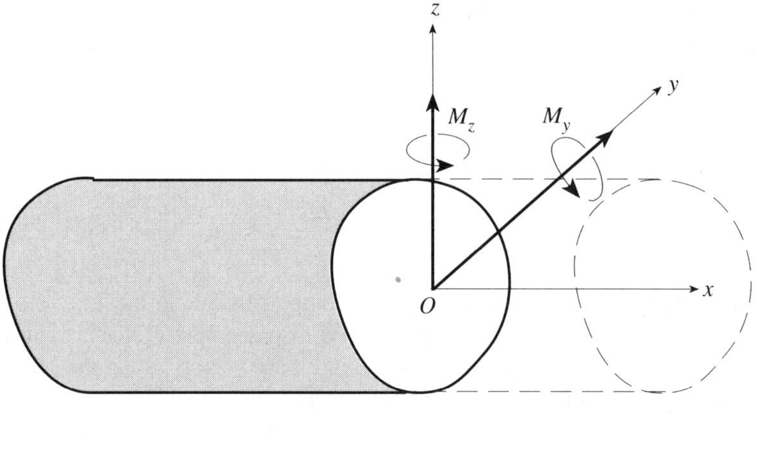

Figure 1.3 Beam in pure bending.

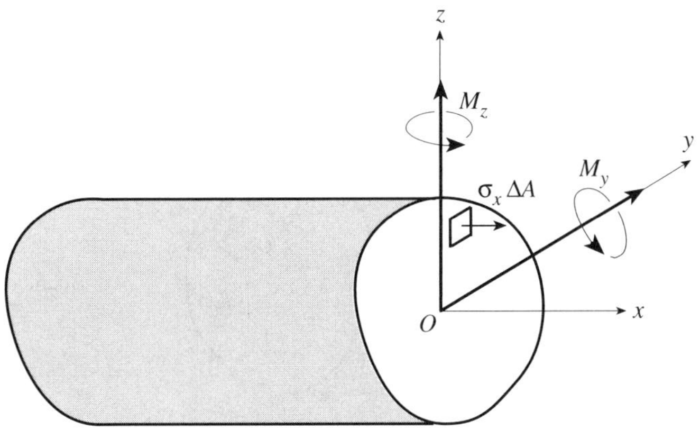

Figure 1.4 Stress resultants on a beam cross section.

where **j**, **k** are the unit vectors parallel to the y, z axes, and the x axis is the beam axis, as shown in Fig. 1.3. In terms of the stress σ_x, the bending moments may be calculated as stress resultants by summing the moments about the origin O of the axes (Fig. 1.4)

$$M_y = \int z\sigma_x \, dA \qquad M_z = -\int y\sigma_x \, dA \tag{1.45}$$

The point about which the moments are taken is arbitrary because the moment of a couple has the same value about any point.

Since in pure bending there is no axial stress resultant,

$$\int \sigma_x \, dA = 0 \tag{1.46}$$

According to the Bernoulli–Euler theory of bending, the cross-sectional planes of the beam remain plane and normal to the beam axis as it deforms. Choose the x axis (i.e. the beam axis) such that it passes through a reference point with coordinates $(x, 0, 0)$. This point is designated by O in Fig. 1.5. The axial displacement u_x of a point on the cross section with coordinates $(x,\ y,\ z)$ can be expressed in terms of the rotations of the cross section about the y and z axes and the axial displacement $u(x)$ of the reference point (Fig. 1.5). Thus

$$u_x(x,\ y,\ z) = u(x) + z\theta_y(x) - y\theta_z(x) \tag{1.47}$$

where θ_y, θ_z are the angles of rotation of the section about the y, z axes. Thus, the displacement u_x at a point on the cross section has been expressed in terms of the beam axis variables u, θ_y, and θ_z. Note that the quantities u, θ_y, and θ_z do not vary over a particular cross section. The terms $z\theta_y$ and $y\theta_z$ vary linearly. Figure 1.6 shows

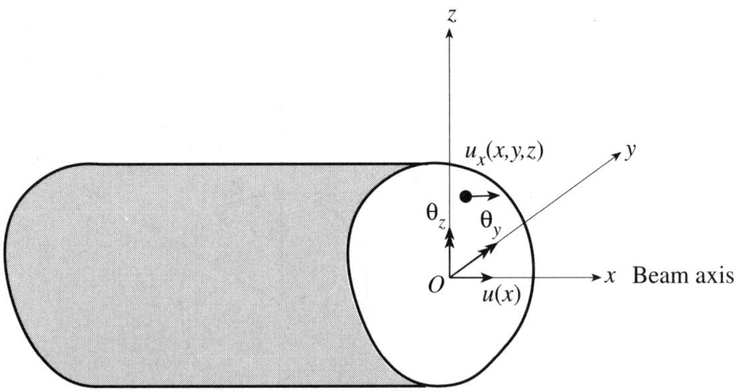

Figure 1.5 Axial displacement. During bending the cross-sectional plane remains plane and normal to the beam axis.

the displacement of a point P of the section with respect to point O as a result of a rotation about the y axis.

The axial strain at the point (x, y, z) is found from the strain–displacement equation (Eq. 1.3)

$$\epsilon_x = \frac{\partial u_x}{\partial x} = \kappa_\epsilon + \kappa_y z - \kappa_z y \tag{1.48}$$

where

$$\kappa_\epsilon = \frac{du}{dx} \qquad \kappa_y = \frac{d\theta_y}{dx} \qquad \kappa_z = \frac{d\theta_z}{dx}$$

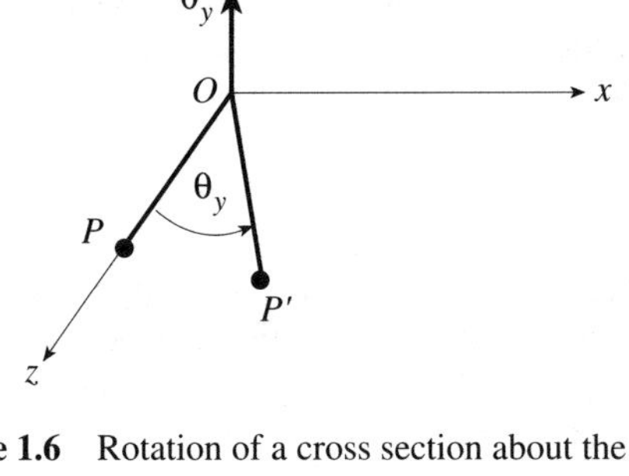

Figure 1.6 Rotation of a cross section about the y axis.

At a given cross section, $\kappa_\epsilon, \kappa_y, \kappa_z$ are constants, so that the normal strain distribution over the section is linear in y and z.

In pure bending, the only nonzero stress is assumed to be the normal stress σ_x, which is given by the material law for linearly elastic isotropic materials as $\sigma_x = E\epsilon_x$, so that

$$\sigma_x = E(\kappa_\epsilon + \kappa_y z - \kappa_z y) \tag{1.49}$$

For a nonhomogeneous beam, the elastic modulus takes on different values over different parts of the section, making E a function of position:

$$E = E(y, z)$$

The stress distribution at a given cross section is then expressed as

$$\sigma_x(y, z) = E(y, z)(\kappa_\epsilon + \kappa_y z - \kappa_z y) \tag{1.50}$$

The stress distribution is statically equivalent to the couple at the section so that the total axial force calculated as a stress resultant is zero and the moments are equal to the bending moments at the section. Thus, from Eqs. (1.45), (1.46), and (1.50),

$$\begin{aligned}
\int \sigma_x \, dA &= \kappa_\epsilon \int E \, dA + \kappa_y \int zE \, dA - \kappa_z \int yE \, dA = 0 \\
\int z\sigma_x \, dA &= \kappa_\epsilon \int zE \, dA + \kappa_y \int z^2 E \, dA - \kappa_z \int yzE \, dA = M_y \\
\int y\sigma_x \, dA &= \kappa_\epsilon \int yE \, dA + \kappa_y \int yzE \, dA - \kappa_z \int y^2 E \, dA = -M_z
\end{aligned} \tag{1.51}$$

Define geometric properties of the cross section as

$$Q_y = \int z \, dA \qquad Q_z = \int y \, dA \tag{1.52a}$$

$$I_y = \int z^2 \, dA \qquad I_z = \int y^2 \, dA \qquad I_{yz} = \int yz \, dA \tag{1.52b}$$

where Q_y and Q_z are first moments of the cross-sectional area and I_y, I_z, and I_{yz} are the second moments of a plane area or the area moments of inertia. Place the definitions of Eqs. (1.52) in Eq. (1.51):

$$\begin{aligned}
\kappa_\epsilon A + \kappa_y Q_y - \kappa_z Q_z &= 0 \\
\kappa_\epsilon Q_y + \kappa_y I_y - \kappa_z I_{yz} &= \frac{M_y}{E} \\
\kappa_\epsilon Q_z + \kappa_y I_{yz} - \kappa_z I_z &= -\frac{M_z}{E}
\end{aligned} \tag{1.53}$$

where the elastic modulus E is assumed to be constant for the cross section.

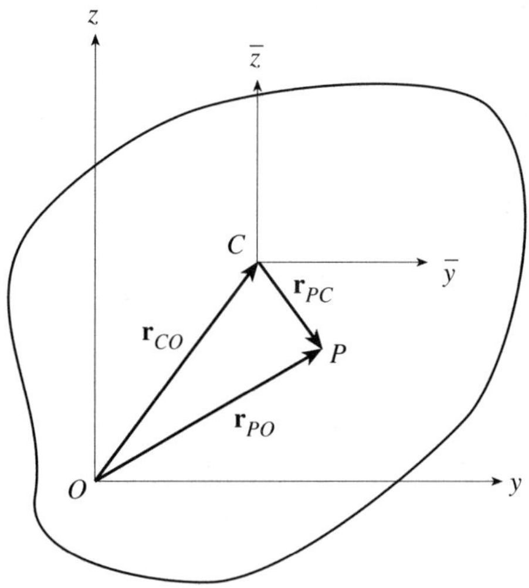

Figure 1.7 Translation of the origin O to the centroid C.

The three simultaneous relations of Eq. (1.53) for the constants κ_ϵ, κ_y, and κ_z become simpler if advantage is taken of the arbitrariness of the choice of origin O. Let a new coordinate system with origin C be defined as shown in Fig. 1.7 by a translation of axes. Figure 1.7 shows that the coordinate transformation equation for any point P of the section is $\mathbf{r}_{PO} = \mathbf{r}_{PC} + \mathbf{r}_{CO}$, or

$$\mathbf{r}_{PC} = \mathbf{r}_{PO} - \mathbf{r}_{CO} \tag{1.54a}$$

The components of this vector equation are

$$\overline{y} = y - y_C \qquad \overline{z} = z - z_C \tag{1.54b}$$

where y and z are the coordinates of P relative to the y, z coordinate axes and y_C and z_C are the coordinates of C relative to the y, z coordinate axes. Choose the origin C such that the first moments of area in the coordinate system $C\overline{y}\,\overline{z}$ are zero:

$$\begin{aligned} Q_{\overline{y}} &= \int \overline{z}\, dA = \int (z - z_C)\, dA = 0 \\ Q_{\overline{z}} &= \int \overline{y}\, dA = \int (y - y_C)\, dA = 0 \end{aligned} \tag{1.55}$$

With the definitions of Eq. (1.52a), these conditions yield the familiar geometric *centroid* of the cross section:

$$y_C = \frac{Q_z}{A} \qquad z_C = \frac{Q_y}{A} \tag{1.56}$$

Transform Eq. (1.53) to the centroidal coordinate system by assuming that Q_y, Q_z, I_y, I_z, and I_{yz} are measured from the centroidal coordinates. Since $Q_{\overline{y}}$ and $Q_{\overline{z}}$ are equal to zero (Eq. 1.55), Eq. (1.53) reduces to

$$\overline{\kappa}_\epsilon A = 0$$

$$\overline{\kappa}_y I_{\overline{y}} - \overline{\kappa}_z I_{\overline{y}\,\overline{z}} = \frac{M_y}{E}$$

$$\overline{\kappa}_y I_{\overline{y}\,\overline{z}} - \overline{\kappa}_z I_{\overline{z}} = -\frac{M_z}{E}$$

where $I_{\overline{y}}$, $I_{\overline{z}}$, and $I_{\overline{yz}}$ are the moments of inertia about the $\overline{y}$, $\overline{z}$ centroidal axes. Solve these expressions for $\overline{\kappa}_\epsilon$, $\overline{\kappa}_y$, and $\overline{\kappa}_z$, and substitute the results into σ_x of Eq. (1.49) expressed in terms of centroidal coordinates [i.e., $\sigma_x = E(\overline{\kappa}_\epsilon + \overline{\kappa}_y \overline{z} - \overline{\kappa}_z \overline{y})$]. This leads to an expression for the normal stress:

$$\sigma_x = -\frac{I_{\overline{y}\,\overline{z}} M_y + I_{\overline{y}} M_z}{I_{\overline{y}} I_{\overline{z}} - I^2_{\overline{y}\,\overline{z}}} \overline{y} + \frac{I_{\overline{z}} M_y + I_{\overline{y}\,\overline{z}} M_z}{I_{\overline{y}} I_{\overline{z}} - I^2_{\overline{y}\,\overline{z}}} \overline{z} \tag{1.57}$$

The *neutral axis* is defined as the line on the cross section for which the normal stress σ_x is zero. This axis is the line of intersection of the neutral surface, which passes through the centroid of the section, and the cross-sectional plane. By equating Eq. (1.57) to zero, we find that the neutral axis is a straight line defined by

$$-(I_{\overline{y}\,\overline{z}} M_y + I_{\overline{y}} M_z)\overline{y} + (I_{\overline{z}} M_y + I_{\overline{y}\,\overline{z}} M_z)\overline{z} = 0$$

or

$$\overline{y} = \frac{I_{\overline{z}} M_y + I_{\overline{y}\,\overline{z}} M_z}{I_{\overline{y}\,\overline{z}} M_y + I_{\overline{y}} M_z} \overline{z} \tag{1.58}$$

If $M_z = 0$, Eq. (1.57) reduces to

$$\sigma_x = M_y \frac{I_{\overline{z}} \overline{z} - I_{\overline{y}\,\overline{z}} \overline{y}}{I_{\overline{y}} I_{\overline{z}} - I^2_{\overline{y}\,\overline{z}}} \tag{1.59}$$

The centroidal coordinates can be located using Eq. (1.56). Sometimes it is convenient to calculate the area moment of inertia first about a judiciously selected coordinate system and then transform them to the centroidal coordinate system. The calculation for $I_{\overline{z}}$, for instance, is

$$\begin{aligned} I_{\overline{z}} &= \int \overline{y}^2 \, dA = \int (y - y_C)^2 \, dA \\ &= \int (y^2 - 2yy_C + y_C^2) \, dA \\ &= I_z - y_C^2 A \end{aligned}$$

From Eq. (1.55), the integral $\int y\, dA$ in this expression is equal to $\int y_C\, dA$. This is one of *Huygens's* or *Steiner's laws* and is referred to as a *parallel axis theorem*. The complete set of equations is

$$\begin{aligned} I_{\overline{y}} &= I_y - z_C^2 A \\ I_{\overline{z}} &= I_z - y_C^2 A \\ I_{\overline{y}\,\overline{z}} &= I_{yz} - y_C z_C A \end{aligned} \tag{1.60}$$

Here I_y, I_z, I_{yz} and $I_{\overline{y}}$, $I_{\overline{z}}$, $I_{\overline{yz}}$ are the moments of inertia about the y, z and $\overline{y}, \overline{z}$ (centroidal) axes, respectively.

Example 1.1 Thin-Walled Cantilevered Beam with an Asymmetrical Cross Section. Find the normal stress distribution for the cantilevered angle shown in Fig. 1.8. The beam is fixed at one end and loaded with a vertical concentrated force $\overline{P}$ at the other end.

SOLUTION. The centroid for this asymmetrical section is found to be located as shown in Fig. 1.9a. Assume that the thickness t is much smaller than the dimension a. Then the moments of inertia can be calculated from Eq. (1.52b) as

$$I_{\overline{y}} = \int \overline{z}^2\, dA = \int_{-4a/3}^{2a/3-t/2} \int_{-a/6-t/2}^{-a/6+t/2} \overline{z}^2\, d\overline{y}\, d\overline{z}$$

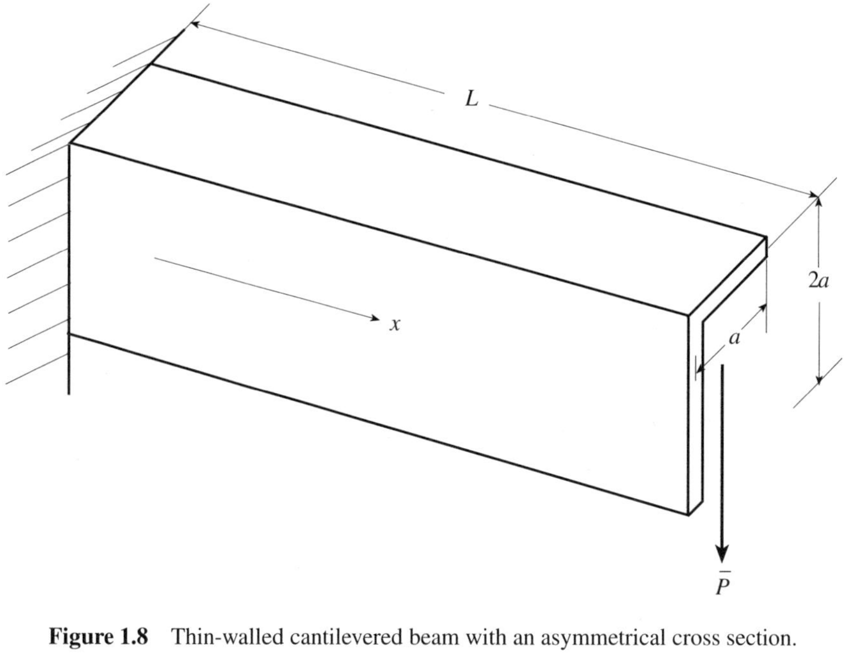

Figure 1.8 Thin-walled cantilevered beam with an asymmetrical cross section.

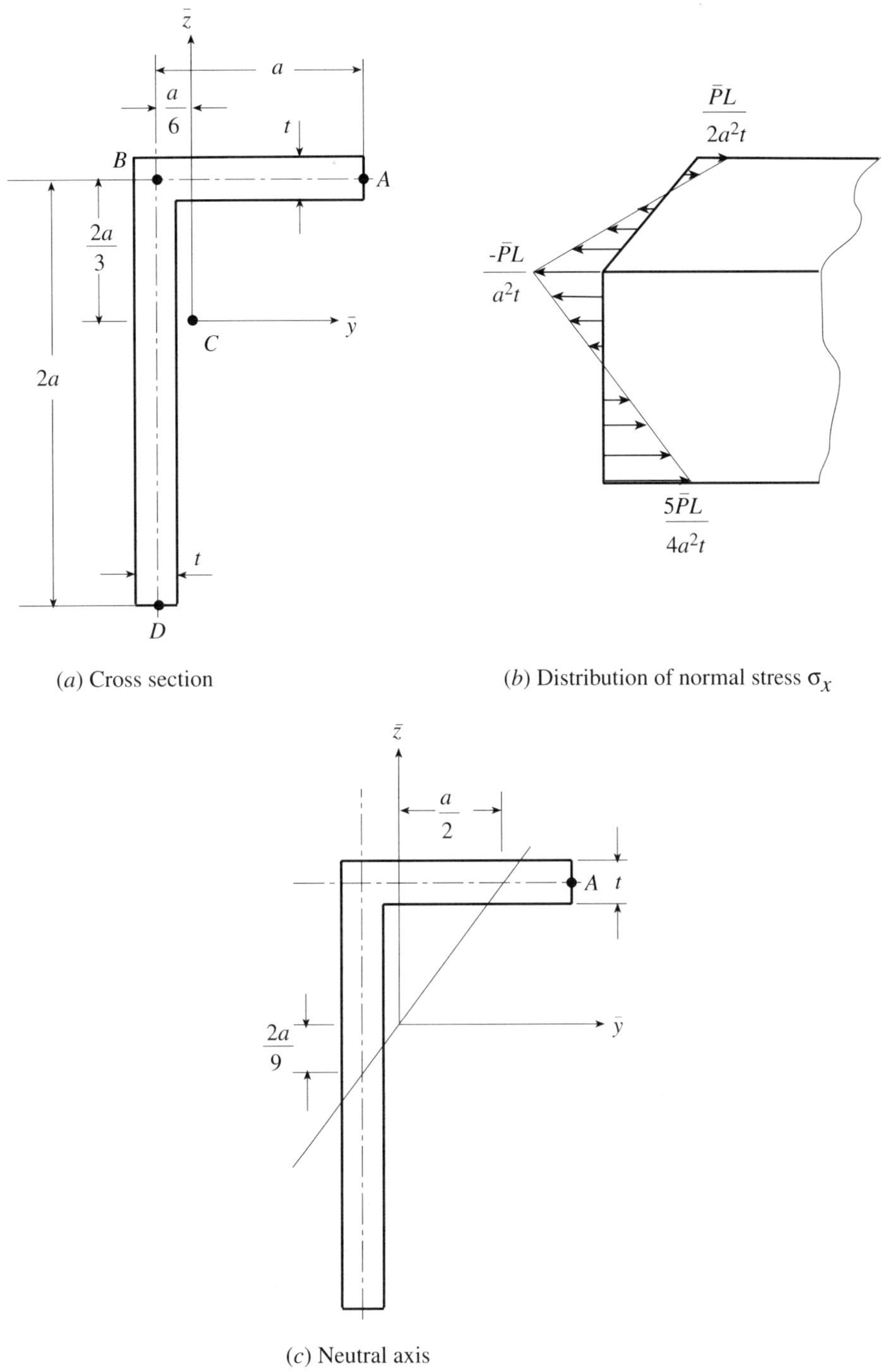

(*a*) Cross section

(*b*) Distribution of normal stress σ_x

(*c*) Neutral axis

Figure 1.9 Normal stress distribution of the thin-walled beam of Example 1.1.

$$+\int_{2a/3-t/2}^{2a/3+t/2}\int_{-a/6-t/2}^{5a/6}\overline{z}^2\,d\overline{y}\,d\overline{z}$$

$$=\tfrac{4}{3}a^3t+\tfrac{1}{4}at^3\approx\tfrac{4}{3}a^3t \tag{1}$$

$$I_{\overline{z}}=\int\overline{y}^2\,dA=\tfrac{1}{4}a^3t+\tfrac{5}{24}at^3\approx\tfrac{1}{4}a^3t$$

$$I_{\overline{y}\overline{z}}=\int\overline{y}\overline{z}\,dA=\tfrac{1}{3}a^3t-\tfrac{5}{48}at^3\approx\tfrac{1}{3}a^3t$$

Numerical values for these and other cross-sectional parameters are given in Table 1.1.

The normal stress σ_x on a cross-sectional face is given by Eq. (1.57). The sign convention for M_y and M_z is detailed in Chapter 2. At a distance L from the free end, $M_y=-\overline{P}L$ and $M_z=0$, so that Eq. (1.57) becomes

$$\begin{aligned}\sigma_x &= -\frac{I_{\overline{y}\overline{z}}M_y+I_{\overline{y}}M_z}{I_{\overline{y}}I_{\overline{z}}-I_{\overline{y}\overline{z}}^2}\overline{y}+\frac{I_{\overline{z}}M_y+I_{\overline{y}\overline{z}}M_z}{I_{\overline{y}}I_{\overline{z}}-I_{\overline{y}\overline{z}}^2}\overline{z}\\ &= -\frac{(a^3t/3)(-\overline{P}L)}{(4a^3t/3)(a^3t/4)-(a^3t/3)^2}\overline{y}+\frac{(a^3t/4)(-\overline{P}L)}{(4a^3t/3)(a^3t/4)-(a^3t/3)^2}\overline{z}\\ &= \frac{3\overline{P}L}{2a^3t}\overline{y}-\frac{9\overline{P}L}{8a^3t}\overline{z}\end{aligned} \tag{2}$$

TABLE 1.1 Part of the Output File for the Computer Program of the Appendixes for the Angle Section of Examples 1.1 and 1.2 with $a=1$ and $t=0.1$[a]

Cross-Sectional Properties		Corresponds to Equation:
Cross-Sectional Area	0.3000	
Y Moment of Area	−0.1999	1.52a
Z Moment of Area	0.0499	1.52a
Y Centroid	0.1663	1.56
Z Centroid	−0.6663	1.56
Moment of Inertia $I_{\overline{y}}$	0.1336	1.52b
Moment of Inertia $I_{\overline{z}}$	0.0252	1.52b
Product of Inertia $I_{\overline{y}\overline{z}}$	0.0332	1.52b
Principal Bending Angle (deg)	−15.7589	1.82 or 1.95
Principal Moment of Inertia (max)	0.1430	1.88
Principal Moment of Inertia (min)	0.0158	1.88

[a] See Fig. 5.26b for coordinate systems.

If the terms involving t^3 in (1) are not neglected, we would find

$$\begin{aligned}\sigma_x = {} & \frac{48\overline{P}L}{at}\frac{16a^2 - 5t^2}{512a^4 + 944a^2t^2 + 95t^4}\overline{y} \\ & - \frac{48\overline{P}L}{at}\frac{12a^2 + 10t^2}{512a^4 + 944a^2t^2 + 95t^4}\overline{z}\end{aligned} \tag{3}$$

Equation (2) is the desired distribution of σ_x on the cross section. The stress at point A of Fig. 1.9a is found by substituting $\overline{z} = 2a/3$ and $\overline{y} = 5a/6$ into (2):

$$(\sigma_x)_A = \frac{\overline{P}L}{2a^2t} \tag{4}$$

At point B, $\overline{y} = -a/6$, $\overline{z} = 2a/3$, and σ_x becomes

$$(\sigma_x)_B = -\frac{\overline{P}L}{a^2t} \tag{5}$$

Finally, at point D, $\overline{y} = -a/6$, $\overline{z} = -4a/3$, and σ_x is found to be

$$(\sigma_x)_D = \frac{5\overline{P}L}{4a^2t} \tag{6}$$

The distribution of the normal stresses is illustrated in Fig. 1.9b.

The neutral axis is defined by Eq. (1.58) as

$$\overline{y} = \frac{I_{\overline{z}}M_y}{I_{\overline{yz}}M_y}\overline{z} = \frac{I_{\overline{z}}}{I_{\overline{yz}}}\overline{z} = \frac{3}{4}\overline{z} \tag{7}$$

This line is plotted in Fig. 1.9*c*. The angle between the $\overline{y}$ axis and the neutral axis is 53.13°.

If the asymmetrical nature of the cross section is ignored, $I_{\overline{yz}}$ would be zero and the normal stress σ_x of Eq. (1.57) would be

$$\sigma_x = \frac{M_y}{I_{\overline{y}}}\overline{z} \tag{8}$$

The maximum stress occurs at point D with $\overline{z} = -4a/3$, so that (8) becomes

$$(\sigma_x)_D = \frac{\overline{P}L}{a^2t} \tag{9}$$

At points A and B, $\overline{z} = 2a/3$ and (8) becomes

$$(\sigma_x)_A = (\sigma_x)_B = -\frac{\overline{P}L}{2a^2t} \tag{10}$$

Note that these values are not consistent with (4), (5), and (6).

The radii of gyration about the centroidal axes $\overline{y}$, $\overline{z}$ are defined by

$$\overline{r}_y = \sqrt{\frac{I_{\overline{y}}}{A}} \qquad \overline{r}_z = \sqrt{\frac{I_{\overline{z}}}{A}} \tag{1.61}$$

The elastic section moduli Y_e, Z_e about the centroidal axes $\overline{y}$, $\overline{z}$ are defined by

$$Y_e = \frac{I_{\overline{y}}}{\overline{z}_{\max}} \qquad Z_e = \frac{I_{\overline{z}}}{\overline{y}_{\max}} \tag{1.62}$$

where $\overline{z}_{\max}$ is the maximum distance between the $\overline{y}$ axis and the material points of the cross section, that is, $\overline{z}_{\max}$ is the distance from the $\overline{y}$ axis to the outermost fiber; $\overline{y}_{\max}$ is the maximum distance between the $\overline{z}$ axis and the material points of the cross section. The polar moment of inertia I_p with respect to the centroid of the section is the sum of the area moments of inertia about the $\overline{y}$ and $\overline{z}$ axes:

$$I_p = I_{\overline{y}} + I_{\overline{z}} \tag{1.63}$$

Modulus-Weighted Properties If the material properties are not homogeneous on the cross section, it is useful to introduce a reference modulus E_r and to define a *modulus-weighted* differential area by

$$d\tilde{A} = \frac{E}{E_r}\, dA \tag{1.64}$$

Then, Eq. (1.51) appears as

$$\begin{aligned} \kappa_\epsilon \tilde{A} + \kappa_y \tilde{Q}_y - \kappa_z \tilde{Q}_z &= 0 \\ \kappa_\epsilon \tilde{Q}_y + \kappa_y \tilde{I}_y - \kappa_z \tilde{I}_{yz} &= \frac{M_y}{E_r} \\ \kappa_\epsilon \tilde{Q}_z + \kappa_y \tilde{I}_{yz} - \kappa_z \tilde{I}_z &= -\frac{M_z}{E_r} \end{aligned} \tag{1.65}$$

In this equation, modulus-weighted section properties of the beam are utilized. The modulus-weighted first moments of area are

$$\tilde{Q}_y = \int z\, d\tilde{A} \qquad \tilde{Q}_z = \int y\, d\tilde{A} \tag{1.66a}$$

The modulus-weighted area moments of inertia are given by

$$\tilde{I}_y = \int z^2 d\tilde{A} \qquad \tilde{I}_z = \int y^2 d\tilde{A} \tag{1.66b}$$

and the modulus-weighted area product of inertia is

$$\tilde{I}_{yz} = \int yz\, d\tilde{A} \tag{1.66c}$$

For a homogeneous beam, the elastic modulus E has the same value at any point of the section, and E_r is chosen equal to E. The modulus-weighted properties then become purely geometric properties of the cross section of Eq. (1.52).

Equation (1.65) is simplified if the relationships are transformed to the centroidal coordinates. For the modulus-weighted case, the components of Eq. (1.54b) are

$$\overline{y} = y - \tilde{y}_C \qquad \overline{z} = z - \tilde{z}_C \tag{1.67}$$

As in the homogeneous case, the origin C is chosen such that the first moments of area in the coordinate system $C\,\overline{yz}$ are zero:

$$\begin{aligned} \tilde{Q}_{\overline{y}} &= \int \overline{z}\, d\tilde{A} = \int (z - \tilde{z}_C)\, d\tilde{A} = 0 \\ \tilde{Q}_{\overline{z}} &= \int \overline{y}\, d\tilde{A} = \int (y - \tilde{y}_C)\, d\tilde{A} = 0 \end{aligned} \tag{1.68}$$

These conditions give

$$\tilde{y}_C = \frac{\tilde{Q}_z}{\tilde{A}} \qquad \tilde{z}_C = \frac{\tilde{Q}_y}{\tilde{A}} \tag{1.69}$$

The point C is the *modulus-weighted centroid* of the cross section. When the material is homogeneous, C becomes the familiar geometric centroid, given by Eq. (1.56).

Transform Eq. (1.65) for the constants κ_ϵ, κ_y, κ_z to the centroidal coordinate system. Introduce Eq. (1.68). Then

$$\begin{aligned} \overline{\kappa}_\epsilon \tilde{A} &= 0 \\ \overline{\kappa}_y \tilde{I}_{\overline{y}} - \overline{\kappa}_z \tilde{I}_{\overline{y}\,\overline{z}} &= \frac{M_y}{E_r} \\ \overline{\kappa}_y \tilde{I}_{\overline{y}\,\overline{z}} - \overline{\kappa}_z \tilde{I}_{\overline{z}} &= -\frac{M_z}{E_r} \end{aligned} \tag{1.70}$$

Solve these equations for $\overline{\kappa}_\epsilon$, $\overline{\kappa}_y$, and $\overline{\kappa}_z$, and substitute the results into σ_x of Eq. (1.50), expressed in terms of $\overline{\kappa}_\epsilon$, $\overline{\kappa}_y$, and $\overline{\kappa}_z$. This leads to the normal stress

$$\sigma_x = \frac{E}{E_r}\left(-\frac{\tilde{I}_{\overline{y}\,\overline{z}} M_y + \tilde{I}_{\overline{y}} M_z}{\tilde{I}_{\overline{y}} \tilde{I}_{\overline{z}} - \tilde{I}^2_{\overline{y}\,\overline{z}}}\overline{y} + \frac{\tilde{I}_{\overline{z}} M_y + \tilde{I}_{\overline{y}\,\overline{z}} M_z}{\tilde{I}_{\overline{y}} \tilde{I}_{\overline{z}} - \tilde{I}^2_{\overline{yz}}}\overline{z} \right) \tag{1.71}$$

The parallel axis theorem transformation equations for the modulus-weighted properties are

$$\tilde{I}_{\overline{y}} = \tilde{I}_y - \tilde{z}_C^2 \tilde{A}$$
$$\tilde{I}_{\overline{z}} = \tilde{I}_z - \tilde{y}_C^2 \tilde{A} \qquad (1.72)$$
$$\tilde{I}_{\overline{y}\,\overline{z}} = \tilde{I}_{yz} - \tilde{y}_C \tilde{z}_C \tilde{A}$$

The radii of gyration about the centroidal axes $\overline{y}$, $\overline{z}$ are defined as

$$\overline{r}_y = \sqrt{\frac{\tilde{I}_{\overline{y}}}{\tilde{A}}} \qquad \overline{r}_z = \sqrt{\frac{\tilde{I}_{\overline{z}}}{\tilde{A}}} \qquad (1.73)$$

and the elastic section moduli about the centroidal axes are

$$Y_e = \frac{\tilde{I}_{\overline{y}}}{\overline{z}_{\max}} \qquad Z_e = \frac{\tilde{I}_{\overline{z}}}{\overline{y}_{\max}} \qquad (1.74)$$

Finally, the polar moment of inertia with respect to the centroid is

$$\tilde{I}_p = \tilde{I}_{\overline{y}} + \tilde{I}_{\overline{z}} \qquad (1.75)$$

1.3 PRINCIPAL BENDING AXES

Figure 1.10 shows centroidal axes $\overline{y}$, $\overline{z}$ and a rotated set of centroidal axes y', z'. The unit vectors **j**, **k** are directed along the $\overline{y}$, $\overline{z}$ axes and the unit vectors **j**$'$, **k**$'$ along the

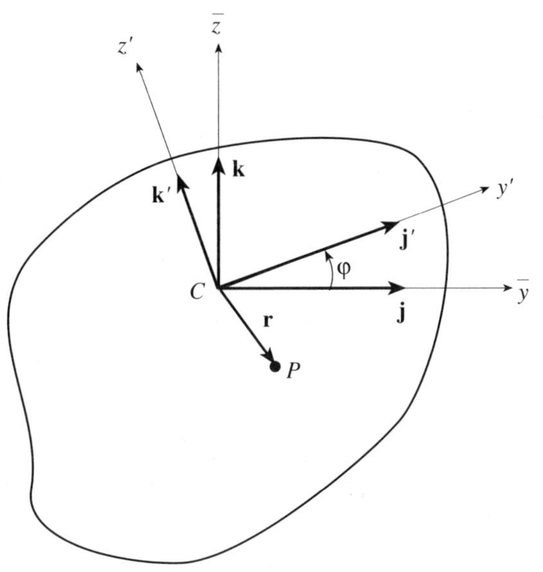

Figure 1.10 Rotated centroidal coordinate system.

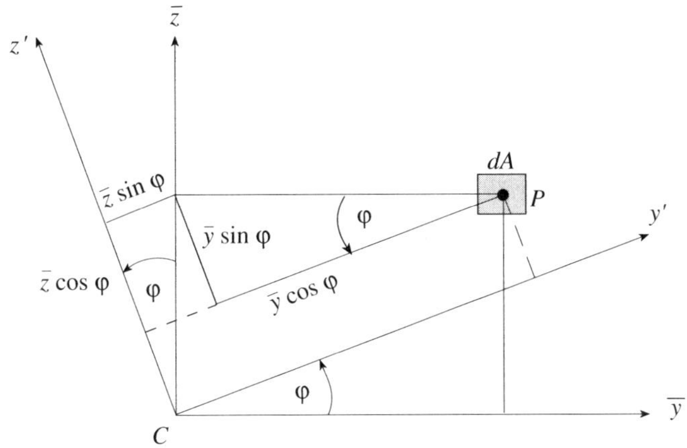

Figure 1.11 Rotation of the centroidal coordinate system.

y', z' axes. The position vector $\mathbf{r}$ of a point P on the cross section may be expressed as

$$\mathbf{r} = \overline{y}\mathbf{j} + \overline{z}\mathbf{k} = y'\mathbf{j}' + z'\mathbf{k}' \tag{1.76}$$

where $\overline{y}$ and $\overline{z}$ are the coordinates of P from the $\overline{y}, \overline{z}$ axes. Similarly, y' and z' are the coordinates of P from the y', z' axes. The y', z' coordinates can be obtained in terms of the $\overline{y}, \overline{z}$ coordinates (Fig. 1.11):

$$\begin{aligned} y' &= \overline{y}\mathbf{j} \cdot \mathbf{j}' + \overline{z}\mathbf{k} \cdot \mathbf{j}' = \overline{y} \cos\varphi + \overline{z} \sin\varphi \\ z' &= \overline{y}\mathbf{j} \cdot \mathbf{k}' + \overline{z}\mathbf{k} \cdot \mathbf{k}' = -\overline{y} \sin\varphi + \overline{z} \cos\varphi \end{aligned} \tag{1.77}$$

Suppose that the differential area dA is located at point P. The second moments of the area (i.e., the area moments of inertia) in the rotated coordinate system are

$$\begin{aligned} I_{y'} &= \int z'^2\, dA = I_{\overline{z}} \sin^2\varphi + I_{\overline{y}} \cos^2\varphi - 2I_{\overline{yz}} \sin\varphi \cos\varphi \\ I_{z'} &= \int y'^2\, dA = I_{\overline{z}} \cos^2\varphi + I_{\overline{y}} \sin^2\varphi + 2I_{\overline{yz}} \sin\varphi \cos\varphi \\ I_{y'z'} &= \int y'z'\, dA = I_{\overline{yz}}(\cos^2\varphi - \sin^2\varphi) + (I_{\overline{y}} - I_{\overline{z}}) \sin\varphi \cos\varphi \end{aligned} \tag{1.78}$$

where the relations of Eq. (1.77) have been utilized. The use of the familiar trigonometric identities $2\cos^2\varphi = 1 + \cos 2\varphi$, $2\sin^2\varphi = 1 - \cos 2\varphi$, $2\sin\varphi\cos\varphi = \sin 2\varphi$, leads to an alternative form:

$$I_{y'} = \frac{I_{\overline{y}} + I_{\overline{z}}}{2} + \frac{I_{\overline{y}} - I_{\overline{z}}}{2} \cos 2\varphi - I_{\overline{yz}} \sin 2\varphi \tag{1.79a}$$

$$I_{z'} = \frac{I_{\overline{y}} + I_{\overline{z}}}{2} - \frac{I_{\overline{y}} - I_{\overline{z}}}{2} \cos 2\varphi + I_{\overline{yz}} \sin 2\varphi \tag{1.79b}$$

$$I_{y'z'} = \frac{I_{\overline{y}} - I_{\overline{z}}}{2} \sin 2\varphi + I_{\overline{yz}} \cos 2\varphi \tag{1.79c}$$

Equations (1.78) and (1.79) provide the area moments of inertia $I_{y'}$, $I_{z'}$, and $I_{y'x'}$ about coordinate axes y', x' at rotation angle φ. These three area moments of inertia as functions of φ are shown in Fig. 1.12. Note that these moments of inertia are bounded. The upper bound for $I_{y'}$ and $I_{z'}$ is $I_{\max} = I_1$ and the lower bound is $I_{\min} = I_2$. Also, for the product of inertia $I_{y'z'}$,

$$-\tfrac{1}{2}(I_1 - I_2) \leq I_{y'z'} \leq +\tfrac{1}{2}(I_1 - I_2) \tag{1.80}$$

The extreme values of the moments of inertia I_1 and I_2 are called *principal moments of inertia* and the corresponding angles define the *principal directions*. In the case shown in Fig. 1.12, both I_1 and I_2 are positive. As observed in Fig. 1.12 by the vertical dashed lines, the product of inertia is zero at the principal directions, which are 90° apart. That is, the two principal directions are perpendicular to each other.

To find the angle φ at which the moment of inertia $I_{y'}$ assumes its maximum value, set $\partial I_{y'}/\partial\varphi$ equal to zero. From Eq. (1.79a) this gives

$$(I_{\overline{y}} - I_{\overline{z}})(-\sin 2\varphi) - 2I_{\overline{yz}} \cos 2\varphi = 0 \tag{1.81}$$

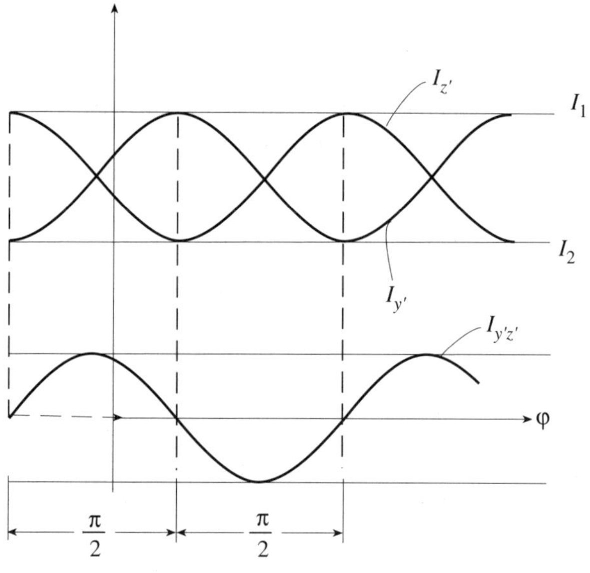

Figure 1.12 Three moments of inertia as a function of the rotation angle φ.

or

$$\tan 2\varphi = \frac{2I_{\overline{yz}}}{I_{\overline{z}} - I_{\overline{y}}} \tag{1.82}$$

This angle φ identifies the so-called *centroidal principal bending axes*. Note that φ of Eq. (1.82) also corresponds to the rotation for which the product of inertia $I_{y'z'}$ is zero. This result, which also was observed in Fig. 1.12, is verified by substituting Eq. (1.81) into Eq. (1.79c). Equation (1.81) determines two values of 2φ that are 180° apart, that is, two values of φ that are 90° apart. At these values, the moments of inertia $I_{y'}$ and $I_{z'}$ assume their maximum or minimum possible values, that is, the principal moments of inertia I_1 and I_2. The magnitudes of I_1 and I_2 can be obtained by substituting φ of Eq. (1.82) into Eq. (1.79a and b). These same values will be obtained below by a different technique. The corresponding directions defined by $\pm\mathbf{j}'$ and $\pm\mathbf{k}'$ are the principal directions. As a particular case, if a cross section is symmetric about an axis, this axis of symmetry is a principal axis.

Consider another approach to finding the magnitudes of the principal moments of inertia I_1 and I_2. It is possible to derive some relationships that are invariant with respect to the rotating coordinate system. It follows from Eq. (1.78) or (1.79) that

$$\begin{aligned} I_{y'} + I_{z'} &= I_{\overline{y}} + I_{\overline{z}} \\ I_{y'}I_{z'} - I_{y'z'}^2 &= I_{\overline{y}}I_{\overline{z}} - I_{\overline{yz}}^2 \end{aligned} \tag{1.83}$$

As noted above, the product of inertia $I_{y'z'}$ is zero at the principal directions and $I_{y'}$ and $I_{z'}$ become I_1 and I_2. Then

$$\begin{aligned} I_{y'} + I_{z'} &= I_{\overline{y}} + I_{\overline{z}} = I_1 + I_2 \\ I_{y'}I_{z'} - I_{y'z'}^2 &= I_{\overline{y}}I_{\overline{z}} - I_{\overline{yz}}^2 = I_1 I_2 \end{aligned} \tag{1.84}$$

The principal moments of inertia I_1 and I_2 can be considered to be the roots of the equation

$$(I - I_1)(I - I_2) = 0 \tag{1.85}$$

Expand Eq. (1.85):

$$I^2 - (I_1 + I_2)I + I_1 I_2 = 0 \tag{1.86}$$

and introduce Eq. (1.84):

$$I^2 - (I_{\overline{y}} + I_{\overline{z}})I + I_{\overline{y}}I_{\overline{z}} - I_{\overline{yz}}^2 = 0 \tag{1.87}$$

The two roots of this equation are the principal moments of inertia

$$
\begin{aligned}
I_1 = I_{\max} &= \frac{I_{\overline{y}} + I_{\overline{z}}}{2} + \Delta \\
I_2 = I_{\min} &= \frac{I_{\overline{y}} + I_{\overline{z}}}{2} - \Delta
\end{aligned}
\tag{1.88}
$$

where

$$
\Delta = \sqrt{\left(\frac{I_{\overline{y}} - I_{\overline{z}}}{2}\right)^2 + I_{\overline{yz}}^2}
$$

Numerical values for some of these parameters are given in Table 1.1 for an angle section.

It is useful to place the transformation relations of Eq. (1.78) or (1.79) in a particular matrix form. Equation (1.87) can be expressed as

$$
(I - I_{\overline{y}})(I - I_{\overline{z}}) - I_{\overline{yz}}^2 = 0 \tag{1.89}
$$

or

$$
\begin{vmatrix} I - I_{\overline{y}} & I_{\overline{yz}} \\ I_{\overline{yz}} & I - I_{\overline{z}} \end{vmatrix} = 0 \tag{1.90}
$$

This determinant is the characteristic equation for the symmetric 2×2 matrix $\mathbf{A}$:

$$
\mathbf{A} = \begin{bmatrix} I_{\overline{y}} & -I_{\overline{yz}} \\ -I_{\overline{yz}} & I_{\overline{z}} \end{bmatrix} \tag{1.91}
$$

With the negative signs on $I_{\overline{yz}}$, $\mathbf{A}$ transforms according to the rotation conventions implied by Fig. 1.10 with φ measured counterclockwise positive from the $\overline{y}$ axis. Define a rotation matrix

$$
\mathbf{R} = \begin{bmatrix} \cos\varphi & \sin\varphi \\ -\sin\varphi & \cos\varphi \end{bmatrix} \tag{1.92}
$$

It may be verified that the transformation

$$
\mathbf{A}' = \begin{bmatrix} I_{y'} & -I_{y'z'} \\ -I_{y'z'} & I_{z'} \end{bmatrix} = \mathbf{RAR}^{-1} = \mathbf{RAR}^{\mathrm{T}} \tag{1.93}
$$

is identical to the rotation transformation equations of Eq. (1.78) or (1.79) derived from Fig. 1.10. If the off-diagonal elements of $\mathbf{A}$ are taken to be $+I_{yz}$, the relationship between $\mathbf{A}$ and $\mathbf{A}'$ no longer matches these equations.

An alternative approach is to base the determination of the principal axes on the diagonalization of matrix $\mathbf{A}$ of Eq. (1.91). When the product of inertia $I_{\overline{yz}}$ is zero, the $\overline{y}$, $\overline{z}$ axes are already the principal axes and no further computation is necessary.

In the special case when $I_{\overline{y}} = I_{\overline{z}}$, any axis is a principal axis. If $I_{\overline{yz}}$ is not zero, the two vectors

$$\begin{aligned} \mathbf{v}_1 &= I_{\overline{yz}}\mathbf{j} + (I_{\overline{y}} - I_1)\mathbf{k} \\ \mathbf{v}_2 &= (I_{\overline{z}} - I_2)\mathbf{j} + I_{\overline{yz}}\mathbf{k} \end{aligned} \tag{1.94}$$

are two orthogonal eigenvectors of $\mathbf{A}$ corresponding to the eigenvalues I_1 and I_2. Some characteristics of eigenvectors are discussed in Chapter 8. The angle φ between the $\overline{y}$ axis and the axis belonging to the larger principal moment of inertia can be computed as the angle between $\pm\mathbf{v}_1$ and $\mathbf{j}$:

$$\varphi = \tan^{-1} \frac{I_{\overline{y}} - I_1}{I_{\overline{yz}}} \tag{1.95}$$

Since the angle between the smaller principal value and the $\overline{y}$ axis is $\varphi + 90°$, the specification of φ is enough to determine both principal axes.

The results of this section apply also to nonhomogeneous beams. It is only necessary to replace all geometric section properties with modulus-weighted ones. If a nonhomogeneous section has an axis of geometric as well as elastic symmetry, it may be concluded that this axis is a principal axis.

Normal Stresses from the Principal Bending Axes If y', z' are the centroidal principal bending axes, Eq. (1.71) simplifies to

$$\sigma_x = \frac{E}{E_r}\left(-\frac{M_z y'}{\tilde{I}_{z'}} + \frac{M_y z'}{\tilde{I}_{y'}}\right) \tag{1.96}$$

For homogeneous materials, Eq. (1.96) reduces to

$$\sigma_x = -\frac{M_z y'}{I_{z'}} + \frac{M_y z'}{I_{y'}} \tag{1.97}$$

In general, the bending moment components are initially calculated in any convenient coordinate system, and when using Eq. (1.96) or (1.97), it is necessary to compute the bending moment components along the principal bending axes.

Example 1.2 Thin-Walled Cantilevered Beam with an Asymmetrical Cross Section. Return to the cantilevered angle of Fig. 1.8 and find the normal stresses using Eq. (1.97), which is based on the principal bending axes.

SOLUTION. From Eq. (1) of Example 1.1 and Eq. (1.88),

$$I_{\overline{y}} = \tfrac{4}{3}a^3t \qquad I_{\overline{z}} = \tfrac{1}{4}a^3t \qquad I_{\overline{yz}} = \tfrac{1}{3}a^3t \tag{1}$$

$$\Delta = \sqrt{\left(\frac{I_{\overline{y}} - I_{\overline{z}}}{2}\right)^2 + I_{\overline{yz}}^2} = \frac{\sqrt{233}}{24}a^3t \tag{2}$$

$$\begin{aligned} I_1 &= \frac{I_{\bar{y}} + I_{\bar{z}}}{2} + \Delta = \frac{a^3 t}{24}(19 + \sqrt{233}) \\ I_2 &= \frac{I_{\bar{y}} + I_{\bar{z}}}{2} - \Delta = \frac{a^3 t}{24}(19 - \sqrt{233}) \end{aligned} \tag{3}$$

The centroidal principal bending axes are located by the angle φ, where (Eq. 1.82)

$$\tan 2\varphi = \frac{2I_{\bar{y}\bar{z}}}{I_{\bar{z}} - I_{\bar{y}}} = -\frac{8}{13} \tag{4}$$

This relationship leads to the two angles $\varphi = -15.8^\circ$ and $\varphi = 74.2^\circ$, one of which corresponds to I_1 and the other to I_2. Further manipulations are necessary to determine which angle corresponds to I_1 and which to I_2. For example, place $\varphi = -15.8^\circ$ into $I_{y'}$ of Eq. (1.78) and find $I_{y'} = (a^3 t/24)(19 + \sqrt{233})$, which is equal to I_1.

The problem of the uncertainty of which value of φ corresponds to I_1 is avoided if Eq. (1.95) is used. In this case,

$$\varphi = \tan^{-1} \frac{I_{\bar{y}} - I_1}{I_{\bar{y}\bar{z}}} = \tan^{-1}\left[\tfrac{1}{8}\left(13 - \sqrt{233}\right)\right] \tag{5}$$

so that $\varphi = -15.8^\circ$ and 164.2°, both of which identify I_1 (Fig. 1.13).

The cross-sectional normal stress σ_x is given by Eq. (1.97). At an axial distance L from the free end, the bending moment components along the principal bending axes are

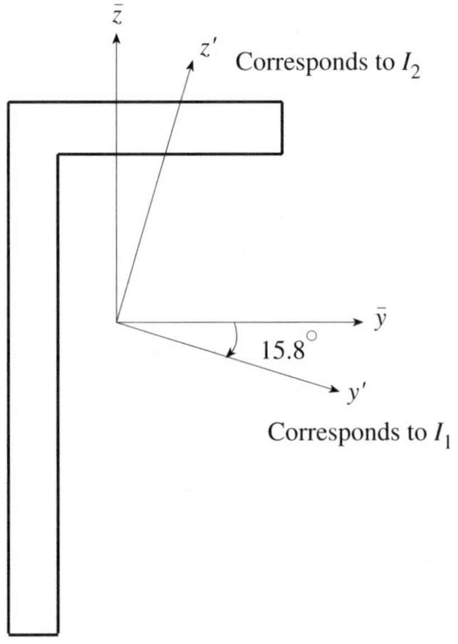

Figure 1.13 Principal bending axes of an asymmetrical cross section.

$$
\begin{aligned}
M_{y'} &= -\overline{P}L\cos(-15.8^\circ) \\
M_{z'} &= -[-\overline{P}L\sin(-15.8^\circ)] = \overline{P}L\sin(-15.8^\circ)
\end{aligned} \tag{6}
$$

The sign convention for these moments is discussed in Chapter 2. Equation (1.97) becomes

$$
\begin{aligned}
\sigma_x &= -\frac{M_z y'}{I_{z'}} + \frac{M_y z'}{I_{y'}} = -\frac{\overline{P}L\sin(-15.8^\circ)y'}{(a^3t/24)(19-\sqrt{233})} + \frac{-\overline{P}L\cos(-15.8^\circ)z'}{(a^3t/24)(19+\sqrt{233})} \\
&= \frac{1.75\overline{P}L}{a^3t}y' - \frac{0.674\overline{P}L}{a^3t}z'
\end{aligned} \tag{7}
$$

At point A of the cross section shown in Fig. 1.9a, $\overline{y} = 5a/6$ and $\overline{z} = 2a/3$, and from Eq. (1.77),

$$
\begin{aligned}
y' &= \frac{5a}{6}\cos(-15.8^\circ) + \frac{2a}{3}\sin(-15.8^\circ) = 0.620a \\
z' &= -\frac{5a}{6}\sin(-15.8^\circ) + \frac{2a}{3}\cos(-15.8^\circ) = 0.869a
\end{aligned} \tag{8}
$$

Substitution of these coordinates into (7) gives $(\sigma_x)_A = \overline{P}L/2a^2t$. At point B, $\overline{y} = -a/6$, $\overline{z} = 2a/3$, and Eq. (1.77) gives $y' = -0.342a$ and $z' = 0.596a$. From (7), $(\sigma_x)_B = -\overline{P}L/a^2t$. At point D, $\overline{y} = -a/6$, $\overline{z} = -4a/3$, $y' = 0.203a$, $z' = -1.328a$, and $(\sigma_x)_D = 5\overline{P}L/4a^2t$. These are the values calculated in Example 1.1.

1.4 AXIAL LOADS

An axial load N_x applied in the x direction at point P of the beam cross section shown in Fig. 1.7 causes additional normal stress. In this case, it is necessary to replace the force at P with its force–couple equivalent at the centroid C. The moment of the equivalent couple is

$$
\mathbf{r}_{PC} \times N_x\mathbf{i} = (z_P\mathbf{k} + y_P\mathbf{j}) \times N_x\mathbf{i} = z_P N_x\mathbf{j} - y_P N_x\mathbf{k} \tag{1.98}
$$

where y_P, z_P are the coordinates of point P in the coordinate system $C\overline{y}\overline{z}$. The additional bending moments due to the axial force are added to the pure bending moments M_y^0, M_z^0 at the section

$$
M_y = M_y^0 + z_P N_x \qquad M_z = M_z^0 - y_P N_x \tag{1.99}
$$

With the inclusion of the axial force, Eqs. (1.57) and (1.71) for normal stress become

$$
\sigma_x = \frac{N_x}{A} - \frac{I_{\overline{y}\overline{z}}M_y + I_{\overline{y}}M_z}{I_{\overline{y}}I_{\overline{z}} - I_{\overline{y}\overline{z}}^2}\overline{y} + \frac{I_{\overline{z}}M_y + I_{\overline{y}\overline{z}}M_z}{I_{\overline{y}}I_{\overline{z}} - I_{\overline{y}\overline{z}}^2}\overline{z} \tag{1.100}
$$

and

$$\sigma_x = \frac{E}{E_r}\left(\frac{N_x}{\tilde{A}} - \frac{\tilde{I}_{\overline{y}\overline{z}}M_y + \tilde{I}_{\overline{y}}M_z}{\tilde{I}_{\overline{y}}\tilde{I}_{\overline{z}} - \tilde{I}^2_{\overline{y}\overline{z}}}\overline{y} + \frac{\tilde{I}_{\overline{z}}M_y + \tilde{I}_{\overline{y}\overline{z}}M_z}{\tilde{I}_{\overline{y}}\tilde{I}_{\overline{z}} - \tilde{I}^2_{\overline{y}\overline{z}}}\overline{z}\right) \tag{1.101}$$

1.5 ELASTICITY SOLUTION FOR PURE BENDING

A beam for which the moments M_y and M_z are constant along the length is said to be in the state of *pure bending*. The elasticity solution for the displacements u_x, u_y, and u_z of a homogeneous beam in pure bending is obtained by assuming a strain field and attempting to satisfy the equations of elasticity. The axes are chosen as shown in Fig. 1.3. The origin O is at the centroid C of the cross section, so that the (x, y, z) and $(\overline{x}, \overline{y}, \overline{z})$ axes coincide. The beam material is assumed to be homogeneous and body forces are assumed to be absent. It follows from the displacement of Eq. (1.47) that a reasonable form of the strains is

$$\begin{aligned}
\epsilon_x &= \kappa_\epsilon + \kappa_y z - \kappa_z y \\
\epsilon_y &= -\nu\epsilon_x \\
\epsilon_z &= -\nu\epsilon_x \\
\gamma_{xy} &= 0 \\
\gamma_{yz} &= 0 \\
\gamma_{zx} &= 0
\end{aligned} \tag{1.102}$$

As shown in Eq. (1.48), the strain ϵ_x is obtained from $\partial u/\partial x$. This strain field identically satisfies the conditions of compatibility of Eq. (1.9). Substitution of the strains of Eq. (1.102) into the Hooke's law formulas of Eq. (1.18) shows that the only nonzero stress is the axial stress:

$$\sigma_x = E\epsilon_x \tag{1.103}$$

The total axial force N_x acting on the cross section is

$$N_x = \int \sigma_x \, dA = E\int (\kappa_\epsilon + \kappa_y z - \kappa_z y)\, dA = E\kappa_\epsilon A \tag{1.104}$$

In this calculation, the factors multiplying κ_y and κ_z, that is, the integrals $E\int z\, dA$ and $E\int y\, dA$, are proportional to the y, z coordinates of the centroid, which are both zero because the centroid C is at the origin O of the coordinates. For pure bending the axial force N_x will be zero (Eq. 1.46). It follows from Eq. (1.104) that the constant κ_ϵ is zero. Then the y, z components of the bending moment as expressed by Eq. (1.45) are

$$
\begin{aligned}
M_y &= \int z\sigma_x \, dA = E(\kappa_y I_y - \kappa_z I_{yz}) \\
M_z &= -\int y\sigma_x \, dA = E(-\kappa_y I_{yz} + \kappa_z I_z)
\end{aligned} \tag{1.105}
$$

where the moments of inertia are given by Eq. (1.52b). Equation (1.105) can be solved for the constants κ_y and κ_z, giving

$$
\kappa_y = \frac{I_z M_y + I_{yz} M_z}{E(I_y I_z - I_{yz}^2)} \qquad \kappa_z = \frac{I_{yz} M_y + I_y M_z}{E(I_y I_z - I_{yz}^2)} \tag{1.106}
$$

It follows from $\sigma_x = E\epsilon_x = E(\kappa_y z - \kappa_z y)$ that the axial stress is again given by Eq (1.57), with $\overline{y} = y$ and $\overline{z} = z$.

The displacements can be obtained from the strain–displacement relations of Eq. (1.3). With the strains given by Eq. (1.102),

$$
\begin{aligned}
\epsilon_x &= \frac{\partial u_x}{\partial x} = \kappa_y z - \kappa_z y & \epsilon_y &= \frac{\partial u_y}{\partial y} = -\nu(\kappa_y z - \kappa_z y) \\
\epsilon_z &= \frac{\partial u_z}{\partial z} = -\nu(\kappa_y z - \kappa_z y) & \gamma_{xy} &= \frac{\partial u_y}{\partial x} + \frac{\partial u_x}{\partial y} = 0 \\
\gamma_{xz} &= \frac{\partial u_z}{\partial x} + \frac{\partial u_x}{\partial z} = 0 & \gamma_{yz} &= \frac{\partial u_z}{\partial y} + \frac{\partial u_y}{\partial z} = 0
\end{aligned} \tag{1.107}
$$

The displacements will be determined from these six equations by direct integration. From the first equation, the axial displacement may be expressed as

$$
u_x = \kappa_y xz - \kappa_z xy + u_{x0}(y, z) \tag{1.108}
$$

where u_{x0} is an unknown function of y and z. The derivatives of u_y and u_z with respect to x are given in terms of u_{x0} by $\gamma_{xy} = 0$ and $\gamma_{xz} = 0$:

$$
\begin{aligned}
\frac{\partial u_y}{\partial x} &= -\frac{\partial u_x}{\partial y} = \kappa_z x - \frac{\partial u_{x0}}{\partial y} \\
\frac{\partial u_z}{\partial x} &= -\frac{\partial u_x}{\partial z} = -\kappa_y x - \frac{\partial u_{x0}}{\partial z}
\end{aligned} \tag{1.109}
$$

from which the displacements u_y and u_z are obtained in the form

$$
\begin{aligned}
u_y &= \kappa_z \frac{x^2}{2} - x\frac{\partial u_{x0}}{\partial y} + u_{y0}(y, z) \\
u_z &= -\kappa_y \frac{x^2}{2} - x\frac{\partial u_{x0}}{\partial z} + u_{z0}(y, z)
\end{aligned} \tag{1.110}
$$

where u_{y0} and u_{z0} are unknown functions of y and z, respectively.

The second and third strain–displacement relations of Eq. (1.107) become

$$\begin{aligned}
\epsilon_y &= \frac{\partial u_y}{\partial y} = -x\frac{\partial^2 u_{x0}}{\partial y^2} + \frac{\partial u_{y0}}{\partial y} = -\nu(\kappa_y z - \kappa_z y) \\
\text{or} \quad & -x\frac{\partial^2 u_{x0}}{\partial y^2} + \frac{\partial u_{y0}}{\partial y} + \nu(\kappa_y z - \kappa_z y) = 0 \\
\epsilon_z &= \frac{\partial u_z}{\partial z} = -x\frac{\partial^2 u_{x0}}{\partial z^2} + \frac{\partial u_{z0}}{\partial z} = -\nu(\kappa_y z - \kappa_z y) \\
\text{or} \quad & -x\frac{\partial^2 u_{x0}}{\partial z^2} + \frac{\partial u_{z0}}{\partial z} + \nu(\kappa_y z - \kappa_z y) = 0
\end{aligned} \tag{1.111}$$

Note the functional form of these equations, with the coordinate x occurring only once as a factor multiplying a second partial derivative of u_{x0}. Since these equations must hold for all values of x,

$$\frac{\partial^2 u_{x0}}{\partial y^2} = 0 \qquad \frac{\partial^2 u_{x0}}{\partial z^2} = 0 \tag{1.112}$$

Consequently, u_{y0} and u_{z0} can be obtained by integration of Eq. (1.111):

$$\begin{aligned}
u_{y0} &= -\nu\left(\kappa_y yz - \kappa_z \frac{y^2}{2}\right) + u_{y1}(z) \\
u_{z0} &= -\nu\left(\kappa_y \frac{z^2}{2} - \kappa_z yz\right) + u_{z1}(y)
\end{aligned} \tag{1.113}$$

The final strain–displacement relation of Eq. (1.107),

$$\gamma_{yz} = \frac{\partial u_z}{\partial y} + \frac{\partial u_y}{\partial z} = 0 \tag{1.114}$$

becomes

$$-2x\frac{\partial^2 u_{x0}}{\partial y\,\partial z} + \frac{\partial u_{z0}}{\partial y} + \frac{\partial u_{y0}}{\partial z} = -2x\frac{\partial^2 u_{x0}}{\partial y\,\partial z} + \frac{du_{z1}}{dy} + \nu\kappa_z z + \frac{du_{y1}}{dz} - \nu\kappa_y y = 0 \tag{1.115}$$

The functional form of this equality shows that the factor multiplying x is zero

$$\frac{\partial^2 u_{x0}}{\partial y\,\partial z} = 0 \tag{1.116}$$

Hence

$$\frac{du_{z1}}{dy} - \nu\kappa_y y + \frac{du_{y1}}{dz} + \nu\kappa_z z = 0 \tag{1.117}$$

By a separation of variables,

$$\frac{du_{z1}}{dy} - \nu\kappa_y y = C_0 \qquad \frac{du_{y1}}{dz} + \nu\kappa_z z = -C_0 \tag{1.118}$$

where C_0 is a constant.

The relations

$$\frac{\partial^2 u_{x0}}{\partial y^2} = 0 \qquad \frac{\partial^2 u_{x0}}{\partial z^2} = 0 \qquad \frac{\partial^2 u_{x0}}{\partial y\, \partial z} = 0 \tag{1.119}$$

of Eqs. (1.112) and (1.116) show that the function u_{x0} has the form

$$u_{x0} = C_1 y + C_2 z + C_3 \tag{1.120}$$

in which the C_k are constants. It follows from Eq. (1.108) that the axial displacement appears as

$$u_x = \kappa_y x z - \kappa_z x y + C_1 y + C_2 z + C_3 \tag{1.121}$$

To find u_{y1}, u_{z1} of Eq. (1.113), integrate Eq. (1.118):

$$\begin{aligned} u_{y1} &= -C_0 z - \nu\kappa_z \frac{z^2}{2} + C_4 \\ u_{z1} &= C_0 y + \nu\kappa_y \frac{y^2}{2} + C_5 \end{aligned} \tag{1.122}$$

so that, from Eq. (1.113),

$$\begin{aligned} u_{y0} &= -\nu\left(\kappa_y yz - \kappa_z \frac{y^2 - z^2}{2}\right) - C_0 z + C_4 \\ u_{z0} &= -\nu\left(\kappa_y \frac{z^2 - y^2}{2} - \kappa_z yz\right) + C_0 y + C_5 \end{aligned} \tag{1.123}$$

The displacements u_y and u_z of Eq. (1.110) may therefore be expressed as

$$\begin{aligned} u_y &= \kappa_z \frac{x^2}{2} - \nu\left(\kappa_y yz - \kappa_z \frac{y^2 - z^2}{2}\right) - C_0 z - C_1 x + C_4 \\ u_z &= -\kappa_y \frac{x^2}{2} - \nu\left(\kappa_y \frac{z^2 - y^2}{2} - \kappa_z yz\right) + C_0 y - C_2 x + C_5 \end{aligned} \tag{1.124}$$

The constants of integration in the expressions derived for the displacements depend on the support conditions. For example, suppose that the centroid at the origin of the coordinates ($x = 0,\ y = 0,\ z = 0$) at the left end ($x = 0$) of the horizontal beam is fixed such that no translational or rotational motion is possible. Then $u_x = 0,\ u_y = 0,\ u_z = 0$ at $x = 0,\ y = 0,\ z = 0$. Also, at $x = 0,\ y = 0,\ z = 0$, there is no rotation in the z direction ($\partial u_z/\partial x = 0$), no rotation in the y direction ($\partial u_y/\partial x = 0$), and no rotation about the x axis ($\partial u_y/\partial z = 0$).

The enforcement of these boundary conditions amounts to restraining the beam at $x = 0,\ y = 0,\ z = 0$ against rigid-body translation and rotation. From Eqs. (1.121) and (1.124) these boundary conditions require that

$$C_0 = 0, \quad C_1 = 0, \quad C_2 = 0, \quad C_3 = 0, \quad C_4 = 0, \quad C_5 = 0 \tag{1.125}$$

The displacements can now be written as (Eqs. 1.121, 1.124, and 1.125)

$$u_x = (\kappa_y z - \kappa_z y)x \tag{1.126a}$$

$$u_y = \kappa_z \frac{x^2}{2} - \nu \left(\kappa_y yz - \kappa_z \frac{y^2 - z^2}{2} \right) \tag{1.126b}$$

$$u_z = -\kappa_y \frac{x^2}{2} - \nu \left(\kappa_y \frac{z^2 - y^2}{2} - \kappa_z yz \right) \tag{1.126c}$$

Consider a special case of a beam with a cross section symmetric about the z axis and for which $M_z = 0$. Then, from Eq. (1.106), $\kappa_z = 0$ and $\kappa_y = M_y/EI_y$.For this case, the displacements of Eq. (1.126) become

$$u_x = \kappa_y zx \tag{1.127a}$$

$$u_y = -\nu \kappa_y yz \tag{1.127b}$$

$$u_z = -\frac{\kappa_y}{2}[x^2 + \nu(z^2 - y^2)] \tag{1.127c}$$

The deflection of the centroidal beam axis is given by Eq. (1.127c) with y and z equal to zero; that is,

$$u_z(x, 0, 0) = w = -\kappa_y \frac{x^2}{2} = -\frac{M_y}{EI_y}\frac{x^2}{2} \tag{1.128}$$

This is the same deflection given by engineering beam theory (Chapter 2) for a cantilevered beam loaded with a concentrated moment at the free end. Some interesting characteristics of beams in bending can be studied by considering the displacements away from the central axis.

To find the axial displacement at a particular cross section, say at $x = a$, consider $u_x(x, y, z)$ of Eq. (1.127a). Thus

$$u_x(a,\ y,\ z) = \kappa_y za \tag{1.129}$$

We see that cross-sectional planes remain planar. This is not surprising since the assumed strain ϵ_x corresponds to a linear variation in the displacements in the y and z directions (Eq. 1.47).

Note from Eq. (1.127a) that beam fibers in the $z = 0$ plane do not displace in the x direction [i.e., $u_x(x, y, 0) = 0$]. Consequently, this plane is referred to as the *neutral plane*. The x axis before deformation is designated as the *neutral axis*.

To illustrate the distortion of the cross-section profile, consider the rectangular section of Fig. 1.14. From Eq. (1.127b), the horizontal displacements u_y of the vertical sides are

$$u_y\left(x, \pm\frac{b}{2}, z\right) = \pm\frac{b}{2}(-\nu\kappa_y z) \tag{1.130}$$

Thus, the vertical sides rotate. The vertical displacements of the top and bottom ($z = \pm h/2$) are (Eq. 1.127c)

$$u_z\left(x, y, \pm\frac{h}{2}\right) = -\frac{\kappa_y}{2}\left[x^2 + \nu\left(\frac{h^2}{4} - y^2\right)\right] = \frac{M}{2EI_y}\left[x^2 + \nu\left(\frac{h^2}{4} - y^2\right)\right] \tag{1.131}$$

where, as seen in Fig. 1.15a, $\kappa_y = M_y/EI_y = -M/EI_y$. This shows that the top and bottom are deformed into parabolic shapes. Assume that $b \gg h$. Note that if the curvature of the longitudinal axis of the beam is concave upward (Fig. 1.15a), the curvature of the top and bottom surfaces are concave downward (Fig. 1.15b). This is referred to as *anticlastic curvature*. For the thin-walled beam of Fig. 1.15, the anticlastic curvature can be significant. In contrast, if the depth and width are of comparable size (Fig. 1.14), the effect is small.

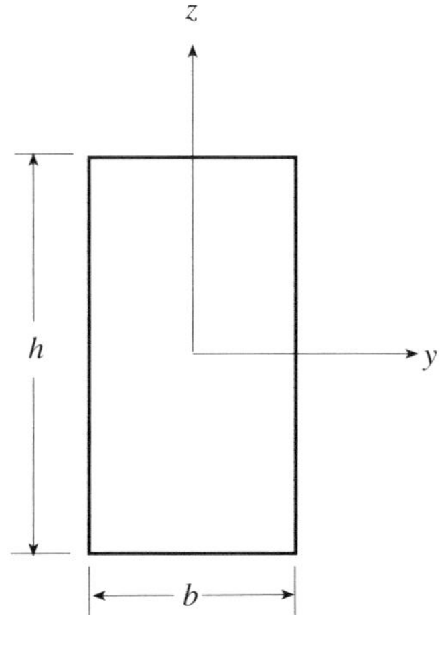

Figure 1.14 Beam cross section.

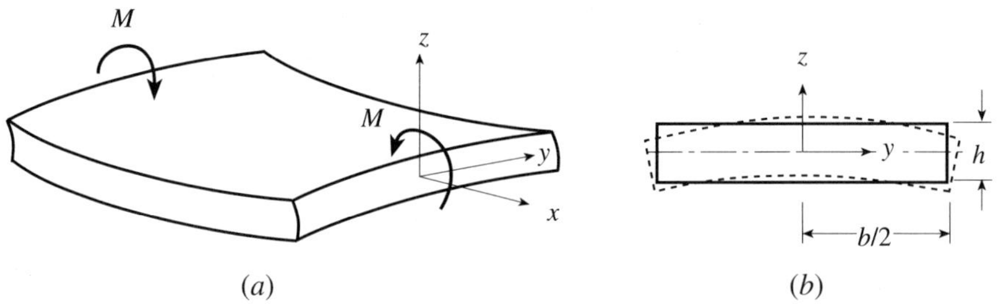

Figure 1.15 Anticlastic curvature.

There is a simple physical interpretation of this behavior in bending. For pure bending as shown in Fig. 1.15a, the upper fibers are in compression and the lower fibers in tension. Strain ϵ_x along the x direction is accompanied by strain $-\nu\epsilon_x$ in the y direction, where ν is Poisson's ratio. It follows that as the upper fibers are compressed in the x direction, they become somewhat longer in the y direction. Conversely, as the lower fibers are extended in the x direction, they shorten in the y direction.

Engineering Beam Theory In contrast to the pure bending assumptions of this section, engineering beam theory, which is presented in Chapter 2, is applied to beams under general lateral loading conditions. The bounding surface of the beam is often not free of stress; body forces are not necessarily zero; the shear force at each section is nonzero; and the bending moment is not constant along the beam. Engineering beam theory neglects the normal stresses σ_y and σ_z, which are much smaller than the axial stress. Also neglected is the influence of Poisson's ratio, so that longitudinal fibers deform independently. For engineering beam theory, the normal stresses and strains are calculated as in the case of pure bending, although the bending moment is no longer constant along the beam axis.

REFERENCES

Atanackovic, T. M., and Guran, A. (2000). *Theory of Elasticity for Scientists and Engineers*, Birkhäuser, Boston.

Boresi, A. P., and Chong, K. P. (1987). *Elasticity in Engineering Mechanics*, Elsevier, Amsterdam, The Netherlands.

Gere, J. M. (2001). *Mechanics of Materials*, 5th ed., PWS, Boston.

Gould, P. L. (1994). *Introduction to Linear Elasticity*, 2nd ed., Springer-Verlag, New York.

Love, A. E. (1944). *A Treatise on the Mathematical Theory of Elasticity*, 4th ed., Dover, New York.

Muskhelishvili, N. I. (1953). *Some Basic Problems of the Mathematical Theory of Elasticity*, Walters-Noordhoff, Groningen, The Netherlands.

Oden, J. T., and Ripperger, E. A. (1981) *Mechanics of Elastic Structures*, 2nd ed., McGraw-Hill, New York.

Pilkey, W. D., and Wunderlich, W. (1994). *Mechanics of Structures; Variational and Computational Methods*, CRC Press, Boca Raton, Fla.

Rivello, R. M. (1969). *Theory and Analysis of Flight Structures*, McGraw-Hill, New York.

Sokolnikoff, I. S. (1956). *Mathematical Theory of Elasticity*, McGraw-Hill, New York.

Uğural, A. C., and Fenster, S. K. (1981) *Advanced Strength and Applied Elasticity*, Elsevier, Amsterdam, The Netherlands.

CHAPTER 2

BEAM ELEMENTS

In Chapter 1, the deflection of a beam subject to pure bending was obtained using the theory of elasticity. Theory of elasticity–based solutions are limited to simple loadings and geometries. The technical or *engineering beam theory* is distinguished from the theory of elasticity approach in that a broad range of problems can be solved. In this chapter, engineering beam theory is employed to find matrices for beam elements that can be used to find the response of the beam. That is, the deflection, moments, and shear forces along the beam can be calculated.

In engineering beam theory, for which the beam can be subject to general loading conditions, the shear force is not zero and the bending moment is not constant along the beam. As in the case of the theory of elasticity, engineering beam theory neglects the normal stresses σ_y and σ_z, which are much smaller than the axial stress σ_x. Engineering beam theory also assumes that Poisson's ratio is zero. Perhaps the most significant assumption is that normal stresses and strains for any loading or geometry may be calculated as in the case of pure bending. That is, although engineering beam theory is established, as is the theory of elasticity approach, for pure bending, it is assumed to apply for arbitrary loading, which in general means that the bending moment is not constant along the beam.

For this chapter the cross sections of the beams are not assumed, in general, to be symmetric with respect to either the y or z axes. The x axis is the axis of centroids of the cross sections, which may have differing shapes or dimensions at different values of x.

2.1 FUNDAMENTAL ENGINEERING THEORY EQUATIONS FOR A STRAIGHT BEAM

2.1.1 Geometry of Deformation

In Eq. (1.47), on the basis of cross-sectional planes remaining plane and normal to the beam axis during bending, the axial displacement of a point on a cross section for bending deformation was taken to be

$$u_x(x, y, z) = u(x) + z\theta_y(x) - y\theta_z(x) \tag{2.1}$$

where θ_y and θ_z are the angles of rotation about the centroidal y and z axes, respectively, and $u(x) = u_x(x, 0, 0)$ is the axial displacement of the centerline of the beam. That is, the beam axis passes through the centroidal x axis of the beam cross sections. These angles of rotation, also referred to as *angles of slope* or just *slopes*, are the angles between the x axis and the tangents to the deflection curve.

In Chapter 1 we set the derivatives $d\theta_y/dx$ and $d\theta_z/dx$ equal to the constants κ_y and κ_z, respectively. It is shown in references such as Gere (2001) that for small rotations, $\kappa_y = d\theta_y/dx$ and $\kappa_z = d\theta_z/dx$ are curvatures of the transverse displacement curve. In Chapter 1, quantities measured with respect to centroidal axes were designated by superscript lines. In this chapter most of the variables are referred to the centroidal axes and hence the superscript lines are dropped. For example, $\overline{y}, \overline{z}, \overline{\kappa}_y$, and $\overline{\kappa}_z$ of Chapter 1 are simply y, z, κ_y, and κ_z here in Chapter 2.

The axial strain corresponding to the displacement of Eq. (2.1) is Eq. (1.3)

$$\epsilon_x = \frac{du_x}{dx} = \frac{du}{dx} + z\frac{d\theta_y}{dx} - y\frac{d\theta_z}{dx} = \kappa_\epsilon + z\kappa_y - y\kappa_z \tag{2.2}$$

where $\kappa_\epsilon = du/dx$. The shear strains γ_{xy} and γ_{xz} take the forms (Eq. 1.3)

$$\gamma_{xy} = \frac{\partial v}{\partial x} + \frac{\partial u_x}{\partial y} = \frac{\partial v}{\partial x} - \theta_z \qquad \gamma_{xz} = \frac{\partial u_x}{\partial z} + \frac{\partial w}{\partial x} = \theta_y + \frac{\partial w}{\partial x} \tag{2.3}$$

where $v = u_y(x, 0, 0)$ and $w = u_z(x, 0, 0)$ are the y and z displacements of the centroidal x axis. For the cross sections to remain planar, it is necessary that these shear strains be zero. In this case, the effects of shear deformation are being neglected. Thus

$$\gamma_{xy} = 0 \quad \text{and} \quad \gamma_{xz} = 0 \tag{2.4}$$

lead to, using ordinary derivatives,

$$\theta_z = \frac{dv}{dx} \qquad \text{and} \quad \theta_y = -\frac{dw}{dx} \tag{2.5}$$

or

$$\kappa_z = \frac{d^2 v}{dx^2} \qquad \text{and} \quad \kappa_y = -\frac{d^2 w}{dx^2} \tag{2.6}$$

These are the strain–displacement relations for bending, since the bending strains are taken to be the curvatures. The displacements v and w are the *deflections* of the beam axis.

Bending in the xz Plane Consider, for the moment, standard planar engineering beam theory for which the cross section is symmetric about the z axis ($I_{yz} = 0$) (Fig. 2.1), only bending about the y axis is taken into account, and the only applied forces are in the z direction. As discussed in Chapter 1, bending such that the product of inertia is zero also occurs for an asymmetrical cross section by transforming (rotating) to the principal bending axes. Since the deflection for a planar beam is usually shown as being downward, the positive z coordinate is chosen to be downward. The relevant strain–displacement relation, referred to the centroidal coordinate system, is $\kappa_y = -d^2 w/dx^2$. Since the strain of the centerline of the beam is $du/dx = \kappa_\epsilon$, the strain–displacement relations in matrix form appear as

$$\begin{bmatrix} \kappa_\epsilon \\ \kappa_y \end{bmatrix} = \begin{bmatrix} d/dx & 0 \\ 0 & -d^2/dx^2 \end{bmatrix} \begin{bmatrix} u \\ w \end{bmatrix} \tag{2.7}$$
$$\boldsymbol{\epsilon} = \mathbf{D}_u \quad \mathbf{u}$$

If shear deformation effects are retained for this standard engineering beam theory, γ_{xz} is not zero and the strain–displacement relations with $\theta_y = \theta$ and $\gamma_{xz} = \gamma$

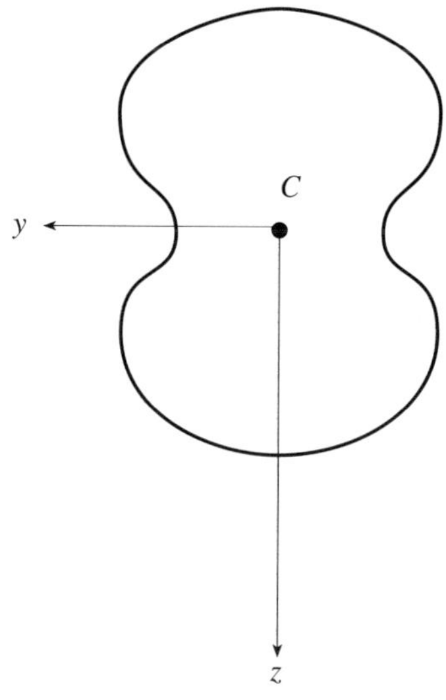

Figure 2.1 Beam cross section with $I_{yz} = 0$.

can be written as

$$\begin{bmatrix} \kappa_\epsilon \\ \gamma \\ \kappa_y \end{bmatrix} = \begin{bmatrix} d/dx & 0 & 0 \\ 0 & d/dx & 1 \\ 0 & 0 & d/dx \end{bmatrix} \begin{bmatrix} u \\ w \\ \theta \end{bmatrix} \qquad (2.8)$$

$$\boldsymbol{\epsilon} = \mathbf{D}_u \; \mathbf{u}$$

2.1.2 Force–Deformation Relations

For the case of a general beam, a material law in the form (Eq. 1.70)

$$\begin{aligned} \kappa_\epsilon \tilde{A} &= 0 \\ \kappa_y \tilde{I}_y - \kappa_z \tilde{I}_{yz} &= \frac{M_y}{E_r} \\ \kappa_y \tilde{I}_{yz} - \kappa_z \tilde{I}_z &= -\frac{M_z}{E_r} \end{aligned} \qquad (2.9a)$$

was derived in Chapter 1. For a homogeneous beam, where the elastic modulus E is the same at any point on the cross section,

$$\begin{aligned} \kappa_\epsilon EA &= 0 \\ \kappa_y EI_y - \kappa_z EI_{yz} &= M_y \\ \kappa_y EI_{yz} - \kappa_z EI_z &= -M_z \end{aligned} \qquad (2.9b)$$

Bending in the xz Plane The material law is much simpler for standard planar engineering beam theory. If there is an axial force N, the first relationship of Eq. (2.9b) becomes

$$\kappa_\epsilon EA = N \qquad (2.10a)$$

and the second relationship of Eq. (2.9b) appears as

$$\kappa_y EI = M \qquad (2.10b)$$

where M_y and I_y have been set equal to M and I, respectively. In matrix notation, Eq. (2.10) can be written as

$$\begin{bmatrix} N \\ M \end{bmatrix} = \begin{bmatrix} EA & 0 \\ 0 & EI \end{bmatrix} \begin{bmatrix} \kappa_\epsilon \\ \kappa_y \end{bmatrix} \qquad (2.11a)$$

$$\mathbf{s} = \mathbf{E} \; \boldsymbol{\epsilon}$$

or

$$\boldsymbol{\epsilon} = \begin{bmatrix} 1/EA & 0 \\ 0 & 1/EI \end{bmatrix} \mathbf{s} \qquad (2.11b)$$

$$\boldsymbol{\epsilon} = \mathbf{E}^{-1} \; \mathbf{s}$$

where

$$\mathbf{s} = \begin{bmatrix} N \\ M \end{bmatrix}$$

These material relationships for the standard planar engineering beam theory can be supplemented with another equation if shear deformation effects are to be taken into account. This new constitutive equation should be a relationship between the shear strain and the shear force on the cross section. From Eq. (1.15), $\tau_{xz} = \tau = G\gamma_{xz} = G\gamma$. Let the shear stress and shear force $V_z = V$ be related as $V = \tau_{\text{average}} A$. Suppose that τ is the shear stress at the centroid of the cross section. Then assume that $\tau_{\text{average}} = k_s \tau$, where k_s is a dimensionless *shear form* or *shear stiffness factor* that depends on the cross-sectional shape. The determination of this factor is treated in Chapter 6. The reciprocal of k_s is the *shear correction factor* or *shear deformation coefficient* that is found in formula books for various cross-sectional shapes. Then the constitutive relationship we seek is

$$V = k_s G A \gamma \tag{2.12}$$

and the material law of Eq. (2.11a) is expanded to become

$$\begin{bmatrix} N \\ V \\ M \end{bmatrix} = \begin{bmatrix} EA & 0 & 0 \\ 0 & k_s GA & 0 \\ 0 & 0 & EI \end{bmatrix} \begin{bmatrix} \kappa_\epsilon \\ \gamma \\ \kappa_y \end{bmatrix} \tag{2.13}$$

$$\mathbf{s} \quad = \quad \mathbf{E} \quad \boldsymbol{\epsilon}$$

or

$$\boldsymbol{\epsilon} = \begin{bmatrix} 1/EA & 0 & 0 \\ 0 & 1/k_s GA & 0 \\ 0 & 0 & 1/EI \end{bmatrix} \mathbf{s}$$

$$\boldsymbol{\epsilon} = \quad \mathbf{E}^{-1} \quad \mathbf{s}$$

where $\mathbf{s}$ is now given as $\mathbf{s} = [N \quad V \quad M]^{\mathrm{T}}$.

2.1.3 Equations of Equilibrium

The equilibrium equations, in terms of axial and shear forces and bending moments, are

$$\frac{dN}{dx} = -\overline{p}_x \qquad \frac{dV_y}{dx} = -\overline{p}_y \qquad \frac{dV_z}{dx} = -\overline{p}_z \tag{2.14}$$

$$\frac{dM_z}{dx} = -V_y \qquad \frac{dM_y}{dx} = V_z \tag{2.15}$$

where $\overline{p}_x$, $\overline{p}_y$, and $\overline{p}_z$ are applied forces per unit length of the beam in the x, y, and z directions, respectively. The derivation of these relationships is given in mechanics-

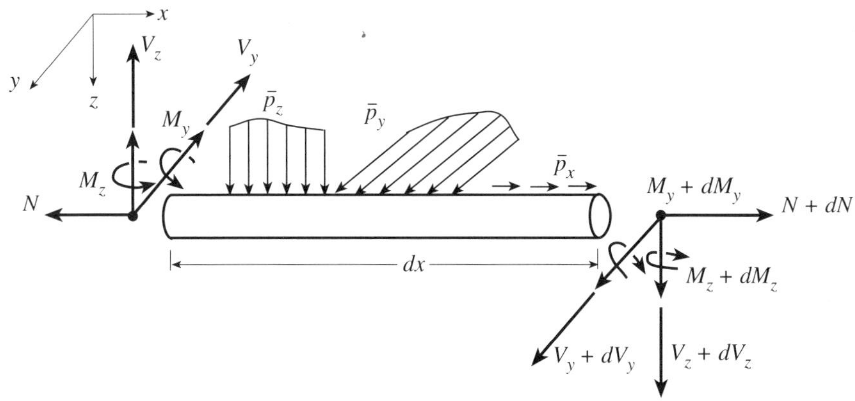

Figure 2.2 Beam element with applied forces per unit length $\overline{p}_x$, $\overline{p}_y$, and $\overline{p}_z$.

of-materials textbooks. These equations are derived from the conditions of equilibrium for a differential beam element of length dx (Fig. 2.2). With a local right-handed coordinate system whose origin is placed at the left end of this beam element, the axial and shear forces $N + dN$, $V_y + dV_y$, and $V_z + dV_z$ on the cross section at the right end are taken to be positive, that is, in the x, y, and z coordinate directions. In terms of the conditions of equilibrium, the axial and shear forces on the cross section at the left end must then be $-N$, $-V_y$, $-V_z$, so that the net differences are dN, dV_y, and dV_z for the differential beam element. Since the applied loads $\overline{p}_x$, $\overline{p}_y$, and $\overline{p}_z$ are positive in the x, y, and z coordinate directions, the force equilibrium conditions for the beam element yield the three relations of Eq. (2.14). Similarly, the bending moments on the cross section at the right end $M_y + dM_y$, $M_z + dM_z$ are taken to be positive if their vectors are in the y, z coordinate directions. The moments on the cross section at the left end are then $-M_y$, $-M_z$, and the moment equilibrium conditions for the beam element yield the two relations of Eq. (2.15).

Equations (2.14) and (2.15) can be combined to form differential equations for the bending moments:

$$\frac{d^2 M_z}{dx^2} = \overline{p}_y \qquad \frac{d^2 M_y}{dx^2} = -\overline{p}_z \tag{2.16}$$

Bending in the xz Plane For the standard planar engineering beam theory, the conditions of equilibrium of Eq. (2.16) with $M_y = M$ reduce to

$$\frac{d^2 M}{dx^2} + \overline{p}_z = 0 \tag{2.17}$$

In matrix form, the equilibrium relations are

$$\begin{aligned} \begin{bmatrix} d/dx & 0 \\ 0 & d^2/dx^2 \end{bmatrix} \begin{bmatrix} N \\ M \end{bmatrix} + \begin{bmatrix} \overline{p}_x \\ \overline{p}_z \end{bmatrix} &= \mathbf{0} \\ \mathbf{D}_s^T \qquad\qquad \mathbf{s} \quad + \quad \overline{\mathbf{p}} \quad &= \mathbf{0} \end{aligned} \tag{2.18}$$

if shear deformation effects are not included. If shear deformation effects are considered, then

$$\begin{bmatrix} d/dx & 0 & 0 \\ 0 & d/dx & 0 \\ 0 & -1 & d/dx \end{bmatrix} \begin{bmatrix} N \\ V \\ M \end{bmatrix} + \begin{bmatrix} \overline{p}_x \\ \overline{p}_z \\ 0 \end{bmatrix} = \mathbf{0}$$
$$\mathbf{D}_s^{\mathrm{T}} \qquad \mathbf{s} + \overline{\mathbf{p}} = \mathbf{0} \tag{2.19}$$

The differential operator $\mathbf{D}_s$ of Eq. (2.18) or (2.19) is functionally the adjoint (Pilkey and Wunderlich, 1994) of $\mathbf{D}_u$ of Eq. (2.7) or (2.8) in the sense that they satisfy the relationship

$$\int_V (\mathbf{D}_u\mathbf{u})^{\mathrm{T}}\mathbf{s}\, dV = \oint_S \mathbf{u}^{\mathrm{T}}\mathbf{N}^{\mathrm{T}}\mathbf{s}\, dS - \int_V \mathbf{u}^{\mathrm{T}}\mathbf{D}_s^{\mathrm{T}}\mathbf{s}\, dV \tag{2.20}$$

where the vectors **u** and **s** are smooth in the domain and $\mathbf{N}^{\mathrm{T}}$ is the matrix of direction cosines of Eq. (1.28) for the tractions on an oblique surface. With appropriate adjustments this adjoint relationship holds for the differential operators for kinematics and equilibrium of the theory of elasticity.

2.1.4 Boundary Conditions

The boundary conditions for a beam occur in-span or on the ends of a beam where either conditions on forces (S_p) or displacements (S_u) can be imposed. For a fixed end, for example, $u = v = w = \theta_y = \theta_z = 0$ on S_u and for a free end, $N = M_y = M_z = V_y = V_z = 0$ on S_p.

Bending in the xz Plane The boundary conditions for the standard planar engineering beam theory are, of course, simpler. For example, for a simply supported end that is not constrained in the axial direction, $w = N = M = 0$ on $S_u + S_p = S$. In general, for this planar beam with

$$\mathbf{s} = \begin{bmatrix} N \\ V \\ M \end{bmatrix} \quad \text{and} \quad \mathbf{u} = \begin{bmatrix} u \\ w \\ \theta \end{bmatrix}$$

the boundary conditions can be summarized as

$$\begin{aligned} N &= \overline{N} \\ V &= \overline{V} \quad \text{or} \quad \mathbf{s} = \overline{\mathbf{s}} \quad \text{on } S_p \\ M &= \overline{M} \end{aligned} \tag{2.21}$$

for static or force conditions and

$$\begin{aligned} u &= \overline{u} \\ w &= \overline{w} \quad \text{or} \quad \mathbf{u} = \overline{\mathbf{u}} \quad \text{on } S_u \\ \theta &= \overline{\theta} \end{aligned} \tag{2.22}$$

for displacement or kinematic conditions. Applied forces or displacement constraints are designated by overbars.

2.1.5 Displacement Form of the Governing Differential Equations

If the kinematic conditions of Eq. (2.6), referred to the centroidal coordinate system, are introduced into the constitutive relations of Eq. (2.9b) for homogeneous materials, we obtain the differential equations for the deflections v, w:

$$\frac{d^2 v}{dx^2} = \frac{I_{yz} M_y + I_y M_z}{E(I_y I_z - I_{yz}^2)} \tag{2.23a}$$

$$\frac{d^2 w}{dx^2} = -\frac{I_z M_y + I_{yz} M_z}{E(I_y I_z - I_{yz}^2)} \tag{2.23b}$$

For a beam whose elastic modulus E and cross-sectional properties I_y, I_z, I_{yz} do not vary with x, these equations can be written, by introduction of Eq. (2.16), in terms of the applied distributed loads $\overline{p}_y$, $\overline{p}_z$:

$$\frac{d^4 v}{dx^4} = \frac{\overline{p}_y I_y - \overline{p}_z I_{yz}}{E(I_y I_z - I_{yz}^2)} \tag{2.24a}$$

$$\frac{d^4 w}{dx^4} = \frac{\overline{p}_z I_z - \overline{p}_y I_{yz}}{E(I_y I_z - I_{yz}^2)} \tag{2.24b}$$

These uncoupled differential equations may be solved one at a time for given loads and boundary conditions. The solutions, which define the elastic curve for a beam, are the displacements of the centroidal line in the y and z directions.

Example 2.1 Bending in the xz Plane of a Cantilevered Beam with a Constant Moment. Suppose that the cross section of the beam of Fig. 2.3a is symmetric about the z axis. Alternatively, for an asymmetrical cross section transform the coordinates to the principal bending planes. Then $I_{yz} = 0$ and the elastic curve in the xz plane is defined by Eq. (2.23b)

$$\frac{d^2 w}{dx^2} = -\frac{M_y}{E I_y} \tag{1}$$

The deflection is found by integrating this equation.

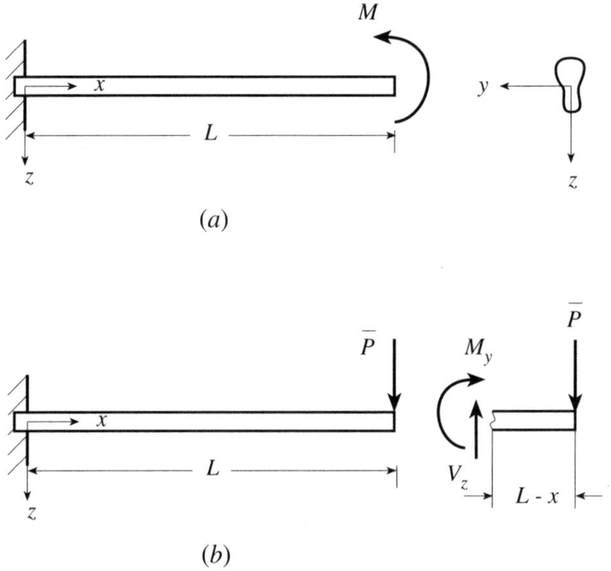

Figure 2.3 Cantilever beams.

From equilibrium requirements, the net moment at any cross section is $M_y = \overline{M}$. Then, from (1), $d^2w/dx^2 = -\overline{M}/EI_y$, so that

$$\frac{dw}{dx} = -\frac{\overline{M}x}{EI_y} + C_1 \tag{2}$$

$$w = -\frac{\overline{M}x^2}{2EI_y} + C_1x + C_2 \tag{3}$$

The displacement boundary conditions are

$$w(0) = 0 \quad \text{and} \quad \frac{dw}{dx}(0) = 0$$

Applied to (2) and (3), these boundary conditions lead to $C_1 = 0$ and $C_2 = 0$. Thus the deflection of the centroidal beam axis is

$$u_z(x, 0, 0) = w = -\frac{\overline{M}x^2}{2EI_y} \tag{4}$$

which corresponds to Eq. (1.128).

Example 2.2 Bending in the xz Plane of a Cantilevered Beam with an Applied Concentrated Force. Repeat the analysis of Example 2.1, except that the applied loading is a concentrated force applied at the free end as shown in Fig. 2.3b.

SOLUTION. The conditions of equilibrium for the small figure of Fig. 2.3b show that

$$M_y = -\overline{P}(L - x) \tag{1}$$

Integrate

$$\frac{d^2w}{dx^2} = -\frac{M_y}{EI_y} = \frac{\overline{P}}{EI_y}(L - x) \tag{2}$$

twice to find the deflection and apply the displacement boundary condition at $x = 0$. This leads to

$$w = \frac{\overline{P}x^2}{EI_y}\left(\frac{L}{2} - \frac{x}{3}\right) \tag{3}$$

Example 2.3 Cantilevered Beam with an Asymmetric Cross Section. For the cantilever beam in Fig. 2.4, the applied loads $\overline{p}_y$ and $\overline{p}_z$ can be taken as zero over the interior of the beam span $0 < x < L$, and the applied load $\overline{P}$ in the z direction at $x = L$ can be treated as a boundary condition. This cross section does not exhibit symmetry about either the y or the z axis. We choose to solve this problem in the yz coordinate system and not transform to the principal bending axes where the product of inertia is zero. For the angled cross section of Example 1.2 (Fig. 1.8), a $-15.8°$ rotation would be required to reach the principal axes.

The fourth-order deflection relations of Eq. (2.24) become

$$\frac{d^4v}{dx^4} = 0 \qquad \frac{d^4w}{dx^4} = 0 \tag{1}$$

The boundary conditions at the fixed end are

$$\begin{aligned} v(0) &= 0 \qquad \theta_z(0) = \frac{dv}{dx}(0) = 0 \\ w(0) &= 0 \qquad \theta_y(0) = -\frac{dw}{dx}(0) = 0 \end{aligned} \tag{2}$$

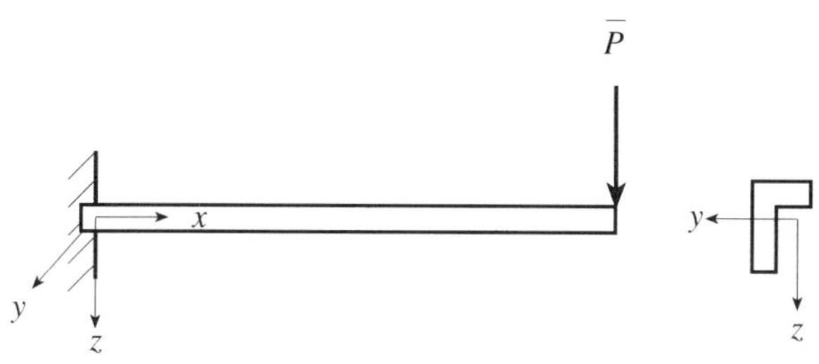

Figure 2.4 Cantilever beam with concentrated applied force in the z direction.

At the free end, there are conditions on shear forces and bending moments

$$
\begin{aligned}
M_y(L) &= 0 \qquad & M_z(L) &= 0 \\
V_y(L) &= 0 \qquad & V_z(L) &= \overline{P}
\end{aligned}
\tag{3}
$$

The deflection equations (1) show that both $v(x)$ and $w(x)$ are polynomials of degree three in x. The general solution $w(x)$ for the deflection in the z direction may be written as

$$
w(x) = D_1 x^3 + D_2 x^2 + D_3 x + D_4 \tag{4}
$$

where D_i, $i = 1, 2, 3, 4$, are constants. The boundary conditions on $w(0) = 0$ and $w'(0) = 0$ at the fixed end show that both D_3 and D_4 are zero. Since both $M_y(L)$ and $M_z(L)$ are zero, Eq. (2.23b) gives the boundary condition

$$
\frac{d^2 w}{dx^2}(L) = 0 \tag{5}
$$

The remaining boundary conditions $V_y(L) = 0$ and $V_z(L) = \overline{P}$ are implemented by manipulating Eq. (2.23b). From Eq. (2.15), Eq. (2.23b) becomes

$$
\frac{d^3 w}{dx^3} = -\frac{I_z V_z - I_{yz} V_y}{E(I_y I_z - I_{yz}^2)} \tag{6}
$$

Substitution into (6) of the conditions $V_y(L) = 0$ and $V_z(L) = \overline{P}$ at $x = L$ gives the relationship

$$
\frac{d^3 w}{dx^3}(L) = -\frac{\overline{P} I_z}{E(I_y I_z - I_{yz}^2)} \tag{7}
$$

In terms of the integration constants of (4), conditions (5) and (7) give

$$
2D_2 + 6D_1 L = 0 \qquad 6D_1 = -\frac{\overline{P} I_z}{E(I_y I_z - I_{yz}^2)} \tag{8}
$$

so that

$$
D_1 = -\frac{\overline{P} I_z}{6E(I_y I_z - I_{yz}^2)} \qquad D_2 = -3D_1 L \tag{9}
$$

and the expression for the deflection $w(x)$ becomes

$$
w(x) = \frac{\overline{P} I_z (3L - x) x^2}{6E(I_y I_z - I_{yz}^2)} \tag{10}
$$

The general solution $v(x)$ for the deflection in the y direction may be written as

$$v(x) = C_1 x^3 + C_2 x^2 + C_3 x + C_4 \tag{11}$$

The boundary conditions on v at the fixed end $x = 0$ show that both C_3 and C_4 are zero. The boundary conditions at the free end $x = L$ imply that

$$\frac{d^2 v}{dx^2}(L) = 0 \qquad \frac{d^3 v}{dx^3}(L) = \frac{\overline{P} I_{yz}}{E(I_y I_z - I_{yz}^2)} \tag{12}$$

from which $v(x)$ is obtained as

$$v(x) = -\frac{\overline{P} I_{yz}(3L - x)x^2}{6E(I_y I_z - I_{yz}^2)} \tag{13}$$

Note that if the product of inertia I_{yz} is nonzero, the deflection $v(x)$ is not identically zero even though there is no external loading in the y direction.

The deflections $v(x)$ and $w(x)$ are components of the total deflection, which is calculated as $(v^2 + w^2)^{1/2}$.

Example 2.4 Statically Indeterminate Beam with an Asymmetric Cross Section. A variety of beams, including statically indeterminate configurations, can be solved as in Example 2.3; that is, $v(x)$ and $w(x)$ can be found one at a time. Sometimes the inclusion of boundary conditions is more involved than in Example 2.3. To illustrate this, find the deflection of the statically indeterminate beam with an asymmetric cross section as shown in Fig. 2.5. Do not transform to the principal bending axes.

SOLUTION. The applied loading can be represented with delta functions (Table 2.1) as

$$\overline{p}_y = \overline{P}\left\langle x - \frac{L}{2}\right\rangle^{-1} \qquad \overline{p}_z = \overline{P}\left\langle x - \frac{L}{2}\right\rangle^{-1} \tag{1}$$

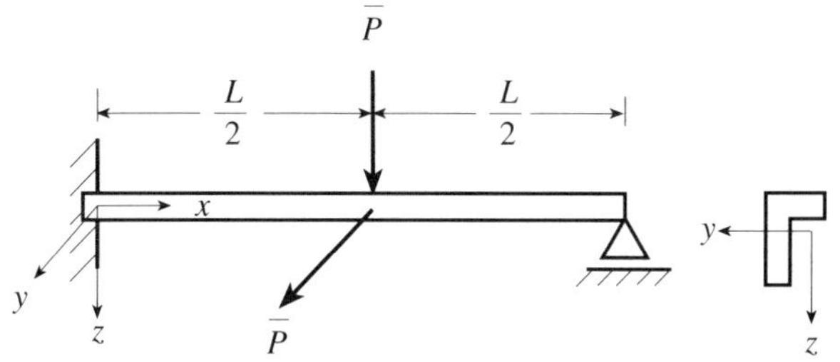

Figure 2.5 Statically indeterminate beam.

TABLE 2.1 Properties of Some Singularity Functions

Name	Pictorial Representation	Definition and Integration Property
Dirac delta function	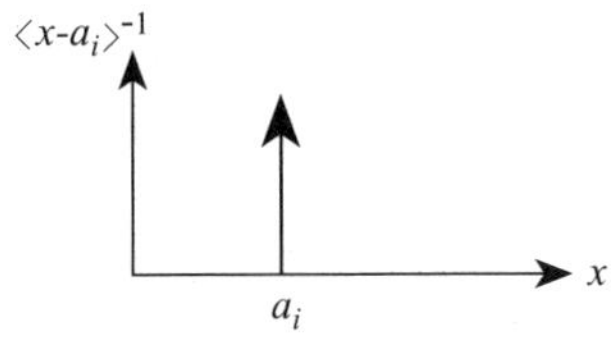	$\langle x-a_i\rangle^{-1}=0 \quad \text{if } x\neq a_i$ $\int_0^x \langle x-a_i\rangle^{-1}\,dx=\langle x-a_i\rangle^{0}$
Unit step function	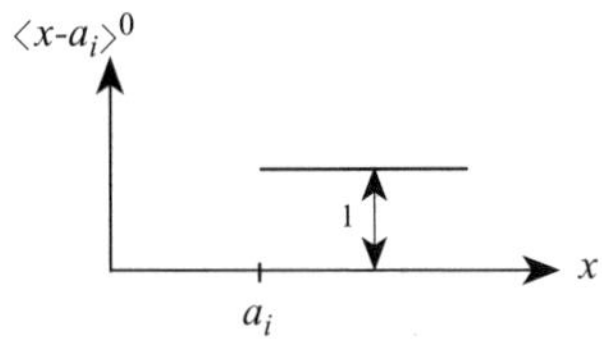	$\langle x-a_i\rangle^{0}=\begin{cases}0 & \text{if } x<a_i\\ 1 & \text{if } x\geq a_i\end{cases}$ $\int_0^x \langle x-a_i\rangle^{0}dx=\langle x-a_i\rangle$
Linear ramp function	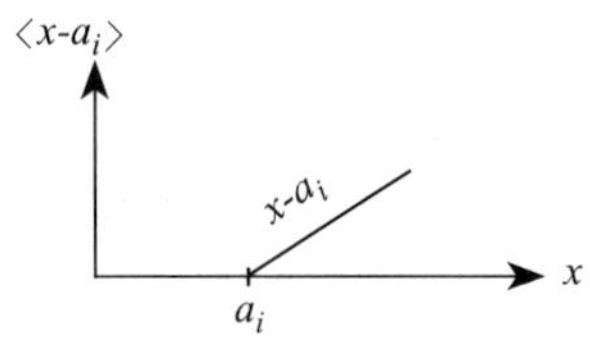	$\langle x-a_i\rangle=\begin{cases}0 & \text{if } x<a_i\\ x-a_i & \text{if } x\geq a_i\end{cases}$ $\int_0^x \langle x-a_i\rangle dx=\dfrac{\langle x-a_i\rangle^2}{2}$
General Macauley notation	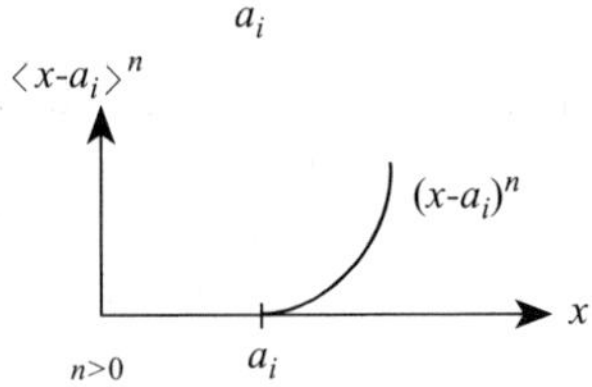	$\langle x-a_i\rangle^{n}=\begin{cases}0 & \text{if } x<a_i\\ (x-a_i)^n & \text{if } x\geq a_i\end{cases}$ $\int_0^x \langle x-a_i\rangle^{n}dx=\dfrac{\langle x-a_i\rangle^{n+1}}{n+1}$

Define

$$\alpha=\frac{I_y}{E(I_yI_z-I_{yz}^2)}\qquad \beta=\frac{-I_{yz}}{E(I_yI_z-I_{yz}^2)}\qquad \gamma=\frac{I_z}{E(I_yI_z-I_{yz}^2)} \tag{2}$$

so that the relations of Eq. (2.24) become

$$\begin{aligned}\frac{d^4v}{dx^4}&=\alpha\overline{p}_y+\beta\overline{p}_z=\alpha\overline{P}\left\langle x-\frac{L}{2}\right\rangle^{-1}+\beta\overline{P}\left\langle x-\frac{L}{2}\right\rangle^{-1}\\ \frac{d^4w}{dx^4}&=\beta\overline{p}_y+\gamma\overline{p}_z=\beta\overline{P}\left\langle x-\frac{L}{2}\right\rangle^{-1}+\gamma\overline{P}\left\langle x-\frac{L}{2}\right\rangle^{-1}\end{aligned} \tag{3}$$

Integrate the first expression

$$\frac{d^3v}{dx^3}=\alpha\overline{P}\left\langle x-\frac{L}{2}\right\rangle^{0}+\beta\overline{P}\left\langle x-\frac{L}{2}\right\rangle^{0}+C_1$$

$$
\begin{aligned}
\frac{d^2v}{dx^2} &= \alpha\overline{P}\left\langle x - \frac{L}{2}\right\rangle^1 + \beta\overline{P}\left\langle x - \frac{L}{2}\right\rangle^1 + C_1x + C_2 \\
\frac{dv}{dx} &= \alpha\overline{P}\frac{\langle x - L/2\rangle^2}{2} + \beta\overline{P}\frac{\langle x - L/2\rangle^2}{2} + C_1\frac{x^2}{2} + C_2x + C_3 \\
v &= \alpha\overline{P}\frac{\langle x - L/2\rangle^3}{3!} + \beta\overline{P}\frac{\langle x - L/2\rangle^3}{3!} + C_1\frac{x^3}{3!} + C_2\frac{x^2}{2} + C_3x + C_4
\end{aligned} \tag{4}
$$

The boundary conditions for $v(x)$ are

$$
v(0) = 0 \qquad \frac{dv}{dx}(0) = 0 \qquad v(L) = 0 \tag{5}
$$

At $x = L$, M_y and M_z are zero, and from Eq. (2.23a),

$$
\frac{d^2v}{dx^2} = -\beta M_y + \alpha M_z \tag{6}
$$

it follows that

$$
\frac{d^2v}{dx^2}(L) = 0 \tag{7}
$$

The four conditions of (5) and (7) can be employed to find the four constants C_1, C_2, C_3, and C_4.

$$
C_1 = -\tfrac{11}{16}\overline{P}(\alpha + \beta) \qquad C_2 = \tfrac{3}{16}\overline{P}L(\alpha + \beta) \qquad C_3 = 0 \qquad C_4 = 0 \tag{8}
$$

From the second expression of (3),

$$
\begin{aligned}
w(x) = {} & \gamma\overline{P}\frac{\langle x - L/2\rangle^3}{3!} + \beta\overline{P}\frac{\langle x - L/2\rangle^3}{3!} - \frac{11}{16}\overline{P}(\beta + \gamma)\frac{x^3}{3!} \\
& + \frac{3}{16}\overline{P}L(\beta + \gamma)\frac{x^2}{2}
\end{aligned} \tag{9}
$$

To complete the solution find the support reactions, that is, the moments and shear forces at the ends of the beam. For the moments, use (6) and the comparable relationship for d^2w/dx^2:

$$
\begin{aligned}
\frac{d^2v}{dx^2}(0) &= C_2 = \frac{3}{16}\overline{P}L(\alpha + \beta) = -\beta M_y(0) + \alpha M_z(0) \\
\frac{d^2w}{dx^2}(0) &= \frac{3}{16}\overline{P}L(\beta + \gamma) = \beta M_z(0) - \gamma M_y(0)
\end{aligned} \tag{10}
$$

These give

$$M_y(0) = -\frac{3\overline{P}L}{16} \qquad M_z(0) = \frac{3\overline{P}L}{16} \tag{11}$$

From Eqs. (2.15) and (2.23)

$$\begin{aligned} \frac{d^3v}{dx^3}(x) &= -\alpha V_y - \beta V_z \\ \frac{d^3w}{dx^3}(x) &= -\gamma V_z - \beta V_y \end{aligned} \tag{12}$$

Then

$$\begin{aligned} \frac{d^3v}{dx^3}(0) &= C_1 = -\frac{11}{16}\overline{P}(\alpha+\beta) = -\alpha V_y(0) - \beta V_z(0) \\ \frac{d^3w}{dx^3}(0) &= -\frac{11}{16}\overline{P}(\beta+\gamma) = -\beta V_y(0) - \gamma V_z(0) \end{aligned} \tag{13}$$

giving

$$V_y(0) = \tfrac{11}{16}\overline{P} = V_z(0) \tag{14}$$

Finally, at $x = L$,

$$\begin{aligned} \frac{d^3v}{dx^3}(L) &= \alpha\overline{P} + \beta\overline{P} + C_1 = -\alpha V_y(L) - \beta V_z(L) \\ \frac{d^3w}{dx^3}(L) &= \gamma\overline{P} + \beta\overline{P} - \frac{11}{16}\overline{P}(\beta+\gamma) = -\beta V_y(L) - \gamma V_z(L) \end{aligned} \tag{15}$$

so that the reactions at $x = L$ are

$$V_y(L) = -\tfrac{5}{16}\overline{P} \qquad V_z(L) = -\tfrac{5}{16}\overline{P} \tag{16}$$

Example 2.5 Beam with In-Span Support. To study the question of when the equations for beams with asymmetric cross sections appear to be coupled, consider the beam of Fig. 2.6.

Let the unknown reactions at the in-span support be R_y and R_z. The loading functions $\overline{p}_y$ and $\overline{p}_z$ can be represented as

$$\begin{aligned} \overline{p}_y(x) &= \overline{P}\,\langle x-3a\rangle^{-1} + R_y\,\langle x-2a\rangle^{-1} \\ \overline{p}_z(x) &= \overline{P}\,\langle x-a\rangle^{-1} + R_z\,\langle x-2a\rangle^{-1} \end{aligned} \tag{1}$$

Retain the definitions for α, β, and γ of Eq. (2) of Example 2.4. From Eq. (2.24),

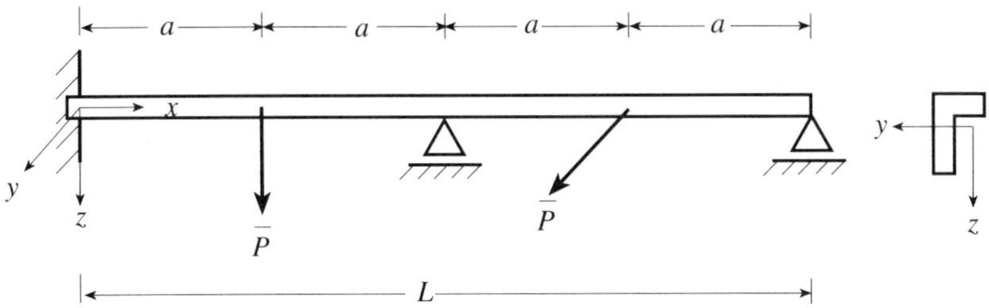

Figure 2.6 Beam with in-span support.

$$\frac{d^4v}{dx^4} = \alpha\overline{p}_y + \beta\overline{p}_z = \alpha\overline{P}\,\langle x - 3a\rangle^{-1} + \alpha R_y\,\langle x - 2a\rangle^{-1} + \beta\overline{P}\,\langle x - a\rangle^{-1} + \beta R_z\,\langle x - 2a\rangle^{-1} \tag{2}$$

Integration leads to

$$\begin{aligned}
\frac{d^3v}{dx^3} &= \alpha\overline{P}\,\langle x - 3a\rangle^0 + \beta\overline{P}\,\langle x - a\rangle^0 + (\alpha R_y + \beta R_z)\,\langle x - 2a\rangle^0 + C_1 \\
\frac{d^2v}{dx^2} &= \alpha\overline{P}\,\langle x - 3a\rangle^1 + \beta\overline{P}\,\langle x - a\rangle^1 + (\alpha R_y + \beta R_z)\,\langle x - 2a\rangle^1 + C_1x + C_2 \\
\frac{dv}{dx} &= \alpha\overline{P}\frac{\langle x - 3a\rangle^2}{2} + \beta\overline{P}\frac{\langle x - a\rangle^2}{2} + (\alpha R_y + \beta R_z)\frac{\langle x - 2a\rangle^2}{2} \\
&\quad + C_1\frac{x^2}{2} + C_2x + C_3 \\
v &= \alpha\overline{P}\frac{\langle x - 3a\rangle^3}{3!} + \beta\overline{P}\frac{\langle x - a\rangle^3}{3!} + (\alpha R_y + \beta R_z)\frac{\langle x - 2a\rangle^3}{3!} \\
&\quad + C_1\frac{x^3}{3!} + C_2\frac{x^2}{2} + C_3x + C_4
\end{aligned} \tag{3}$$

The displacement boundary conditions are

$$v(0) = 0 \qquad v(2a) = 0 \qquad v(4a) = 0 \tag{4}$$

Since M_y and M_z are zero at $x = 0$ and $x = 4a$,

$$\frac{d^2v}{dx^2} = -\beta M_y + \alpha M_z \tag{5}$$

gives the conditions

$$\frac{d^2v}{dx^2}(0) = 0 \qquad \frac{d^2v}{dx^2}(4a) = 0 \tag{6}$$

There are five conditions from (4) and (6) and six unknown constants

$$C_1, C_2, C_3, C_4, R_y, \text{ and } R_z \tag{7}$$

It will be necessary to use information from beam analysis for deflection in the xz plane in order to find the six unknowns. However, the five conditions can be used to calculate the five quantities C_1, C_2, C_3, C_4, and $\alpha R_y + \beta R_z$. Thus

$$\begin{aligned} &C_4 = 0 \qquad C_2 = 0 \qquad C_1 = \frac{\overline{P}}{32}(3\alpha - 13\beta) \\ &C_3 = \frac{a^2\overline{P}}{16}(-\alpha + 3\beta) \qquad \alpha R_y + \beta R_z = -\frac{11}{16}\overline{P}(\alpha + \beta) \end{aligned} \tag{8}$$

For the displacement $w(x)$,

$$\begin{aligned} \frac{d^4 w}{dx^4} &= \gamma \overline{p}_z + \beta \overline{p}_y = \gamma \overline{P} \langle x - a \rangle^{-1} + \gamma R_z \langle x - 2a \rangle^{-1} \\ &\quad + \beta \overline{P} \langle x - 3a \rangle^{-1} + \beta R_y \langle x - 2a \rangle^{-1} \\ \frac{d^2 w}{dx^2} &= \gamma \overline{P} \langle x - a \rangle^{1} + \beta \overline{P} \langle x - 3a \rangle^{1} + (\gamma R_z + \beta R_y) \langle x - 2a \rangle^{1} + D_1 x + D_2 \\ w &= \gamma \overline{P} \frac{\langle x - a \rangle^{3}}{3!} + \beta \overline{P} \frac{\langle x - 3a \rangle^{3}}{3!} + (\gamma R_z + \beta R_y) \frac{\langle x - 2a \rangle^{3}}{3!} \\ &\quad + D_1 \frac{x^3}{3!} + D_2 \frac{x^2}{2} + D_3 x + D_4 \end{aligned} \tag{9}$$

The boundary conditions for $w(x)$ are

$$w(0) = 0 \qquad w(2a) = 0 \qquad w(4a) = 0 \qquad \frac{d^2 w}{dx^2}(0) = 0 \qquad \frac{d^2 w}{dx^2}(4a) = 0 \tag{10}$$

These five conditions can be used to find the four constants D_1, D_2, D_3, and D_4, as well as the combination of reactions $(\gamma R_z + \beta R_y)$.

$$\begin{aligned} &D_4 = 0 \qquad D_2 = 0 \qquad D_1 = \frac{\overline{P}}{32}(3\beta - 13\gamma) \\ &D_3 = \frac{a^2\overline{P}}{16}(-\beta + 3\gamma) \qquad \gamma R_z + \beta R_y = -\frac{11}{16}\overline{P}(\beta + \gamma) \end{aligned} \tag{11}$$

The final relationships of (8) and (11),

$$
\begin{aligned}
-\frac{11}{16}\overline{P}(\alpha+\beta) &= \alpha R_y + \beta R_z \\
-\frac{11}{16}\overline{P}(\beta+\gamma) &= \beta R_y + \gamma R_z
\end{aligned}
\tag{12}
$$

can be solved to find the two reactions

$$
R_y = -\frac{11\overline{P}}{16} \qquad R_z = -\frac{11\overline{P}}{16}
\tag{13}
$$

It is seen that the solution for this beam is obtained by finding the responses in the two planes (xy and xz) separately, with the solution being coupled through the boundary conditions.

The shear forces are obtained by considering the response in both planes.

$$
\begin{aligned}
\frac{d^3v}{dx^3} &= -\alpha V_y - \beta V_z \\
\frac{d^3w}{dx^3} &= -\gamma V_z - \beta V_y
\end{aligned}
\tag{14}
$$

From (3) and (9) at $x = 0$,

$$
\begin{aligned}
\frac{d^3v}{dx^3}(0) &= C_1 = \frac{\overline{P}}{32}(3\alpha - 13\beta) \\
\frac{d^3w}{dx^3}(0) &= D_1 = \frac{\overline{P}}{32}(3\beta - 13\gamma)
\end{aligned}
\tag{15}
$$

Evaluate (14) at $x = 0$ and use (15). Solve the resulting equations to find

$$
V_y(0) = -\tfrac{3}{32}\overline{P} \qquad V_z(0) = \tfrac{13}{32}\overline{P}
\tag{16}
$$

The shear reactions at $x = 4a$ are found in a similar fashion. At $x = 4a$,

$$
\begin{aligned}
\frac{d^3v}{dx^3}(4a) &= \alpha\overline{P} + \beta\overline{P} + (\alpha R_y + \beta R_z) + C_1 \\
\frac{d^3w}{dx^3}(4a) &= \gamma\overline{P} + \beta\overline{P} + (\beta R_y + \gamma R_z) + D_1
\end{aligned}
\tag{17}
$$

From (14) and (17),

$$
V_y(4a) = -\tfrac{13}{32}\overline{P} \qquad V_z(4a) = \tfrac{3}{32}\overline{P}
\tag{18}
$$

Summation of forces and moments quickly shows that the solution found above satisfies the conditions of equilibrium.

Engineering Beam Theory for Bending in a Single Plane For standard planar engineering beam theory, the fundamental equations, with shear deformation included, are Eqs. (2.8), (2.13), and (2.19)

$$\boldsymbol{\epsilon} = \mathbf{D}_u \mathbf{u} \qquad \mathbf{s} = \mathbf{E}\boldsymbol{\epsilon} \qquad \mathbf{D}_s^{\mathrm{T}} \mathbf{s} + \overline{\mathbf{p}} = 0 \tag{2.25}$$

The more common forms of the beam equations are obtained by eliminating $\boldsymbol{\epsilon}$ from these relationships.

The most familiar form of the beam equations is the *Bernoulli–Euler beam*, for which axial and shear deformations are ignored. Let $M_y = M$, $V_z = V$, $\theta_y = \theta$, and $I_y = I$. From the material law of Eq. (2.10b) and the kinematical relationships of $\kappa_y = d\theta/dx$ and Eq. (2.5),

$$M = EI\kappa_y = EI\frac{d\theta}{dx} = -EI\frac{d^2 w}{dx^2} \tag{2.26}$$

Substitution of this relationship into the equilibrium conditions of Eqs. (2.14) and (2.15) leads to

$$V = \frac{dM}{dx} = -\frac{d}{dx}EI\frac{d^2 w}{dx^2} \tag{2.27a}$$

$$-\frac{dV}{dx} = \frac{d^2}{dx^2}EI\frac{d^2 w}{dx^2} = \overline{p}_z \tag{2.27b}$$

In summary, the complete set of differential equations is

$$\frac{d^2}{dx^2}EI\frac{d^2 w}{dx^2} = \overline{p}_z \tag{2.28a}$$

$$V = -\frac{d}{dx}EI\frac{d^2 w}{dx^2} \tag{2.28b}$$

$$M = -EI\frac{d^2 w}{dx^2} \tag{2.28c}$$

$$\theta = -\frac{dw}{dx} \tag{2.28d}$$

The inclusion of shear deformation effects leads to a more general form of the displacement equations for the standard engineering beam theory. Substitute the strain–displacement relations of Eq. (2.8), $\gamma = dw/dx + \theta$ and $\kappa_y = d\theta/dx$, into the constitutive relations of Eq. (2.13), $V = k_s GA\gamma$ and $M = EI\kappa_y$. Place the resulting force–displacement relations in the conditions of equilibrium of Eq. (2.19), $-dV/dx = \overline{p}_z$ and $dM/dx = V$.

$$-\frac{d}{dx}\left[k_s GA\left(\frac{dw}{dx} + \theta\right)\right] = \overline{p}_z \tag{2.29a}$$

$$\frac{d}{dx}\left(EI\frac{d\theta}{dx}\right) - k_s GA\left(\frac{dw}{dx} + \theta\right) = 0 \tag{2.29b}$$

2.1.6 Mixed Form of the Governing Differential Equations

A useful form of the governing differential equations for standard engineering beam theory is written in terms of the *state variables:* displacements v and w, slopes θ_y and θ_z, moments M_y and M_z, and shear forces V_y and V_z. This is referred to as a *mixed form*, as both displacement and force variables are involved. The mixed equations are obtained by assembling the equilibrium equations of Eqs. (2.14) and (2.15), the kinematical relations of Eq. (2.5), and the differential equations of Eq. (2.23) in the form

$$\frac{d^2v}{dx^2} = \frac{d\theta_z}{dx} = \frac{I_{yz}M_y + I_y M_z}{E(I_y I_z - I_{yz}^2)} \qquad -\frac{d^2w}{dx^2} = \frac{d\theta_y}{dx} = \frac{I_z M_y + I_{yz} M_z}{E(I_y I_z - I_{yz}^2)} \tag{2.30}$$

along with the axial relationship $du/dx = N/AE$. Then

$$\begin{aligned}
\frac{du}{dx} &= \frac{N}{AE} \\
\frac{dv}{dx} &= \theta_z \\
\frac{dw}{dx} &= -\theta_y \\
\frac{d\theta_y}{dx} &= \frac{I_z M_y + I_{yz} M_z}{E(I_y I_z - I_{yz}^2)} \\
\frac{d\theta_z}{dx} &= \frac{I_{yz} M_y + I_y M_z}{E(I_y I_z - I_{yz}^2)} \\
\frac{dN}{dx} &= -\overline{p}_x \\
\frac{dV_y}{dx} &= -\overline{p}_y \\
\frac{dV_z}{dx} &= -\overline{p}_z \\
\frac{dM_y}{dx} &= V_z \\
\frac{dM_z}{dx} &= -V_y
\end{aligned} \tag{2.31}$$

or, in matrix notation,

$$\frac{d\mathbf{z}}{dx} = \mathbf{A}\mathbf{z} + \overline{\mathbf{P}} \tag{2.32a}$$

where

$$
\mathbf{z} = \begin{bmatrix} u \\ v \\ w \\ \theta_y \\ \theta_z \\ N \\ V_y \\ V_z \\ M_y \\ M_z \end{bmatrix} \quad
\mathbf{A} = \begin{bmatrix}
0 & 0 & 0 & 0 & 0 & 1/EA & 0 & 0 & 0 & 0 \\
0 & 0 & 0 & 0 & 1 & 0 & 0 & 0 & 0 & 0 \\
0 & 0 & 0 & -1 & 0 & 0 & 0 & 0 & 0 & 0 \\
0 & 0 & 0 & 0 & 0 & 0 & 0 & 0 & I_z/K & I_{yz}/K \\
0 & 0 & 0 & 0 & 0 & 0 & 0 & 0 & I_{yz}/K & I_y/K \\
0 & 0 & 0 & 0 & 0 & 0 & 0 & 0 & 0 & 0 \\
0 & 0 & 0 & 0 & 0 & 0 & 0 & 0 & 0 & 0 \\
0 & 0 & 0 & 0 & 0 & 0 & 0 & 0 & 0 & 0 \\
0 & 0 & 0 & 0 & 0 & 0 & 0 & 1 & 0 & 0 \\
0 & 0 & 0 & 0 & 0 & 0 & -1 & 0 & 0 & 0
\end{bmatrix} \quad
\overline{\mathbf{P}} = \begin{bmatrix} 0 \\ 0 \\ 0 \\ 0 \\ 0 \\ -\overline{p}_x \\ -\overline{p}_y \\ -\overline{p}_z \\ 0 \\ 0 \end{bmatrix} \tag{2.32b}
$$

with $K = E(I_y I_z - I_{yz}^2)$.

In contrast to the displacement form of the governing equations, these mixed or state variable equations do not involve derivatives with respect to x of the geometrical or material parameters and all derivatives are of first order. These first-order equations are often the most useful form of the governing equations, as numerical integration algorithms frequently utilize first-order derivatives only, and equations with higher-order derivatives must initially be reduced to a system with first-order derivatives.

For standard planar engineering beam theory the mixed equations are found by eliminating the strains $\boldsymbol{\epsilon}$ from Eq. (2.25) and ignoring axial motion terms,

$$\frac{dw}{dx} = -\theta + \frac{V}{k_s GA} \tag{2.33a}$$

$$\frac{d\theta}{dx} = \frac{M}{EI} \tag{2.33b}$$

$$\frac{dV}{dx} = -\overline{p}_z \tag{2.33c}$$

$$\frac{dM}{dx} = V \tag{2.33d}$$

or in matrix notation,

$$\frac{d\mathbf{z}}{dx} = \mathbf{A}\mathbf{z} + \overline{\mathbf{P}} \tag{2.34a}$$

where

$$
\mathbf{z} = \begin{bmatrix} w \\ \theta \\ V \\ M \end{bmatrix} \quad
\mathbf{A} = \begin{bmatrix} 0 & -1 & 1/k_s GA & 0 \\ 0 & 0 & 0 & 1/EI \\ 0 & 0 & 0 & 0 \\ 0 & 0 & 1 & 0 \end{bmatrix} \quad
\overline{\mathbf{P}} = \begin{bmatrix} 0 \\ 0 \\ -\overline{p}_z \\ 0 \end{bmatrix} \tag{2.34b}
$$

If axial deformation terms are included, Eq. (2.34a) still applies, but the matrices are defined as

$$\mathbf{z} = \begin{bmatrix} u \\ w \\ \theta \\ N \\ V \\ M \end{bmatrix} \qquad \mathbf{A} = \begin{bmatrix} 0 & 0 & 0 & 1/EA & 0 & 0 \\ 0 & 0 & -1 & 0 & 1/k_s GA & 0 \\ 0 & 0 & 0 & 0 & 0 & 1/\mathrm{EI} \\ 0 & 0 & 0 & 0 & 0 & 0 \\ 0 & 0 & 0 & 0 & 0 & 0 \\ 0 & 0 & 0 & 0 & 1 & 0 \end{bmatrix}$$

$$\overline{\mathbf{P}} = \begin{bmatrix} 0 \\ 0 \\ 0 \\ -\overline{p}_x \\ -\overline{p}_z \\ 0 \end{bmatrix} \tag{2.35}$$

2.1.7 Principle of Virtual Work: Integral Form of the Governing Equations

The principle of virtual work represents, in a sense, the global (integral) form of the conditions of equilibrium. Consult a structural mechanics textbook such as Pilkey and Wunderlich (1994) for a thorough discussion of the fundamentals of the principle of virtual work. Virtual work is expressed in terms of a small, admissible change (virtual) in displacements or strains. These virtual quantities, which correspond to variations of functions as in the calculus of variations, are indicated by δ.

The principle of virtual work is embodied in the relationship

$$\delta W = \delta W_{\text{int}} + \delta W_{\text{ext}} = 0 \tag{2.36}$$

where δW is the total virtual work, δW_{int} the internal virtual work, and δW_{ext} the virtual work of the applied loadings. The principle can be stated as: *A deformable solid is in equilibrium if the sum of the internal virtual work and the external virtual work is zero for virtual displacements that satisfy the strain–displacement equations and the displacement boundary conditions* (i.e., the virtual displacements are *kinematically admissible*). The fundamental unknowns for the principle of virtual work are the displacements, and the variations are taken on the displacements. This principle is also called the *principle of virtual displacements.*

The dimensions of a bar transverse to its axial (longitudinal) coordinate are very small relative to the length of the bar. Hence, the influence of the strains associated with the transverse directions will be neglected. The principle-of-virtual-work relationship $-\delta W = -\delta W_{\text{int}} - \delta W_{\text{ext}} = 0$ can be expressed in terms of stresses and strains as

$$
\overbrace{\int_V (\sigma_x\,\delta\epsilon_x + \tau_{xy}\,\delta\gamma_{xy} + \tau_{xz}\,\delta\gamma_{xz} + \underbrace{\sigma_y\,\delta\epsilon_y + \sigma_z\,\delta\epsilon_z + \tau_{yz}\,\delta\gamma_{yz})}_{\text{neglected}}\quad dV}^{-\delta W_{\text{int}}}
$$

$$
\overbrace{-\int_{S_p} (\delta u_x \overline{p}_x + \delta u_y \overline{p}_y + \delta u_z \overline{p}_z)\, dS}^{-\delta W_{\text{ext}}} \quad = 0 \tag{2.37}
$$

where the virtual work due to body forces has been ignored.

Kinematical Relations The three displacements $[u_x \quad u_y \quad u_z]^{\mathrm{T}} = \mathbf{u}$ describe the movement of a point. As discussed in Section 2.1.1, the description of the bending and axial motion of a point on the axis of a beam involves more quantities. These are the three translations u, v, w and the two rotations θ_y, θ_z. The displacements u_x, u_y, u_z of a point on the cross section of a beam are related to the beam axis variables $u, v, w, \theta_y, \theta_z$ by Eq. (2.1)

$$
u_x = u + z\theta_y - y\theta_z \qquad u_y = v \qquad u_z = w \tag{2.38}
$$

Strain–Displacement Relations Substitute the displacements of Eq. (2.38) into the strain–displacement relations of Eq. (1.3) that are needed in the virtual work expression of Eq. (2.37) [Eqs. (2.2) and (2.3), repeated]:

$$
\begin{aligned}
\epsilon_x &= \frac{\partial u_x}{\partial x} = \frac{du}{dx} + z\frac{d\theta_y}{dx} - y\frac{d\theta_z}{dx} \\
&= \kappa_\epsilon + z\kappa_y - y\kappa_z && \text{(2.39a)}
\end{aligned}
$$

$$
\gamma_{xy} = \frac{\partial u_y}{\partial x} + \frac{\partial u_x}{\partial y} = \frac{dv}{dx} - \theta_z \tag{2.39b}
$$

$$
\gamma_{xz} = \frac{\partial u_z}{\partial x} + \frac{\partial u_x}{\partial z} = \frac{dw}{dx} + \theta_y \tag{2.39c}
$$

where the curvatures are defined $\kappa_\epsilon = du/dx$, $\kappa_y = d\theta_y/dx$, and $\kappa_z = d\theta_z/dx$.

From Eq. (2.37), the internal virtual work is expressed as

$$
-\delta W_{\text{int}} = \int_V (\sigma_x\,\delta\epsilon_x + \tau_{xy}\,\delta\gamma_{xy} + \tau_{xz}\,\delta\gamma_{xz})\, dV \tag{2.40}
$$

Substitute the strain expressions of Eq. (2.39) into Eq. (2.40):

$$
\begin{aligned}
-\delta W_{\text{int}} = \int_x \int_A \Bigg[& \sigma_x\,\delta\left(\frac{du}{dx} + z\frac{d\theta_y}{dx} - y\frac{d\theta_z}{dx}\right) + \tau_{xy}\,\delta\left(\frac{dv}{dx} - \theta_z\right) \\
& + \tau_{xz}\,\delta\left(\frac{dw}{dx} + \theta_y\right)\Bigg]\, dA\, dx && \text{(2.41a)}
\end{aligned}
$$

$$= \int_x \Bigg[\underbrace{\left(\int_A \sigma_x \, dA \right)}_{N} \delta \frac{du}{dx} + \underbrace{\left(\int_A \sigma_x z \, dA \right)}_{M_y} \delta \frac{d\theta_y}{dx} + \underbrace{\left(- \int_A \sigma_x y \, dA \right)}_{M_z} \delta \frac{d\theta_z}{dx}$$

$$+ \underbrace{\left(\int_A \tau_{xy} \, dA \right)}_{V_y} \delta \left(\frac{dv}{dx} - \theta_z \right) + \underbrace{\left(\int_A \tau_{xz} \, dA \right)}_{V_z} \delta \left(\frac{dw}{dx} + \theta_y \right) \Bigg] \, dx \tag{2.41b}$$

$$= \int_x \left(N \, \delta\kappa_\epsilon + M_y \, \delta\kappa_y + M_z \, \delta\kappa_z + V_y \, \delta\gamma_{xy} + V_z \, \delta\gamma_{xz} \right) \, dx \tag{2.41c}$$

where stress resultants are defined by (Eqs. 1.45 and 1.46)

$$\begin{aligned} N &= \int_A \sigma_x \, dA \qquad M_y = \int_A \sigma_x z \, dA \qquad M_z = - \int_A \sigma_x y \, dA \\ V_y &= \int_A \tau_{xy} \, dA \qquad V_z = \int_A \tau_{xz} \, dA \end{aligned} \tag{2.41d}$$

Kinematic Assumptions to Simplify the Virtual Work Bernoulli's hypothesis maintains that plane cross sections remain plane and orthogonal to the beam axis during bending. This corresponds to neglecting the direct shear strains (Eq. 2.4):

$$\gamma_{xy} = \frac{dv}{dx} - \theta_z = 0 \qquad \gamma_{xz} = \frac{dw}{dx} + \theta_y = 0 \tag{2.42}$$

Thus Eq. (2.5)

$$\theta_z = \frac{dv}{dx} \quad \text{and} \quad \theta_y = -\frac{dw}{dx} \tag{2.43}$$

when shear deformation is ignored.

The internal virtual work expression of Eq. (2.41) can now be expressed as

$$-\delta W_{\text{int}} = \int_x \left(N \, \delta \frac{du}{dx} - M_y \, \delta \frac{d^2 w}{dx^2} + M_z \, \delta \frac{d^2 v}{dx^2} \right) \, dx \tag{2.44}$$

Introduction of the Material Law If the transverse normal stresses σ_y and σ_z are ignored and the constitutive relations of Eq. (1.12) are introduced into Eq. (2.40), we obtain

$$-\delta W_{\text{int}} = \int_V \left(E\epsilon_x \, \delta\epsilon_x + G\gamma_{xy} \, \delta\gamma_{xy} + G\gamma_{xz} \, \delta\gamma_{xz} \right) \, dV \tag{2.45}$$

From Eq. (2.39) with the shear strains γ_{xy} and γ_{xz} ignored,

$$-\delta W_{\text{int}} = \int_x \left[\int_A E(\kappa_\epsilon + z\kappa_y - y\kappa_z)\, \delta(\kappa_\epsilon + z\kappa_y - y\kappa_z)\, dA \right] dx \tag{2.46}$$

If E is constant over the cross section, this becomes

$$\begin{aligned}
-\,\delta W_{\text{int}} = \int_x E \Bigg\{ & \Bigg[\underbrace{\left(\int_A dA \right)}_{A} \kappa_\epsilon + \underbrace{\left(\int_A z\, dA \right)}_{Q_y} \kappa_y - \underbrace{\left(\int_A y\, dA \right)}_{Q_z} \kappa_z \Bigg] \delta\kappa_\epsilon \\
& + \Bigg[\underbrace{\left(\int_A z\, dA \right)}_{Q_y} \kappa_\epsilon + \underbrace{\left(\int_A z^2\, dA \right)}_{I_y} \kappa_y - \underbrace{\left(\int_A zy\, dA \right)}_{I_{yz}} \kappa_z \Bigg] \delta\kappa_y \\
& - \Bigg[\underbrace{\left(\int_A y\, dA \right)}_{Q_z} \kappa_\epsilon + \underbrace{\left(\int_A yz\, dA \right)}_{I_{yz}} \kappa_y - \underbrace{\left(\int_A y^2 dA \right)}_{I_z} \kappa_z \Bigg] \delta\kappa_z \Bigg\}\, dx
\end{aligned} \tag{2.47}$$

where the geometric properties of Eq. (1.52) have been utilized. To study bending only, ignore the strain in the axial direction. Also, since the centroid of the cross section is the origin of the coordinates, the first moments Q_y and Q_z are zero. Then

$$-\delta W_{\text{int}} = \int_x E(I_y\kappa_y\, \delta\kappa_y - I_{yz}\kappa_z\, \delta\kappa_y - I_{yz}\kappa_y\, \delta\kappa_z + I_z\kappa_z\, \delta\kappa_z)\, dx \tag{2.48}$$

The terms with the products of inertia I_{yz} can be avoided by moving to the centroidal principal bending axes. For this orientation (Chapter 1) $I_{y'z'} = 0$ and $I_{y'}$ and $I_{z'}$ are the principal moments of inertia. In this case

$$-\delta W_{\text{int}} = \int_x E(I_{y'}\kappa_y\, \delta\kappa_y + I_{z'}\kappa_z\, \delta\kappa_z)\, dx \tag{2.49}$$

Insertion of this expression into the principle of virtual work of Eq. (2.37) leads to

$$\begin{aligned}
-\delta W &= -\delta W_{\text{int}} - \delta W_{\text{ext}} \\
&= \int_x E(I_y\kappa_y\, \delta\kappa_y + I_z\kappa_z\, \delta\kappa_z)\, dx - \int_{S_p} (\delta u_y \overline{p}_y + \delta u_z \overline{p}_z)\, dS \\
&= 0
\end{aligned} \tag{2.50}$$

where the primes on y' and z' have been dropped and the applied axial force $\overline{p}_x$ has been ignored.

2.2 RESPONSE OF BEAM ELEMENTS

2.2.1 First-Order Form of the Governing Equations

There are a variety of effective analytical and numerical methods for solving a system of first-order equations of the form of Eq. (2.32). The first-order-form solutions are readily converted to stiffness matrix form in case the beam is a member of a framework. In this section several analytical solution techniques are discussed.

Exponential Expansion A familiar solution to the first-order scalar differential equation $dz/dx = Az + \overline{p}$ is

$$z = z_0 e^{Ax} + e^{Ax} \int_0^x e^{-A\tau} \overline{p}(\tau)\, d\tau$$

Similarly, integration of the matrix relations of Eq. (2.32) ($d\mathbf{z}/dx = \mathbf{Az} + \overline{\mathbf{P}}$) for a constant coefficient matrix $\mathbf{A}$ gives

$$\mathbf{z} = e^{\mathbf{A}x}\mathbf{z}_a + e^{\mathbf{A}x} \int_{x=a}^{x} e^{-\mathbf{A}\tau} \overline{\mathbf{P}}(\tau)\, d\tau \tag{2.51}$$

for an element beginning at $x = a$. For element e of length ℓ extending from $x = a$ to $x = b$,

$$\mathbf{z}_b = \mathbf{U}^e \left[\mathbf{z}_a + \int_a^b (\mathbf{U}^e)^{-1} \overline{\mathbf{P}}\, d\tau \right] = \mathbf{U}^e \mathbf{z}_a + \overline{\mathbf{z}}^e \tag{2.52}$$

where

$$\overline{\mathbf{z}}^e = \overline{\mathbf{z}}_b^e = \mathbf{U}^e \int_0^\ell [\mathbf{U}^e(\tau)]^{-1} \overline{\mathbf{P}}(\tau)\, d\tau \tag{2.53}$$

and as indicated above for a constant coefficient matrix $\mathbf{A}$,

$$\mathbf{U}^e = \mathbf{U}^e(\ell) = e^{\mathbf{A}(b-a)} = e^{\mathbf{A}\ell} \tag{2.54}$$

with $b - a = \ell$.

As is evident in Eq. (2.52), the matrix $\mathbf{U}^e$ transfers the state variables $\mathbf{z}$ from $x = a$ to $x = b$. Hence $\mathbf{U}^e$ is referred to as a *transfer matrix*.

The transfer matrix of Eq. (2.54) can be expanded as

$$\mathbf{U}^e = e^{\mathbf{A}\ell} = \mathbf{I} + \frac{\mathbf{A}\ell}{1!} + \frac{\mathbf{A}^2\ell^2}{2!} + \cdots = \sum_{s=0}^{\infty} \frac{\mathbf{A}^s \ell^s}{s!} \tag{2.55}$$

where $\mathbf{I}$ is the identity matrix, a diagonal matrix with diagonal values of unity.

Since $[\mathbf{U}^e(x)]^{-1} = e^{-\mathbf{A}x}$ for constant $\mathbf{A}$, it follows that

$$[\mathbf{U}^e(x)]^{-1} = \mathbf{U}^e(-x) \tag{2.56}$$

Then the loading expression of Eq. (2.53), with the help of Eq. (2.56), can be written as

$$\overline{\mathbf{z}}^e = e^{\mathbf{A}(b-a)} \int_a^b e^{-\mathbf{A}(\tau-a)} \overline{\mathbf{P}}\, d\tau = e^{\mathbf{A}\ell} \int_0^\ell e^{-\mathbf{A}\tau} \overline{\mathbf{P}}\, d\tau \tag{2.57}$$

Example 2.6 Transfer Matrix for a Bernoulli–Euler Beam. To illustrate the use of Eq. (2.55) to compute a transfer matrix, consider a Bernoulli–Euler beam with [Eq. (2.34b) with no shear deformation terms]

$$\mathbf{A} = \begin{bmatrix} 0 & -1 & 0 & 0 \\ 0 & 0 & 0 & 1/EI \\ 0 & 0 & 0 & 0 \\ 0 & 0 & 1 & 0 \end{bmatrix} \tag{1}$$

To use Eq. (2.55), the powers of $\mathbf{A}$ are needed. We find that

$$\mathbf{A}^2 = \mathbf{A}\mathbf{A} = \begin{bmatrix} 0 & 0 & 0 & -1/EI \\ 0 & 0 & 1/EI & 0 \\ 0 & 0 & 0 & 0 \\ 0 & 0 & 0 & 0 \end{bmatrix} \tag{2}$$

$$\mathbf{A}^3 = \mathbf{A}\mathbf{A}^2 = \begin{bmatrix} 0 & 0 & -1/EI & 0 \\ 0 & 0 & 0 & 0 \\ 0 & 0 & 0 & 0 \\ 0 & 0 & 0 & 0 \end{bmatrix} \tag{3}$$

$$\mathbf{A}^4 = \mathbf{A}\mathbf{A}^3 = \mathbf{0} \tag{4}$$

It is apparent that $\mathbf{A}^s = \mathbf{0}$ for s greater than 3. This result is expected since the analytical solution for this beam is a polynomial of order 3. Substitution of (2), (3), and (4) into Eq. (2.55) leads to the correct transfer matrix:

$$\mathbf{U}^e = \begin{bmatrix} 1 & -\ell & -\ell^3/6EI & -\ell^2/2EI \\ 0 & 1 & \ell^2/2EI & \ell/EI \\ 0 & 0 & 1 & 0 \\ 0 & 0 & \ell & 1 \end{bmatrix} \tag{5}$$

Numerous methods are available for computing transfer matrices; consult a reference such as Pilkey and Wunderlich (1994). The number of terms needed in an expansion of $e^{\mathbf{A}\ell}$ can be reduced by using a Padé approximation. The Laplace transform is one of the feasible solution techniques that can be used. Another approach is to use the Cayley–Hamilton theorem (i.e., the minimal polynomial) to represent the solution $e^{\mathbf{A}\ell}$ as a matrix polynomial. This requires knowledge of the eigenvalues of $\mathbf{A}$, which can be difficult to obtain for a large matrix $\mathbf{A}$.

For some structural members (e.g., circular plates), $\mathbf{A}$ is not constant. In such cases, numerical integration techniques can be used to find the transfer matrix, as can Picard iteration.

Example 2.7 Deflection of a Cantilevered Beam with an Asymmetrical Cross Section. For the cantilevered beam of Example 2.3 with an asymmetrical cross section, the applied loads $\overline{p}_y$ and $\overline{p}_z$ can be taken as zero over the interior of the beam span $0 < x < L$, and the applied load $\overline{P}$ in the z direction at $x = L$ can be treated as a boundary condition. Ignore $du/dx = N/AE$ and $dN/dx = -\overline{p}_x$. From Eq. (2.32b)

$$\mathbf{A} = \begin{bmatrix} 0 & 0 & 0 & 1 & 0 & 0 & 0 & 0 \\ 0 & 0 & -1 & 0 & 0 & 0 & 0 & 0 \\ 0 & 0 & 0 & 0 & 0 & 0 & I_z/K & I_{yz}/K \\ 0 & 0 & 0 & 0 & 0 & 0 & I_{yz}/K & I_y/K \\ 0 & 0 & 0 & 0 & 0 & 0 & 0 & 0 \\ 0 & 0 & 0 & 0 & 0 & 0 & 0 & 0 \\ 0 & 0 & 0 & 0 & 0 & 1 & 0 & 0 \\ 0 & 0 & 0 & 0 & -1 & 0 & 0 & 0 \end{bmatrix} \qquad \mathbf{z} = \begin{bmatrix} v \\ w \\ \theta_y \\ \theta_z \\ V_y \\ V_z \\ M_y \\ M_z \end{bmatrix} \tag{1}$$

where

$$K = E(I_y I_z - I_{yz}^2)$$

The transfer matrix is found by applying Eq. (2.55). Note that $\mathbf{A}^s = 0$ for s greater than 3.

$$\mathbf{U}^e = \begin{bmatrix} 1 & 0 & 0 & L & -I_y L^3/6K & I_{yz} L^3/6K & I_{yz} L^2/2K & I_y L^2/2K \\ 0 & 1 & -L & 0 & I_{yz} L^3/6K & -I_z L^3/6K & -I_z L^2/2K & -I_{yz} L^2/2K \\ 0 & 0 & 1 & 0 & -I_{yz} L^2/2K & I_z L^2/2K & I_z L/K & I_{yz} L/K \\ 0 & 0 & 0 & 1 & -I_y L^2/2K & I_{yz} L^2/2K & I_{yz} L/K & I_y L/K \\ 0 & 0 & 0 & 0 & 1 & 0 & 0 & 0 \\ 0 & 0 & 0 & 0 & 0 & 1 & 0 & 0 \\ 0 & 0 & 0 & 0 & 0 & L & 1 & 0 \\ 0 & 0 & 0 & 0 & -L & 0 & 0 & 1 \end{bmatrix} \tag{2}$$

Since there are no applied loads on the interior of the beam, $\overline{\mathbf{z}}^e$ is zero and Eq. (2.52) reduces to

$$\mathbf{z}_L = \mathbf{U}^e \mathbf{z}_0 \tag{3}$$

The initial parameter vector $\mathbf{z}_0$ is defined by applying boundary conditions to (3). The boundary conditions at the fixed end are

$$\begin{aligned} v(0) &= 0 \qquad \theta_z(0) = \frac{dv}{dx}(0) = 0 \\ w(0) &= 0 \qquad \theta_y(0) = -\frac{dw}{dx}(0) = 0 \end{aligned} \tag{4}$$

At the free end there are conditions on the shear forces and bending moments:

$$\begin{aligned} M_y(L) &= 0 \qquad M_z(L) = 0 \\ V_y(L) &= 0 \qquad V_z(L) = \overline{P} \end{aligned} \tag{5}$$

Equation (3) can be rewritten as

$$\begin{bmatrix} v_L \\ w_L \\ \theta_{yL} \\ \theta_{zL} \\ V_{yL} = 0 \\ V_{zL} = \overline{P} \\ M_{yL} = 0 \\ M_{zL} = 0 \end{bmatrix} = \begin{bmatrix} \vdots & \vdots & \vdots & \vdots & \cdots & \cdots & \cdots & \cdots \\ \vdots & \vdots & \vdots & \vdots & \cdots & \cdots & \cdots & \cdots \\ \vdots & \vdots & \vdots & \vdots & \cdots & \cdots & \cdots & \cdots \\ \vdots & \vdots & \vdots & \vdots & \cdots & \cdots & \cdots & \cdots \\ \vdots & \vdots & \vdots & \vdots & 1 & 0 & 0 & 0 \\ \vdots & \vdots & \vdots & \vdots & 0 & 1 & 0 & 0 \\ \vdots & \vdots & \vdots & \vdots & 0 & L & 1 & 0 \\ \vdots & \vdots & \vdots & \vdots & -L & 0 & 0 & 1 \end{bmatrix} \begin{bmatrix} v_0 = 0 \\ w_0 = 0 \\ \theta_{y0} = 0 \\ \theta_{z0} = 0 \\ V_{y0} \\ V_{z0} \\ M_{y0} \\ M_{z0} \end{bmatrix} \tag{6}$$

where columns 1 through 4 have been canceled because the corresponding displacements at $x = 0$ are zero, and rows 1 through 4 have been temporarily ignored because the displacements at $x = L$ are unknown.

The unknown shear forces and bending moments at $x = 0$ can be calculated using the reduced transfer matrix of (6):

$$\begin{bmatrix} V_{y0} \\ V_{zo} \\ M_{y0} \\ M_{zo} \end{bmatrix} = \begin{bmatrix} 1 & 0 & 0 & 0 \\ 0 & 1 & 0 & 0 \\ 0 & L & 1 & 0 \\ -L & 0 & 0 & 1 \end{bmatrix}^{-1} \begin{bmatrix} 0 \\ \overline{P} \\ 0 \\ 0 \end{bmatrix} = \begin{bmatrix} 0 \\ \overline{P} \\ -L\overline{P} \\ 0 \end{bmatrix} \tag{7}$$

Of course, these are the reactions that can be obtained from the conditions of equilibrium. With all the elements of $\mathbf{z}_0$ now known, $\mathbf{z}_L$ can be calculated using Eq. (3)

$$\mathbf{z}_L = \begin{bmatrix} 1 & 0 & 0 & L & -\dfrac{I_y L^3}{6K} & \dfrac{I_{yz} L^3}{6K} & \dfrac{I_{yz} L^2}{2K} & \dfrac{I_y L^2}{2K} \\ 0 & 1 & -L & 0 & \dfrac{I_{yz} L^3}{6K} & -\dfrac{I_z L^3}{6K} & -\dfrac{I_z L^2}{2K} & -\dfrac{I_{yz} L^2}{2K} \\ 0 & 0 & 1 & 0 & -\dfrac{I_{yz} L^2}{2K} & \dfrac{I_z L^2}{2K} & \dfrac{I_z L}{K} & \dfrac{I_{yz} L}{K} \\ 0 & 0 & 0 & 1 & -\dfrac{I_y L^2}{2K} & \dfrac{I_{yz} L^2}{2K} & \dfrac{I_{yz} L}{K} & \dfrac{I_y L}{K} \\ 0 & 0 & 0 & 0 & 1 & 0 & 0 & 0 \\ 0 & 0 & 0 & 0 & 0 & 1 & 0 & 0 \\ 0 & 0 & 0 & 0 & 0 & L & 1 & 0 \\ 0 & 0 & 0 & 0 & -L & 0 & 0 & 1 \end{bmatrix} \begin{bmatrix} 0 \\ 0 \\ 0 \\ 0 \\ 0 \\ 0 \\ -L\overline{P} \\ 0 \end{bmatrix}$$

$$= \begin{bmatrix} -\dfrac{I_{yz} L^3 \overline{P}}{3K} \\ \dfrac{I_z L^3 \overline{P}}{3K} \\ -\dfrac{I_z L^2 \overline{P}}{2K} \\ -\dfrac{I_{yz} L^2 \overline{P}}{2K} \\ 0 \\ \overline{P} \\ 0 \\ 0 \end{bmatrix} \tag{8}$$

These are the values obtained in Example 2.3.

Example 2.8 Transfer Matrix for a General Beam Element Based on Standard Planar Engineering Beam Theory. Standard planar engineering beam theory for beam elements with such effects as elastic foundations and axial loads remains as a fourth-order problem in that the governing equations are four first-order equations or a single fourth-order equation. For example, for a beam with compressive axial load N, displacement (Winkler) foundation modulus k_w, rotary foundation modulus k_θ, mass per unit length ρ, radius of gyration $r_y^2 = I/A$, applied moment per unit length $\overline{c}$, frequency of harmonic motion ω, shear correction factor $\alpha_z = 1/k_s$, and thermal moment $M_T = \int_A E\alpha \, \Delta T z \, dA$ (α is the coefficient of thermal expansion and ΔT is the temperature change), the first-order differential equations for the response are given by

$$
\begin{aligned}
\frac{dw}{dx} &= -\theta + \frac{V}{k_s GA} \\
\frac{d\theta}{dx} &= \frac{M}{EI} + \frac{M_T}{EI} \\
\frac{dV}{dx} &= k_w w - \rho\omega^2 w - \overline{p}_z(x) \\
\frac{dM}{dx} &= V + (k_\theta - N)\theta - \rho r_y^2 \omega^2 \theta - \overline{c}(x)
\end{aligned}
\tag{1}
$$

or, in matrix notation,

$$
\frac{d\mathbf{z}}{dx} = \mathbf{A}\mathbf{z} + \overline{\mathbf{P}} \tag{2}
$$

where

$$
\mathbf{z} = \begin{bmatrix} w \\ \theta \\ V \\ M \end{bmatrix} \quad \mathbf{A} = \begin{bmatrix} 0 & -1 & 1/k_s GA & 0 \\ 0 & 0 & 0 & 1/EI \\ k_w - \rho\omega^2 & 0 & 0 & 0 \\ 0 & k_\theta - N - \rho r_y^2 \omega^2 & 1 & 0 \end{bmatrix} \quad \overline{\mathbf{P}} = \begin{bmatrix} 0 \\ M_T/EI \\ -\overline{p}_z \\ -\overline{c} \end{bmatrix}
$$

If the applied loadings are set to zero, (1) or (2) can be combined into the fourth-order equation

$$
\frac{d^4 w}{dx^4} + (\zeta - \eta)\frac{d^2 w}{dx^2} + (\lambda - \zeta\eta)w = 0 \tag{3}
$$

where

$$
\begin{aligned}
\zeta &= \frac{1}{EI}(N - k_\theta + \rho r_y^2 \omega^2) \\
\eta &= \frac{1}{k_s GA}(k_w - \rho\omega^2) \\
\lambda &= \frac{1}{EI}(k_w - \rho\omega^2)
\end{aligned}
$$

A number of techniques are available to solve the first-order equations of (2) or the fourth-order equation of (3). One of these, an exponential expansion, was presented earlier in this section. Other solution methods are detailed in Pilkey (1994) and in Pilkey and Wunderlich (1994). The transfer matrix solution for a general beam element is given in Table 2.2. This table includes applied loading functions for a variety of types of loading.

Note that in Table 2.2 the transfer matrix has been "extended" by introducing a fifth column to include the particular solutions F_w, F_θ, F_V, F_M due to applied loading. These correspond to the responses w, θ, V, M, respectively. In addition to

TABLE 2.2A Transfer and Stiffness Matrices for a General Beam Element

Notation

$\lambda = (k_w - \rho\omega^2)/EI$
$\xi = EI/(k_s AG)$
k_θ = rotary foundation modulus
ℓ = element length
G = shear modulus of elasticity
$k_s A$ = equivalent shear area, $k_s = 1/\alpha_z$
$\eta = (k_w - \rho\omega^2)/(k_s AG)$
k_w = displacement (Winkler) foundation modulus
ω = natural frequency
I = moment of inertia
r_y = radius of gyration
α_z = shear correction factor (Chapter 6)
$\zeta = (N - k_\theta + \rho r_y^2 \omega^2)/EI$
ρ = mass per unit length
E = elastic modulus, Young's modulus
N = axial force compessive: replace by $-N$ for tensile axial force

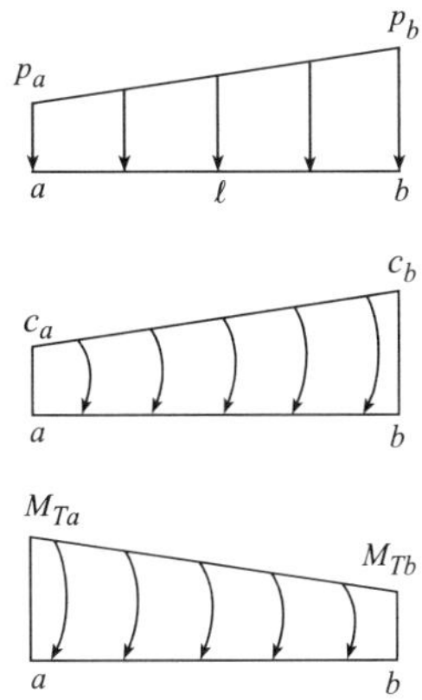

To use the matrices, follow the steps:

1. Calculate the three parameters γ, ζ, η. If shear deformation is not to be considered, set $1/k_s AG = 0$.
2. Compare the magnitude of these parameters and look up the appropriate e_i using the definitions of e_i given below
3. Substitute these expressions in the matrices below.

(continued)

the fifth column, the transfer matrix of Table 2.2 is extended with a fifth row, and a fifth element in the vector $\mathbf{z}_a$ has been added.

Example 2.9 Reduction of a General Beam Element Solution to the Transfer Matrix for a Bernoulli–Euler Beam. To demonstrate the use of Table 2.2 to find a transfer matrix, consider a Bernoulli–Euler beam. First, evaluate the constants λ, η, and ζ of Table 2.2. Since for a Bernoulli–Euler beam k_w, ρ, N, and k_θ are zero, we find that

$$\lambda = 0 \qquad \eta = 0 \qquad \zeta = 0 \tag{1}$$

From the column in Table 2.2 for $\lambda = 0$, $\eta = 0$, $\zeta = 0$, we find that

$$e_0 = 0 \qquad e_1 = 1 \qquad e_2 = \ell \qquad e_3 = \frac{\ell^2}{2} \qquad e_4 = \frac{\ell^3}{6} \tag{2}$$

Insertion of these values for e_i, $i = 0, 1, \ldots, 4$, into the transfer matrix at the top of Table 2.2 leads immediately to the correct transfer matrix. This is the same matrix derived in Example 2.6.

TABLE 2.2A (*Continued*)

Transfer Matrix (Sign Convention 1)

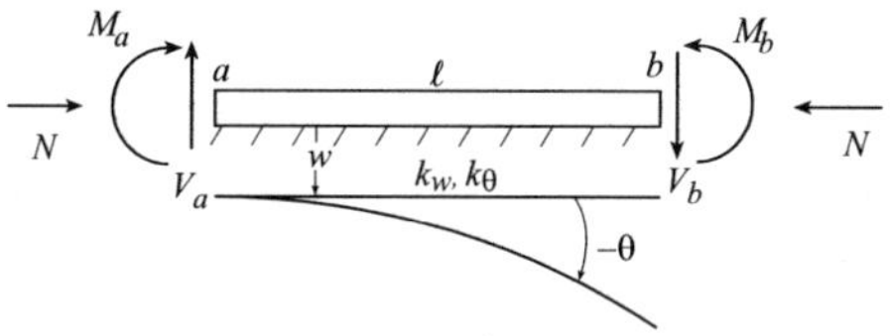

$$\mathbf{z}_b = \mathbf{U}^e \mathbf{z}_a$$

$$\underbrace{\begin{bmatrix} e_1 + \zeta e_3 & -e_2 & -\dfrac{e_4}{EI} + \dfrac{(e_2 + \zeta e_4)}{k_s AG} & -\dfrac{e_3}{EI} & F_w \\ \lambda e_1 & e_1 - \eta e_3 & \dfrac{e_3}{EI} & \dfrac{(e_2 - \eta e_4)}{EI} & F_\theta \\ \lambda EI(e_2 + \zeta e_4) & -\lambda EI e_3 & e_1 + \zeta e_3 & -\lambda e_4 & F_V \\ \lambda EI e_3 & EI(e_0 - \eta e_2) & e_2 & e_1 - \eta e_3 & F_M \\ 0 & 0 & 0 & 0 & 1 \end{bmatrix}}_{\mathbf{U}^e} \underbrace{\begin{bmatrix} w_a \\ \theta_a \\ V_a \\ M_a \\ 1 \end{bmatrix}}_{\mathbf{z}_a}$$

$$F_w = [p_a(e_5 - e_6/\ell) + p_b e_6/\ell + c_a(e_4 - e_5/\ell) + c_b e_5/\ell - M_{Ta}(e_3 - e_4/\ell)$$
$$- M_{Tb} e_4/\ell]/EI - \{p_a[e_3 + \zeta e_5 - (e_4 + \zeta e_6)/\ell] + p_b(e_4 + \zeta e_6)/\ell\}/k_s AG$$

$$F_\theta = \{p_a(-e_4 + e_5/\ell) - p_b e_5/\ell + c_a[-e_3 + \eta e_5 + (e_4 - \eta e_6)/\ell] - c_b(e_4 - \eta e_6)/\ell$$
$$+ M_{Ta}[(e_2 - \eta e_4) - (e_3 - \eta e_5)/\ell] + M_{Tb}(e_3 - \eta e_5)/\ell\}/EI$$

$$F_V = p_a\{-(e_2 + \zeta e_4) + (e_3 + \zeta e_5)/\ell\} - p_b(e_3 + \zeta e_5)/\ell + \lambda[c_a(e_5 - e_6/\ell) + c_b e_6/\ell$$
$$+ M_{Ta}(-e_4 + c_5/\ell) - M_{Tb} e_5/\ell]$$

$$F_M = p_a(-e_3 + e_4/\ell) - p_b e_4/\ell + c_a[(-e_2 + e_3/\ell) + \eta(e_4 - e_5/\ell)] - c_b(e_3 - \eta e_5)/\ell$$
$$+ M_{Ta}[(e_1 - 1 - e_2/\ell) + \eta(-e_3 + e_4/\ell)] + M_{Tb}(e_2 - \ell - \eta e_4)/\ell$$

2.2.2 Sign Conventions for Beams

The sign convention for internal forces employed thus far in this book is in common use for analytical formulas of structural members. We refer to this as *Sign Convention 1*. The face of a beam element with its outward normal along a positive direction of a coordinate axis is said to be a positive face. A face with its normal in the opposite direction is defined to be a negative face. As indicated in Fig. 2.7, where only bending in the xz plane is displayed, for Sign Convention 1 force and moment components on the positive face of a beam element are positive when acting along the positive direction of the coordinates. Components are defined to be positive on the negative face when acting in the negative direction of the coordinate axis.

Another sign convention, which will be called *Sign Convention 2*, is more appropriate for use with the stiffness method of analysis of a network of beams. In the

TABLE 2.2A ***(Continued)***

Stiffness Matrix (Sign Convention 2)

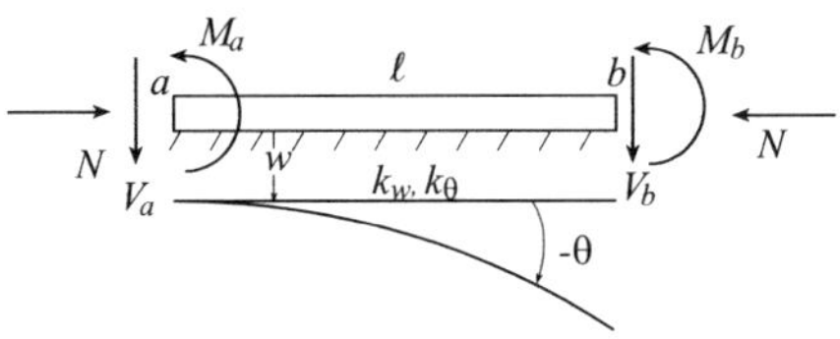

$\mathbf{p}^e = \mathbf{k}^e\mathbf{v}^e + \mathbf{p}^{eo}$

$$\begin{bmatrix} V_a \\ M_a \\ V_b \\ M_b \end{bmatrix} = \begin{bmatrix} k_{11} & k_{12} & k_{13} & k_{14} \\ k_{21} & k_{22} & k_{23} & k_{24} \\ k_{31} & k_{32} & k_{33} & k_{34} \\ k_{41} & k_{42} & k_{43} & k_{44} \end{bmatrix} \begin{bmatrix} w_a \\ \theta_a \\ w_b \\ \theta_b \end{bmatrix} + \mathbf{p}^{e0}$$

$$\mathbf{p}^e \quad = \qquad \mathbf{k}^e \qquad \mathbf{v}^e \quad + \mathbf{p}^{e0}$$

$$k_{11} = [(e_2 - \eta e_4)(e_1 + \zeta e_3) + \lambda e_3 e_4]EI/\Delta$$
$$k_{12} = [e_3(e_1 - \eta e_3) - e_2(e_2 - \eta e_4)]EI/\Delta$$
$$k_{13} = -(e_2 - \eta e_4)EI/\Delta$$
$$k_{14} = -e_3 EI/\Delta$$
$$k_{21} = k_{12}$$
$$k_{22} = \{-(e_1 - \eta e_3)[e_4 - \xi(e_2 + \zeta e_4)] + e_2 e_3\}EI/\Delta$$
$$k_{23} = e_3 EI/\Delta = -k_{14}$$
$$k_{24} = [e_4 - \xi(e_2 + \zeta e_4)]EI/\Delta$$
$$k_{31} = k_{13}, k_{41} = k_{14}, k_{42} = k_{24}$$
$$k_{32} = k_{23}, k_{43} = k_{34}$$
$$k_{33} = [(e_1 + \zeta e_3)(e_2 - \eta e_4) + \lambda e_3 e_4]EI/\Delta = k_{11}$$
$$k_{34} = \{(e_1 + \zeta e_3)e_3 + \lambda e_4[e_4 - \xi(e_2 + \zeta e_4)]\}EI/\Delta$$
$$k_{44} = \{e_2 e_3 - (e_1 - \eta e_3)[e_4 - \xi(e_2 + \zeta e_4)]\}EI/\Delta = k_{22}$$
$$\Delta = e_3^2 - (e_2 - \eta e_4)[e_4 - \xi(e_2 + \zeta e_4)]$$

$$\mathbf{p}^{e0} = \begin{bmatrix} V_a^0 \\ M_a^0 \\ V_b^0 \\ M_b^0 \end{bmatrix}$$

$$V_a^0 = [(e_2 - \eta e_4)F_w + e_3 F_\theta]EI/\Delta$$
$$M_a^0 = -\{e_3 F_w + [e_4 - \xi(e_2 + \zeta e_4)]F_\theta\}EI/\Delta$$
$$V_b^0 = F_V - \{[(e_1 + \zeta e_3)(e_2 - \eta e_4) + \lambda e_3 e_4]F_w + [(e_1 + \zeta e_3)e_3 + \lambda e_4[e_4 - \xi(e_2 + \zeta e_4)]]F_\theta\}EI/\Delta$$
$$M_b^0 = F_M - \{[(e_1 + \zeta e_3)e_3 + \xi e_4[e_4 - \xi(e_2 + \zeta e_4)]]F_w + [e_2 e_3 - (e_1 - \eta e_3)[e_4 - \xi(e_2 + \zeta e_4)]]F_\theta\}EI/\Delta$$

TABLE 2.2B

Definitions for $e_i\,(i = 0, 1, 2, \ldots, 6)$						
		$\lambda = 0, \lambda - \zeta\eta = 0$			$\lambda > 0, \lambda - \zeta\eta > 0$	
	1. $\lambda < 0$	2. $\zeta = \eta = 0$	3. $\eta = 0, \zeta \neq 0$	4. $\lambda - \zeta\eta = \frac{1}{4}(\zeta - \eta)^2$	5. $\lambda - \zeta\eta < \frac{1}{4}(\zeta - \eta)^2$, $\zeta - \eta \neq 0$	6. $\lambda - \zeta\eta > \frac{1}{4}(\zeta - \eta)^2$
e_0	$\frac{1}{g}(d^3C - q^3D)$	0	$-\zeta B$	$-\frac{\zeta - \eta}{4}(3C + A\ell)$	$-\frac{1}{g}(q^3D - d^3C)$	$-(\lambda - \zeta\eta)e_4 - (\zeta - \eta)e_2$
e_1	$\frac{1}{g}(d^2A + q^2B)$	1	A	$\frac{1}{2}(2A - B\ell)$	$\frac{p}{g}(q^2B - d^2A)$	$AB - \frac{q^2 - d^2}{2dq}CD$
e_2	$\frac{1}{g}(dC + qD)$	ℓ	B	$\frac{1}{2}(C + A\ell)$	$\frac{p}{g}(qD - dC)$	$\frac{1}{2dq}(dAD + qBC)$
e_3	$\frac{1}{g}(A - B)$	$\frac{\ell^2}{2}$	$\frac{1}{\zeta}(1 - A)$	$\frac{C\ell}{2}$	$\frac{1}{g}(A - B)$	$\frac{1}{2dq}CD$
e_4	$\frac{1}{g}\left(\frac{C}{d} - \frac{D}{q}\right)$	$\frac{\ell^3}{6}$	$\frac{1}{\zeta}(\ell - B)$	$\frac{1}{(\zeta - \eta)}(C - A\ell)$	$\frac{1}{g}\left(\frac{C}{d} - \frac{D}{q}\right)$	$\frac{1}{2(d^2 + q^2)}\left(\frac{AD}{q} - \frac{BC}{d}\right)$
e_5	$\frac{1}{g}\left(\frac{A}{d^2} + \frac{B}{q^2}\right) - \frac{1}{d^2q^2}$	$\frac{\ell^4}{24}$	$\frac{1}{\zeta}\left(\frac{\ell^2}{2} - e_3\right)$	$\frac{2}{(\zeta - \eta)^2}(-2A - B\ell + 2)$	$\frac{p}{g}\left(\frac{B}{q^2} - \frac{A}{d^2}\right) + \frac{1}{d^2q^2}$	$\frac{1 - e_1}{\lambda - \zeta\eta} - \frac{\zeta - \eta}{\lambda - \zeta\eta}e_3$
e_6	$\frac{1}{g}\left(\frac{C}{d^3} + \frac{D}{q^3}\right) - \frac{\ell}{d^2q^2}$	$\frac{\ell^5}{120}$	$\frac{1}{\zeta}\left(\frac{\ell^3}{6} - e_4\right)$	$\frac{2}{(\zeta - \eta)^2}(-3C + A\ell + 2\ell)$	$\frac{p}{g}\left(\frac{D}{q^3} - \frac{C}{d^3}\right) + \frac{\ell}{d^2q^2}$	$\frac{\ell - e_2}{\lambda - \zeta\eta} - \frac{\zeta - \eta}{\lambda - \zeta\eta}e_4$

(continued)

TABLE 2.2B (*Continued*)

Definitions for A, B, C, D, g, d, q				
		$\lambda > 0, \lambda - \zeta\eta > 0$		
$\lambda < 0$	$\lambda = 0, \lambda - \zeta\eta = 0$	$\lambda - \zeta\eta = \frac{1}{4}(\zeta - \eta)^2$	$\lambda - \zeta\eta < \frac{1}{4}(\zeta - \eta)^2, \zeta - \eta \neq 0$	$\lambda - \zeta\eta > \frac{1}{4}(\zeta - \eta)^2$
1. $A = \cosh d\ell$ $B = \cos q\ell$ $C = \sinh d\ell$ $D = \sin q\ell$ $g = d^2 + q^2$	2. $\zeta > 0$: $\alpha^2 = \zeta$ $A = \cos \alpha\ell$ $B = (\sin \alpha\ell)/\alpha$	4. $\zeta - \eta > 0$: $\beta^2 = \frac{1}{2}(\zeta - \eta)$ $A = \cos \beta\ell$, $B = \beta \sin \beta\ell$ $C = (\sin \beta\ell)/\beta$	6. $\zeta - \eta > 0$: $g = q^2 - d^2$, $p = 1$ $A = \cos d\ell$, $B = \cos q\ell$ $C = \sin d\ell$, $D = \sin q\ell$ $d^2 = \frac{1}{2}(\zeta - \eta) - \sqrt{\frac{1}{4}(\zeta + \eta)^2 - \lambda}$ $q^2 = \frac{1}{2}(\zeta - \eta) + \sqrt{\frac{1}{4}(\zeta + \eta)^2 - \lambda}$	8. $A = \cosh d\ell$, $B = \cos q\ell$ $C = \sinh d\ell$, $D = \sin q\ell$ $d^2 = \frac{1}{2}\sqrt{\lambda - \zeta\eta} - \frac{1}{4}(\zeta - \eta)$ $q^2 = \frac{1}{2}\sqrt{\lambda - \zeta\eta} + \frac{1}{4}(\zeta - \eta)$
$d^2 = \sqrt{\beta^4 + \frac{1}{4}(\zeta + \eta)^2} - \frac{1}{2}(\zeta - \eta)$ $q^2 = \sqrt{\beta^4 + \frac{1}{4}(\zeta + \eta)^2} + \frac{1}{2}(\zeta - \eta)$ $\beta^4 = -\lambda$	3. $\zeta < 0$: $\alpha^2 = -\zeta$ $A = \cosh \alpha\ell$ $B = (\sinh \alpha\ell)/\alpha$	5. $\zeta - \eta < 0$: $\beta^2 = -\frac{1}{2}(\zeta - \eta)$ $A = \cosh \beta\ell$, $B = -\beta \sinh \beta\ell$ $C = (\sinh \beta\ell)/\beta$	7. $\zeta - \eta < 0$: $g = d^2 - q^2$, $p = -1$ $A = \cosh d\ell$, $B = \cosh q\ell$ $C = \sinh d\ell$, $D = \sinh q\ell$ $d^2 = -\frac{1}{2}(\zeta - \eta) + \sqrt{\frac{1}{4}(\zeta + \eta)^2 - \lambda}$ $q^2 = -\frac{1}{2}(\zeta - \eta) - \sqrt{\frac{1}{4}(\zeta + \eta)^2 - \lambda}$	

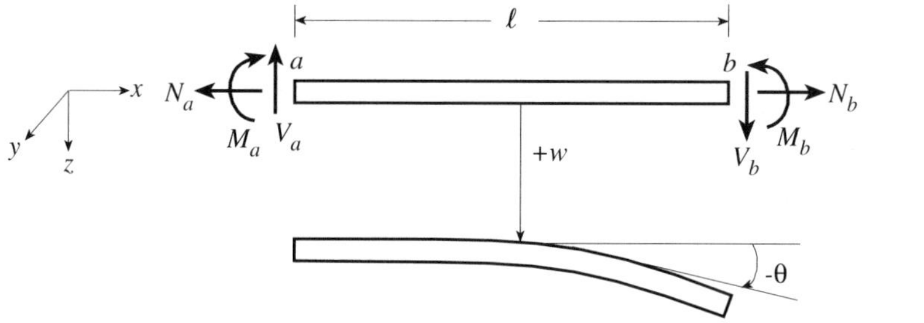

(*a*) Sign Convention 1: positive forces, moments, slopes, and displacements are shown. Used for analytical formulas of beam elements.

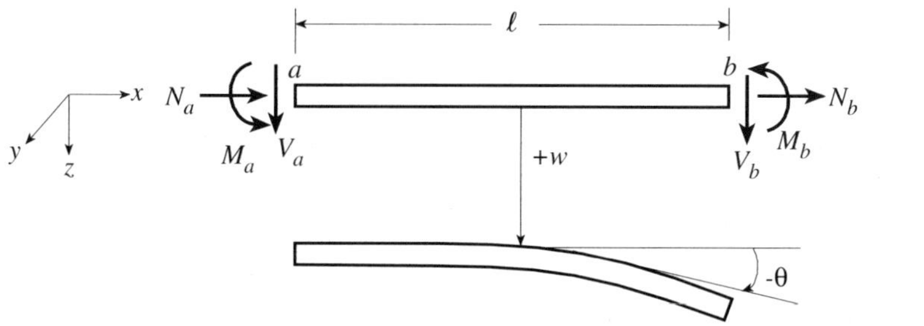

(*b*) Sign Convention 2: positive forces, moments, slopes, and displacements are shown. Positive deflection and slope are the same for as for Sign Convention 1. Sign Convention 2 is convenient to use in the matrix analysis of beam networks.

Figure 2.7 Sign conventions for beam element.

case of Sign Convention 2, the force and moment components on both ends of the beam element are positive along the positive coordinate directions. From Fig. 2.7 it is evident that the difference between the two sign conventions occurs at the left end, $x = a$, where the force and moment components differ in sign.

2.2.3 Definition of Stiffness Matrices

A stiffness matrix provides a relationship between the forces $\mathbf{p}^e$ on both ends of a beam element and the displacements $\mathbf{v}^e$ on both ends. For the eth beam element, the stiffness matrix $\mathbf{k}^e$ is defined as

$$\mathbf{p}^e = \mathbf{k}^e \mathbf{v}^e \tag{2.58}$$

Let the forces (actually, forces and moments) at the left end ($x = a$) of a beam element be $\mathbf{p}_a$. For example, for standard planar engineering beam theory

$$\mathbf{p}_a = \begin{bmatrix} V_a \\ M_a \end{bmatrix} \tag{2.59}$$

Also, designate the forces and moments at the right end ($x = b$) as $\mathbf{p}_b$, the displacements (deflections and slopes) at the left end as $\mathbf{v}_a = [w_a \quad \theta_a]^T$, and displacements at the right end as $\mathbf{v}_b$. Then Eq. (2.58) can be written as

$$\begin{aligned} \begin{bmatrix} \mathbf{p}_a \\ \mathbf{p}_b \end{bmatrix} &= \begin{bmatrix} \mathbf{k}_{aa} & \mathbf{k}_{ab} \\ \mathbf{k}_{ba} & \mathbf{k}_{bb} \end{bmatrix} \begin{bmatrix} \mathbf{v}_a \\ \mathbf{v}_b \end{bmatrix} \\ \mathbf{p}^e &= \qquad \mathbf{k}^e \qquad\quad \mathbf{v}^e \end{aligned} \tag{2.60}$$

where $\mathbf{k}_{jk}$ are matrices.

2.2.4 Determination of Stiffness Matrices

Conversion of Transfer Matrices Recall that the state vector $\mathbf{z}$ of the transfer matrix solution ($\mathbf{z}_b = \mathbf{U}^e\mathbf{z}_a + \overline{\mathbf{z}}^e$) is composed of the displacements and forces at the end $x = a$ and at the end $x = b$ of member e. Thus, if the displacements at a are $\mathbf{v}_a$ and the forces are $\mathbf{p}_a$, the state vector appears as

$$\mathbf{z}_a = \begin{bmatrix} \mathbf{v}_a \\ \mathbf{p}_a \end{bmatrix} \tag{2.61}$$

Similarly, at the right end ($x = b$),

$$\mathbf{z}_b = \begin{bmatrix} \mathbf{v}_b \\ \mathbf{p}_b \end{bmatrix} \tag{2.62}$$

Also, subdivide the applied force vector $\overline{\mathbf{z}}^e$ of Eq. (2.57) as

$$\overline{\mathbf{z}}^e = \begin{bmatrix} \mathbf{F}_v \\ \mathbf{F}_p \end{bmatrix} \tag{2.63}$$

Then the transfer matrix solution can be written as

$$\begin{aligned} \begin{bmatrix} \mathbf{v}_b \\ \mathbf{p}_b \end{bmatrix} &= \begin{bmatrix} \mathbf{U}_{vv} & \mathbf{U}_{vp} \\ \mathbf{U}_{pv} & \mathbf{U}_{pp} \end{bmatrix} \begin{bmatrix} \mathbf{v}_a \\ \mathbf{p}_a \end{bmatrix} + \begin{bmatrix} \mathbf{F}_v \\ \mathbf{F}_p \end{bmatrix} \\ \mathbf{z}_b &= \qquad \mathbf{U}^e \qquad\quad \mathbf{z}_a \;\; + \;\; \overline{\mathbf{z}}^e \end{aligned} \tag{2.64}$$

where the transfer matrix has been subdivided into the square submatrices $\mathbf{U}_{vv}$, $\mathbf{U}_{vp}$, $\mathbf{U}_{pv}$, $\mathbf{U}_{pp}$. It is assumed that the forces are defined according to Sign Convention 2.

It is apparent by comparing Eqs. (2.60) and (2.64) that the stiffness matrix can be obtained by reorganizing the transfer matrix. From (Eq. 2.64)

$$\mathbf{p}_b = \mathbf{U}_{pv}\mathbf{v}_a + \mathbf{U}_{pp}\mathbf{p}_a + \mathbf{F}_p \qquad \mathbf{v}_b = \mathbf{U}_{vv}\mathbf{v}_a + \mathbf{U}_{vp}\mathbf{p}_a + \mathbf{F}_v$$

it follows that

$$\mathbf{p}_a = \mathbf{U}_{vp}^{-1}\mathbf{v}_b - \mathbf{U}_{vp}^{-1}\mathbf{U}_{vv}\mathbf{v}_a - \mathbf{U}_{vp}^{-1}\mathbf{F}_v$$

and

$$\mathbf{p}_b = \mathbf{U}_{pv}\mathbf{v}_a + \mathbf{U}_{pp}\mathbf{p}_a + \mathbf{F}_p = (\mathbf{U}_{pv} - \mathbf{U}_{pp}\mathbf{U}_{vp}^{-1}\mathbf{U}_{vv})\mathbf{v}_a + \mathbf{U}_{pp}\mathbf{U}_{vp}^{-1}\mathbf{v}_b + \mathbf{F}_p - \mathbf{U}_{pp}\mathbf{U}_{vp}^{-1}\mathbf{F}_v$$

In matrix form

$$\begin{bmatrix} \mathbf{p}_a \\ \\ \mathbf{p}_b \end{bmatrix} = \begin{bmatrix} -\mathbf{U}_{vp}^{-1}\mathbf{U}_{vv} & \vdots & \mathbf{U}_{vp}^{-1} \\ \cdots\cdots\cdots\cdots\cdots & \vdots & \cdots\cdots\cdots \\ \mathbf{U}_{pv} - \mathbf{U}_{pp}\mathbf{U}_{vp}^{-1}\mathbf{U}_{vv} & \vdots & \mathbf{U}_{pp}\mathbf{U}_{vp}^{-1} \end{bmatrix} \begin{bmatrix} \mathbf{v}_a \\ \\ \mathbf{v}_b \end{bmatrix} + \begin{bmatrix} -\mathbf{U}_{vp}^{-1}\mathbf{F}_v \\ \\ \mathbf{F}_p - \mathbf{U}_{pp}\mathbf{U}_{vp}^{-1}\mathbf{F}_v \end{bmatrix}$$

$$\mathbf{p}^e \quad = \quad \mathbf{k}^e \quad \mathbf{v}^e \quad + \quad \mathbf{p}^{e0} \tag{2.65}$$

The force vector $\mathbf{p}^{e0}$ of Eq. (2.65) can be used to include the influence of distributed load between the nodes.

Example 2.10 Stiffness Matrix for a Bernoulli–Euler Beam. The transfer matrix of Eq. (5) of Example 2.6 can be converted to a stiffness matrix using Eq. (2.65). First change the signs of V_a and M_a to make the adjustment from Sign Convention 1 to Sign Convention 2. From Eq. (5),

$$\mathbf{U}_{vv} = \begin{bmatrix} 1 & -\ell \\ 0 & 1 \end{bmatrix} \qquad \mathbf{U}_{vp} = \begin{bmatrix} \ell^3/6EI & \ell^2/2EI \\ -\ell^2/2EI & -\ell/EI \end{bmatrix}$$
$$\mathbf{U}_{pv} = \begin{bmatrix} 0 & 0 \\ 0 & 0 \end{bmatrix} \qquad \mathbf{U}_{pp} = \begin{bmatrix} -1 & 0 \\ -\ell & -1 \end{bmatrix} \tag{1}$$

The inverse matrix $\mathbf{U}_{vp}^{-1}$ is needed in Eq. (2.65). We find that

$$\mathbf{U}_{vp}^{-1} = \begin{bmatrix} -12EI/\ell^3 & -6EI/\ell^2 \\ 6EI/\ell^2 & 2EI/\ell \end{bmatrix} \tag{2}$$

From Eq. (2.65) with $\mathbf{p}^{e0}$ equal to zero

$$\begin{bmatrix} V_a \\ M_a \\ \\ V_b \\ M_b \end{bmatrix} = \begin{bmatrix} 12EI/\ell^3 & -6EI/\ell^2 & \vdots & -12EI/\ell^3 & -6EI/\ell^2 \\ -6EI/\ell^2 & 4EI/\ell & \vdots & 6EI/\ell^2 & 2EI/\ell \\ \cdots & \cdots & \vdots & \cdots & \cdots \\ -12EI/\ell^3 & 6EI/\ell^2 & \vdots & 12EI/\ell^3 & 6EI/\ell^2 \\ -6EI/\ell^2 & 2EI/\ell & \vdots & 6EI/\ell^2 & 4EI/\ell \end{bmatrix} \begin{bmatrix} w_a \\ \theta_a \\ \\ w_b \\ \theta_b \end{bmatrix} \tag{3}$$

$$= \begin{bmatrix} \mathbf{k}_{aa}^e & \mathbf{k}_{ab}^e \\ \mathbf{k}_{ba}^e & \mathbf{k}_{bb}^e \end{bmatrix} \mathbf{v}^e$$

$$\mathbf{p}^e = \quad \mathbf{k}^e \quad \mathbf{v}^e$$

where

$$\mathbf{k}_{aa}^e = \begin{bmatrix} 12EI/\ell^3 & -6EI/\ell^2 \\ -6EI/\ell^2 & 4EI/\ell \end{bmatrix} \quad \mathbf{k}_{ab}^e = \begin{bmatrix} -12EI/\ell^3 & -6EI/\ell^2 \\ 6EI/\ell^2 & 2EI/\ell \end{bmatrix}$$

$$\mathbf{k}_{ba}^e = \begin{bmatrix} -12EI/\ell^3 & 6EI/\ell^2 \\ -6EI/\ell^2 & 2EI/\ell \end{bmatrix} \quad \mathbf{k}_{bb}^e = \begin{bmatrix} 12EI/\ell^3 & 6EI/\ell^2 \\ 6EI/\ell^2 & 4EI/\ell \end{bmatrix} \tag{4}$$

Common notation for a stiffness matrix of element e is

$$\begin{bmatrix} V_a \\ M_a \\ V_b \\ M_b \end{bmatrix} = \begin{bmatrix} k_{11} & k_{12} & k_{13} & k_{14} \\ k_{21} & k_{22} & k_{23} & k_{24} \\ k_{31} & k_{32} & k_{33} & k_{34} \\ k_{41} & k_{42} & k_{43} & k_{44} \end{bmatrix} \begin{bmatrix} w_a \\ \theta_a \\ w_b \\ \theta_b \end{bmatrix} \tag{5}$$
$$\mathbf{p}^e \quad = \quad \mathbf{k}^e \quad \mathbf{v}^e$$

where, for example,

$$\mathbf{k}_{aa}^e = \begin{bmatrix} k_{11} & k_{12} \\ k_{21} & k_{22} \end{bmatrix} \tag{6}$$

If $\mathbf{p}^{e0}$ of Eq. (2.65) is included, it would appear as

$$\mathbf{p}^{e0} = \begin{bmatrix} \mathbf{p}_a^{e0} & \mathbf{p}_b^{e0} \end{bmatrix}^T = \begin{bmatrix} V_a^0 & M_a^0 & V_b^0 & M_b^0 \end{bmatrix}^{eT} = \begin{bmatrix} -\mathbf{U}_{vp}^{-1}\mathbf{F}_v & \mathbf{F}_p - \mathbf{U}_{pp}\mathbf{U}_{vp}^{-1}\mathbf{F}_v \end{bmatrix}^T \tag{7}$$

The stiffness components can be interpreted physically. From (5) observe that the stiffness element k_{ij} (e.g., $k_{12} = -6EI/\ell^2$) is the force developed at coordinate i due to a unit displacement at coordinate j, with all other displacements set equal to zero. The displacements at the ends of an element are often referred to as *degrees of freedom* (DOFs), which are usually defined as the independent coordinates (displacements) necessary to characterize the location of a point in a structure undergoing deformation.

The stiffness matrix of (3) is often scaled so that

$$\begin{bmatrix} V_a \\ M_a/\ell \\ V_b \\ M_b/\ell \end{bmatrix} = \frac{EI}{\ell^3} \begin{bmatrix} 12 & -6 & -12 & -6 \\ -6 & 4 & 6 & 2 \\ -12 & 6 & 12 & 6 \\ -6 & 2 & 6 & 4 \end{bmatrix} \begin{bmatrix} w_a \\ \ell\theta_a \\ w_b \\ \ell\theta_b \end{bmatrix} \tag{8}$$
$$\mathbf{p}^e \quad = \quad \mathbf{k}^e \quad \mathbf{v}^e$$

Example 2.11 Stiffness Matrix for a General Beam Element Using Standard Engineering Beam Theory. Apply the conversion formula of Eq. (2.65) to convert the general transfer matrix of Table 2.2 to a general stiffness matrix. The resulting stiffness matrix is shown in Table 2.2.

Example 2.12 Reduction of a General Beam Element Solution to the Stiffness Matrix for a Bernoulli–Euler Beam. As shown in Example 2.9 for a Bernoulli–Euler beam, the constants λ, η, and ζ needed to utilize Table 2.2 are zero. This leads to the values given in Example 2.9 for $e_i, i = 0, 1, \ldots, 4$. When these values are placed in the stiffness matrix of Table 2.2, the Bernoulli–Euler stiffness matrix of Eq. (3) of Example 2.10 results.

Direct Evaluation Standard textbooks describe a variety of methods for determining stiffness matrices. Often, a direct evaluation is introduced using the governing differential equations. Consider Eq. (5) of Example 2.10. Use the configuration of Fig. 2.8a to calculate $k_{i1}, i = 1, 2, 3, 4$, which correspond to the first column of the stiffness matrix. For displacements $w_a = 1, \theta_a = 0, w_b = 0$, and $\theta_b = 0$, Eq. (5) of Example 2.10 becomes

$$k_{11} = V_a \qquad k_{21} = M_a \qquad k_{31} = V_b \qquad k_{41} = M_b$$

The fixed–fixed beam element of Fig. 2.8b, which corresponds to the configuration of Fig. 2.8a with $w_a = 1, \theta_a = 0, w_b = 0$, and $\theta_b = 0$, can be used to compute V_a, M_a, V_b, and M_b. For Sign Convention 2, the conditions of equilibrium give $M(x) = -M_a - V_a x$. Integration of Eq. (2.28c), $d^2w/dx^2 = -M/EI$, leads to

$$\frac{dw}{dx} = \frac{1}{EI}\left(M_a x + V_a \frac{x^2}{2}\right) + C_1 = -\theta \qquad w = \frac{1}{EI}\left(M_a \frac{x^2}{2} + V_a \frac{x^3}{6}\right) + C_1 x + C_2$$

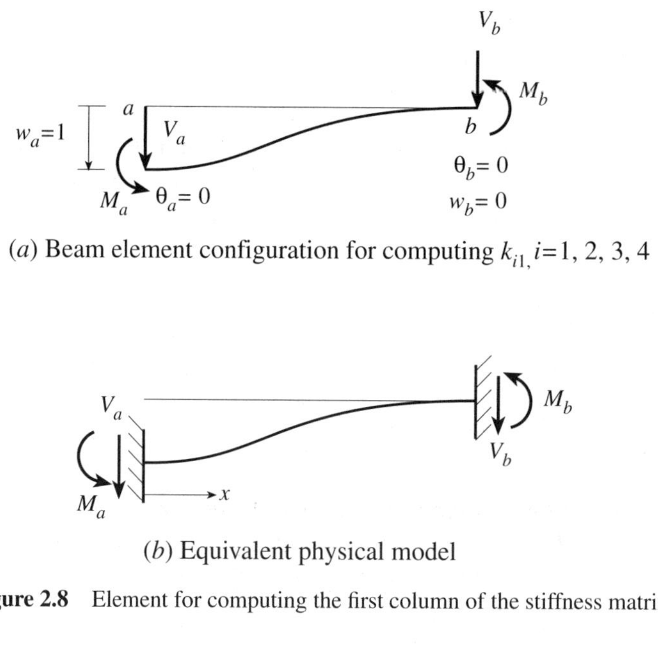

(a) Beam element configuration for computing k_{i1}, i=1, 2, 3, 4

(b) Equivalent physical model

Figure 2.8 Element for computing the first column of the stiffness matrix.

At the left and right ends, $x = x_a = 0$ and $x = x_b = \ell$, respectively. The boundary conditions $\theta_a = 0$ and $w_b = 0$ give $C_1 = 0$ and $C_2 = -M_a\ell^2/2EI - V_a\ell^3/6EI$. Apply $\theta_b = 0$ to the first equation and $w_a = 1$ to the second. These show that

$$V_a = \frac{12EI}{\ell^3} = k_{11} \qquad M_a = -\frac{6EI}{\ell^2} = k_{21}$$

The force M_b and V_b can be obtained by imposing the conditions of equilibrium on the beam of Fig. 2.8a.

$$V_b = -V_a = -\frac{12EI}{\ell^3} = k_{31} \qquad M_b = -V_a\ell - M_a = -\frac{6EI}{\ell^2} = k_{41}$$

The second, third, and fourth columns of the stiffness matrix of Eq. (3) of Example 2.10 are computed in the same fashion.

Approximation by Trial Function In the preceding section, an exact beam theory solution was placed in the form of a stiffness matrix. It is possible to do this for beams, but often not for other types of structural elements. In such cases, an *assumed* or *trial function solution* can be employed to obtain an approximate stiffness matrix. This is the approach that is commonly used to find stiffness matrices for the finite element method.

Shape Functions We assume that the deflection of a beam element can be approximated by a polynomial

$$w = C_0 + C_1x + C_2x^2 + C_3x^3 \cdots = \hat{w}_0 + \hat{w}_1x + \hat{w}_2x^2 + \hat{w}_3x^3 \cdots \tag{2.66}$$

where the coefficients of the polynomial $C_j = \hat{w}_j$, $j = 0, 1, 2, \ldots$ are unknown constants. These coefficients are often referred to as *generalized displacements*. This polynomial is a *trial function* or *basis function*. Retain only the first four terms of this series:

$$w = \begin{bmatrix} 1 & x & x^2 & x^3 \end{bmatrix} \begin{bmatrix} \hat{w}_0 \\ \hat{w}_1 \\ \hat{w}_2 \\ \hat{w}_3 \end{bmatrix} = \mathbf{N}_u\hat{\mathbf{w}} = \hat{\mathbf{w}}^{\mathrm{T}}\mathbf{N}_u^{\mathrm{T}} \tag{2.67}$$

where

$$\mathbf{N}_u = \begin{bmatrix} 1 & x & x^2 & x^3 \end{bmatrix}$$

We choose to express this series as an *interpolation function* by transforming the coefficients $\hat{w}_j$ so that the series is written in terms of the unknown displacements at the ends of the beam element. Express these unknown displacements for the eth element as

$$\mathbf{v}^e = \begin{bmatrix} w_a \\ \theta_a = -dw_a/dx \\ w_b \\ \theta_b = -dw_b/dx \end{bmatrix} \tag{2.68}$$

The derivative of w with respect to x is

$$\frac{dw}{dx} = \frac{d\mathbf{N}_u}{dx}\hat{\mathbf{w}} = \hat{\mathbf{w}}^{\mathrm{T}} \left(\frac{d\mathbf{N}_u}{dx}\right)^{\mathrm{T}}$$

where

$$\frac{d\mathbf{N}_u}{dx} = \begin{bmatrix} 0 & 1 & 2x & 3x^2 \end{bmatrix}$$

Also, properties of a transpose of a scalar [i.e., $(dw/dx)^{\mathrm{T}} = dw/dx$] and a matrix product [i.e., $(\mathbf{AB})^{\mathrm{T}} = \mathbf{B}^{\mathrm{T}}\mathbf{A}^{\mathrm{T}}$] were utilized.

To construct a relationship between the coefficients $\hat{\mathbf{w}}$ and the unknown end displacements $\mathbf{v}^e$, evaluate $w = \mathbf{N}_u\hat{\mathbf{w}}$ and $\theta = -dw/dx = -(d\mathbf{N}_u/dx)\hat{\mathbf{w}}$ at $x = x_a = 0$ and $x = x_b = \ell$

$$\underset{\mathbf{v}^e}{\begin{bmatrix} w_a \\ \theta_a \\ w_b \\ \theta_b \end{bmatrix}} = \begin{bmatrix} w(0) \\ -dw(0)/dx \\ w(\ell) \\ -dw(\ell)/dx \end{bmatrix} = \underset{\hat{\mathbf{N}}_u}{\begin{bmatrix} 1 & 0 & 0 & 0 \\ 0 & -1 & 0 & 0 \\ 1 & \ell & \ell^2 & \ell^3 \\ 0 & -1 & -2\ell & -3\ell^2 \end{bmatrix}} \underset{\hat{\mathbf{w}}}{\begin{bmatrix} \hat{w}_0 \\ \hat{w}_1 \\ \hat{w}_2 \\ \hat{w}_3 \end{bmatrix}}$$

It follows that the coefficients $\hat{\mathbf{w}}$ can be expressed in terms of the discrete, unknown nodal displacements $\mathbf{v}^e$ as

$$\hat{\mathbf{w}} = \hat{\mathbf{N}}_u^{-1}\mathbf{v}^e = \mathbf{G}\mathbf{v}^e \tag{2.69}$$

where

$$\mathbf{G} = \hat{\mathbf{N}}_u^{-1} = \begin{bmatrix} 1 & 0 & 0 & 0 \\ 0 & -1 & 0 & 0 \\ -3/\ell^2 & 2/\ell & 3/\ell^2 & 1/\ell \\ 2/\ell^3 & -1/\ell^2 & -2/\ell^3 & -1/\ell^2 \end{bmatrix}$$

The desired relationship between w and the unknown nodal displacements $\mathbf{v}^e$ is

$$w = \mathbf{N}_u\hat{\mathbf{w}} = \mathbf{N}_u\mathbf{G}\mathbf{v}^e = \mathbf{N}\mathbf{v}^e \tag{2.70}$$

or

$$w(x) = \left(1 - 3\frac{x^2}{\ell^2} + 2\frac{x^3}{\ell^3}\right) w_a + \left(-x + 2\frac{x^2}{\ell} - \frac{x^3}{\ell^2}\right)\theta_a$$
$$+ \left(3\frac{x^2}{\ell^2} - 2\frac{x^3}{\ell^3}\right) w_b + \left(\frac{x^2}{\ell} - \frac{x^3}{\ell^2}\right)\theta_b$$

where $\mathbf{N} = \mathbf{N}_u\mathbf{G}$.

This final expression is the interpolation form of the assumed deflection polynomial. The components of this expression are often referred to as *shape*, *basis*, or *interpolation functions*. Correspondingly, the matrix $\mathbf{N}$ is often called a *shape function matrix*. Equation (2.70) represents an interpolation of the displacement w between the nodes.

Interpolation Functions Based on a Normalized Coordinate Various interpolation functions are tabulated in mathematical handbooks. When derivatives of the displacements at the nodes are involved, as with the beams of this section, the Hermitian interpolation polynomial (Fig. 2.9) can be employed. These can be obtained from Eq. (2.70) using the normalized coordinate $\xi = x/\ell$. From Eq. (2.70),

$$w(\xi) = \begin{bmatrix} 1 & \xi & \xi^2 & \xi^3 \end{bmatrix} \begin{bmatrix} 1 & 0 & 0 & 0 \\ 0 & -1 & 0 & 0 \\ -3 & 2 & 3 & 1 \\ 2 & -1 & -2 & -1 \end{bmatrix} \begin{bmatrix} w_a \\ \ell\theta_a \\ w_b \\ \ell\theta_b \end{bmatrix}$$

$$= \underbrace{\begin{bmatrix} 1 & \xi & \xi^2 & \xi^3 \end{bmatrix}}_{\mathbf{N}_u(\xi)} \underbrace{\begin{bmatrix} 1 & 0 & 0 & 0 \\ 0 & -\ell & 0 & 0 \\ -3 & 2\ell & 3 & \ell \\ 2 & -\ell & -2 & -\ell \end{bmatrix}}_{\mathbf{G}} \underbrace{\begin{bmatrix} w_a \\ \theta_a \\ w_b \\ \theta_b \end{bmatrix}}_{\mathbf{v}^e}$$

$$= \mathbf{N}_u(\xi)\mathbf{G}\mathbf{v}^e = \mathbf{N}\mathbf{v}^e = \mathbf{N}(\xi)\mathbf{v}^e \tag{2.71a}$$

or

$$w(\xi) = \left(1 - 3\xi^2 + 2\xi^3\right) w_a + \left(-\xi + 2\xi^2 - \xi^3\right)\theta_a\ell$$
$$+ \left(3\xi^2 - 2\xi^3\right) w_b + \left(\xi^2 - \xi^3\right)\theta_b\ell \tag{2.71b}$$

The bracketed quantities are Hermitian polynomials. It should be noted that $\mathbf{G}$ of Eq. (2.69) differs from $\mathbf{G}$ of Eq. (2.71a).

The polynomial trial functions of Eq. (2.70) or (2.71b) can be used with the principle of virtual work to generate stiffness matrices.

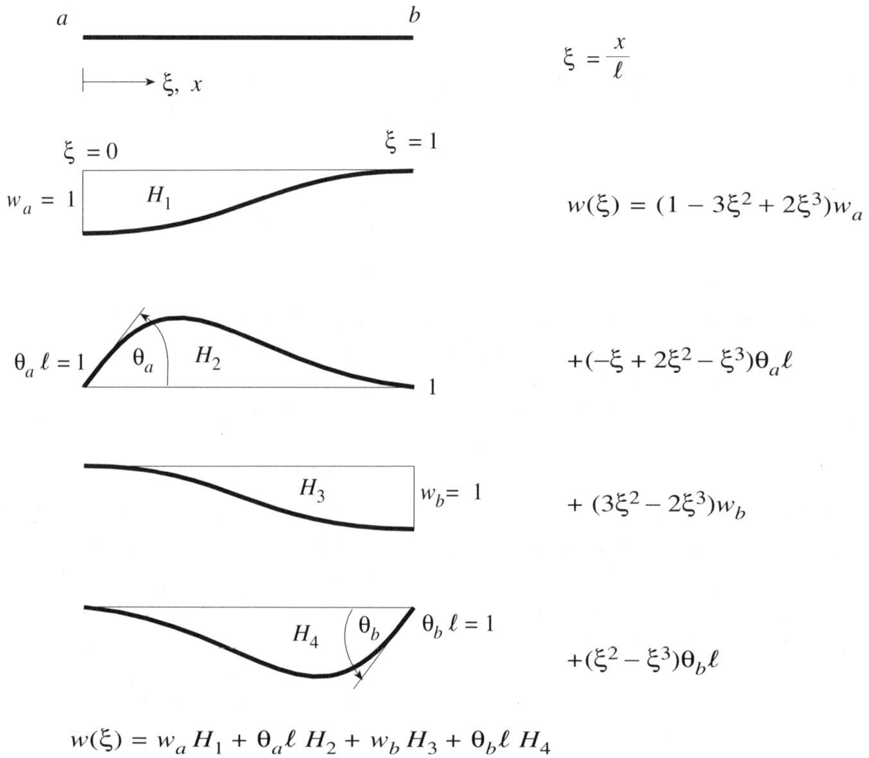

Figure 2.9 Hermitian interpolation polynomials as shape functions.

Principle of Virtual Work for Beams As indicated in Section 2.1.7, the principle of virtual work can be represented as

$$\delta W_{\text{int}} + \delta W_{\text{ext}} = 0 \tag{2.72}$$

From the "Strain–Displacement Relations" subsection of Section 2.1.7, for Bernoulli–Euler bending in the *xz* plane

$$\begin{aligned} -\delta W_{\text{int}} &= \int_x \delta\kappa_y M_y \, dx = \int_x \delta \frac{d\theta_y}{dx} M_y \, dx \\ &= -\int_x \delta \frac{d^2 w}{dx^2} M_y \, dx = \int_x EI \, \delta\kappa_y \, \kappa_y \, dx \\ &= \int_x \left(\delta \frac{d^2 w}{dx^2} \right) EI \frac{d^2 w}{dx^2} dx \end{aligned} \tag{2.73}$$

The virtual work due to the applied loads follows from Eq. (2.50) in the form

$$-\delta W_{\text{ext}} = -\int_x \delta w \, \overline{p}_z \, dx \tag{2.74}$$

The sum of the virtual works for element e from $x = a$ to $x = b$ is

$$-(\delta W_{\text{int}} + \delta W_{\text{ext}})^e = \int_a^b \left(\delta \frac{d^2 w}{dx^2}\right) EI \frac{d^2 w}{dx^2} \, dx - \int_a^b \delta w \, \overline{p}_z \, dx \tag{2.75}$$

and for the entire system, the principle of virtual work (Eq. 2.72) becomes

$$\begin{aligned} -\delta W &= -(\delta W_{\text{int}} + \delta W_{\text{ext}}) \\ &= \sum_{\substack{e \\ \text{all elements}}} \left[\int_a^b \left(\delta \frac{d^2 w}{dx^2}\right) EI \frac{d^2 w}{dx^2} \, dx - \int_a^b \delta w \, \overline{p}_z \, dx \right] = 0 \end{aligned} \tag{2.76}$$

Stiffness Matrix from the Principle of Virtual Work The procedure to be described is generally applicable in the sense that it can be used to derive stiffness matrices for many types of elements. To find the beam element stiffness matrix, substitute the assumed polynomial for w Eq. (2.70) or (2.71) into Eq. (2.75). Begin by finding the variational quantities δw and $\delta w''$ expressed in terms of the trial series, where $w'' = d^2 w/dx^2$. In Eq. (2.70) or (2.71), $\mathbf{G}$ is constant and $\mathbf{N}_u$ is a function of the axial coordinate of x or ξ. Then

$$\delta w = \delta(\mathbf{N}\mathbf{v}^e) = \mathbf{N}\,\delta\mathbf{v}^e = \delta\mathbf{v}^{e\text{T}}\mathbf{N}^{\text{T}} \tag{2.77}$$

where $\delta\mathbf{v}^{e\text{T}}$ are the virtual end displacements, and

$$\begin{aligned} w'' &= \mathbf{N}_u''\mathbf{G}\mathbf{v}^e = \mathbf{B}_u\mathbf{G}\mathbf{v}^e = \mathbf{N}''\mathbf{v}^e = \mathbf{B}\,\mathbf{v}^e \\ \delta w'' &= \mathbf{B}\,\delta\mathbf{v}^e = \delta\mathbf{v}^{e\text{T}}\mathbf{B}^{\text{T}} \end{aligned} \tag{2.78}$$

with $\mathbf{N}_u'' = \mathbf{B}_u$, $\mathbf{B} = \mathbf{N}'' = \mathbf{B}_u\mathbf{G}$, and $\mathbf{N}_u''(x) = [0 \quad 0 \quad 2 \quad 6x]$.

Substitute these expressions into Eq. (2.75):

$$\begin{aligned} -(\delta W_{\text{int}} + \delta W_{\text{ext}})^e &= \int_a^b \overbrace{\delta\mathbf{v}^{e\text{T}}\mathbf{B}^{\text{T}}}^{\delta w''} \, EI \, \overbrace{\mathbf{B}\mathbf{v}^e}^{w''} \, dx - \int_a^b \overbrace{\delta\mathbf{v}^{e\text{T}}\mathbf{N}^{\text{T}}}^{\delta w} \, \overline{p}_z \, dx \\ &= \delta\mathbf{v}^{e\text{T}} \left[\underbrace{\int_a^b \mathbf{B}^{\text{T}}(x) EI \mathbf{B}(x) \, dx}_{\mathbf{k}^e} \quad \underbrace{\mathbf{v}^e}_{\mathbf{v}^e} \quad \underbrace{- \int_a^b \mathbf{N}^{\text{T}}(x) \, \overline{p}_z(x) \, dx}_{\mathbf{p}^{e0}} \right]^e \end{aligned} \tag{2.79}$$

Since $\mathbf{B}(x) = \mathbf{N}''(x) = \mathbf{B}_u(x)\mathbf{G}$, $\mathbf{N}(x) = \mathbf{N}_u(x)\mathbf{G}$, $\mathbf{B}^{\text{T}} = \mathbf{G}^{\text{T}}\mathbf{B}_u^{\text{T}}$, $\mathbf{N}^{\text{T}} = \mathbf{G}^{\text{T}}\mathbf{N}_u^{\text{T}}$:

$$
\begin{aligned}
&-(\delta W_{\text{int}} + \delta W_{\text{ext}})^e \\
&= \delta \mathbf{v}^{e\mathrm{T}} \left[\underbrace{\mathbf{G}^{\mathrm{T}} \int_a^b \mathbf{B}_u^{\mathrm{T}}(x) EI \mathbf{B}_u(x)\, dx\, \mathbf{G}}_{\mathbf{k}^e} \quad \underbrace{\mathbf{v}^e}_{\mathbf{v}^e} \quad \underbrace{-\mathbf{G}^{\mathrm{T}} \int_a^b \mathbf{N}_u^{\mathrm{T}}(x) \overline{p}_z(x)\, dx}_{\mathbf{p}^{e0}} \right]^e
\end{aligned}
\tag{2.80}
$$

Finally,

$$
-(\delta W_{\text{int}} + \delta W_{\text{ext}})^e = \delta \mathbf{v}^{e\mathrm{T}} \left(\mathbf{k}^e \mathbf{v}^e + \mathbf{p}^{e0} \right) \tag{2.81}
$$

The virtual work for element e can be expressed from a different standpoint in terms the displacements $\mathbf{v}^e$ and internal forces (moments and shear forces) $\mathbf{p}^e$ at the ends $x = a$ and $x = b$ of the element. In this case, the virtual work for an element is expressed as $\delta \mathbf{v}^{e\mathrm{T}} \mathbf{p}^e$; that is,

$$
-(\delta W_{\text{int}} + \delta W_{\text{ext}})^e = \delta \mathbf{v}^{e\mathrm{T}} \mathbf{p}^e \tag{2.82}
$$

Equate the two expressions for virtual work of Eqs. (2.81) and (2.82). This gives

$$
\delta \mathbf{v}^{e\mathrm{T}} \left(\mathbf{k}^e \mathbf{v}^e + \mathbf{p}^{e0} \right) = \delta \mathbf{v}^{e\mathrm{T}} \mathbf{p}^e
$$

or

$$
\mathbf{p}^e = \mathbf{k}^e \mathbf{v}^e + \mathbf{p}^{e0} \tag{2.83}
$$

This fundamental relationship corresponds to Eq. (2.65).

In summary, the vectors $\mathbf{p}^e$, $\mathbf{v}^e$, $\mathbf{p}^{e0}$ and the stiffness matrix $\mathbf{k}^e$ of the relationship $\mathbf{p}^e = \mathbf{k}^e \mathbf{v}^e + \mathbf{p}^{e0}$ are defined as

$$
\begin{aligned}
&\mathbf{p}^e = \begin{bmatrix} \mathbf{p}_a \\ \mathbf{p}_b \end{bmatrix} = \begin{bmatrix} V_a \\ M_a \\ V_b \\ M_b \end{bmatrix} \qquad \mathbf{v}^e = \begin{bmatrix} \mathbf{v}_a \\ \mathbf{v}_b \end{bmatrix} = \begin{bmatrix} w_a \\ \theta_a \\ w_b \\ \theta_b \end{bmatrix} \\
&\mathbf{p}^{e0} = -\int_a^b \mathbf{N}^{\mathrm{T}}(x) \overline{p}_z\, dx = -\mathbf{G}^{\mathrm{T}} \int_a^b \mathbf{N}_u^{\mathrm{T}}(x) \overline{p}_z\, dx \\
&\mathbf{k}^e = \int_a^b \mathbf{B}^{\mathrm{T}}(x) EI \mathbf{B}(x)\, dx = \mathbf{G}^{\mathrm{T}} \int_a^b \mathbf{B}_u^{\mathrm{T}}(x) EI \mathbf{B}_u(x)\, dx \mathbf{G}
\end{aligned}
\tag{2.84}
$$

where

$\mathbf{k}^e$ is the stiffness matrix for element e.

$\mathbf{v}^e$ is the column matrix of displacements on the ends of element e.

$\mathbf{p}^e$ is the column matrix of forces on the ends of element e.

$\mathbf{p}^{e0}$ is the column matrix of the influence of the applied loads on element e.

$\mathbf{N}(x)$ is the row matrix of the assumed displacements for element e (Eq. (2.70). The components of $\mathbf{N}$ are referred to as *shape functions* or *interpolation functions*.

$\mathbf{N}_u(x)$ is the row matrix defined by Eq. (2.67).

$\mathbf{G}$ is a matrix defined in Eq. (2.69).

$\mathbf{B}$ and $\mathbf{B}_u$ are equal to $\mathbf{N}''$ and $\mathbf{N}''_u$, respectively.

For constant EI, the stiffness matrix of Eq. (2.84) becomes

$$\mathbf{k}^e = EI \int_a^b \mathbf{B}^{\mathrm{T}}\mathbf{B}\, dx \tag{2.85}$$

Recall that the principle of virtual work corresponds (under prescribed displacement conditions) to the equations of equilibrium. For a framework composed of beam elements including element e, Eq. (2.80) represents the contribution of the eth element to the equilibrium of the entire framework. This contribution is expressed as the virtual work of the eth element that belongs to the virtual work of the entire framework.

Example 2.13 Evaluation of the Stiffness Matrix for a Beam. Find the stiffness matrix for a Bernoulli–Euler beam element. Assume that the cross-sectional area is constant along the beam.

SOLUTION. Introduce the trial function

$$w(x) = \mathbf{N}_u(x)\mathbf{G}\mathbf{v}^e = \mathbf{N}\mathbf{v}^e \tag{1}$$

with Eq. (2.67):

$$\mathbf{N}_u = \begin{bmatrix} 1 & x & x^2 & x^3 \end{bmatrix} \tag{2}$$

We wish to evaluate the stiffness matrix $\mathbf{k}^e$ of $\mathbf{p}^e = \mathbf{k}^e\mathbf{v}^e$ defined as Eq. (2.80)

$$\mathbf{k}^e = \mathbf{G}^{\mathrm{T}} \int_0^\ell \left\{ \mathbf{B}_u^{\mathrm{T}}(x) EI \mathbf{B}_u(x)\, dx \right\} \mathbf{G} \tag{3}$$

with

$$\mathbf{v}^e = [w_a \quad \theta_a \quad w_b \quad \theta_b]^{e\mathrm{T}}$$

and

$$\mathbf{p}^e = [V_a \quad M_a \quad V_b \quad M_b]^{e\mathrm{T}}$$

Also,

$$\mathbf{B}_u(x) = \mathbf{N}_u''(x) = \frac{d^2\mathbf{N}_u(x)}{dx^2} = [0 \quad 0 \quad 2 \quad 6x] \tag{4}$$

and

$$\mathbf{B}_u^{\mathrm{T}}(x) = \begin{bmatrix} 0 \\ 0 \\ 2 \\ 6x \end{bmatrix} \tag{5}$$

For constant EI,

$$\int_0^\ell \mathbf{B}_u^{\mathrm{T}} EI \mathbf{B}_u \, dx = EI \int_0^\ell \begin{bmatrix} 0 & 0 & 0 & 0 \\ 0 & 0 & 0 & 0 \\ 0 & 0 & 4 & 12x \\ 0 & 0 & 12x & 36x^2 \end{bmatrix} dx$$

$$= EI \begin{bmatrix} 0 & 0 & 0 & 0 \\ 0 & 0 & 0 & 0 \\ 0 & 0 & 4\ell & 6\ell^2 \\ 0 & 0 & 6\ell^2 & 12\ell^3 \end{bmatrix} \tag{6}$$

Finally, using Eq. (2.69), we obtain

$$\mathbf{G} = \begin{bmatrix} 1 & 0 & 0 & 0 \\ 0 & -1 & 0 & 0 \\ -3/\ell^2 & 2/\ell & 3/\ell^2 & 1/\ell \\ 2/\ell^3 & -1/\ell^2 & -2/\ell^3 & -1/\ell^2 \end{bmatrix} \tag{7}$$

Inserting these values into (3) leads to

$$\mathbf{k}^e = \frac{EI}{\ell^3} \begin{bmatrix} 12 & -6\ell & -12 & -6\ell \\ -6\ell & 4\ell^2 & 6\ell & 2\ell^2 \\ -12 & 6\ell & 12 & 6\ell \\ -6\ell & 2\ell^2 & 6\ell & 4\ell^2 \end{bmatrix} \tag{8}$$

Note that use of the cubic polynomial of (2) to represent w results in the exact rather than an approximate stiffness matrix. That is, $\mathbf{k}^e$ of (8) is the same stiffness matrix obtained by "exact" approaches earlier in this section.

Example 2.14 Stiffness Matrix Based on Normalized Coordinate. Calculate the stiffness matrix for a Bernoulli–Euler beam element using the normalized coordinate $\xi = x/\ell$.

SOLUTION. From Eq. (2.71a), the displacement assumed is

$$w(\xi) = \mathbf{N}_u(\xi)\mathbf{G}\mathbf{v}^e \tag{1}$$

with $\xi = x/\ell$, $\mathbf{G}$ given in Eq. (2.71a), and

$$\mathbf{N}_u(\xi) = \begin{bmatrix} 1 & \xi & \xi^2 & \xi^3 \end{bmatrix} \tag{2}$$

$$\mathbf{v}^e = [w_a \quad \theta_a \quad w_b \quad \theta_b]^{eT} \tag{3}$$

The corresponding force vector is

$$\mathbf{p}^e = [V_a \quad M_a \quad V_b \quad M_b]^{eT} \tag{4}$$

In terms of the coordinate x, the stiffness matrix $\mathbf{k}^e$ of $\mathbf{p}^e = \mathbf{k}^e\mathbf{v}^e$ is calculated using Eq. (2.84):

$$\begin{aligned} \mathbf{k}^e &= \int_a^b \mathbf{B}^T(x)EI\mathbf{B}(x)\,dx = \int_0^\ell \mathbf{B}^T(x)EI\mathbf{B}(x)\,dx \\ &= \mathbf{G}^T \int_0^\ell \mathbf{B}_u^T(x)EI\mathbf{B}_u(x)\,dx\mathbf{G} \end{aligned} \tag{5}$$

with $\mathbf{G}$ from Eq. (2.69). For the normalized coordinate, $dx = \ell\, d\xi$ and

$$\mathbf{B}_u(\xi) = \mathbf{N}_u''(\xi) = \frac{d^2}{dx^2}\mathbf{N}_u(\xi) = \frac{d^2\mathbf{N}_u(\xi)}{\ell^2 d\xi^2} = \frac{1}{\ell^2}[0 \quad 0 \quad 2 \quad 6\xi] \tag{6}$$

$$\mathbf{B}_u^T(\xi) = \mathbf{N}_u''^T(\xi) = \frac{1}{\ell^2}\begin{bmatrix} 0 \\ 0 \\ 2 \\ 6\xi \end{bmatrix} \tag{7}$$

Then

$$\mathbf{k}^e = \mathbf{G}^T \int_0^\ell \mathbf{B}_u^T(x)EI\mathbf{B}_u(x)\,dx\,\mathbf{G} \qquad = \qquad \mathbf{G}^T \int_0^1 \mathbf{B}_u^T(\xi)EI\mathbf{B}_u(\xi)\ell\,d\xi\mathbf{G}$$

$\mathbf{G}$ from Eq. (2.69) — $\mathbf{G}$ from Eq. (2.71a)

$\mathbf{B}_u(x)$ from Example 2.13 — $\mathbf{B}_u(\xi)$ from (6) of this example

$$= \mathbf{G}^T EI \int_0^1 \frac{1}{\ell^4}\begin{bmatrix} 0 & 0 & 0 & 0 \\ 0 & 0 & 0 & 0 \\ 0 & 0 & 4 & 12\xi \\ 0 & 0 & 12\xi & 36\xi^2 \end{bmatrix} \ell\,d\xi\ \mathbf{G} = \mathbf{G}^T \left[\begin{array}{cc:cc} 0 & 0 & 0 & 0 \\ 0 & 0 & 0 & 0 \\ \hdashline 0 & 0 & 4 & 6 \\ 0 & 0 & 6 & 12 \end{array}\right] \frac{EI}{\ell^3}\mathbf{G} \tag{8}$$

Finally, with $\mathbf{G}$ of Eq. (2.71a),

$$\mathbf{k}^e = \mathbf{G}^{\mathrm{T}} \int_0^1 \mathbf{B}_u^{\mathrm{T}}(\xi) EI \mathbf{B}_u(\xi) \ell \, d\xi \mathbf{G} = \underbrace{\begin{bmatrix} 12 & -6\ell & \vdots & -12 & -6\ell \\ -6\ell & 4\ell^2 & \vdots & 6\ell & 2\ell^2 \\ \cdots & \cdots & \vdots & \cdots & \cdots \\ -12 & 6\ell & \vdots & 12 & 6\ell \\ -6\ell & 2\ell^2 & \vdots & 6\ell & 4\ell^2 \end{bmatrix} \frac{EI}{\ell^3}}_{\mathbf{k}^e} \tag{9}$$

which, as expected, is the same matrix obtained in Example 2.13.

Example 2.15 Loading Vector. The loading vector $\mathbf{p}^{e0}$ is calculated using Eq. (2.84). If the normalized coordinate $\xi = x/\ell$ is introduced,

$$\mathbf{p}^{e0} = -\int_a^b \mathbf{N}^{\mathrm{T}}(x)\overline{p}_z(x)\, dx = -\mathbf{G}^{\mathrm{T}} \int_0^\ell \mathbf{N}_u^{\mathrm{T}}(x)\overline{p}_z(x)\, dx$$

$\mathbf{G}$ from Eq. (2.69)

$\mathbf{N}_u$ from Eq. (2.67)

$$= -\ell \mathbf{G}^{\mathrm{T}} \int_0^1 \mathbf{N}_u^{\mathrm{T}}(\xi)\overline{p}_z(\xi)\, d\xi$$

$\mathbf{N}_u$ and $\mathbf{G}$ from Eq. (2.71a)

$$= -\ell \int_0^1 \mathbf{N}^{\mathrm{T}}(\xi)\overline{p}_z(\xi)\, d\xi = -\ell \int_0^1 \begin{bmatrix} 1 - 3\xi^2 + 2\xi^3 \\ (-\xi + 2\xi^2 - \xi^3)\ell \\ 3\xi^2 - 2\xi^3 \\ (\xi^2 - \xi^3)\ell \end{bmatrix} \overline{p}_z(\xi)\, d\xi \tag{1}$$

For $\overline{p}_z$ of constant magnitude p_0,

$$\mathbf{p}^{e0} = p_0 \ell \begin{bmatrix} -\frac{1}{2} \\ \frac{1}{12}\ell \\ -\frac{1}{2} \\ -\frac{1}{12}\ell \end{bmatrix} \tag{2}$$

For hydrostatic loading with $\overline{p}_z$ varying linearly $\overline{p}_z = p_0\xi$ from $\xi = 0$ to $\xi = 1$:

$$\mathbf{p}^{e0} = \frac{p_0 \ell}{60}\begin{bmatrix} -9 \\ 2\ell \\ -21 \\ -3\ell \end{bmatrix} \tag{3}$$

For the more general linear case with $\overline{p}_z$ taking on the value p_0 at $\xi = 0$ and p_1 at $\xi = 1$,

$$\overline{p}_z = p_0 + (p_1 - p_0)\xi \tag{4}$$

$$\mathbf{p}^{e0} = \frac{p_0 \ell}{60}\begin{bmatrix} -21 \\ 3\ell \\ -9 \\ -2\ell \end{bmatrix} + \frac{p_1 \ell}{60}\begin{bmatrix} -9 \\ 2\ell \\ -21 \\ -3\ell \end{bmatrix} \tag{5}$$

Example 2.16 Stiffness Matrix for Axial Extension and Bending. The stiffness matrix for axial extension of a bar from $x = a$ to $x = b$ is

$$\begin{aligned} \begin{bmatrix} N_a \\ N_b \end{bmatrix}^e &= \begin{bmatrix} EA/\ell & -EA/\ell \\ -EA/\ell & EA/\ell \end{bmatrix}^e \begin{bmatrix} u_a \\ u_b \end{bmatrix}^e + \begin{bmatrix} N_a^0 \\ N_b^0 \end{bmatrix}^e \\ \mathbf{p}^e &= \qquad\qquad \mathbf{k}^e \qquad\qquad\quad \mathbf{v}^e \;\; + \;\; \mathbf{p}^{e0} \end{aligned} \tag{1}$$

where for constant distributed axial load p_{x0},

$$\begin{bmatrix} N_a^0 \\ N_b^0 \end{bmatrix}^e = -p_{x0}\frac{\ell}{2}\begin{bmatrix} 1 \\ 1 \end{bmatrix}$$

If this stiffness matrix is combined with that for the bending of a bar

$$\begin{aligned} \begin{bmatrix} N_a \\ V_a \\ M_a \\ N_b \\ V_b \\ M_b \end{bmatrix}^e &= \begin{bmatrix} \frac{EA}{\ell} & 0 & 0 & -\frac{EA}{\ell} & 0 & 0 \\ 0 & \frac{12EI}{\ell^3} & -\frac{6EI}{\ell^2} & 0 & -\frac{12EI}{\ell^3} & -\frac{6EI}{\ell^2} \\ 0 & -\frac{6EI}{\ell^2} & \frac{4EI}{\ell} & 0 & \frac{6EI}{\ell^2} & \frac{2EI}{\ell} \\ -\frac{EA}{\ell} & 0 & 0 & \frac{EA}{\ell} & 0 & 0 \\ 0 & -\frac{12EI}{\ell^3} & \frac{6EI}{\ell^2} & 0 & \frac{12EI}{\ell^3} & \frac{6EI}{\ell^2} \\ 0 & -\frac{6EI}{\ell^2} & \frac{2EI}{\ell} & 0 & \frac{6EI}{\ell^2} & \frac{4EI}{\ell} \end{bmatrix}^e \begin{bmatrix} u_a \\ w_a \\ \theta_a \\ u_b \\ w_b \\ \theta_b \end{bmatrix}^e \\ \mathbf{p}^e &= \qquad\qquad\qquad\qquad \mathbf{k}^e \qquad\qquad\qquad\qquad\qquad \mathbf{v}^e \end{aligned} \tag{2}$$

These stiffness matrices adhere to sign convention 2, as do all of the stiffness matrices of this chapter.

Example 2.17 Stiffness Matrix for a Bar in Space. The stiffness matrix for a bar for which forces and displacements occur along three coordinate directions is obtained by continuing the sort of uncoupled superposition used in Example 2.16. This leads to the stiffness matrix of (1) on page 93 with the symbols defined in Fig. 2.10.

The bar stiffness relationship is of the form $\mathbf{p}^e = \mathbf{k}^e\mathbf{v}^e + \mathbf{p}^{e0}$. The relationship of (1) represents extension in the x direction, bending in the xz plane (in-plane), bending in the xy plane (out-of-plane), and torsion about the x axis. Torsion is discussed in Chapter 5. The torsion in (1) is for "simple" torsion with no consideration of restrained warping. For a bar with this torsion only, the stiffness matrix appears as

$$\begin{bmatrix} M_{xa} \\ M_{xb} \end{bmatrix}^e = \begin{bmatrix} \dfrac{GJ}{\ell} & -\dfrac{GJ}{\ell} \\ -\dfrac{GJ}{\ell} & \dfrac{GJ}{\ell} \end{bmatrix}^e \begin{bmatrix} \theta_{xa} \\ \theta_{xb} \end{bmatrix}^e + \begin{bmatrix} M^0_{xa} \\ M^0_{xb} \end{bmatrix}^e \tag{2}$$

$$\mathbf{p}^e \quad = \quad \mathbf{k}^e \quad \mathbf{v}^e \quad + \quad \mathbf{p}^{e0}$$

where G is the shear modulus of elasticity and J is the torsional constant, which is equal to the polar moment of inertia for bars of circular cross section. As shown in

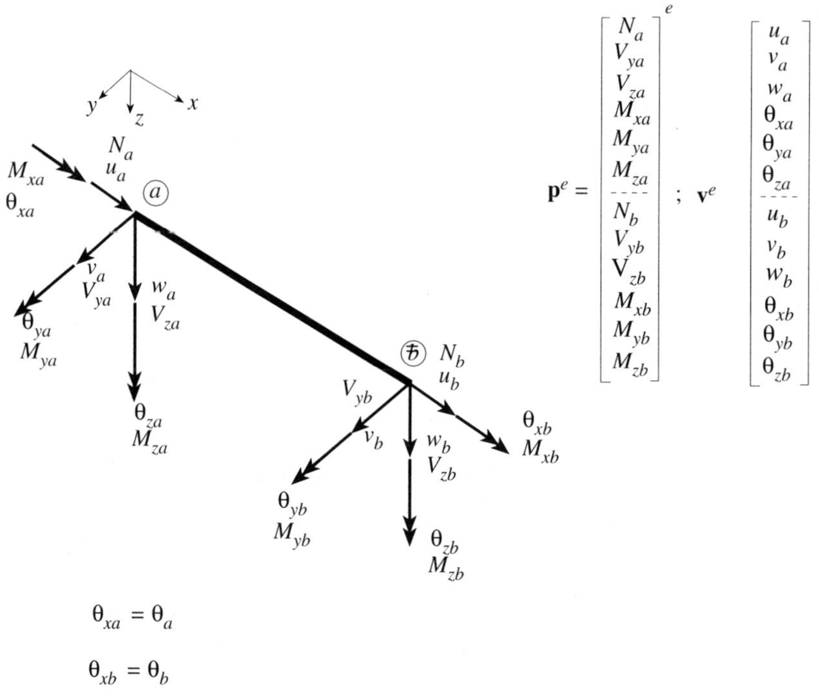

Figure 2.10 Local coordinate system, forces, and displacements on the ends of a bar in space. Sign Convention 2.

$$
\begin{array}{c}
\begin{bmatrix} u_a & v_a & w_a & \theta_{xa} & \theta_{ya} & \theta_{za} & & u_b & v_b & w_b & \theta_{xb} & \theta_{yb} & \theta_{zb} \end{bmatrix} \\
\begin{bmatrix} N_a \\ V_{ya} \\ V_{za} \\ M_{xa} \\ M_{ya} \\ M_{za} \\ \cdots \\ N_b \\ V_{yb} \\ V_{zb} \\ M_{xb} \\ M_{yb} \\ M_{zb} \end{bmatrix}^e
=
\begin{bmatrix}
\frac{EA}{\ell} & 0 & 0 & 0 & 0 & 0 & \vdots & -\frac{EA}{\ell} & 0 & 0 & 0 & 0 & 0 \\
0 & \frac{12EI_y}{\ell^3} & 0 & 0 & 0 & \frac{6EI_y}{\ell^2} & \vdots & 0 & -\frac{12EI_y}{\ell^3} & 0 & 0 & 0 & \frac{6EI_y}{\ell^2} \\
0 & 0 & \frac{12EI_z}{\ell^3} & 0 & -\frac{6EI_z}{\ell^2} & 0 & \vdots & 0 & 0 & -\frac{12EI_z}{\ell^3} & 0 & -\frac{6EI_z}{\ell^2} & 0 \\
0 & 0 & 0 & \frac{GJ}{\ell} & 0 & 0 & \vdots & 0 & 0 & 0 & -\frac{GJ}{\ell} & 0 & 0 \\
0 & 0 & -\frac{6EI_z}{\ell^2} & 0 & \frac{4EI_z}{\ell} & 0 & \vdots & 0 & 0 & \frac{6EI_z}{\ell^2} & 0 & \frac{2EI_z}{\ell} & 0 \\
0 & \frac{6EI_y}{\ell^2} & 0 & 0 & 0 & \frac{6EI_y}{\ell} & \vdots & 0 & -\frac{6EI_y}{\ell^2} & 0 & 0 & 0 & \frac{2EI_y}{\ell} \\
\cdots & \cdots & \cdots & \cdots & \cdots & \cdots & \vdots & \cdots & \cdots & \cdots & \cdots & \cdots & \cdots \\
-\frac{EA}{\ell} & 0 & 0 & 0 & 0 & 0 & \vdots & \frac{EA}{\ell} & 0 & 0 & 0 & 0 & 0 \\
0 & -\frac{12EI_y}{\ell^3} & 0 & 0 & 0 & -\frac{6EI_y}{\ell^2} & \vdots & 0 & \frac{12EI_y}{\ell^3} & 0 & 0 & 0 & -\frac{6EI_y}{\ell^2} \\
0 & 0 & -\frac{12EI_z}{\ell^3} & 0 & \frac{6EI_z}{\ell^2} & 0 & \vdots & 0 & 0 & \frac{12EI_z}{\ell^3} & 0 & \frac{6EI_z}{\ell^2} & 0 \\
0 & 0 & 0 & -\frac{GJ}{\ell} & 0 & 0 & \vdots & 0 & 0 & 0 & \frac{GJ}{\ell} & 0 & 0 \\
0 & 0 & -\frac{6EI_z}{\ell^2} & 0 & \frac{2EI_z}{\ell} & 0 & \vdots & 0 & 0 & \frac{6EI_z}{\ell^2} & 0 & \frac{4EI_z}{\ell} & 0 \\
0 & \frac{6EI_y}{\ell^2} & 0 & 0 & 0 & \frac{2EI_y}{\ell} & \vdots & 0 & \frac{6EI_y}{\ell^2} & 0 & 0 & 0 & \frac{4EI_y}{\ell}
\end{bmatrix}^e
\begin{bmatrix} u_a \\ v_a \\ w_a \\ \theta_{xa} \\ \theta_{ya} \\ \theta_{za} \\ \cdots \\ u_b \\ v_b \\ w_b \\ \theta_{xb} \\ \theta_{yb} \\ \theta_{zb} \end{bmatrix}^e
+
\begin{bmatrix} N_a^0 \\ V_{ya}^0 \\ V_{za}^0 \\ M_{xa}^0 \\ M_{ya}^0 \\ M_{za}^0 \\ \cdots \\ N_b^0 \\ V_{yb}^0 \\ V_{zb}^0 \\ M_{xb}^0 \\ M_{yb}^0 \\ M_{zb}^0 \end{bmatrix}^e \\
\mathbf{p}^e = \mathbf{k}^e \, \mathbf{v}^e + \mathbf{p}^{e0}
\end{array}
\tag{1}
$$

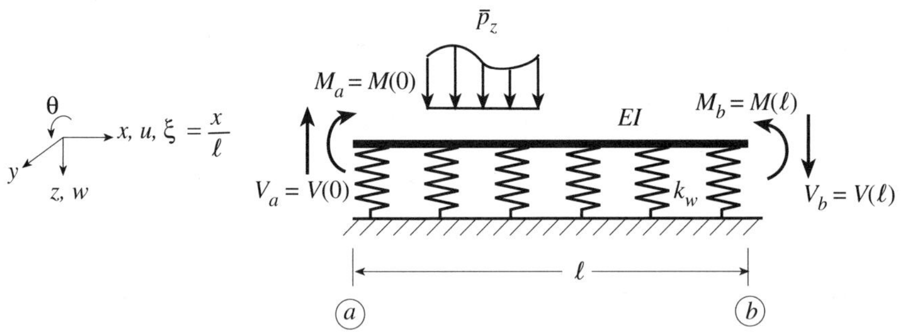

Figure 2.11 Beam element on elastic foundation. Sign Convention 1.

Chapter 5, the governing equations for torsion are the same as those for extension of a bar with a change of variables $E \rightarrow G$, $A \rightarrow J$, $u \rightarrow \theta_x(\phi)$, $N \rightarrow M_x$. Hence, with this adjustment in notation the stiffness matrix of (2) is obtained from the stiffness matrix for extension of Example 2.16.

Example 2.18 Stiffness Matrix for a Beam Element on an Elastic Foundation. Determine the stiffness matrix for a beam element on an elastic (Winkler) foundation.

SOLUTION. The methods described earlier in this chapter can be employed to derive an exact stiffness matrix for a beam on an elastic foundation. Although the third-order polynomial trial function of this section led to an exact stiffness matrix for a simple Bernoulli–Euler beam element, the same polynomial leads to an approximate stiffness matrix for a beam on elastic foundation element.

Notation for the beam element of Fig. 2.11 is:

E is the modulus of elasticity for the beam (force/length^2).

I is the moment of inertia of the cross section about the centroidal y axis (length^4).

ℓ is length of the beam element (length).

k_w is the modulus of the elastic (Winkler) foundation (force/length^2).

κ_y is the curvature of the beam element.

$L = \sqrt[4]{4EI/k_w} = 1/\lambda$ is the *characteristic length* of the beam on an elastic foundation element (length).

Differential Relationships The displacement or Winkler elastic foundation imposes a force of magnitude $k_w w$ on the beam element. This is introduced as an equivalent distributed load (force/length), opposite in sign to $\overline{p}_z$ (Fig. 2.12). The fundamental differential equations governing the motion of this beam element are derived using the procedure described in Chapter 1 for the simple beam element.

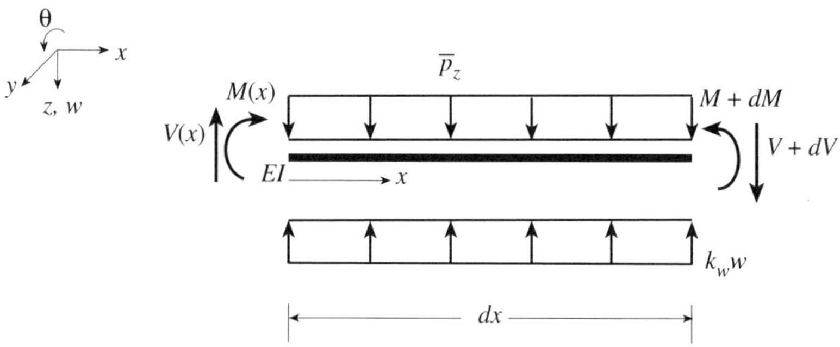

Figure 2.12 Differential element for beam on elastic foundation. Sign Convention 1.

Conditions of equilibrium:

$$\frac{dV}{dx} + \overline{p}_z - k_w w = 0 \qquad \frac{dM}{dx} - V = 0 \tag{1}$$

Material law:

$$M = EI\kappa_y \qquad V = GA_s\gamma \tag{2}$$

Kinematics:

$$\kappa_y = \frac{d\theta}{dx} \qquad \gamma = \theta + \frac{dw}{dx} \tag{3}$$

If shear deformation is not taken into account,

$$\gamma = 0 \quad \text{or} \quad \theta = -\frac{dw}{dx} \qquad \kappa_y = -\frac{d^2w}{dx^2} \tag{4}$$

Combine (1), (2), (3), and (4) to obtain the differential equation

$$\frac{d^2}{dx^2}\left(EI\frac{d^2w}{dx^2}\right) + k_w w - \overline{p}_z = 0 \tag{5}$$

The exact solution of this rather simple fourth-order differential equation is readily established for constant EI and k_w (Pilkey, 1994). This solution is of the form

$$w(x) = C_0 e^{\lambda x}\cos\lambda x + C_1 e^{\lambda x}\sin\lambda x + C_3 e^{-\lambda x}\cos\lambda x + C_4 e^{-\lambda x}\sin\lambda x \tag{6}$$

where $\lambda = \sqrt[4]{k_w/4EI}$.

The exact stiffness matrix (which is provided in Table 2.2) is readily obtained from this solution using the techniques discussed earlier in this chapter. However, here we choose to utilize the polynomial trial function and the principle of virtual work to derive an approximate stiffness matrix.

Approximate Stiffness Matrix from the Principle of Virtual Work The principle of virtual work expression of Eq. (2.76) should be supplemented with an extra term for the elastic foundation

$$
\begin{aligned}
-\delta W &= -(\delta W_{\text{int}} + \delta W_{\text{ext}}) \\
&= \sum_{\substack{e \\ \text{all elements}}} \left[\int_a^b \left(\delta \frac{d^2 w}{dx^2} \right) EI \frac{d^2 w}{dx^2}\, dx - \int_a^b \delta w\, \overline{p}_z\, dx - \int_a^b \delta w(-k_w w)\, dx \right] \\
&= 0
\end{aligned} \tag{7}
$$

where it is recognized that an elastic foundation of modulus k_w has the same effect as a distributed load $\overline{p}_z$, but of opposite sign.

Choose as a trial function the same third-order polynomial of Eq. (2.67) used for the simple Bernoulli–Euler beam element. Substitute $w = \mathbf{N}_u \mathbf{G} \mathbf{v}^e = \mathbf{N} \mathbf{v}^e$ into (7), where

$$
\mathbf{N}_u = \begin{bmatrix} 1 & x & x^2 & x^3 \end{bmatrix} \tag{8}
$$

and $\mathbf{G}$ is given in Eq. (2.69). With $\mathbf{N}_u'' = d^2\mathbf{N}_u/dx^2$ the virtual work for element e is

$$
\begin{aligned}
\delta W^e = {} & \delta \mathbf{v}^{e\mathrm{T}} \underbrace{\mathbf{G}^{\mathrm{T}} \int_a^b \mathbf{N}_u''^{\mathrm{T}}(x) EI \mathbf{N}_u''(x)\, dx \mathbf{G}}_{\mathbf{k}_B^e} \mathbf{v}^e \\
& + \delta \mathbf{v}^{e\mathrm{T}} \underbrace{\mathbf{G}^{\mathrm{T}} \int_a^b \mathbf{N}_u^{\mathrm{T}}(x) k_w \mathbf{N}_u(x)\, dx \mathbf{G}}_{\mathbf{k}_w^e} \mathbf{v}^e \\
& - \delta \mathbf{v}^{e\mathrm{T}} \underbrace{\mathbf{G}^{\mathrm{T}} \int_a^b \mathbf{N}_u^{\mathrm{T}}(x) \overline{p}_z(x)\, dx}_{-\mathbf{p}^{e0}}
\end{aligned} \tag{9}
$$

where

- $\mathbf{k}_B^e$ is the stiffness matrix for bending of the element [Eq. (8) of Example 2.13].
- $\mathbf{k}_w^e$ is the element stiffness matrix for the elastic foundation.
- $\mathbf{p}^{e0}$ is the loading vector.

Introduce the familiar notation $\mathbf{N}_u\mathbf{G} = \mathbf{N}$, $\mathbf{B} = \mathbf{N}''$. Then

$$\delta W^e = \delta \mathbf{v}^{e\mathrm{T}} \underbrace{\int_a^b \mathbf{B}^{\mathrm{T}} EI \, \mathbf{B} \, dx}_{\mathbf{k}_B^e} \; \mathbf{v}^e + \delta \mathbf{v}^{e\mathrm{T}} \underbrace{\int_a^b \mathbf{N}^{\mathrm{T}} k_w \mathbf{N} \, dx}_{\mathbf{k}_w^e} \mathbf{v}^e \qquad (10)$$

$$- \delta \mathbf{v}^{e\mathrm{T}} \underbrace{\int_a^b \mathbf{N}^{\mathrm{T}} \overline{p}_z \, dx}_{-\mathbf{p}^{e0}}$$

The total virtual work of (7) can be expressed as

$$\delta W = \sum_{\text{elements}} \delta \mathbf{v}^{e\mathrm{T}} \left[\left(\mathbf{k}_B^e + \mathbf{k}_w^e \right) \mathbf{v}^e + \mathbf{p}^{e0} \right] \qquad (11)$$

Calculation of the Element Stiffness Matrix $\mathbf{k}_w^e$ *for the Elastic Foundation* The matrices $\mathbf{k}_B^e$ and $\mathbf{p}^{e0}$ of (11) were developed in earlier examples. The stiffness matrix $\mathbf{k}_w^e$ for the elastic foundation will be calculated here. We wish to find

$$\mathbf{k}_w^e = \mathbf{G}^{\mathrm{T}} \int_o^\ell \left(\mathbf{N}_u^{\mathrm{T}} k_w \mathbf{N}_u \, dx \right) \mathbf{G} \qquad (12)$$

with Eq. (2.69)

$$\mathbf{G} = \hat{\mathbf{N}}_u^{-1} = \begin{bmatrix} 1 & 0 & 0 & 0 \\ 0 & -1 & 0 & 0 \\ -3/\ell^2 & 2/\ell & 3/\ell^2 & 1/\ell \\ 2/\ell^3 & -1/\ell^2 & -2/\ell^3 & -1/\ell^2 \end{bmatrix} \qquad (13)$$

Assume that k_ω, the modulus for the foundation, is constant. Then

$$\int_o^\ell \mathbf{N}_u^{\mathrm{T}} k_w \mathbf{N}_u \, dx = k_w \int_o^\ell \begin{bmatrix} 1 & x & x^2 & x^3 \\ x & x^2 & x^3 & x^4 \\ x^2 & x^3 & x^4 & x^5 \\ x^3 & x^4 & x^5 & x^6 \end{bmatrix} dx$$

$$= k_w \begin{bmatrix} \ell & \ell^2/2 & \ell^3/3 & \ell^4/4 \\ \ell^2/2 & \ell^3/3 & \ell^4/4 & \ell^5/5 \\ \ell^3/3 & \ell^4/4 & \ell^5/5 & \ell^6/6 \\ \ell^4/4 & \ell^5/5 & \ell^6/6 & \ell^7/7 \end{bmatrix} \qquad (14)$$

so that

$$\mathbf{k}_w^e = \mathbf{G}^{\mathrm{T}} \int_o^\ell \left(\mathbf{N}_u^{\mathrm{T}} k_w \mathbf{N}_u \, dx \right) \mathbf{G}$$

$$= \frac{k_w \ell}{420} \begin{bmatrix} 156 & -22\ell & 54 & 13\ell \\ -22\ell & 4\ell^2 & -13\ell & -3\ell^2 \\ 54 & -13\ell & 156 & 22\ell \\ 13\ell & -3\ell^2 & 22\ell & 4\ell^2 \end{bmatrix} \tag{15}$$

Although the third-order polynomial led to an exact matrix $\mathbf{k}_B^e$, the same polynomial gives an approximate stiffness matrix $\mathbf{k}_w^e$ for the elastic foundation.

2.2.5 Development of an Element by Mapping from a Reference Element

It is common practice with the finite-element method to develop elements using a mapping from the simple geometry of a reference element to the geometry of the real element. In general, reference elements are used to simplify calculations with elements of complex shape of the sort that occur often in later chapters. A reference element Ω_r is defined with a very simple geometry. The geometry of Ω_r is then mapped to the geometry of the real element Ω by a transformation. This transformation defines the coordinates of each point of the real domain Ω in terms of coordinates of the corresponding point in the reference domain Ω_r. The geometrical transformation has the following properties.

1. It is a bijective mapping; that is, for each point of Ω_r, there is precisely one point of Ω, and any two distinct points of Ω_r are mapped to two distinct points of Ω.
2. The geometrical nodes of Ω_r are mapped to the geometrical nodes of Ω.
3. Every part of the boundary of Ω_r, defined by the geometrical nodes of that boundary, corresponds to the part the boundary of Ω defined by the corresponding nodes.

Shape Functions We choose to use the polynomials of third degree employed earlier in this chapter in developing shape function for a straight beam in bending (i.e., we will use cubic line elements). The reference element of Fig. 2.13a has the domain $-1 \le \xi \le 1$, and the corresponding physical element domain of Fig. 2.13b is $a \le x \le b$. The physical element domain could also be represented as $x_a \le x \le x_b$. These elements have two geometrical nodes, and each node has two displacement variables, a deflection and a slope. If $w(x)$ is a function (the deflection) to be interpolated over element e, the nodal vector $\mathbf{v}^e$ in the physical domain for a Bernoulli–Euler

(*a*) Reference element (*b*) Physical element

Figure 2.13 Beam bending elements in the reference and physical domains.

beam element is

$$\mathbf{v}^e = [w_a \quad \theta_a \quad w_b \quad \theta_b]^{\mathrm{T}} = \left[w_a \quad -\frac{dw_a}{dx} \quad w_b \quad -\frac{dw_b}{dx} \right]^{\mathrm{T}} \tag{2.86}$$

where the subscript a indicates values at node a and the subscript b is for values at node b. The nodal vector for the reference element is different because the derivatives are taken with respect to ξ:

$$\mathbf{v}_r^e = \left[w_a \quad -\frac{dw_a}{d\xi} \quad w_b \quad -\frac{dw_b}{d\xi} \right]^{\mathrm{T}} \tag{2.87}$$

Points on the line in the reference domain with two nodes can be transformed to points in the physical domain using

$$x(\xi) = a_1 + a_2\xi$$

The coefficients a_1, a_2 are found in terms of the deflections at the nodes a and b in the physical domain by evaluating this equation at the nodes ($\xi = -1$ and $\xi = +1$)

$$a = a_1 - a_2$$

$$b = a_1 + a_2$$

which, upon solution for a_1 and a_2, shows that the geometric transformation is

$$x(\xi) = \frac{a+b}{2} + \frac{b-a}{2}\xi \tag{2.88}$$

The Jacobian J of an element is defined by

$$\frac{dw}{d\xi} = \frac{dw}{dx}\frac{dx}{d\xi} = J\frac{dw}{dx} \tag{2.89a}$$

and the value of J is determined by the physical node coordinate relationship of Eq. (2.88):

$$J = \frac{dx}{d\xi} = \frac{b-a}{2} = \frac{\ell}{2} \tag{2.89b}$$

where ℓ is the length of the physical line element. The inverse Jacobian defined by

$$\frac{dw}{dx} = J^{-1}\frac{dw}{d\xi} \tag{2.89c}$$

is the reciprocal of the Jacobian

$$J^{-1} = \frac{1}{J} = \frac{2}{\ell} \tag{2.89d}$$

The interpolation formula for a function $w(\xi)$ defined on the reference element is written as a linear combination of basis functions

$$w(\xi) = \mathbf{N}_u(\xi)\hat{\mathbf{w}}$$

where $\hat{\mathbf{w}} = \begin{bmatrix}\hat{w}_0 & \hat{w}_1 & \hat{w}_2 & \hat{w}_3\end{bmatrix}^{\mathrm{T}}$ is the vector of generalized displacements. The basis functions for the cubic element are

$$\mathbf{N}_u(\xi) = \begin{bmatrix}1 & \xi & \xi^2 & \xi^3\end{bmatrix} \tag{2.90}$$

The nodal vector for the function $w(\xi)$ on the reference element is

$$\mathbf{v}_r^e = \begin{bmatrix} w(\xi_a) \\ -\dfrac{dw}{d\xi}(\xi_a) \\ w(\xi_b) \\ -\dfrac{dw}{d\xi}(\xi_b) \end{bmatrix} = \begin{bmatrix} \mathbf{N}_u(\xi_a) \\ -\dfrac{d\mathbf{N}_u}{d\xi}(\xi_a) \\ \mathbf{N}_u(\xi_b) \\ -\dfrac{d\mathbf{N}_u}{d\xi}(\xi_b) \end{bmatrix} \hat{\mathbf{w}}$$

in which $\xi_a = -1$ and $\xi_b = 1$. When the preceding equation is evaluated, the nodal vector is expressed in terms of the coefficient vector $\hat{\mathbf{w}}$ as

$$\mathbf{v}_r^e = \begin{bmatrix} 1 & -1 & 1 & -1 \\ 0 & -1 & 2 & -3 \\ 1 & 1 & 1 & 1 \\ 0 & -1 & -2 & -3 \end{bmatrix} \hat{\mathbf{w}} = \hat{\mathbf{N}}_u^* \hat{\mathbf{w}}$$

The matrix $\hat{\mathbf{N}}_u^*$ is invertible, so that

$$\hat{\mathbf{w}} = \hat{\mathbf{N}}_u^{*-1}\mathbf{v}_r^e = \frac{1}{4}\begin{bmatrix} 2 & -1 & 2 & 1 \\ -3 & 1 & 3 & 1 \\ 0 & 1 & 0 & -1 \\ 1 & -1 & -1 & -1 \end{bmatrix} \mathbf{v}_r^e \tag{2.91a}$$

The results of the preceding paragraph give a nodal interpolation formula using the reference element nodal vector

$$w(\xi) = \mathbf{N}_u(\xi)\hat{\mathbf{w}} = \mathbf{N}_u(\xi)\hat{\mathbf{N}}_u^{*-1}\mathbf{v}_r^e \tag{2.91b}$$

We wish to convert this expression to a nodal interpolation formula for the element in the physical domain

$$w(\xi) = \mathbf{N}^*(\xi)\mathbf{v}^e \tag{2.91c}$$

The relationship between $\mathbf{v}_r^e$, the nodal vector for the reference element Eq. (2.87), and $\mathbf{v}^e$, the nodal vector in the physical domain (Eq. (2.86), can be formed as

$$\mathbf{v}_r^e = \mathbf{Y}\mathbf{v}^e \tag{2.91d}$$

where, from Eq. (2.89), the transformation matrix $\mathbf{Y}$ is given by

$$\mathbf{Y} = \begin{bmatrix} 1 & 0 & 0 & 0 \\ 0 & \ell/2 & 0 & 0 \\ 0 & 0 & 1 & 0 \\ 0 & 0 & 0 & \ell/2 \end{bmatrix} \tag{2.92}$$

From Eq. (2.91b–d), the nodal interpolation formula for the function $w(\xi)$ becomes

$$w(\xi) = \mathbf{N}_u(\xi)\hat{\mathbf{N}}_u^{*\,-1}\mathbf{v}_r^e = \mathbf{N}_u(\xi)\hat{\mathbf{N}}_u^{*\,-1}\mathbf{Y}\mathbf{v}^e = \mathbf{N}^*(\xi)\mathbf{v}^e = \mathbf{v}^{e\mathrm{T}}\mathbf{N}^{*\mathrm{T}}(\xi) \tag{2.93}$$

Thus, the four shape functions N_i^* for the cubic element are defined by

$$\mathbf{N}^*(\xi) = \begin{bmatrix} N_1^* & N_2^* & N_3^* & N_4^* \end{bmatrix} = \mathbf{N}_u(\xi)\hat{\mathbf{N}}_u^{*-1}\mathbf{Y} \tag{2.94}$$

The shape functions and their derivatives are listed in Table 2.3. These shape functions depend on the length of the physical line element and are not determined by the reference line element alone. The advantage of defining the shape functions in this way is that the nodal interpolation formula uses the nodal vector $\mathbf{v}^e$, in which the derivatives are with respect to the physical domain coordinate x, rather than the nodal vector $\mathbf{v}_r^e$, in which the derivatives are with respect to the reference domain coordinate ξ.

Stiffness Matrix The element stiffness matrix $\mathbf{k}^e$ is obtained from the internal virtual work in the principle of virtual work of Eq. (2.76). From this expression the internal virtual work for element e is given by

$$-\delta W_{\mathrm{int}}^e = \int_a^b \delta\frac{d^2w}{dx^2} EI \frac{d^2w}{dx^2}\, dx \tag{2.95}$$

TABLE 2.3 Shape Functions and Derivatives for the Cubic Line Element

i	$4N_i^*$	$4\dfrac{dN_i^*}{d\xi}$	$4\dfrac{d^2N_i^*}{d\xi^2}$	$4\dfrac{d^3N_i^*}{d\xi^3}$	$4\dfrac{d^4N_i^*}{d\xi^4}$
1	$(1-\xi)^2(2+\xi)$	$-3(1-\xi^2)$	6ξ	6	0
2	$-(1-\xi)^2(1+\xi)\ell/2$	$(1-\xi)(1+3\xi)\ell/2$	$(1-3\xi)\ell$	-3ℓ	0
3	$(1+\xi)^2(2-\xi)$	$3(1-\xi^2)$	-6ξ	-6	0
4	$(1+\xi)^2(1-\xi)\ell/2$	$(1+\xi)(1-3\xi)\ell/2$	$(1+3\xi)\ell$	3ℓ	0

To express this integral in the reference domain, use

$$dx = \frac{\ell}{2} d\xi \qquad \text{(from Eq. 2.89b)}$$

$$\xi = \begin{cases} -1 & \text{at } x = a \\ +1 & \text{at } x = b \end{cases} \qquad \text{(from Eq. 2.88)}$$

$$\frac{d^2w}{dx^2} = \frac{d}{dx}\frac{dw}{dx} = \frac{d}{dx} J^{-1} \frac{dw}{d\xi} = J^{-2}\frac{d^2w}{d\xi^2} = \frac{4}{\ell^2}\frac{d^2w}{d\xi^2} \qquad \text{(from Eq. 2.89c and d)}$$

Then

$$\begin{aligned} -\delta W^e_{\text{int}} &= \int_a^b \delta \frac{d^2w}{dx^2} EI \frac{d^2w}{dx^2}\, dx = \frac{16EI}{\ell^4} \int_{-1}^{1} \delta \frac{d^2w}{d\xi^2}\frac{d^2w}{d\xi^2}\frac{\ell}{2}\, d\xi \\ &= \frac{8EI}{\ell^3} \int_{-1}^{1} \delta \frac{d^2w}{d\xi^2}\frac{d^2w}{d\xi^2} d\xi = \frac{8EI}{\ell^3}\int_{-1}^{1} \delta \mathbf{v}^{e\mathrm{T}} \frac{d^2\mathbf{N}^{*\mathrm{T}}}{d\xi^2}\frac{d^2\mathbf{N}^*}{d\xi^2}\mathbf{v}^e d\xi \end{aligned}$$

where Eq. (2.93) has been introduced. Also,

$$\frac{d^2w}{d\xi^2} = \frac{d^2\mathbf{N}^*}{d\xi^2}\mathbf{v}^e = \mathbf{v}^{e\mathrm{T}}\frac{d^2\mathbf{N}^{*\mathrm{T}}}{d\xi^2}$$

Then the stiffness matrix is given by Eq. (2.79)

$$\mathbf{k}^e = \frac{8EI}{\ell^3}\int_{-1}^{1} \frac{d^2\mathbf{N}^{*\mathrm{T}}}{d\xi^2}\frac{d^2\mathbf{N}^*}{d\xi^2} d\xi \tag{2.96}$$

With the shape functions of Table 2.3, $\mathbf{k}^e$ is found to be the exact stiffness matrix of Eq. (3) of Example 2.10.

2.3 MASS MATRICES FOR DYNAMIC PROBLEMS

Consider an undamped structure undergoing dynamic motion. The governing equations of motion for a beam moving dynamically are obtained using D'Alembert's principle by including an acceleration times mass term as a new force. Thus, in the governing differential equation or in the principle of virtual work expression, include a term $\overline{p}_z(x) = -\rho \ddot{w}(x)$, where ρ is the mass per unit length and $\ddot{w} = \partial^2 w/\partial t^2$ is the acceleration at axial location x.

In the case of the governing differential equations of Eq. (2.28), the first relationship becomes the partial differential equation

$$\frac{\partial}{\partial x^2} EI \frac{\partial^2 w}{\partial x^2} + \rho \frac{\partial^2 w}{\partial t^2} = \overline{p}_z \tag{2.97a}$$

and the remaining relationships are

$$V = -\frac{\partial}{\partial x} EI \frac{\partial^2 w}{\partial x^2} \tag{2.97b}$$

$$M = -EI \frac{\partial^2 w}{\partial x^2} \tag{2.97c}$$

$$\theta = -\frac{\partial w}{\partial x} \tag{2.97d}$$

Direct general solutions of these equations are difficult to achieve. Some solution techniques are introduced in Chapter 3. These solution methods involve a *mass matrix* to represent the inertia properties of a beam element. Element mass matrices for beams are discussed in this section.

2.3.1 Consistent Mass Matrices

The introduction of the inertia term to the integral governing equations (Eq. 2.75) of the principle of virtual work relation for the eth element leads to

$$-(\delta W_{\text{int}} + \delta W_{\text{ext}})^e = \int_a^b \left(\delta \frac{\partial^2 w}{\partial x^2}\right) EI \frac{\partial^2 w}{\partial x^2}\, dx + \int_a^b \delta w\, \rho(x) \frac{\partial^2 w}{\partial t^2}\, dx - \int_a^b \delta w\, \overline{p}_z\, dx \tag{2.98}$$

In contrast to static responses, the degrees of freedom for dynamic responses include not only the nodal deflections and slopes, but also their time derivatives. To find an approximate solution, we choose to employ trial solutions for the displacements and accelerations. Often, the same trial solution is chosen for both. For beam elements, the trial solutions are usually cubic polynomials. For sinusoidal vibrations (displacements), the acceleration also responds sinusoidally and use of the same approximate solution for displacements and accelerations does not introduce an additional approximation.

For the displacement and acceleration of the eth element, use the shape functions

$$w(x) = \mathbf{N}(x)\mathbf{v}^e = \mathbf{N}_u(x)\mathbf{G}\mathbf{v}^e \tag{2.99a}$$

$$\ddot{w}(x) = \mathbf{N}(x)\ddot{\mathbf{v}}^e = \mathbf{N}_u(x)\mathbf{G}\ddot{\mathbf{v}}^e \tag{2.99b}$$

where $\mathbf{N}_u(x)$ and $\mathbf{G}$ are given by Eqs. (2.67) and (2.69), respectively, and

$$\ddot{\mathbf{v}}^e = \frac{\partial^2 \mathbf{v}^e}{\partial t^2}$$

In the expression of Eq. (2.99a) the function $\mathbf{N}_u$ depends only on the coordinate x; hence the derivatives with respect to time affect only the unknown nodal displacements. Introduce the trial functions of Eq. (2.99) into the principle of virtual work relation of Eq. (2.98)

$$
\begin{aligned}
\delta W^e &= \delta \mathbf{v}^{e\mathrm{T}} \underbrace{\mathbf{G}^{\mathrm{T}} \int_a^b \mathbf{N}_u''^{\mathrm{T}}(x) EI \mathbf{N}_u''(x)\, dx \mathbf{G}}_{\mathbf{k}^e \text{ stiffness matrix}} \mathbf{v}^e \\
&+ \delta \mathbf{v}^{e\mathrm{T}} \underbrace{\mathbf{G}^{\mathrm{T}} \int_a^b \mathbf{N}_u^{\mathrm{T}}(x)\, \rho(x)\, \mathbf{N}_u(x)\, dx \mathbf{G}}_{\mathbf{m}^e \text{ mass matrix}} \ddot{\mathbf{v}}^e \\
&+ \delta \mathbf{v}^{e\mathrm{T}} \underbrace{\left[-\mathbf{G}^{\mathrm{T}} \int_a^b \mathbf{N}_u^{\mathrm{T}}(x)\, \overline{p}_z(x)\, dx \right]}_{\mathbf{p}^{e0} \text{ loading vector}}
\end{aligned}
\tag{2.100}
$$

where $\mathbf{N}_u'' = d^2\, \mathbf{N}_u / dx^2$. The mass matrix is identified as

$$
\mathbf{m}^e = \int_a^b \rho \mathbf{N}^{\mathrm{T}} \mathbf{N}\, dx = \mathbf{G}^{\mathrm{T}} \int_a^b \mathbf{N}_u^{\mathrm{T}}(x)\, \rho(x)\, \mathbf{N}_u(x)\, dx \mathbf{G} \tag{2.101}
$$

If the same $\mathbf{N}$ is substituted into Eq. (2.101) for the mass matrix $\mathbf{m}^e$ as was employed to develop the stiffness matrix $\mathbf{k}^e$, the mass matrix $\mathbf{m}^e$ is said to be *consistent.* If $\mathbf{N}_u$ of Eq. (2.67) is used, both $\mathbf{N}_u$ and $\mathbf{N}$ contain up to cubic terms. Substitution of $\mathbf{N}$ of Eq. (2.70) into Eq. (2.101) gives for constant mass density ρ

$$
\mathbf{m}^e = \frac{\rho \ell}{420} \begin{bmatrix} 156 & -22\ell & 54 & 13\ell \\ -22\ell & 4\ell^2 & -13\ell & -3\ell^2 \\ 54 & -13\ell & 156 & 22\ell \\ 13\ell & -3\ell^2 & 22\ell & 4\ell^2 \end{bmatrix} \tag{2.102}
$$

Note that this mass matrix $\mathbf{m}^e$ is symmetric. It often leads to computationally efficient solutions. A "more exact" mass matrix can be obtained by employing a "more exact" $\mathbf{N}$ in Eq. (2.101). This can result in ρ appearing inside the mass matrix in transcendental terms and the corresponding analysis will be less efficient but more accurate.

From Eq. (2.100), the element equations, comparable to $\mathbf{p}^e = \mathbf{k}^e \mathbf{v}^e$ for static responses, become

$$
\begin{aligned}
\begin{bmatrix} V_a \\ M_a \\ V_b \\ M_b \end{bmatrix} &= \frac{\rho \ell}{420} \begin{bmatrix} 156 & -22\ell & 54 & 13\ell \\ -22\ell & 4\ell^2 & -13\ell & -3\ell^2 \\ 54 & -13\ell & 156 & 22\ell \\ 13\ell & -3\ell^2 & 22\ell & 4\ell^2 \end{bmatrix} \begin{bmatrix} \ddot{w}_a \\ \ddot{\theta}_a \\ \ddot{w}_b \\ \ddot{\theta}_b \end{bmatrix} \\
\mathbf{p}^e \quad &= \qquad\qquad \mathbf{m}^e \qquad\qquad \ddot{\mathbf{v}}^e
\end{aligned}
$$

$$
+\frac{EI}{\ell^3}\begin{bmatrix} 12 & -6\ell & -12 & -6\ell \\ -6\ell & 4\ell^2 & 6\ell & 2\ell^2 \\ -12 & 6\ell & 12 & 6\ell \\ -6\ell & 2\ell^2 & 6\ell & 4\ell^2 \end{bmatrix}\begin{bmatrix} w_a \\ \theta_a \\ w_b \\ \theta_b \end{bmatrix} + \begin{bmatrix} V_a^0 \\ M_a^0 \\ V_b^0 \\ M_b^0 \end{bmatrix}
$$

$$
+ \qquad \mathbf{k}^e \qquad \mathbf{v}^e + \mathbf{p}^{e0} \tag{2.103}
$$

2.3.2 Lumped Mass Matrices

A common approach to model mass is to represent the continuous mass as concentrated masses, that is, assume that the distributed mass can be modeled as lumps of mass at particular locations along an element. This lumped mass modeling provides an alternative to consistent mass matrix discretization of the mass $\rho(x)$. For structural elements such as beams, the mass can be lumped by moving the mass surrounding a node to that node. For example, in the case of the beam element of Fig. 2.14, part of the distributed mass of the element is lumped at node a and part at node b. If half of the mass is lumped at each node,

$$
m_a = m_b = \frac{\rho\ell}{2} \tag{2.104}
$$

If these masses are considered to contribute to translational motion, the mass matrix would appear as

$$
\mathbf{m}^e = \begin{bmatrix} m_a & & & \\ & 0 & & \\ & & m_b & \\ & & & 0 \end{bmatrix} = \frac{\rho\ell}{2}\begin{bmatrix} 1 & & & \\ & 0 & & \\ & & 1 & \\ & & & 0 \end{bmatrix} \tag{2.105}
$$

Terms to take into account the effects of rotational inertia are usually placed in the two diagonal coefficients of Eq. (2.105), which now contain zeros (Pilkey and Wunderlich, 1994). The diagonal character of this matrix can be useful in terms of computational implementation of a solution.

A number of alternatives for the formation of mass matrices have been proposed. For example, a linear combination of the consistent and lumped mass matrices is sometimes utilized. This can take the form

$$
\mathbf{m} = \alpha\mathbf{m}_{\text{consistent}} + \beta\mathbf{m}_{\text{lumped}} \tag{2.106}
$$

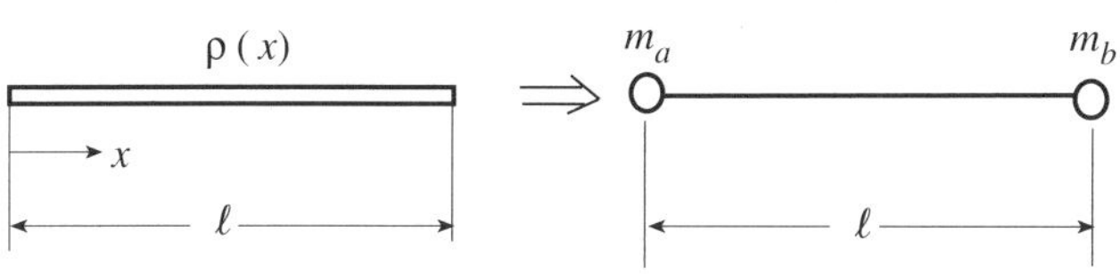

Figure 2.14 Lumped mass assumption.

where α and β are weighting constants. This combined mass model is referred to as a *nonconsistent* or *high-order mass matrix.*

Several other variations for mass matrices are discussed in the literature (Pilkey and Wunderlich, 1994). For example, there are schemes for diagonalizing consistent mass matrices.

2.3.3 Exact Mass and Dynamic Stiffness Matrices

The stiffness matrix is often calculated by substituting the trial solution $w = \mathbf{N}\mathbf{v}^e$ into the internal virtual work relationship Eq. (2.73)

$$-\delta W_{\text{int}} = \int_x \left(\delta \frac{\partial^2 w}{\partial x^2} \right) EI \frac{\partial^2 w}{\partial x^2} \, dx \tag{2.107}$$

giving (Eq. 2.79)

$$\mathbf{k}^e = \int_a^b \frac{\partial^2 \mathbf{N}^{\mathrm{T}}}{\partial x^2} EI \frac{\partial^2 \mathbf{N}}{\partial x^2} \, dx \tag{2.108}$$

The accuracy of beam system analysis depends in part on the accuracy of the beam element stiffness matrix, which depends on the trial function $\mathbf{N}$. In general, the closer $\mathbf{N}$ is to the exact solution for the beam response, the better the stiffness matrix $\mathbf{k}^e$. The familiar stiffness matrix is used to find the static response of a structure. Thus far in this chapter, the polynomials employed as the shape functions of $\mathbf{N}$ in Eq. (2.108) are functions of the axial coordinate x and do not represent time-dependent responses. In some cases these polynomials in x have been the exact solution of the governing differential equations for the static response of a beam and hence they have led to an exact stiffness matrix $\mathbf{k}^e$. In this section it is shown that the effect of dynamic response can be included in the stiffness matrix and that this matrix can be related to the mass matrix. As indicated in Chapter 3, the improved accuracy of the dynamic response of a structure concomitant with the use of more accurate stiffness and mass matrices is usually obtained at the expense of an increase in computational effort.

For a beam, the governing equations of motion including dynamic motion appear as

$$\frac{\partial w}{\partial x} = -\theta + \frac{V}{k_s GA} \tag{2.109a}$$

$$\frac{\partial \theta}{\partial x} = \frac{M}{EI} + \frac{M_T}{EI} \tag{2.109b}$$

$$\frac{\partial V}{\partial x} = k_w w + \rho \frac{\partial^2 w}{\partial t^2} - \overline{p}_z(x, t) \tag{2.109c}$$

$$\frac{\partial M}{\partial x} = V + (k_\theta - N)\theta + \rho r_y^2 \frac{\partial^2 \theta}{\partial t^2} - \overline{c}(x, t) \tag{2.109d}$$

where the terms are defined in Example 2.8 and Table 2.2. The differential equations of Eq. (2.109), including the effects of shear deformation and rotary inertia as well as bending, are the *Timoshenko beam equations.* These equations reduce to those for a *Rayleigh beam* (bending and rotary inertia) by setting the shear deformation term $1/k_s GA$ equal to zero. A *shear beam* models bending with shear deformation and is obtained by setting the rotary inertia term $\rho r_y^2 \partial^2\theta/\partial t^2$ equal to zero. The familiar Bernoulli–Euler beam is obtained by setting both the shear deformation and the rotary inertia terms to zero. For free vibrations, let the motion be harmonic [e.g., set $w(x,t) = w(x)\sin\omega t$, and let the applied loadings M_T, $\overline{p}_z$, and $\overline{c}$ be equal to zero]. The free vibration equations are given in Example 2.8. As indicated earlier in this chapter, the solution of these governing differential equations, including free vibration effects, is given in Table 2.2.

Dynamic Stiffness Matrices The differential equations for harmonic motion of a beam of Example 2.8 can be solved exactly, giving the exact stiffness matrix provided in Table 2.2. Since this stiffness matrix includes the effects of inertia, it is often referred to as a *dynamic stiffness matrix* and is designated as $\mathbf{k}^e_{\text{dyn}}$. This exact stiffness matrix, which is obtained in Table 2.2 by the simple transformation from a transfer matrix Eq. (2.65), can also be computed by using the exact solution of the harmonic motion differential equations of Example 2.8 to compose $w = \mathbf{N}\mathbf{v}^e$ and forming the dynamic stiffness matrix as Eq. (2.108)

$$\mathbf{k}^e_{\text{dyn}} = \int_a^b \frac{\partial^2 \mathbf{N}^T}{\partial x^2} EI \frac{\partial^2 \mathbf{N}}{\partial x^2}\, dx \qquad \mathbf{N} = \mathbf{N}_{\text{exact}} \tag{2.110}$$

Exact Mass Matrices A *frequency-dependent mass matrix* can be formed in a fashion similar to that employed to set up the dynamic stiffness matrix of Eq. (2.110). Thus, place the trial solution $\mathbf{N}$ formed of exact shape functions in the integral expression for a mass matrix Eq. (2.101):

$$\mathbf{m}^e = \int_a^b \rho \mathbf{N}^T \mathbf{N}\, dx \qquad \mathbf{N} = \mathbf{N}_{\text{exact}} \tag{2.111}$$

which is an *exact mass matrix.*

Relationship between Stiffness and Mass Matrices The exact frequency-dependent mass and dynamic stiffness matrices can be shown to be related (Pilkey and Wunderlich, 1994). For example, the exact mass matrix of Eq. (2.111) can be obtained by differentiating the element dynamic stiffness matrix of Eq. (2.110) by the frequency parameter ω^2. That is,

$$\mathbf{m}^e = -\frac{\partial \mathbf{k}^e_{\text{dyn}}}{\partial \omega^2} \tag{2.112}$$

It can be useful to economize somewhat in the calculation of a frequency-dependent mass matrix by forming a "quasistatic" mass matrix $\tilde{\mathbf{m}}^e$, defined as

$$\tilde{\mathbf{m}}^e = \int_a^b \rho \mathbf{N}_0^{\mathrm{T}} \mathbf{N}\, dx \tag{2.113}$$

where $\mathbf{N} = \mathbf{N}_{\text{exact}}$ and $\mathbf{N}_0$ contains static shape functions such as the $\mathbf{N}^*$ of Table 2.3. With this notation, using the static trial function designated as $\mathbf{N}_0$, the consistent mass matrix of Section 2.3.1 is given by

$$\int_a^b \rho \mathbf{N}_0^{\mathrm{T}} \mathbf{N}_0\, dx \tag{2.114}$$

For computational convenience, as discussed in Chapter 3, it is useful to define a frequency-dependent stiffness matrix $\mathbf{k}^e(\omega)$ in terms of the dynamic stiffness matrix $\mathbf{k}^e_{\text{dyn}}(\omega)$ and a frequency-dependent mass matrix $\mathbf{m}^e(\omega)$ as

$$\mathbf{k}^e_{\text{dyn}} = \mathbf{k}^e(\omega) - \omega^2 \mathbf{m}^e(\omega) \tag{2.115}$$

for the eth element. Symbolic manipulation software can be quite helpful in implementing the mathematical operations necessary to form these frequency-dependent matrices for beam elements. For computational purposes it is useful to bring the frequencies ω outside of the frequency-dependent matrices. This can be accomplished by expanding the mass and stiffness matrices in Taylor series in ω^2. Thus,

$$\begin{aligned} \mathbf{m}^e &= \sum_{n=0}^{\infty} \mathbf{m}_n \omega^{2n} \qquad & \tilde{\mathbf{m}}^e &= \sum_{n=0}^{\infty} \tilde{\mathbf{m}}_n \omega^{2n} \\ \mathbf{k}^e &= \sum_{n=0}^{\infty} \mathbf{k}_n \omega^{2n} \qquad & \mathbf{k}^e_{\text{dyn}} &= \sum_{n=0}^{\infty} \left(\mathbf{k}_{\text{dyn}}\right)_n \omega^{2n} \end{aligned} \tag{2.116}$$

where the matrices with subscript n are not functions of the frequency ω.

Typically, for computational purposes only a few terms in these expansions are needed. It can be shown that the expansion terms are related by

$$(n+1)\mathbf{k}_{n+1} = n\mathbf{m}_n = n(n+1)\tilde{\mathbf{m}}_n = -n(n+1)(\mathbf{k}_{\text{dyn}})_{n+1} \qquad n \geq 1 \tag{2.117}$$

Furthermore,

$$\begin{aligned} (\mathbf{k}_{\text{dyn}})_0 &= \mathbf{k}_0 \qquad && \text{traditional stiffness matrix for static responses} \\ (\mathbf{k}_{\text{dyn}})_1 &= -\mathbf{m}_0 \qquad && \text{traditional consistent mass matrix} \\ \mathbf{k}_1 &= 0 && \end{aligned} \tag{2.118}$$

2.4 GEOMETRIC STIFFNESS MATRICES FOR BEAMS WITH AXIAL LOADING

Axial forces in beams are introduced into an analysis in a fashion similar to the treatment of mass. The governing differential equations of motion of Eq. (2.109), which are partial differential equations, and the harmonic motion version of the governing differential equations of Example 2.8, which are ordinary differential equations, include terms that account for the axial force N. It is essential to distinguish between the axial force N of Eq. (2.109), which represents the effect of axial forces on bending motion, and the same axial force N of Example 2.16, where the axial deformation is of concern. It is the effect of N on bending that is considered in this subsection. Table 2.2, which models this effect on bending of the axial force, contains the exact solution of the governing differential equations of Example 2.8.

The principle of virtual work can be used to create a matrix that accounts, approximately, for the bending effects of the axial force N, similar to the consistent mass matrix for the inclusion of inertia in an analysis. The introduction of axial force N to the integral form of the governing equations Eq. (2.118) of the principle of virtual work relation for the eth element gives

$$-(\delta W_{\text{int}} + \delta W_{\text{ext}})^e = \int_a^b \left(\delta \frac{d^2 w}{dx^2}\right) EI \frac{d^2 w}{dx^2}\, dx - \int_a^b \left(\delta \frac{dw}{dx}\right) N \frac{dw}{dx}\, dx - \int_a^b \delta w\, \overline{p}_z\, dx \tag{2.119}$$

where N is in compression. The second integral on the right-hand side, when N is factored out and $w(x) = \mathbf{N}(x)\mathbf{v}^e = \mathbf{N}_u(x)\mathbf{G}\mathbf{v}^e$ is introduced, leads to the stiffness matrix

$$\mathbf{k}_\sigma^e = \int_a^b \mathbf{N}'^{\mathrm{T}}\mathbf{N}'\, dx = \mathbf{G}^{\mathrm{T}} \int_a^b \mathbf{N}_u'^{T}\mathbf{N}_u'\, dx\mathbf{G} \tag{2.120}$$

where $\mathbf{N}' = d\mathbf{N}/dx$. This matrix is referred to as the *geometric, differential*, or *stress stiffness matrix*. Matrix $\mathbf{k}_\sigma^e$ is used in the study of structural stability, that is, in the analysis for the buckling load of beams or beam systems. The phenomenon of a tensile axial force N increasing the bending stiffness is referred to as *stress stiffening*: hence the terminology *stress stiffness matrix*.

Use of the same displacement trial function to form $\mathbf{k}_\sigma^e$ that is used in deriving $\mathbf{k}^e$ leads to a geometric stiffness matrix that is said to be *consistent*. Substitution of the cubic $\mathbf{N}_u = \begin{bmatrix} 1 & x & x^2 & x^3 \end{bmatrix}^{\mathrm{T}}$ into Eq. (2.120) gives

$$\mathbf{k}_\sigma^e = \frac{1}{30\ell} \begin{bmatrix} 36 & -3\ell & -36 & -3\ell \\ -3\ell & 4\ell^2 & 3\ell & -\ell^2 \\ -36 & 3\ell & 36 & 3\ell \\ -3\ell & -\ell^2 & 3\ell & 4\ell^2 \end{bmatrix} \tag{2.121}$$

This symmetric matrix is a commonly used approximate geometric stiffness matrix for beam elements.

There is no requirement that the matrices $\mathbf{k}^e$ and $\mathbf{k}_\sigma^e$ be based on the same displacement functions. Note that only first-order derivatives of $\mathbf{N}$ occur in the $\mathbf{k}_\sigma^e$ of Eq. (2.120), whereas second-order derivatives appear in the $\mathbf{k}^e$ term. Hence a "simpler" displacement function is sometimes employed in calculating $\mathbf{k}_\sigma^e$.

Use of a more accurate $\mathbf{N}$ in Eq. (2.120) can give more accurate but computationally less favorable geometric stiffness matrices. An exact geometric stiffness matrix is obtained by using $N = N_{\text{exact}}$ in Eq. (2.120). The exact geometric stiffness matrix can also be calculated from the dynamic stiffness matrix using

$$\mathbf{k}_\sigma^e = \frac{\partial \mathbf{k}_{dyn}}{\partial N} \tag{2.122}$$

2.5 THERMOELASTIC ANALYSIS

The solutions presented thus far in this chapter taking into account temperature changes ΔT apply when $\Delta T \neq \Delta T(y, z)$. As indicated in Section 1.2, the beam theory here is based on the assumption that the transverse normal strains and stresses (σ_y and σ_z) are small. For large transverse normal stresses and strains, a systematic solution of a beam with general thermal loading is available (Copper, 1993). This approach couples a plane strainlike solution to the strength-of-material thermal response equations for a beam. It is found that if the quantity of $\nu(\sigma_y + \sigma_z)$ where ν is Poisson's ratio, is large, it is important to be careful in using the traditional strength of materials formulas to account for thermal loading. Rather, it may well be necessary to introduce the coupled approach.

Copper (1993) gives several examples to illustrate that if the change in temperature varies with the coordinates y and z, the stresses and deflection as calculated by traditional strength of materials can be quite inaccurate. In the case of a square cross section with the change in temperature varying cubically with z, the use of strength of materials formulas tends to underestimate both the stress and the deflection. Another example from the same reference is that of a free–free hollow tube with a change in temperature that varies with the radial and axial coordinates. It is shown that the error in calculating normal stress σ_x and the shear stress τ_{rx} using traditional strength of materials formulas is $\nu \times 100\%$. Thus, as $\nu \to \frac{1}{2}$, the error approaches 50%.

REFERENCES

Copper, C. D. (1993). Thermoelastic solutions for beams, Ph.D. dissertation, University of Virginia, Charlottesville, Va.

Copper, C. D., and Pilkey, W. D. (2002). Thermoelasticity solutions for straight beams, *J. Appl. Mech.*, Vol. 70.

Gere, J. M. (2001). *Mechanics of Materials*, 5th ed., PWS, Boston.

Johnson, T. (1993). Accurate thermal stresses for beams, Ph.D. dissertation, University of Virginia, Charlottesville, Va.

Pilkey, W. D. (1994). *Formulas for Stress, Strain, and Structural Matrices*, Wiley, New York.

Pilkey, W. D., and Liu, Y. (1989). Thermal bending stresses on beam cross sections, *Finite Elements Anal. Des.*, Vol. 6, pp. 23–31.

Pilkey, W. D., and Wunderlich, W. (1994). *Mechanics of Structures: Variational and Computational Methods*, CRC Press, Boca Raton, Fla.

CHAPTER 3

BEAM SYSTEMS

In Chapter 2, element matrices for beams were developed. In this chapter we study methods of analyzing complex beams or systems of beams using the element matrices of Chapter 2. The static, dynamic, and unstable responses can be computed. The responses include deflections, moments, shear forces along the beam system, as well as natural frequencies and buckling loads.

In practice, most structural analyses are performed using general-purpose analysis software based on the *displacement method.* Alternatively, the *force method* or the *transfer matrix method* can be employed. The displacement and force methods lead to large systems of equations, the number of equations depending on the complexity of the structure. With the transfer matrix method, the size of the system equations does not depend on the complexity of the structure. However, the transfer matrix is restricted to linelike structures, such as a single bay framework. Here the discussion is directed primarily to the displacement method, with a brief description of the transfer matrix method. See a textbook on structural mechanics for a thorough presentation of the force and transfer matrix methods.

Most of the treatment in this chapter is limited to structures with concentrated applied loads. Books on structural mechanics deal with various methods of handling applied distributed loading.

In addition to the static analysis of beams and frameworks, in this chapter we also discuss dynamic responses and stability problems. The computation of the eigenvalues (i.e., natural frequencies and buckling loads) is described.

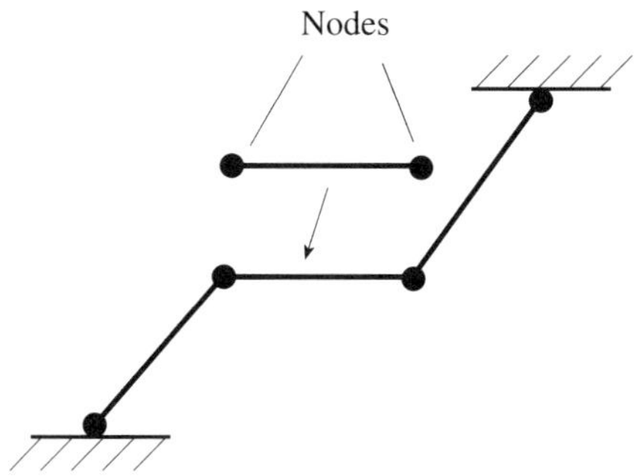

Figure 3.1 Spatially discretized framework with beam elements connected at the nodes.

3.1 STRUCTURAL SYSTEMS

3.1.1 Coordinate System and Degrees of Freedom

Beam structures are usually modeled as a finite number of beam elements connected at joints (*nodes*). This is a spatially discretized model (Fig. 3.1) where the forces and displacements at the nodes are the unknowns.

The locations of the nodes are defined in a global coordinate system (X, Y, Z). After the unknown forces and displacements at the nodes are computed, the internal forces [e.g., moments and shear forces in a local coordinate system (x, y, z)] and displacements between the nodes of the beam elements can be calculated. These internal forces are then employed in computing the cross-sectional distribution of stresses. Most of the rest of this book deals with the problem of computing normal and shear stresses on the beam cross section, along with cross-sectional properties.

As mentioned previously, the *degrees of freedom* (DOF) of a node are the independent coordinates (i.e., displacements) essential for the complete description of the motion of the node. For a beam element in bending with motion in a plane, each node can have three DOF: one transverse displacement, one axial displacement, and one rotation. In terms of forces, one shear force, one axial force, and one bending moment correspond to these DOF.

3.1.2 Transformation of Forces and Displacements

For a beam element of a planar framework, the forces and displacements can be treated in the local x, y, z coordinate system as shown in Fig. 3.2a or in the global X, Y, Z coordinate system of Fig. 3.2b. In both cases, the motion is in-plane; that is, the bending deflection is in the xz plane. The local reference frame x, y, z is aligned in a natural direction with the x axis along the element. The components of the forces and displacements at the ends of beam element e are shown in both coordinate

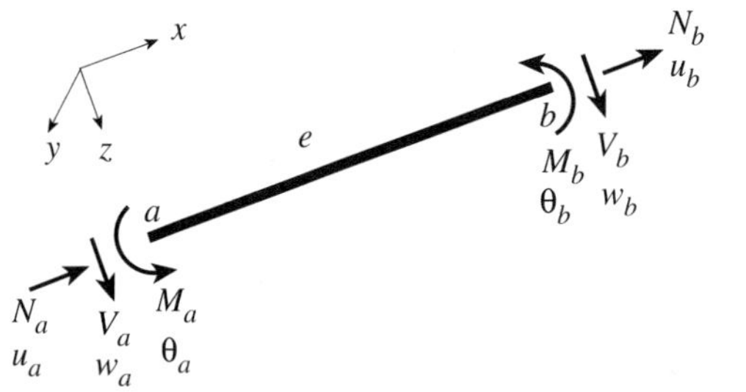

$$\mathbf{p}^e = \begin{bmatrix} N_a \\ V_a \\ M_a \\ N_b \\ V_b \\ M_b \end{bmatrix}^e \quad ; \quad \mathbf{v}^e = \begin{bmatrix} u_a \\ w_a \\ \theta_a \\ u_b \\ w_b \\ \theta_b \end{bmatrix}^e$$

(a) Element force (including shear force, axial force, and bending moment) and displacement (including deflection, axial displacement, and rotation about y) components in the local x, y, z coordinate system; motion in the xz plane

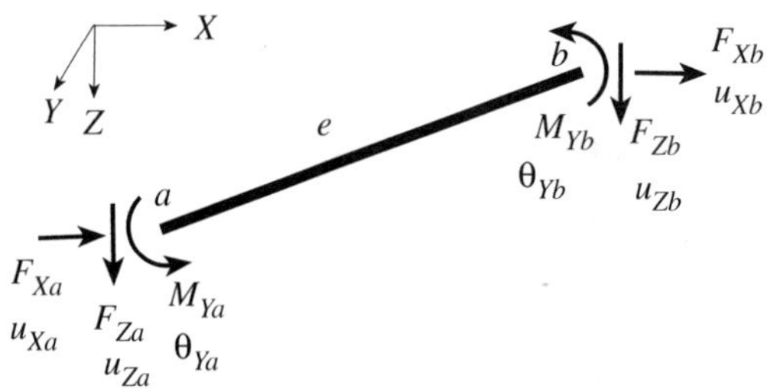

$$\mathbf{p}_G^e = \begin{bmatrix} F_{Xa} \\ F_{Za} \\ M_{Ya} \\ F_{Xb} \\ F_{Zb} \\ M_{Yb} \end{bmatrix}^e \quad ; \quad \mathbf{v}_G^e = \begin{bmatrix} u_{Xa} \\ u_{Za} \\ \theta_{Ya} \\ u_{Xb} \\ u_{Zb} \\ \theta_{Yb} \end{bmatrix}^e$$

(b) Element force and displacement components in the global X, Y, Z coordinate system

Figure 3.2 Representations of forces and displacements in the local and global coordinate systems on the ends of beam element e with in-plane motion. Sign Convention 2.

systems in Fig. 3.2. In the local coordinate system, the forces and displacements are $\mathbf{p}^e$, $\mathbf{v}^e$, and in the global system, the forces and displacements are designated by $\mathbf{p}_G^e$, $\mathbf{v}_G^e$.

In analyzing structures it is customary to transform the components of force and displacement vectors for all elements to the same coordinate system, here the global coordinate system. Thus, these components are transformed from the local to the global coordinate systems. The equations for the transformation of vector components are shown in Fig. 3.3. The equations in this figure are applicable to either node a or b.

To place the local–global relationships of Fig. 3.3 in matrix notation, define the transformation matrix $\mathbf{T}^e$:

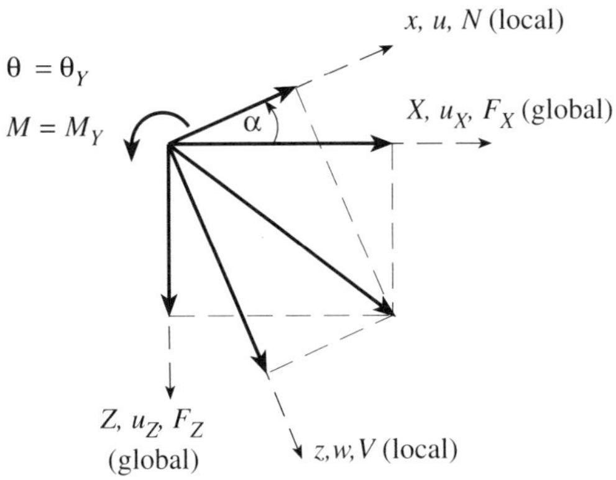

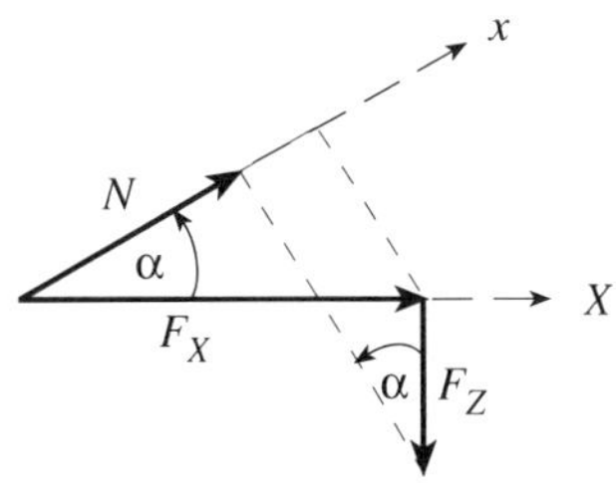

	Local ← Global	Global ← Local
Displacements	$u = u_X \cos\alpha - u_Z \sin\alpha$ $w = u_X \sin\alpha + u_Z \cos\alpha$ $\theta = \theta_Y$	$u_X = u\cos\alpha + w\sin\alpha$ $u_Z = -u\sin\alpha + w\cos\alpha$ $\theta_Y = \theta$
Forces	$N = F_X \cos\alpha - F_Z \sin\alpha$ $V = F_X \sin\alpha + F_Z \cos\alpha$ $M = M_Y$	$F_X = N\cos\alpha + V\sin\alpha$ $F_Z = -N\sin\alpha + V\cos\alpha$ $M_Y = M$

Figure 3.3 Transformation to and from local and global coordinates of the components of the force and displacement vectors at a or b. The right-hand-rule vector corresponding to a is positive along the positive y direction. Angle a is shown positive. It is measured from the global to the local coordinate systems.

$$\mathbf{T}^e = \begin{bmatrix} \cos\alpha & -\sin\alpha & 0 & \vdots & 0 & 0 & 0 \\ \sin\alpha & \cos\alpha & 0 & \vdots & 0 & 0 & 0 \\ 0 & 0 & 1 & \vdots & 0 & 0 & 0 \\ \cdots & \cdots & \cdots & \cdots & \cdots & \cdots & \cdots \\ 0 & 0 & 0 & \vdots & \cos\alpha & -\sin\alpha & 0 \\ 0 & 0 & 0 & \vdots & \sin\alpha & \cos\alpha & 0 \\ 0 & 0 & 0 & \vdots & 0 & 0 & 1 \end{bmatrix}^e$$

$$= \begin{bmatrix} \mathbf{T}_{aa} & \vdots & \mathbf{T}_{ab} \\ \cdots & \cdots & \cdots \\ \mathbf{T}_{ba} & \vdots & \mathbf{T}_{bb} \end{bmatrix}^e \tag{3.1}$$

From the definitions of $\mathbf{v}^e$, $\mathbf{v}_G^e$, $\mathbf{p}^e$, $\mathbf{p}_G^e$ of Fig. 3.2 the transformations can be written as follows:

Displacement variables (displacement components on the ends of a bar element):

$$\begin{array}{ll} \text{local} \leftarrow \text{global} & \text{global} \leftarrow \text{local} \\ \mathbf{v}^e = \mathbf{T}^e \mathbf{v}_G^e & \mathbf{v}_G^e = \mathbf{T}^{e\mathrm{T}} \mathbf{v}^e \end{array} \tag{3.2a}$$

Force variables (force components on the ends of a bar element):

$$\begin{array}{ll} \text{local} \leftarrow \text{global} & \text{global} \leftarrow \text{local} \\ \mathbf{p}^e = \mathbf{T}^e \mathbf{p}_G^e & \mathbf{p}_G^e = \mathbf{T}^{e\mathrm{T}} \mathbf{p}^e \end{array} \tag{3.2b}$$

The superscript T indicates the transpose of a matrix.

The transformation matrix $\mathbf{T}^e$ was established above for a bar element displacing in the x, z plane. The matrix $\mathbf{T}^e$ varies for other planes or spaces as well as for other members (e.g., truss members). Figure 3.4 shows the local and global force and displacement components for bars undergoing out-of-plane motion (xy plane). The transformation matrix $\mathbf{T}^e$ for out-of-plane motion is also shown.

Note that $\mathbf{T}_{jj}^e$, $j = a$ or b, of Eq. (3.1) satisfies

$$\mathbf{T}_{jj}^{e\mathrm{T}} \mathbf{T}_{jj}^e = \mathbf{I} \qquad \text{also} \quad \mathbf{T}^{e\mathrm{T}} \mathbf{T}^e = \mathbf{I} \tag{3.3}$$

where $\mathbf{I}$ is the unit diagonal matrix. Hence, the transformation is orthogonal. Since $(\mathbf{T}^e)^{-1} \mathbf{T}^e = \mathbf{I}$, it follows that

$$(\mathbf{T}^e)^{-1} = \mathbf{T}^{e\mathrm{T}} \tag{3.4}$$

It is important to be able to transform the stiffness matrix from one coordinate system to another. Let $\mathbf{k}^e$ of $\mathbf{p}^e = \mathbf{k}^e \mathbf{v}^e + \mathbf{p}^{e0}$ be the stiffness matrix for the displacements and forces referred to the local coordinate system. Similarly, $\mathbf{k}_G^e$ of $\mathbf{p}_G^e = \mathbf{k}_G^e \mathbf{v}_G^e + \mathbf{p}_G^{e0}$ is the stiffness matrix for which the displacement and force components are referred to the global coordinate system. To find a relationship between $\mathbf{k}_G^e$ and $\mathbf{k}^e$, form

$$\mathbf{T}^{e\mathrm{T}} \mathbf{p}^e = \mathbf{T}^{e\mathrm{T}} \mathbf{k}^e \mathbf{v}^e + \mathbf{T}^{e\mathrm{T}} \mathbf{p}^{e0} = \mathbf{T}^{e\mathrm{T}} \mathbf{k}^e \mathbf{T}^e \mathbf{v}_G^e + \mathbf{T}^{e\mathrm{T}} \mathbf{p}^{e0} = \mathbf{k}_G^e \mathbf{v}_G^e + \mathbf{p}_G^{e0} \tag{3.5}$$

Thus the stiffness matrix has transformed according to

$$\mathbf{k}_G^e = \mathbf{T}^{e\mathrm{T}} \mathbf{k}^e \mathbf{T}^e \tag{3.6a}$$

and the applied loading vector as

$$\mathbf{p}_G^{e0} = \mathbf{T}^{e\mathrm{T}} \mathbf{p}^{e0} \tag{3.6b}$$

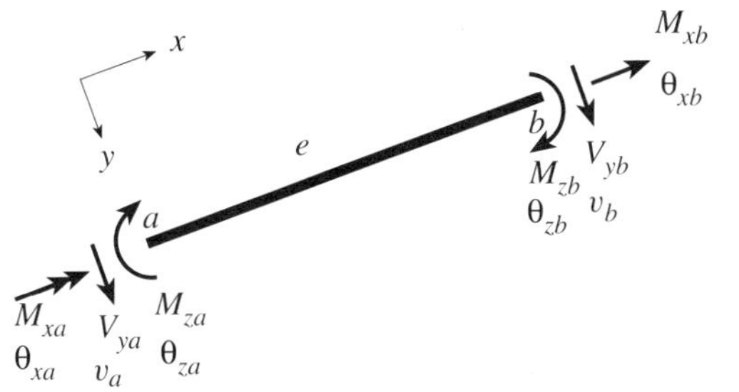

$$\mathbf{p}^e = \begin{bmatrix} V_{ya} \\ M_{xa} \\ M_{za} \\ V_{yb} \\ M_{xb} \\ M_{zb} \end{bmatrix}^e ; \quad \mathbf{v}^e = \begin{bmatrix} v_a \\ \theta_{xa} \\ \theta_{za} \\ v_b \\ \theta_{xb} \\ \theta_{zb} \end{bmatrix}^e$$

Local forces and displacements

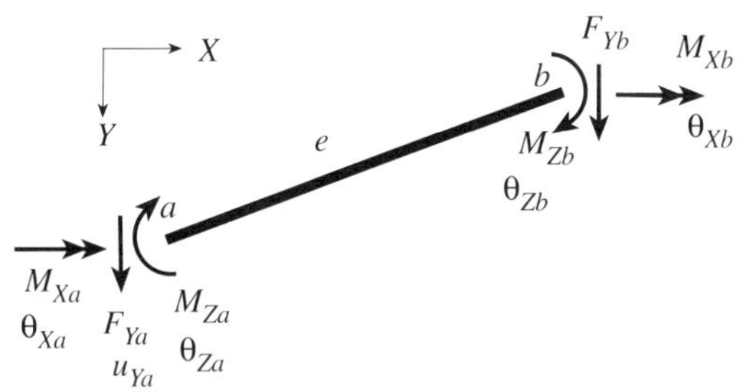

$$\mathbf{p}_G^e = \begin{bmatrix} F_{Ya} \\ M_{Xa} \\ M_{Za} \\ F_{Yb} \\ M_{Xb} \\ M_{Zb} \end{bmatrix}^e ; \quad \mathbf{v}_G^e = \begin{bmatrix} u_{Za} \\ \theta_{Xa} \\ \theta_{Za} \\ u_{Zb} \\ \theta_{Xb} \\ \theta_{Zb} \end{bmatrix}^e$$

Global forces and displacements

$$\mathbf{T}^e = \left[\begin{array}{ccc:ccc} 1 & 0 & 0 & 0 & 0 & 0 \\ 0 & \cos\alpha & \sin\alpha & 0 & 0 & 0 \\ 0 & -\sin\alpha & \cos\alpha & 0 & 0 & 0 \\ \hdashline 0 & 0 & 0 & 1 & 0 & 0 \\ 0 & 0 & 0 & 0 & \cos\alpha & \sin\alpha \\ 0 & 0 & 0 & 0 & -\sin\alpha & \cos\alpha \end{array}\right]^e$$

Figure 3.4 Forces, displacements, and transformation for out-of-plane motion. Sign Convention 2. The right-hand rule vector corresponding to α is positive along the positive z direction. Measure a from the global to local coordinate systems.

A product of the form $\mathbf{T}^{e\mathrm{T}}\mathbf{k}^e\mathbf{T}^e$is referred to as a *congruent transformation*. Under a congruent transformation $\mathbf{k}_G^e$ will be a symmetric matrix if $\mathbf{k}^e$ is symmetric (which it is).

3.2 DISPLACEMENT METHOD OF ANALYSIS

The displacement method can be formulated directly from the conditions of equilibrium, with the imposition of compatibility of displacements at the nodes, in which case it is called the *direct stiffness method*, or it can be considered to be a variationally based approach since it follows from the principle of virtual work.

3.2.1 Direct Stiffness Method

For the displacement method the equations modeling the response of the structural system represent the conditions of equilibrium for the nodal forces. The nodal equilibrium conditions for the forces of a simple framework are shown in Fig 3.5a. The displacements (at the nodes) need to be compatible. Nodal displacement compatibility conditions are shown in Fig. 3.5a.

The equations of equilibrium and compatibility of Fig. 3.5 deserve explanation. Heretofore, the ends of a beam element have been labeled with the letters a and b. In this chapter, where several beam elements are assembled into a structure, the ends of each element will be labeled with numbers. A framework formed of two beam elements is shown in Fig. 3.5. The two elements are identified with numbers enclosed by squares. There are three nodes, the numbers for which are enclosed by circles. Superscripts designate the elements. Subscripts designate nodes or ends of elements. The subscript G means that the components are referred to the global coordinate system.

Thus,

$(\mathbf{p}_j^e)_G, e = 1, 2;\ j = 1, 2, 3$ are the force components for element e at node j. The components are expressed in the global coordinate system. In the notation of Fig. 3.2b for in-plane (X, Z) motion

$$(\mathbf{p}_j^e)_G = \begin{bmatrix} F_X \\ F_Z \\ M_Y \end{bmatrix}_j^e \tag{3.7}$$

$(\mathbf{v}_j^e)_G, e = 1, 2;\ j = 1, 2, 3$ are the displacements for element e at node j. These displacements are the components along the global coordinates. From Fig. 3.2b

$$(\mathbf{v}_j^e)_G = \begin{bmatrix} u_X \\ u_Z \\ \theta_Y \end{bmatrix}_j^e \tag{3.8}$$

$\overline{\mathbf{P}}_j^*, j = 1, 2, 3$ are the globally aligned concentrated system forces applied at node j. Included in this vector are the reactions.

$\mathbf{V}_j, j = 1, 2, 3$ are the system displacements, along the global coordinates, of node j.

The notation needed to represent the forces and displacements in the two (local and global) coordinate systems may appear to be rather intimidating. Initially, the

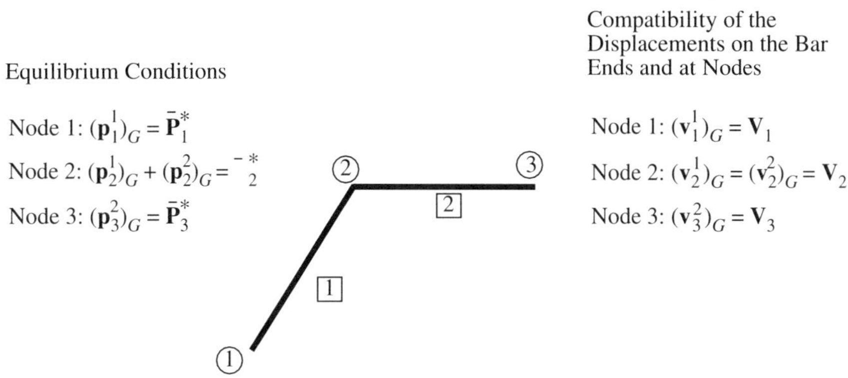

(a) Conditions for the three nodes of a two-element framework

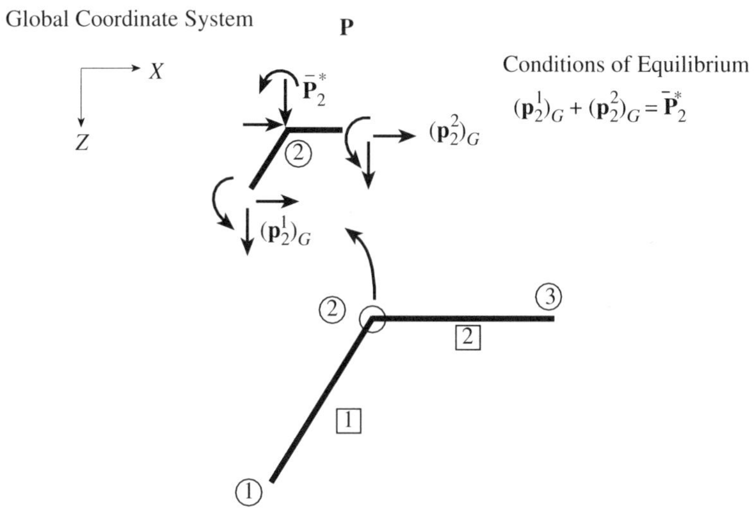

(b) Forces at node 2

Figure 3.5 Conditions of equilibrium for forces and displacement compatibility requirements at the nodes.

displacement and force components for element e were represented in the local coordinate systems as $\mathbf{v}^e$ and $\mathbf{p}^e$. Then these components were transformed to the global coordinate system as $(\mathbf{v}^e)_G$ and $(\mathbf{p}^e)_G$. Finally, since the equations for the response of the whole structure are to be expressed in terms of the displacements and forces at the nodes, another notation system for the nodal variables needs to be adopted. As indicated above, $\mathbf{V}_j$ represents the displacements at node j of the system. The components of these unknown nodal displacements are designated as U_X, U_Z, and

Θ_Y, or

$$\mathbf{V}_j = \begin{bmatrix} U_X \\ U_Z \\ \Theta_Y \end{bmatrix}_j \tag{3.9}$$

Prescribed displacements are defined as $\overline{\mathbf{V}}_j$. The global aligned concentrated system reactions or forces applied at node j are $\overline{\mathbf{P}}_j^*$. The applied loading components of $\overline{\mathbf{P}}_j^*$ are designated as $\overline{P}_X$, $\overline{P}_Z$, and $\overline{M}_Y$, so that

$$\overline{\mathbf{P}}_j^* = \begin{bmatrix} \overline{P}_X \\ \overline{P}_Z \\ \overline{M}_Y \end{bmatrix}_j \tag{3.10}$$

We choose to represent the unknown reactions in $\overline{\mathbf{P}}_j^*$ as

$$\overline{\mathbf{P}}_j^* = \begin{bmatrix} R_X \\ R_Z \\ M_R \end{bmatrix}_j \tag{3.11}$$

The equations for equilibrium and compatibility conditions at the nodes are shown in Fig. 3.5. The equilibrium conditions are simply the familiar conditions of statics applied to each node. For example, for node 2 the forces at the ends of elements 1 and 2 are shown in Fig. 3.5b as $(\mathbf{p}_2^1)_G$ and $(\mathbf{p}_2^2)_G$, while the applied external loads are designated as $\overline{\mathbf{P}}_2^*$. The equilibrium conditions at node 2 are then $(\mathbf{p}_2^1)_G + (\mathbf{p}_2^2)_G = \overline{\mathbf{P}}_2^*$. The compatibility conditions at a particular node prescribe that the local displacements in the global coordinate system are equal to the global displacements. Thus, in the case of node 2, where elements 1 and 2 are attached to each other, the element end displacements $(\mathbf{v}_2^1)_G$ and $(\mathbf{v}_2^2)_G$ are the same as the displacement $\mathbf{V}_2$ of the node. That is, $(\mathbf{v}_2^1)_G = (\mathbf{v}_2^2)_G = \mathbf{V}_2$. In most structural analyses, all local force and displacement components are referred to the global coordinate system and hence it is convenient to drop the use of the subscript G.

We are now in the position to establish equations of equilibrium for the entire structural system. In doing so, the compatibility conditions for the nodes are to be imposed. Let us continue to consider the two-element, three-node plane frame of Fig. 3.5. For elements 1 and 2, which span from nodes 1 to 2 and 2 to 3, respectively, the stiffness equations of Example 2.10, Eqs. (4) and (6), appear as

$$\begin{bmatrix} \mathbf{p}_1^1 \\ \mathbf{p}_2^1 \end{bmatrix} = \begin{bmatrix} \mathbf{k}_{11}^1 & \mathbf{k}_{12}^1 \\ \mathbf{k}_{21}^1 & \mathbf{k}_{22}^1 \end{bmatrix} \begin{bmatrix} \mathbf{v}_1^1 \\ \mathbf{v}_2^1 \end{bmatrix} + \begin{bmatrix} \mathbf{p}_1^{10} \\ \mathbf{p}_2^{10} \end{bmatrix} \tag{3.12}$$

$$\mathbf{p}^1 \quad = \quad \mathbf{k}^1 \quad \mathbf{v}^1 \quad + \quad \mathbf{p}^{10}$$

$$\begin{bmatrix} \mathbf{p}_2^2 \\ \mathbf{p}_3^2 \end{bmatrix} = \begin{bmatrix} \mathbf{k}_{22}^2 & \mathbf{k}_{23}^2 \\ \mathbf{k}_{32}^2 & \mathbf{k}_{33}^2 \end{bmatrix} \begin{bmatrix} \mathbf{v}_2^2 \\ \mathbf{v}_3^2 \end{bmatrix} + \begin{bmatrix} \mathbf{p}_2^{20} \\ \mathbf{p}_3^{20} \end{bmatrix} \tag{3.13}$$

$$\mathbf{p}^2 \quad = \quad \mathbf{k}^2 \quad \mathbf{v}^2 + \mathbf{p}^{20}$$

where, for example,

$$\mathbf{p}_2^e = (\mathbf{p}_2^e)_G = \begin{bmatrix} F_X \\ F_Z \\ M_Y \end{bmatrix}_2^e \qquad \mathbf{v}_2^e = (\mathbf{v}_2^e)_G = \begin{bmatrix} u_X \\ u_Z \\ \theta_Y \end{bmatrix}_2^e \tag{3.14}$$

Recall that the subscript G has been intentionally dropped in expressing Eqs. (3.12) and (3.13).

Introduce the compatibility equations of Fig. 3.5a, that is, $\mathbf{v}_1^1 = \mathbf{V}_1, \mathbf{v}_2^1 = \mathbf{v}_2^2 = \mathbf{V}_2$, and $\mathbf{v}_3^2 = \mathbf{V}_3$, into the stiffness relations of Eqs. (3.12) and (3.13):

$$\begin{aligned} \begin{bmatrix} \mathbf{p}_1^1 \\ \mathbf{p}_2^1 \end{bmatrix} &= \begin{bmatrix} \mathbf{k}_{11}^1 & \mathbf{k}_{12}^1 \\ \mathbf{k}_{21}^1 & \mathbf{k}_{22}^1 \end{bmatrix} \begin{bmatrix} \mathbf{V}_1 \\ \mathbf{V}_2 \end{bmatrix} + \begin{bmatrix} \mathbf{p}_1^{10} \\ \mathbf{p}_2^{10} \end{bmatrix} \\ \begin{bmatrix} \mathbf{p}_2^2 \\ \mathbf{p}_3^2 \end{bmatrix} &= \begin{bmatrix} \mathbf{k}_{22}^2 & \mathbf{k}_{23}^2 \\ \mathbf{k}_{32}^2 & \mathbf{k}_{33}^2 \end{bmatrix} \begin{bmatrix} \mathbf{V}_2 \\ \mathbf{V}_3 \end{bmatrix} + \begin{bmatrix} \mathbf{p}_2^{20} \\ \mathbf{p}_3^{20} \end{bmatrix} \end{aligned} \tag{3.15}$$

Substitute these relationships into the conditions of nodal equilibrium of Fig. 3.5a; that is,

$$\begin{aligned} \mathbf{p}_1^1 &= \overline{\mathbf{P}}_1^* \\ \mathbf{p}_2^1 + \mathbf{p}_2^2 &= \overline{\mathbf{P}}_2^* \\ \mathbf{p}_3^2 &= \overline{\mathbf{P}}_3^* \end{aligned} \tag{3.16}$$

Thus, from Eqs. (3.15) and (3.16),

$$\begin{aligned} \mathbf{p}_1^1 &= \mathbf{k}_{11}^1 \mathbf{V}_1 + \mathbf{k}_{12}^1 \mathbf{V}_2 + \mathbf{p}_1^{10} = \overline{\mathbf{P}}_1^* \\ \mathbf{p}_2^1 + \mathbf{p}_2^2 &= \mathbf{k}_{21}^1 \mathbf{V}_1 + \mathbf{k}_{22}^1 \mathbf{V}_2 + \mathbf{p}_2^{10} + \mathbf{k}_{22}^2 \mathbf{V}_2 + \mathbf{k}_{23}^2 \mathbf{V}_3 + \mathbf{p}_2^{20} = \overline{\mathbf{P}}_2^* \\ &= \mathbf{k}_{21}^1 \mathbf{V}_1 + (\mathbf{k}_{22}^1 + \mathbf{k}_{22}^2) \mathbf{V}_2 + \mathbf{k}_{23}^2 \mathbf{V}_3 + \mathbf{p}_2^{10} + \mathbf{p}_2^{20} = \overline{\mathbf{P}}_2^* \\ \mathbf{p}_3^2 &= \mathbf{k}_{32}^2 \mathbf{V}_2 + \mathbf{k}_{33}^2 \mathbf{V}_3 + \mathbf{p}_3^{20} = \overline{\mathbf{P}}_3^* \end{aligned} \tag{3.17}$$

These system equations in matrix form appear as

$$\begin{bmatrix} \mathbf{k}^1_{11} & \mathbf{k}^1_{12} & \\ \mathbf{k}^1_{21} & \mathbf{k}^1_{22}+\mathbf{k}^2_{22} & \mathbf{k}^2_{23} \\ & \mathbf{k}^2_{32} & \mathbf{k}^2_{33} \end{bmatrix} \begin{bmatrix} \mathbf{V}_1 \\ \mathbf{V}_2 \\ \mathbf{V}_3 \end{bmatrix} = \begin{bmatrix} \overline{\mathbf{P}}^*_1 \\ \overline{\mathbf{P}}^*_2 \\ \overline{\mathbf{P}}^*_3 \end{bmatrix} - \begin{bmatrix} \mathbf{p}^{10}_1 \\ \mathbf{p}^{10}_2 + \mathbf{p}^{20}_2 \\ \mathbf{p}^{20}_3 \end{bmatrix} = \begin{bmatrix} \overline{\mathbf{P}}_1 \\ \overline{\mathbf{P}}_2 \\ \overline{\mathbf{P}}_3 \end{bmatrix}$$

$$\mathbf{K} \qquad \mathbf{V} = \overline{\mathbf{P}}^* - \overline{\mathbf{P}}^\circ = \overline{\mathbf{P}} \tag{3.18}$$

The construction of equilibrium conditions at each node has resulted in this system of equations that represents the global statement of equilibrium

$$\mathbf{KV} = \overline{\mathbf{P}}^* - \overline{\mathbf{P}}^\circ \quad \text{or} \quad \mathbf{KV} = \overline{\mathbf{P}} \tag{3.19}$$

where

$\mathbf{V}$ is the vector of unknown nodal displacements (including nodal translations and rotations).

$\overline{\mathbf{P}}^*$ is the vector of concentrated reactions or loads applied at the nodes. Whereas the applied loads are known quantities, the reactions are usually unknown.

$\overline{\mathbf{P}}^\circ$ is the vector of nodal quantities accounting for loading applied between the nodes.

$\overline{\mathbf{P}}$ is equal to $\overline{\mathbf{P}}^* - \overline{\mathbf{P}}^\circ$.

$\mathbf{K}$ is the global or *system stiffness matrix*. This matrix is singular, a characteristic that is removed when the boundary conditions are introduced.

The system stiffness equations of Eq. (3.19) can be solved for the nodal displacements. Further processing of the solution leads to other characteristics of the response of the structure, such as reactions and stresses.

From the form of the stiffness matrix components of $\mathbf{K}$ of Eq. (3.18) we can observe how the global system of equations is "assembled":

$$\mathbf{K} = \begin{bmatrix} \mathbf{K}_{11} & \mathbf{K}_{12} & \mathbf{K}_{13} \\ \mathbf{K}_{21} & \mathbf{K}_{22} & \mathbf{K}_{23} \\ \mathbf{K}_{31} & \mathbf{K}_{32} & \mathbf{K}_{33} \end{bmatrix} = \begin{bmatrix} \mathbf{k}^1_{11} & \mathbf{k}^1_{12} & \\ \mathbf{k}^1_{21} & \mathbf{k}^1_{22}+\mathbf{k}^2_{22} & \mathbf{k}^2_{23} \\ & \mathbf{k}^2_{32} & \mathbf{k}^2_{33} \end{bmatrix}$$

$=$ 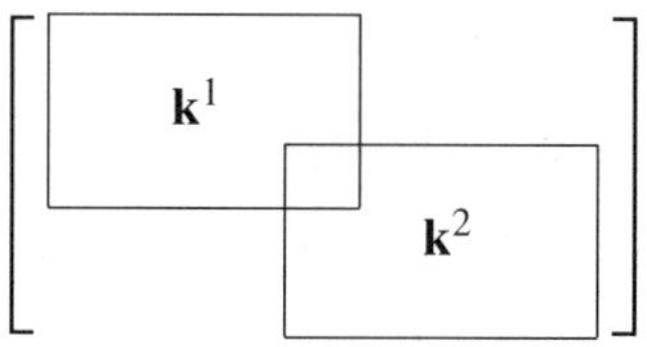

(3.20)

In the assembled equations, each unknown displacement component occurs only once in the displacement vector $\mathbf{V}$. It is apparent from Eq. (3.20) that the global stiffness matrix is assembled by adding appropriate stiffness coefficients which represent elements that are connected to each other. In Eq. (3.20), this addition is achieved by summing, $\mathbf{K}_{jk} = \mathbf{k}^1_{jk} + \mathbf{k}^2_{jk}$ (e.g., $\mathbf{K}_{11} = \mathbf{k}^1_{11}$ and $\mathbf{K}_{22} = \mathbf{k}^1_{22} + \mathbf{k}^2_{22}$). This special summation process is possible only because the element stiffness matrices were fit into a unique global node numbering system. That is, the assembly process is one of the addition of elements of the local stiffness matrices that have the same global address. Other techniques that do not involve the explicit renumbering of element stiffness matrices for assembling $\mathbf{K}$ are available in the literature.

Example 3.1 Alternative Assembly Notation. As is to be expected, it is not necessary to represent the stiffness matrices in partitioned form; for example

$$\mathbf{k}^m = \begin{bmatrix} \mathbf{k}^m_{ii} & \mathbf{k}^m_{ij} \\ \mathbf{k}^m_{ji} & \mathbf{k}^m_{jj} \end{bmatrix} \tag{1}$$

to use the superposition procedure above for assembling the global stiffness matrix. If the local stiffness matrices of the framework of Fig. 3.5 are expressed as [Eq. (5) of Example 2.10]

$$\mathbf{k}^1 = \begin{bmatrix} k^1_{11} & k^1_{12} & k^1_{13} & k^1_{14} \\ k^1_{21} & k^1_{22} & k^1_{23} & k^1_{24} \\ k^1_{31} & k^1_{32} & k^1_{33} & k^1_{34} \\ k^1_{41} & k^1_{42} & k^1_{43} & k^1_{44} \end{bmatrix} \qquad \mathbf{k}^2 = \begin{bmatrix} k^2_{33} & k^2_{34} & k^2_{35} & k^2_{36} \\ k^2_{43} & k^2_{44} & k^2_{45} & k^2_{46} \\ k^2_{53} & k^2_{54} & k^2_{55} & k^2_{56} \\ k^2_{63} & k^2_{64} & k^2_{65} & k^2_{66} \end{bmatrix} \tag{2}$$

where the displacements and forces at the ends of elements 1 and 2 are shown in Fig. 3.6. The assembly procedure of adding stiffness coefficients of common global

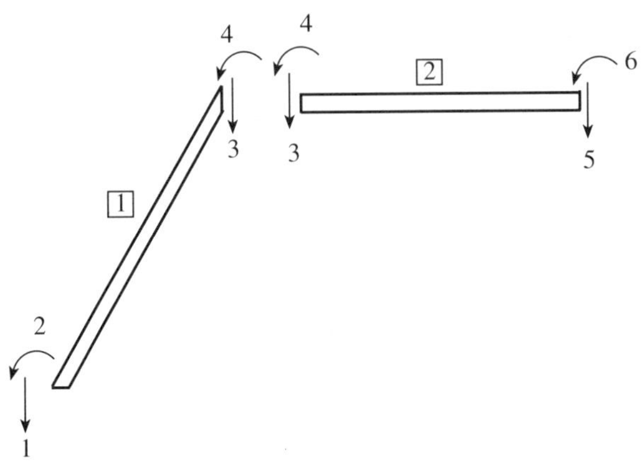

Figure 3.6 Notation for the stiffness matrices of Eq. (2) of Example 3.1.

addresses leads to the global stiffness matrix

$$\mathbf{K} = \begin{bmatrix} k_{11}^1 & k_{12}^1 & k_{13}^1 & k_{14}^1 & 0 & 0 \\ k_{21}^1 & k_{22}^1 & k_{23}^1 & k_{24}^1 & 0 & 0 \\ k_{31}^1 & k_{32}^1 & k_{33}^1 + k_{33}^2 & k_{34}^1 + k_{34}^2 & k_{35}^2 & k_{36}^2 \\ k_{41}^1 & k_{42}^1 & k_{43}^1 + k_{43}^2 & k_{44}^1 + k_{44}^2 & k_{45}^2 & k_{46}^2 \\ 0 & 0 & k_{53}^2 & k_{54}^2 & k_{55}^2 & k_{56}^2 \\ 0 & 0 & k_{63}^2 & k_{64}^2 & k_{65}^2 & k_{66}^2 \end{bmatrix} \tag{3}$$

This is, of course, the global stiffness matrix given in Eq. (3.18).

Example 3.2 Framework. Find the forces and displacements along the frame of Fig. 3.7.

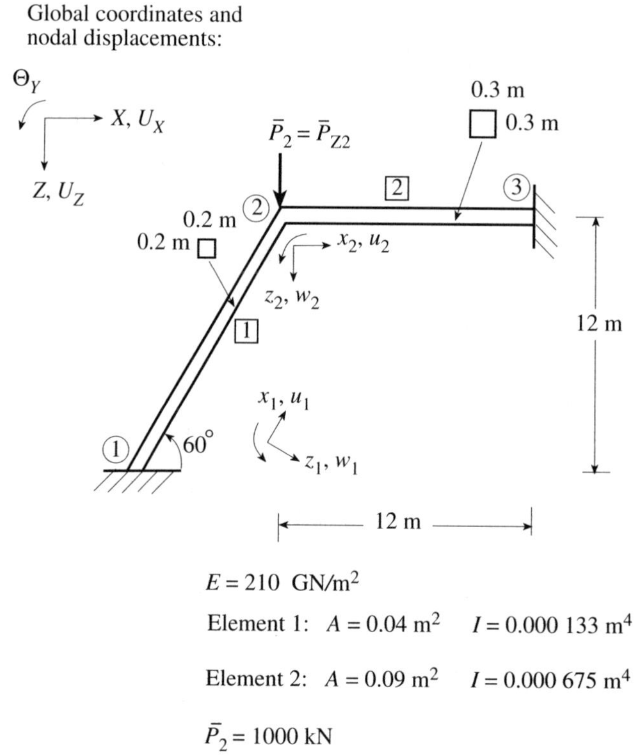

Figure 3.7 Framework with nodal loading.

SOLUTION. **Local Element Stiffness Matrices.** The element stiffness matrix for a bar element undergoing axial extension and bending is given by Eq. (2) of Example 2.16.

$$\mathbf{k}^e = \begin{bmatrix} EA/\ell & 0 & 0 & -EA/\ell & 0 & 0 \\ 0 & 12EI/\ell^3 & -6EI/\ell^2 & 0 & -12EI/\ell^3 & -6EI/\ell^2 \\ 0 & -6EI/\ell^2 & 4EI/\ell & 0 & 6EI/\ell^2 & 2EI/\ell \\ -EA/\ell & 0 & 0 & EA/\ell & 0 & 0 \\ 0 & -12EI/\ell^3 & 6EI/\ell^2 & 0 & 12EI/\ell^3 & 6EI/\ell^2 \\ 0 & -6EI/\ell^2 & 2EI/\ell & 0 & 6EI/\ell^2 & 4EI/\ell \end{bmatrix} \tag{1}$$

This local stiffness matrix for the eth element belongs to the relationship

$$\begin{bmatrix} N_a^e \\ V_a^e \\ M_a^e \\ N_b^e \\ V_b^e \\ M_b^e \end{bmatrix} = \mathbf{k}^e \begin{bmatrix} u_a^e \\ w_a^e \\ \theta_a^e \\ u_b^e \\ w_b^e \\ \theta_b^e \end{bmatrix} \tag{2}$$

$$\mathbf{p}^e \quad = \mathbf{k}^e \quad \mathbf{v}^e$$

Element 1: The length of element one is 13.85 m. In the local coordinate system, (1) with this length is given by

$$\mathbf{k}^1 = 10^6 \begin{bmatrix} 606.218 & 0 & 0 & -606.218 & 0 & 0 \\ 0 & 0.126 & -0.875 & 0 & -0.126 & -0.875 \\ 0 & -0.875 & 8.083 & 0 & 0.875 & 4.041 \\ -606.218 & 0 & 0 & 606.218 & 0 & 0 \\ 0 & -0.126 & 0.875 & 0 & 0.126 & 0.875 \\ 0 & -0.875 & 4.041 & 0 & 0.875 & 8.083 \end{bmatrix} \tag{3}$$

Element 2: For element 2, $\ell = 12$ m and the stiffness matrix in local coordinates is given by

$$\mathbf{k}^2 = 10^6 \begin{bmatrix} 1575 & 0 & 0 & -1575 & 0 & 0 \\ 0 & 0.984 & -5.906 & 0 & -0.984 & -5.906 \\ 0 & -5.906 & 47.25 & 0 & 5.906 & 23.625 \\ -1575 & 0 & 0 & 1575 & 0 & 0 \\ 0 & -0.984 & 5.906 & 0 & 0.984 & 5.906 \\ 0 & -5.906 & 23.625 & 0 & 5.906 & 47.25 \end{bmatrix} \tag{4}$$

Transformation of Element Variables from Local to Global Coordinate Systems The stiffness matrix $\mathbf{k}^e$ needs to be transformed from the local to the global coordinate systems using (Eq. 3.6a) $\mathbf{k}_G^e = \mathbf{T}^{e\mathrm{T}}\mathbf{k}^e\mathbf{T}^e$, with $\mathbf{T}^e$ given by Eq. (3.1).

Element 1: Equation (3.1), with $\alpha = 60°$, gives

$$\mathbf{T}^e = \begin{bmatrix} 0.5000 & -0.8660 & 0.0000 & 0.0000 & 0.0000 & 0.0000 \\ 0.8660 & 0.5000 & 0.0000 & 0.0000 & 0.0000 & 0.0000 \\ 0.0000 & 0.0000 & 1.0000 & 0.0000 & 0.0000 & 0.0000 \\ 0.0000 & 0.0000 & 0.0000 & 0.5000 & -0.8660 & 0.0000 \\ 0.0000 & 0.0000 & 0.0000 & 0.8660 & 0.5000 & 0.0000 \\ 0.0000 & 0.0000 & 0.0000 & 0.0000 & 0.0000 & 1.0000 \end{bmatrix} \tag{5}$$

Then, the element stiffness matrix $\mathbf{k}_G^e$ referred to the global coordinate system becomes

$$\begin{aligned} \mathbf{k}_G^1 &= \mathbf{T}^{1\mathrm{T}}\mathbf{k}^1\mathbf{T}^1 \\ &= 10^6 \begin{bmatrix} 151.649 & -262.445 & -0.758 & -151.649 & 262.445 & -0.758 \\ -262.445 & 454.695 & -0.438 & 262.445 & -454.695 & -0.438 \\ -0.758 & -0.438 & 8.083 & 0.758 & 0.438 & 4.041 \\ -151.649 & 262.445 & 0.758 & 151.649 & -262.445 & 0.758 \\ 262.445 & -454.695 & 0.438 & -262.445 & 454.695 & 0.438 \\ -0.758 & -0.438 & 4.041 & 0.758 & 0.438 & 8.083 \end{bmatrix} \\ &= \begin{bmatrix} \mathbf{k}_{11G}^1 & \mathbf{k}_{12G}^1 \\ \mathbf{k}_{21G}^1 & \mathbf{k}_{22G}^1 \end{bmatrix} \end{aligned} \tag{6}$$

where $\mathbf{k}_{jkG}^1$ are 3×3 submatrices.

Element 2: The local and global coordinate systems coincide for element 2. The angle α is equal to zero and $\mathbf{k}_G^2 = \mathbf{k}^2$ of (4). In partitioned form,

$$\mathbf{k}_G^2 = \begin{bmatrix} \mathbf{k}_{11G}^2 & \mathbf{k}_{12G}^2 \\ \mathbf{k}_{21G}^2 & \mathbf{k}_{22G}^2 \end{bmatrix}$$

Assembly of the System Equations The system equations $\mathbf{KV} = \overline{\mathbf{P}}$ for the frame of Fig. 3.7 are assembled as explained in Section 3.2.1. The nodal displacements $\mathbf{V}$ are defined by Eq. (3.9):

$$\mathbf{V} = \begin{bmatrix} \mathbf{V}_1 \\ \mathbf{V}_2 \\ \mathbf{V}_3 \end{bmatrix} \qquad \mathbf{V}_1 = \begin{bmatrix} U_{X1} \\ U_{Z1} \\ \Theta_{Y1} \end{bmatrix} \qquad \mathbf{V}_2 = \begin{bmatrix} U_{X2} \\ U_{Z2} \\ \Theta_{Y2} \end{bmatrix} \qquad \mathbf{V}_3 = \begin{bmatrix} U_{X3} \\ U_{Z3} \\ \Theta_{Y3} \end{bmatrix} \tag{7}$$

where $U_{Xj}, U_{Zj}, \Theta_{Yj}, j = 1, 2, 3$ are the nodal displacement components. Express Eq. (3.18) as

$$\underset{\mathbf{K}}{\begin{bmatrix} \mathbf{k}_G^1 & \\ & \mathbf{k}_G^2 \end{bmatrix}} \underset{\mathbf{V}}{\begin{bmatrix} \mathbf{V}_1 \\ \mathbf{V}_2 \\ \mathbf{V}_3 \end{bmatrix}} = \underset{\overline{\mathbf{P}}^*}{\begin{bmatrix} \overline{\mathbf{P}}_1^* \\ \overline{\mathbf{P}}_2^* \\ \overline{\mathbf{P}}_3^* \end{bmatrix}} - \underset{\overline{\mathbf{P}}^0}{\begin{bmatrix} \mathbf{p}_{1G}^{10} \\ \mathbf{p}_{2G}^{10} + \mathbf{p}_{2G}^{20} \\ \mathbf{p}_{3G}^{20} \end{bmatrix}} = \underset{\overline{\mathbf{P}}}{\begin{bmatrix} \overline{\mathbf{P}}_1 \\ \overline{\mathbf{P}}_2 \\ \overline{\mathbf{P}}_3 \end{bmatrix}} \tag{8}$$

or

$$\begin{bmatrix} \mathbf{k}_{11G}^1 & \mathbf{k}_{12G}^1 & \\ \mathbf{k}_{21G}^1 & \mathbf{k}_{22G}^1 + \mathbf{k}_{22G}^2 & \mathbf{k}_{23G}^2 \\ & \mathbf{k}_{32G}^2 & \mathbf{k}_{33G}^2 \end{bmatrix} \begin{bmatrix} \mathbf{V}_1 \\ \mathbf{V}_2 \\ \mathbf{V}_3 \end{bmatrix} = \begin{bmatrix} \overline{\mathbf{P}}_1 \\ \overline{\mathbf{P}}_2 \\ \overline{\mathbf{P}}_3 \end{bmatrix}$$

where the subscript G has been included to emphasize that all the variables are referred to the global coordinate system. Since there are no distributed loads between the nodes, $\overline{\mathbf{P}}^0 = \mathbf{0}$ and $\overline{\mathbf{P}} = \overline{\mathbf{P}}^*$.

In their present form, the equations of (8) apply to the two-element framework of Fig. 3.7. The fixed boundaries at nodes 1 and 3 have not yet been introduced to the solution. The global stiffness matrix $\mathbf{K}$ of (8) is a 9×9 matrix, which is singular before the boundary conditions are taken into account. If the boundary conditions are applied to (8), the resulting reduced stiffness matrix will be nonsingular and (8) will be reduced to a set of equations that can be solved for the nodal displacements.

The procedure described in this example for assembling and storing the global stiffness matrix, applying the boundary conditions, and solving the resulting system of equations is appropriate for a structural system with a limited number of elements, but is thought to be somewhat cumbersome for a large structural system. There is a sizable literature covering the problem of efficiently implementing this process of developing the system equations.

Nodes 1 and 3 are fixed so that the boundary conditions are the constraints on the displacements

$$\mathbf{V}_1 = \begin{bmatrix} U_{X1} \\ U_{Z1} \\ \Theta_{Y1} \end{bmatrix} = \mathbf{0} \quad \text{and} \quad \mathbf{V}_3 = \begin{bmatrix} U_{X3} \\ U_{Z3} \\ \Theta_{Y3} \end{bmatrix} = \mathbf{0} \tag{9}$$

Then, in the vector $\mathbf{V}$ of (8), the nodal displacements $\mathbf{V}_1$ and $\mathbf{V}_3$ are set equal to zero. The unknown nodal displacements U_{X2}, U_{Z2}, and Θ_{Y2} occur at node 2, and the vector $\mathbf{V}_2$ is equal to

$$\mathbf{V}_2 = [U_{X2} \quad U_{Z2} \quad \Theta_{Y2}]^{\mathrm{T}} \tag{10}$$

Then the complete nodal displacement vector is

$$\mathbf{V} = [\mathbf{V}_1 \quad \mathbf{V}_2 \quad \mathbf{V}_3]^{\mathrm{T}} = [0 \quad 0 \quad 0 \quad U_{X2} \quad U_{Z2} \quad \Theta_{Y2} \quad 0 \quad 0 \quad 0]^{\mathrm{T}} \tag{11}$$

For each displacement component that is zero at nodes 1 and 3, there will be a corresponding reaction, which is unknown. In the vector $\overline{\mathbf{P}} = \overline{\mathbf{P}}^* = \left[\overline{\mathbf{P}}_1 \quad \overline{\mathbf{P}}_2 \quad \overline{\mathbf{P}}_3\right]^{\mathrm{T}}$ of (8), the reactions correspond to $\overline{\mathbf{P}}_1$ and $\overline{\mathbf{P}}_3$. Hence, we will replace the nodal force vectors $\overline{\mathbf{P}}_1$and $\overline{\mathbf{P}}_3$ by the unknown reactions. The load vector $\overline{\mathbf{P}}_2$ is equal $\overline{\mathbf{P}}_2^*$ of Eq. (3.18). Recall that the distributed load vector $\overline{\mathbf{P}}^\circ$ is ignored, as there are no distributed loads applied between the nodes of the framework of Fig. 3.7. We find that

$$\overline{\mathbf{P}}_2 = \begin{bmatrix} \overline{P}_{X2} \\ \overline{P}_{Z2} \\ \overline{M}_{Y2} \end{bmatrix} = 10^3 \begin{bmatrix} 0.0 \\ 1000.0 \\ 0.0 \end{bmatrix} \tag{12}$$

where the forces are in newtons and the moments in newton-meters. Then the complete loading vector $\overline{\mathbf{P}}$ will appear as

$$\begin{aligned} \overline{\mathbf{P}} = \overline{\mathbf{P}}^* &= \left[\overline{\mathbf{P}}_1 \quad \overline{\mathbf{P}}_2 \quad \overline{\mathbf{P}}_3\right]^{\mathrm{T}} \\ &= \left[R_{X1} \quad R_{Z1} \quad M_{R1} \quad 0.0 \quad (1000)10^3 \quad 0.0 \quad R_{X3} \quad R_{Z3} \quad M_{R3}\right]^{\mathrm{T}} \end{aligned} \tag{13}$$

Now introduce the boundary conditions to (8). The displacement boundary conditions $\mathbf{V}_1 = \mathbf{0}$ and $\mathbf{V}_3 = \mathbf{0}$ lead to the cancellation of the columns in $\mathbf{K}$ corresponding to $\mathbf{V}_1$ and $\mathbf{V}_3$. For the framework of Fig. 3.7, this means cancel columns 1, 2, 3 and 7, 8, 9 of $\mathbf{K}$. Also, in $\mathbf{K}$ temporarily ignore the rows corresponding to $\overline{\mathbf{P}}_1 = \overline{\mathbf{P}}_1^* = [R_{X1} \quad R_{Z1} \quad M_{R1}]^{\mathrm{T}}$ and $\overline{\mathbf{P}}_3 = \overline{\mathbf{P}}_3^* = [R_{X3} \quad R_{Z3} \quad M_{R3}]^{\mathrm{T}}$. Thus, ignore rows 1, 2, 3 and 7, 8, 9. These manipulations lead to a square matrix that can be solved for the unknown displacements $\mathbf{V}_2$:

$$\left[\mathbf{k}_{22G}^1 + \mathbf{k}_{22G}^2\right] \mathbf{V}_2 = \overline{\mathbf{P}}_2 \tag{14}$$

where the displacements at node 2 are

$$\mathbf{V}_2 = \begin{bmatrix} U_{X2} \\ U_{Z2} \\ \Theta_{Y2} \end{bmatrix} \tag{15}$$

and the loading vector $\overline{\mathbf{P}}_2$ is

$$\overline{\mathbf{P}}_2 = 10^3 \begin{bmatrix} 0.0 \\ 1000.0 \\ 0.0 \end{bmatrix} \tag{16}$$

Thus, (14) appears as

$$10^6 \begin{bmatrix} 1726.649 & -262.445 & 0.758 \\ -262.445 & 455.679 & -5.469 \\ 0.758 & -5.469 & 55.333 \end{bmatrix} \begin{bmatrix} U_{X2} \\ U_{Z2} \\ \Theta_{Y2} \end{bmatrix} = 10^3 \begin{bmatrix} 0.0 \\ 1000.0 \\ 0.0 \end{bmatrix} \tag{17}$$

$$\widetilde{\mathbf{K}} \qquad \widetilde{\mathbf{V}} = \widetilde{\overline{\mathbf{P}}}$$

where $\widetilde{\mathbf{K}}$, $\widetilde{\mathbf{V}}$, and $\widetilde{\overline{\mathbf{P}}}$ are reduced matrices.

This is a set of simultaneous linear algebraic equations for U_{X2}, U_{Z2}, and Θ_{Y2}, the nodal displacements. Symbolically, $\widetilde{\mathbf{V}} = \widetilde{\mathbf{K}}^{-1}\widetilde{\overline{\mathbf{P}}}$. We find that

$$\mathbf{V}_2 = \begin{bmatrix} U_{X2} \\ U_{Z2} \\ \Theta_{Y2} \end{bmatrix} = 10^{-4} \begin{bmatrix} 3.659 \\ 24.081 \\ 2.33 \end{bmatrix} \tag{18}$$

where the two translations are in meters and the rotation in radians.

Postprocessing to Find Element Forces and Displacements The reactions at nodes 1 and 3 (the fixed ends of the framework) are obtained from (8), where the entries $\overline{\mathbf{P}}_1$ and $\overline{\mathbf{P}}_3$ are the unknown reactions corresponding to the constraints on the boundaries (i.e., $\mathbf{V}_1 = 0$ and $\mathbf{V}_3 = 0$). Thus, in (8) we will use

$$\mathbf{V} = \begin{bmatrix} \mathbf{V}_1 \\ \mathbf{V}_2 \\ \mathbf{V}_3 \end{bmatrix} \tag{19}$$

where

$$\mathbf{V}_1 = \begin{bmatrix} 0 \\ 0 \\ 0 \end{bmatrix} \qquad \mathbf{V}_3 = \begin{bmatrix} 0 \\ 0 \\ 0 \end{bmatrix} \qquad \mathbf{V}_2 = 10^{-4} \begin{bmatrix} 3.659 \\ 24.081 \\ 2.33 \end{bmatrix} \tag{20}$$

and

$$\overline{\mathbf{P}} = \begin{bmatrix} \overline{\mathbf{P}}_1 \\ \overline{\mathbf{P}}_2 \\ \overline{\mathbf{P}}_3 \end{bmatrix} \tag{21}$$

where

$$\overline{\mathbf{P}}_1 = \begin{bmatrix} R_{X1} \\ R_{Z1} \\ M_{R1} \end{bmatrix} \qquad \overline{\mathbf{P}}_2 = 10^3 \begin{bmatrix} 0.0 \\ 1000.0 \\ 0.0 \end{bmatrix} \qquad \overline{\mathbf{P}}_3 = \begin{bmatrix} R_{X3} \\ R_{Z3} \\ M_{R3} \end{bmatrix} \tag{22}$$

From (8), the reactions are found to be

$$\overline{\mathbf{P}}_1 = \begin{bmatrix} R_X \\ R_Z \\ M_R \end{bmatrix}_1 = 10^5 \begin{bmatrix} 5.763 \\ -9.99 \\ 0.023 \end{bmatrix}$$

and

$$\overline{\mathbf{P}}_3 = \begin{bmatrix} R_X \\ R_Z \\ M_R \end{bmatrix}_3 = 10^5 \begin{bmatrix} -5.763 \\ -9.944 \times 10^{-3} \\ -0.087 \end{bmatrix} \tag{23}$$

The displacements in the local coordinates for element 1 are calculated using

$$\mathbf{v}^1 = \mathbf{T}^1 \mathbf{v}_G^1 = \mathbf{T}^1 \begin{bmatrix} (\mathbf{v}_1^1)_G \\ (\mathbf{v}_2^1)_G \end{bmatrix} \tag{24}$$

where (Fig. 3.5) $(\mathbf{v}_1^1)_G = \mathbf{V}_1 = \mathbf{0}$ and $(\mathbf{v}_2^1)_G = \mathbf{V}_2$. Then using the notation of Fig. 3.2,

$$\mathbf{v}^1 = \begin{bmatrix} u_1 \\ w_1 \\ \theta_1 \\ u_2 \\ w_2 \\ \theta_2 \end{bmatrix} = \mathbf{T}^1 \mathbf{v}_G^1 = \mathbf{T}^1 \begin{bmatrix} (\mathbf{v}_1^1)_G \\ (\mathbf{v}_2^1)_G \end{bmatrix} = \mathbf{T}^1 \begin{bmatrix} u_{X1} \\ w_{Z1} \\ \theta_{Y1} \\ u_{X2} \\ w_{Z2} \\ \theta_{Y2} \end{bmatrix} = \mathbf{T}^1 \times 10^{-4} \begin{bmatrix} 0 \\ 0 \\ 0 \\ 3.659 \\ 24.081 \\ 2.33 \end{bmatrix}$$

$$= 10^{-4} \begin{bmatrix} 0 \\ 0 \\ 0 \\ -19.025 \\ 15.209 \\ 2.33 \end{bmatrix} \tag{25}$$

The forces in the local coordinate system for element 1 can be computed using $\mathbf{p}^1 = \mathbf{k}^1 \mathbf{v}^1$, where $\mathbf{k}^1$ is taken from (3) and $\mathbf{v}^1$ from (25). These forces will be expressed with Sign Convention 2. We find that

$$\mathbf{p}^1 = \mathbf{k}^1 \mathbf{v}^1 = 10^4 \begin{bmatrix} 115.332 \\ -0.04 \\ 0.227 \\ -115.332 \\ 0.04 \\ 0.321 \end{bmatrix} \tag{26}$$

Alternatively, these forces can be found by using

$$\mathbf{p}^1 = \mathbf{T}^1 \mathbf{p}_G^1 \quad \text{with} \quad \mathbf{p}_G^1 = \mathbf{k}_G^1 \mathbf{v}_G^1 \tag{27}$$

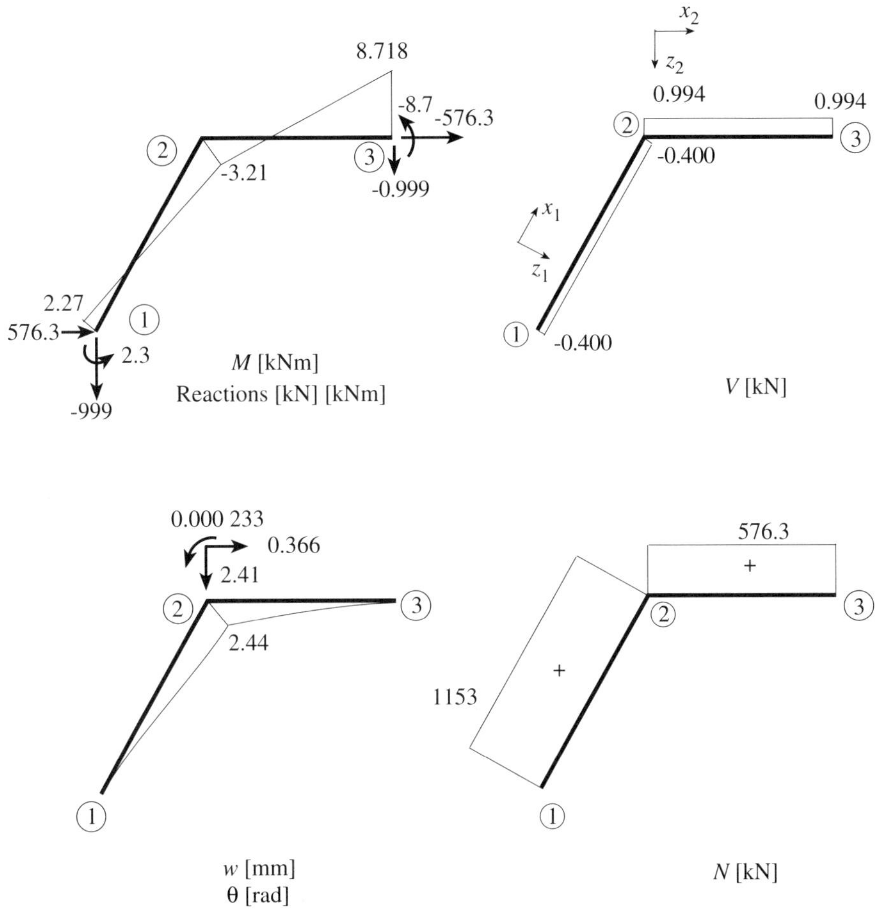

Figure 3.8 Displacements and forces for the framework of Fig. 3.6. Sign Convention 1.

For element 2, the displacements and forces are the same in the global and local coordinate systems; that is, $\mathbf{v}^2 = \mathbf{v}_G^2$ and $\mathbf{p}^2 = \mathbf{p}_G^2$. The reactions at nodes 1 and 3 are equal to the forces and moments at the appropriate ends of elements 1 and 2.

Figure 3.8 shows the distribution of forces and displacements for this framework. The forces are displayed in Sign Convention 1. These results correspond to the values that would be found in strength of materials or elasticity textbooks.

At this stage of an analysis it would be wise to verify that the reactions are in equilibrium with the applied loads. Also, are the forces at each node in equilibrium and are the conditions of equilibrium and compatibility satisfied for each element?

Example 3.3 Beam on Elastic Foundation. Determine the deflection, shear forces, and bending moments along the beam on elastic foundation of Fig. 3.9.

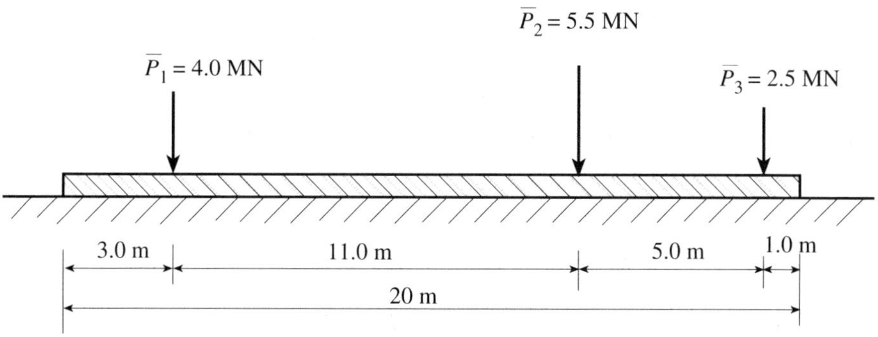

$E = 200\ \text{GN/m}^2$, $I = 0.001\ \text{m}^4$, $k_w = 30\ \text{MN/m}^2$

Figure 3.9 Beam on elastic foundation.

SOLUTION. In using the displacement method, the exact stiffness matrix of Table 2.2 or the approximate stiffness matrix of Eq. (15) of Example 2.18 can be employed. The exact stiffness matrix for a Bernoulli–Euler beam on a Winkler foundation is taken from Table 2.2 using $\lambda = k_w/EI, \eta = 0, \zeta = 0, \xi = 0$. This leads to

$$\mathbf{k}_w^e = \begin{bmatrix} k_{11} & k_{12} & k_{13} & k_{14} \\ k_{21} & k_{22} & k_{23} & k_{24} \\ k_{31} & k_{32} & k_{33} & k_{34} \\ k_{41} & k_{42} & k_{43} & k_{44} \end{bmatrix}^e \tag{1}$$

where

$$\begin{aligned} \Delta &= e_3^2 - e_2 e_4 & & \\ k_{11} &= \frac{(e_2 e_1 + \lambda e_3 e_4)EI}{\Delta} & q^2 &= \tfrac{1}{2}\sqrt{\lambda} \\ k_{12} = k_{21} &= \frac{(e_3 e_1 - e_2^2)EI}{\Delta} & e_0 &= -\lambda e_4 \\ k_{13} = k_{31} &= -\frac{e_2 EI}{\Delta} & e_1 &= AB \\ k_{14} = k_{41} &= -\frac{e_3 EI}{\Delta} & e_2 &= \frac{1}{2q}(AD + BC) \\ k_{22} &= \frac{(-e_1 e_4 + e_2 e_3)EI}{\Delta} & e_3 &= \frac{1}{2q^2} CD \\ k_{23} = k_{32} &= -k_{14} & e_4 &= \frac{1}{4q^3}(AD - BC) \end{aligned} \tag{2}$$

$$k_{24} = k_{42} = \frac{e_4 EI}{\Delta} \qquad e_5 = \frac{1 - e_1}{\lambda}$$

$$k_{33} = k_{11} \qquad e_6 = \frac{\ell - e_2}{\lambda}$$

$$k_{34} = k_{43} = \frac{(e_1 e_3 + \lambda e_4^2) EI}{\Delta}$$

$$k_{44} = k_{22}$$

$$A = \cosh q\ell \qquad B = \cos q\ell$$

$$C = \sinh q\ell \qquad D = \sin q\ell$$

Equations (1) and (2) define the exact stiffness matrix for a beam element on an elastic foundation. The approximate stiffness matrix based on a third-degree polynomial [Eq. (15) of Example 2.18] is

$$\mathbf{k}_w^e = \frac{k_w \ell}{420} \begin{bmatrix} 156 & -22\ell & 54 & 13\ell \\ -22\ell & 4\ell^2 & -13\ell & -3\ell^2 \\ 54 & -13\ell & 156 & 22\ell \\ 13\ell & -3\ell^2 & 22\ell & 4\ell^2 \end{bmatrix} \tag{3}$$

The global stiffness matrix can be assembled and the stiffness equations

$$\mathbf{KV} = \overline{\mathbf{P}}^* - \overline{\mathbf{P}}^\circ \tag{4}$$

formed. For the displacement method, the displacement boundary conditions are applied to (4). However, for a beam resting on an elastic foundation, there are no prescribed displacement conditions on the ends of the beam system. Hence, (4) is solved without reduction. Due to the support provided by the elastic foundation, there is no rigid-body motion and the stiffness matrix **K** is not singular. At each node there will be two unknown displacements, the deflection w and the slope θ.

With the displacement method the vector **V** is assembled for a straight beam such that the deflection w and slope $\theta = -w'$ are continuous at a node for each element connected at the node. In contrast, however, the displacement method of analysis does not assure that the moment M and shear force V are continuous at a node. Indeed, the displacement method will, in general, lead to M and V that are not continuous across a node, even though for the actual beam M and V may be continuous.

The exact deflection w corresponding to Table 2.2 with the definitions of (2) is

$$w(x) = A_0 e^{\lambda^* x} \cos \lambda^* x + A_1 e^{\lambda^* x} \sin \lambda^* x + A_2 e^{-\lambda^* x} \cos \lambda^* x + A_3 e^{-\lambda^* x} \sin \lambda^* x \tag{5}$$

where A_i, $i = 0, 1, 2, 3$, are constants and $\lambda^* = \sqrt[4]{k_w/4EI}$. The deflection w corresponding to the approximate solution given by $\mathbf{k}_w^e$ of (3) is the cubic polynomial

$$w(x) = C_0 + C_1 x + C_2 x^2 + C_3 x^3 \tag{6}$$

where C_i, $i = 0, 1, 2, 3$, are constants. Both of these representations of $w(x)$ are continuous. From the relationship $\theta(x) = -w'(x)$, (6) gives the approximate solution for the slope

$$\theta(x) = -C_1 - 2C_2x - 3C_3x^2 \tag{7}$$

whereas the exact solution $\theta(x) = -w'(x)$, with $w(x)$ taken from (5), is of a form similar to (5). From (6) and $M(x) = -EIw''$, we find the approximate expression for the bending moment:

$$M(x) = -2EIC_2 - 6EIC_3x \tag{8}$$

which is linear and obviously varies from the exact solution $M(x) = -EIw''$, which appears similar to (5). A third derivative of (6) leads to the shear force [i.e., $V =$

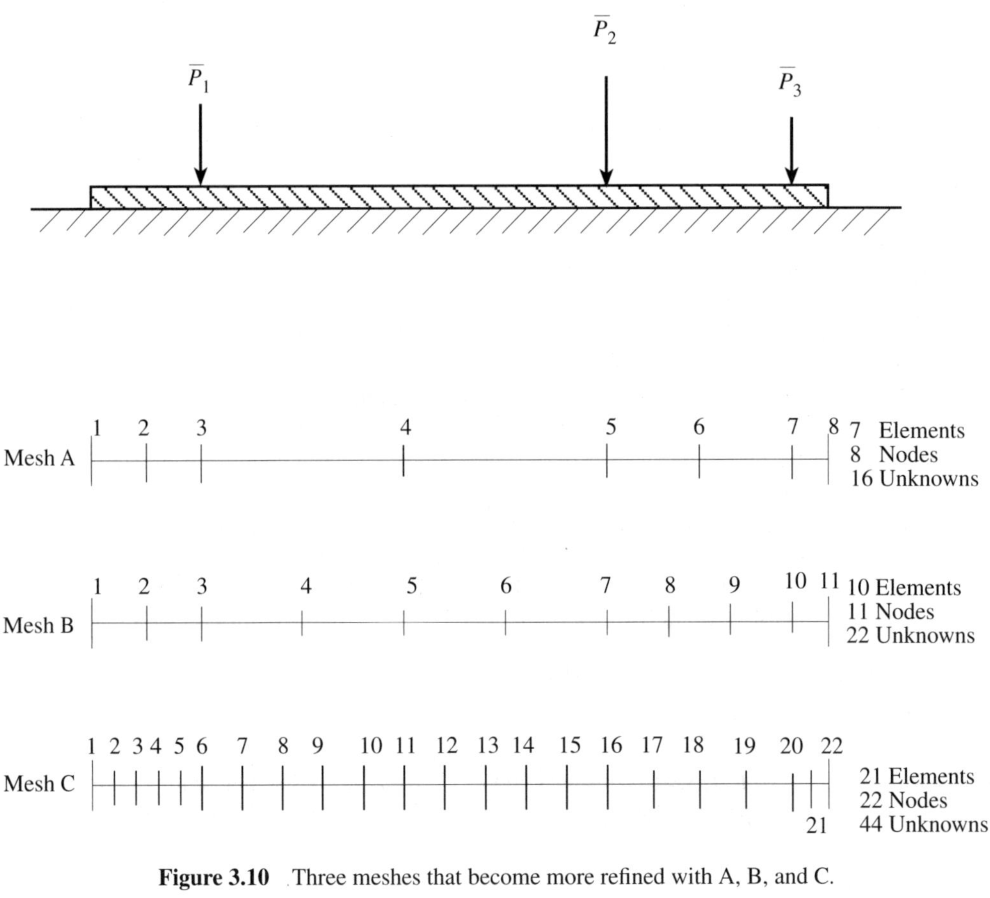

Figure 3.10 Three meshes that become more refined with A, B, and C.

$-EIw'''(x)]$. This approximate V is constant,

$$V(x) = -6EIC_3 \tag{9}$$

Clearly this is different from the exact solution for V, which with $V = -EIw'''(x)$ is similar to (5). These results indicate that it may be problematic if forces (or stresses) are computed using the derivatives of the assumed approximate displacement of (6).

We compute the solution using the displacement method with the stiffness matrix of (3) for the three mesh refinements shown in Fig. 3.10. In particular, we study the response at the point of application of the force $\overline{P}_3$. The moment M at the load $\overline{P}_3$ can be computed using $\mathbf{p}^e = \mathbf{k}^e\mathbf{v}^e$ or $M = -EIw''$. Similarly, the shear force can be calculated using $\mathbf{p}^e = \mathbf{k}^e\mathbf{v}^e$ or $V = -EIw'''$. From the results of Fig. 3.11, it is apparent that the accuracy improves as the number of elements increases. Note that the two different calculation techniques give different values for the moment and shear forces. The exact values can be obtained by analyzing the beam with the displacement method using the exact stiffness matrix of (1) and (2). An even simpler analysis would be to superimpose three solutions for a beam of length L on an infinite foundation with a single concentrated load. This solution for a single load is available in formula books (e.g., Pilkey, 1994).

Bending Moment Distributions The bending moment distributions for mesh refinements A, B, and C, as well as for the exact solution, are shown in Fig. 3.12. The moments for the approximate solution are from $\mathbf{p}^e = \mathbf{k}^e\mathbf{v}^e$. This expression provides the moments at the nodes only. We have chosen to connect the values of the moments at the nodes with straight lines.

Figure 3.13 shows the solution for moments for the same three mesh refinements, but $M = -EIw''$ is employed to calculate the moment distributions. As indicated above, the second derivatives of the assumed $w(x)$ leads to a linear expression, for the moment distribution. Jumps in the values of the moments at the nodes can occur. The approximation for Fig. 3.13, involving derivatives of the deflection, is not as good as the approximate moments of Fig. 3.12.

Shear Force Distributions Figures 3.14 and 3.15 show shear force distributions, plotted similar to the moment distributions of Figs. 3.12 and 3.13. Consider the shear forces distributions determined using $V = -EIw'''$ shown in Fig. 3.15. As indicated in (9), the third derivative of the deflection assumption (cubic polynomial) is constant. This method leads to a piecewise constant distribution of shear force between the nodes. Jumps in shear force can occur at the element boundaries. The use of derivatives of the shape function to compute the shear force results in a poorer approximation than that shown in Fig. 3.14.

3.2.2 Characteristics of the Displacement Method

Introduction of Boundary Conditions Displacement boundary conditions are imposed on the system stiffness equations when using the displacement method. This

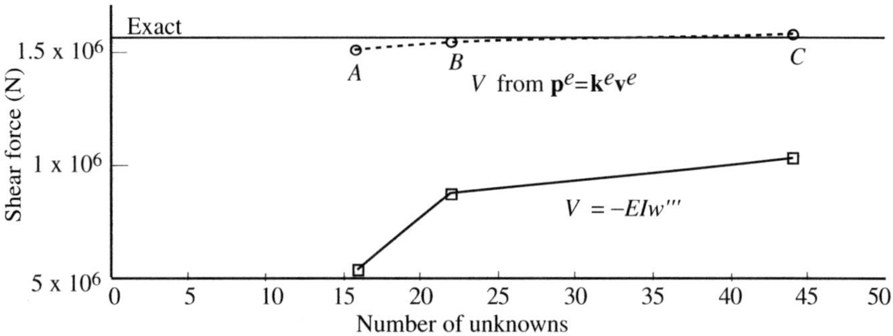

(a) Shear force under load $\overline{P}_3$

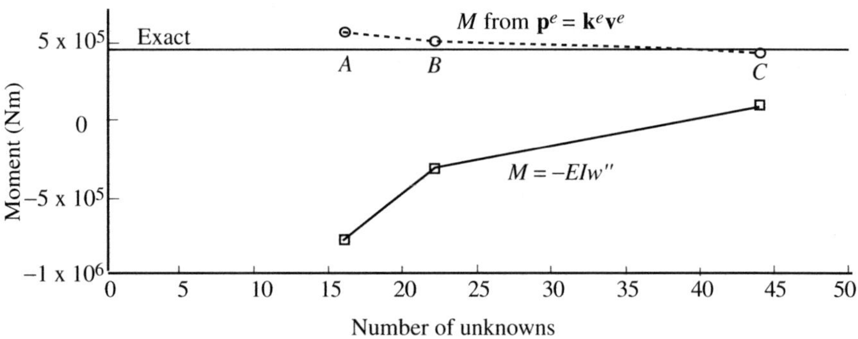

(b) Moment under load $\overline{P}_3$

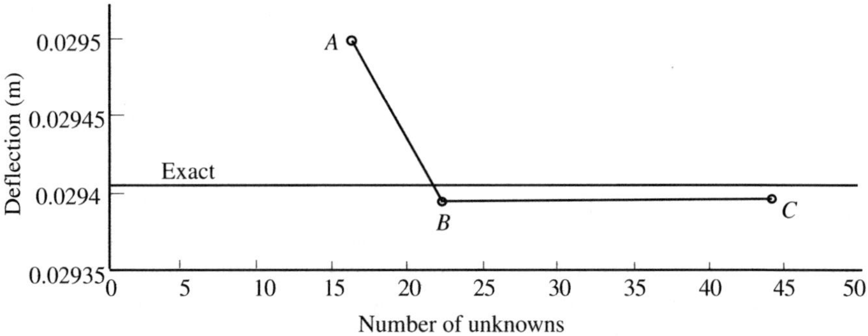

(c) Deflection under load $\overline{P}_3$

Figure 3.11 Moment, shear, deflection under load $\overline{P}_3$ for mesh refinements A, B, and C for beam on elastic foundation.

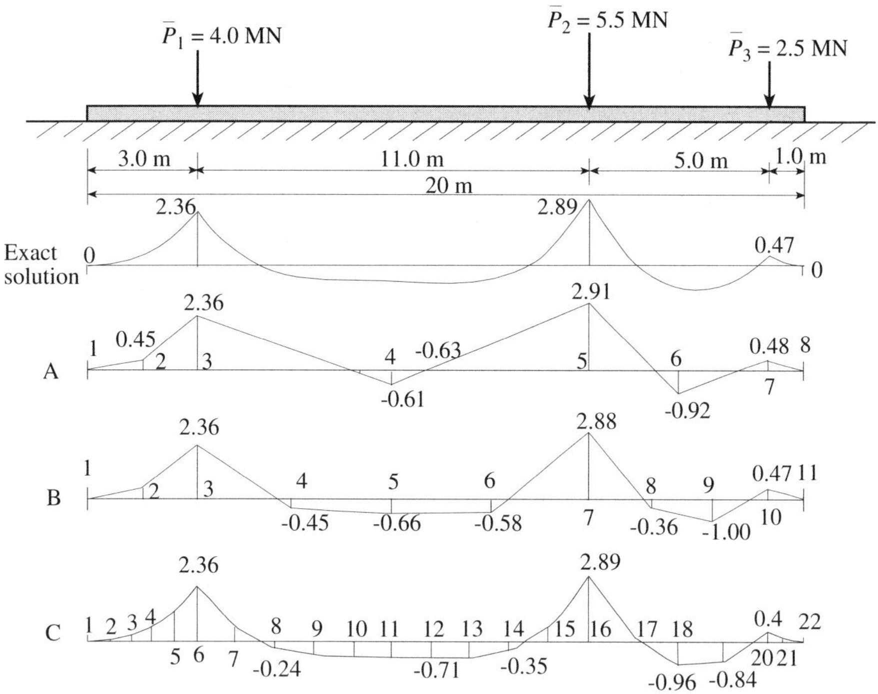

Figure 3.12 Moment distributions for the three mesh refinements A, B, and C, using nodal forces $\mathbf{p}^e = \mathbf{k}^e \mathbf{v}^e$. The nodal values are arbitrarily connected with straight lines. The exact moment distribution is also shown.

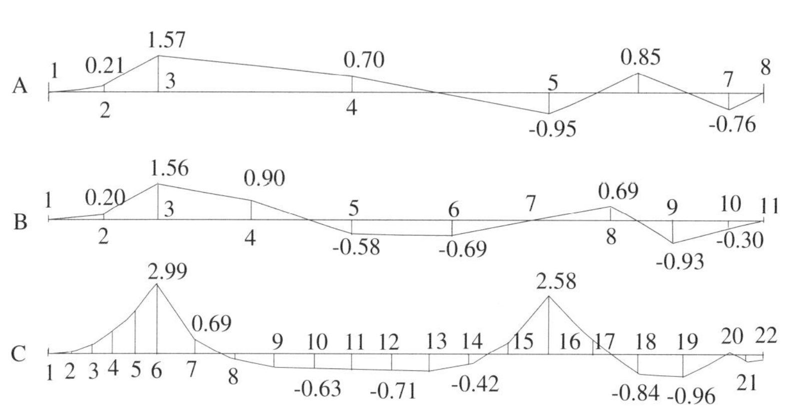

Figure 3.13 Moment distributions, for the three mesh refinements based on $M = -EIw''$.

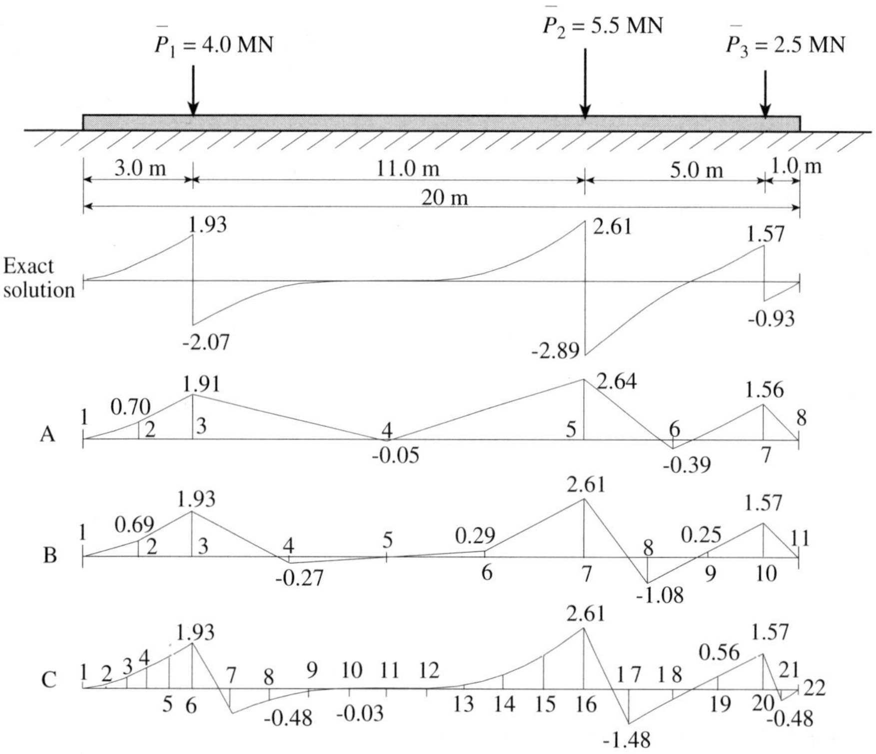

Figure 3.14 Shear force distributions for the three mesh refinements A, B, and C, using nodal forces $\mathbf{p}^e = \mathbf{k}^e \mathbf{v}^e$. The nodal values are arbitrarily connected with straight lines. The exact shear force distribution is also shown.

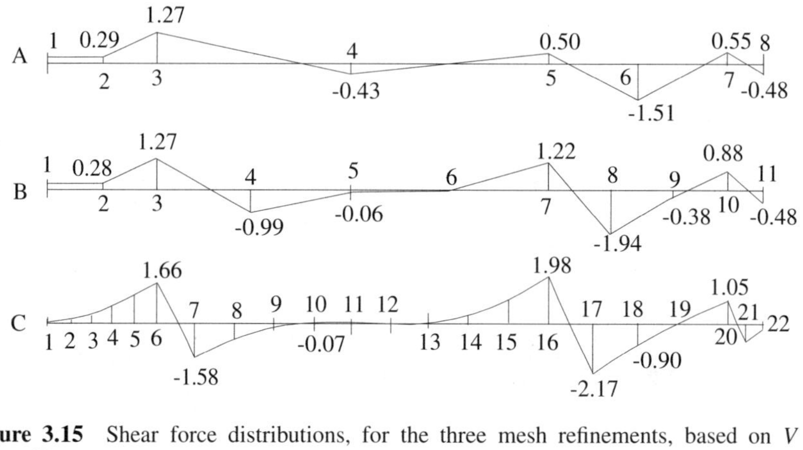

Figure 3.15 Shear force distributions, for the three mesh refinements, based on $V = -EIw'''$.

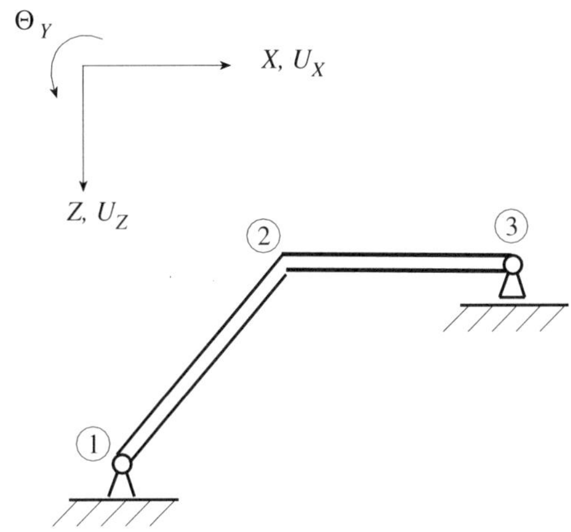

Figure 3.16 Framework with global nodal displacements U_X, U_Z, Θ_Y.

was illustrated in Example 3.2, Fig. 3.7, where all of the displacements at the fixed ends of nodes 1 and 3 were set equal to zero.

Consider the framework of Fig. 3.16, where the node 1 end is free to rotate, but is constrained against translation in the X and Z directions. The node 3 end can rotate about Y as well as translate in the X direction. However, it is restrained against translation in the Z direction. The displacement boundary conditions are

$$U_{X1} = 0, \qquad U_{Z1} = 0, \qquad U_{Z3} = 0 \tag{3.21}$$

Let the unknown reaction forces be R_{X1}, R_{Z1}, and R_{Z3}. The system stiffness equations $\mathbf{KV} = \overline{\mathbf{P}}$ are given by Eq. (3.18) with

$$\mathbf{V} = \begin{bmatrix} \mathbf{V}_1 \\ \mathbf{V}_2 \\ \mathbf{V}_3 \end{bmatrix} = \begin{bmatrix} U_{X1} \\ U_{Z1} \\ \Theta_{Y1} \\ U_{X2} \\ U_{Z2} \\ \Theta_{Y2} \\ U_{X3} \\ U_{Z3} \\ \Theta_{Y3} \end{bmatrix} = \begin{bmatrix} 0 \\ 0 \\ \Theta_{Y1} \\ U_{X2} \\ U_{Z2} \\ \Theta_{Y2} \\ U_{X3} \\ 0 \\ \Theta_{Y3} \end{bmatrix} \qquad \overline{\mathbf{P}} = \begin{bmatrix} \overline{\mathbf{P}}_1 \\ \overline{\mathbf{P}}_2 \\ \overline{\mathbf{P}}_3 \end{bmatrix} = \begin{bmatrix} R_{X1} \\ R_{Z1} \\ \overline{M}_{Y1} \\ \overline{P}_{X2} \\ \overline{P}_{Z2} \\ \overline{M}_{Y2} \\ \overline{P}_{X3} \\ R_{Z3} \\ \overline{M}_{Y3} \end{bmatrix} \tag{3.22}$$

where $\overline{M}_{Y1}$, $\overline{P}_{X2}$, $\overline{P}_{Z2}$, $\overline{M}_{Y2}$, $\overline{P}_{X3}$, and $\overline{M}_{Y3}$ are loads applied at the nodes. Note that the components in $\overline{\mathbf{P}}$ include the forces applied at the nodes and the reactions generated at nodes due to constraints on certain displacements imposed at some of these nodes. The zero components in the displacement vector $\mathbf{V}$ of Eq. (3.22) essentially cancel columns 1, 2, and 8 of $\mathbf{K}$ of Eq. (3.18). To establish a square matrix that can be solved for the unknown displacements, we will temporarily ignore the rows in Eq. (3.18) corresponding to the three reactions R_{X1}, R_{Z1}, and R_{Z3}. Equation (3.18) now appears as

$$\underbrace{\begin{bmatrix} k_{33}^1 & k_{34}^1 & k_{35}^1 & k_{36}^1 & 0 & 0 \\ k_{43}^1 & k_{44}^1+k_{44}^2 & k_{45}^1+k_{45}^2 & k_{46}^1+k_{46}^2 & k_{47}^2 & k_{49}^2 \\ k_{53}^1 & k_{54}^1+k_{54}^2 & k_{55}^1+k_{55}^2 & k_{56}^1+k_{56}^2 & k_{57}^2 & k_{59}^2 \\ k_{63}^1 & k_{64}^1+k_{64}^2 & k_{65}^1+k_{65}^2 & k_{66}^1+k_{66}^2 & k_{67}^2 & k_{69}^2 \\ 0 & k_{74}^2 & k_{75}^2 & k_{76}^2 & k_{77}^2 & k_{79}^2 \\ 0 & k_{94}^2 & k_{95}^2 & k_{96}^2 & k_{97}^2 & k_{99}^2 \end{bmatrix}}_{\widetilde{\mathbf{K}}} \underbrace{\begin{bmatrix} \Theta_{Y1} \\ U_{X2} \\ U_{Z2} \\ \Theta_{Y2} \\ U_{X3} \\ \Theta_{Y3} \end{bmatrix}}_{\widetilde{\mathbf{V}}} = \underbrace{\begin{bmatrix} \overline{M}_{Y1} \\ \overline{P}_{X2} \\ \overline{P}_{Z2} \\ \overline{M}_{Y2} \\ \overline{P}_{X3} \\ \overline{M}_{Y3} \end{bmatrix}}_{\widetilde{\overline{\mathbf{P}}}} \tag{3.23}$$

where $\widetilde{\mathbf{K}}$, $\widetilde{\mathbf{V}}$, and $\widetilde{\overline{\mathbf{P}}}$ are the reduced matrices. Thus, in this example, the displacement boundary conditions, along with discarding the rows corresponding to the reactions, have reduced the 9×9 singular stiffness matrix to a 6×6 nonsingular matrix. Equation (3.23) can be solved for the displacements

$$\widetilde{\mathbf{V}} = [\Theta_{Y1} \quad U_{X2} \quad U_{Z2} \quad \Theta_{Y2} \quad U_{X3} \quad \Theta_{Y3}]^T \tag{3.24}$$

Once the displacements have been calculated, they can be employed to find the reactions R_{X1}, R_{Z1}, and R_{Z3} using the three rows (first, second, and eighth) that were ignored.

$$\begin{bmatrix} k_{13}^1 & k_{14}^1 & k_{15}^1 & k_{16}^1 & 0 & 0 \\ k_{23}^1 & k_{24}^1 & k_{25}^1 & k_{26}^1 & 0 & 0 \\ 0 & k_{84}^2 & k_{85}^2 & k_{86}^2 & k_{87}^2 & k_{89}^2 \end{bmatrix} \begin{bmatrix} \Theta_{Y1} \\ U_{X2} \\ U_{Z2} \\ \Theta_{Y2} \\ U_{X3} \\ \Theta_{Y3} \end{bmatrix} = \begin{bmatrix} R_{X1} \\ R_{Z1} \\ R_{Z3} \end{bmatrix} \tag{3.25}$$

Properties of the System Stiffness Matrix The two-element, three-node framework of this chapter can be used to illustrate that the system stiffness matrix $\mathbf{K}$ is singular prior to the introduction of the boundary conditions. The system stiffness equations of Eqs. (3.18) and (3.20),

$$\underbrace{\begin{bmatrix} \mathbf{K}_{11} & \mathbf{K}_{12} & \mathbf{K}_{13} \\ \mathbf{K}_{21} & \mathbf{K}_{22} & \mathbf{K}_{23} \\ \mathbf{K}_{31} & \mathbf{K}_{32} & \mathbf{K}_{33} \end{bmatrix}}_{\mathbf{K}} \underbrace{\begin{bmatrix} \mathbf{V}_1 \\ \mathbf{V}_2 \\ \mathbf{V}_3 \end{bmatrix}}_{\mathbf{V}} = \underbrace{\begin{bmatrix} \overline{\mathbf{P}}_1 \\ \overline{\mathbf{P}}_2 \\ \overline{\mathbf{P}}_3 \end{bmatrix}}_{\overline{\mathbf{P}}} \tag{3.26}$$

correspond to the structure of Fig. 3.5. Since displacement constraints have not been introduced, the structure can undergo arbitrary rigid-body motion as well as relative nodal motions which correspond to elastic displacements. For example, rigid-body motion can cause nodal displacement components in $\mathbf{V}_1$ to have arbitrary values, while the displacement components in $\mathbf{V}_2$ can depend on the relative motions of nodes 1 and 3. From Cramer's rule for the solution of a system of linear equations, it follows that if the solutions $\mathbf{V}_1$, $\mathbf{V}_2$, and $\mathbf{V}_3$ are not unique, the determinant of matrix $\mathbf{K}$ must be zero. If $|\mathbf{K}| = 0$, then $\mathbf{K}$ is singular. Unique solutions $\mathbf{V}_1$, $\mathbf{V}_2$, and $\mathbf{V}_3$ exist only after proper displacement constraints such as those of Fig. 3.16 are imposed and rigid-body motion is prevented. Then $\widetilde{\mathbf{K}}$, the reduced matrix, is not singular.

As with element stiffness matrices, system stiffness matrices are symmetric. Introduction of proper boundary constraints eliminates rigid-body motion and results in a positive-definite stiffness matrix. The symmetric and positive definite conditions imply that the matrix has a unique inverse and that the eigenvalues are real and positive. For the matrix $\mathbf{K}$ to be positive definite,

$$\mathbf{a}^{\mathrm{T}}\mathbf{K}\mathbf{a} > 0$$

for all nonzero vectors $\mathbf{a}$. Although a singular matrix can be positive semi-definite, $\mathbf{a}^{\mathrm{T}}\mathbf{K}\mathbf{a} \geq 0$, it is not positive definite.

Each row of a system of stiffness equations $\mathbf{KV} = \overline{\mathbf{P}}$ corresponds to a force at a node. The nonzero terms in the stiffness matrix on the main diagonal correspond to DOF at the node and the remaining nonzero terms in the row correspond to DOF of other nodes of elements, connected to the node. The stiffness matrix $\mathbf{K}$ is usually *sparse* since for a framework with many nodes and elements, the matrix usually contains mostly zero terms. As part of the solution procedure for the equations, the stiffness matrix is often rearranged to achieve a banded matrix of minimum width.

Other System Matrix Assembly Techniques Many variations of the assembly procedure described here for system matrices are available. In Chapters 4 and 5 we discuss a technique to make this procedure more systematic.

3.3 TRANSFER MATRIX METHOD OF ANALYSIS

In Section 2.2.1, the transfer matrix $\mathbf{U}^e$ was obtained as the solution to the first-order form of the governing differential equations for a beam. The transfer matrix relates to the state variables $\mathbf{z}$ at the left end a of element e to the state variables at the right end b of the element:

$$\mathbf{z}_b = \mathbf{U}^e\mathbf{z}_a + \overline{\mathbf{z}}^e \tag{3.27}$$

The primary use of transfer matrices in this book is as an important intermediate step in deriving stiffness matrices for beam elements. However, for linelike structural systems, such as structures made of beam elements, responses can be calculated directly from the transfer matrices. In contrast to the displacement method, for which the equations represent the conditions of equilibrium, the transfer matrix method solution for a whole structure satisfies the equilibrium conditions, kinematic admissibility, and the material law simultaneously.

From Chapter 2, the transfer matrix for a Bernoulli–Euler beam element with sign convention 1 appears as

$$\begin{bmatrix} w \\ \theta \\ V \\ M \end{bmatrix}_b = \begin{bmatrix} 1 & -\ell & -\ell^3/6EI & -\ell^2/2EI \\ 0 & 1 & \ell^2/2EI & \ell/EI \\ 0 & 0 & 1 & 0 \\ 0 & 0 & \ell & 1 \end{bmatrix} \begin{bmatrix} w \\ \theta \\ V \\ M \end{bmatrix}_a + \begin{bmatrix} F_w \\ F_\theta \\ F_V \\ F_M \end{bmatrix} \tag{3.28}$$

$$\mathbf{z}_b \quad = \quad \mathbf{U}^e \quad \mathbf{z}_a \quad + \quad \overline{\mathbf{z}}^e$$

It is frequently useful, as noted in Table 2.2, to include the loading terms $\overline{\mathbf{z}}^e$ in the transfer matrix $\mathbf{U}^e$. This leads to an expression that is equivalent to Eq. (3.28), utilizing an *extended state vector* $\mathbf{z}$ and an *extended transfer matrix* $\mathbf{U}^e$:

$$\begin{bmatrix} w \\ \theta \\ V \\ M \\ 1 \end{bmatrix}_b = \begin{bmatrix} 1 & -\ell & -\ell^3/6EI & -\ell^2/2EI & F_w \\ 0 & 1 & \ell^2/2EI & \ell/EI & F_\theta \\ 0 & 0 & 1 & 0 & F_V \\ 0 & 0 & \ell & 1 & F_M \\ 0 & 0 & 0 & 0 & 1 \end{bmatrix} \begin{bmatrix} w \\ \theta \\ V \\ M \\ 1 \end{bmatrix}_a \tag{3.29}$$

$$\mathbf{z}_b \quad = \quad \mathbf{U}^e \quad \mathbf{z}_a$$

The structure shown in Fig. 3.17 has four beam elements for which there are four extended transfer matrices:

$$\begin{aligned} \mathbf{z}_2 &= \mathbf{U}^1\mathbf{z}_1 \\ \mathbf{z}_3 &= \mathbf{U}^2\mathbf{z}_2 \\ \mathbf{z}_4 &= \mathbf{U}^3\mathbf{z}_3 \\ \mathbf{z}_5 &= \mathbf{U}^4\mathbf{z}_4 \end{aligned} \tag{3.30}$$

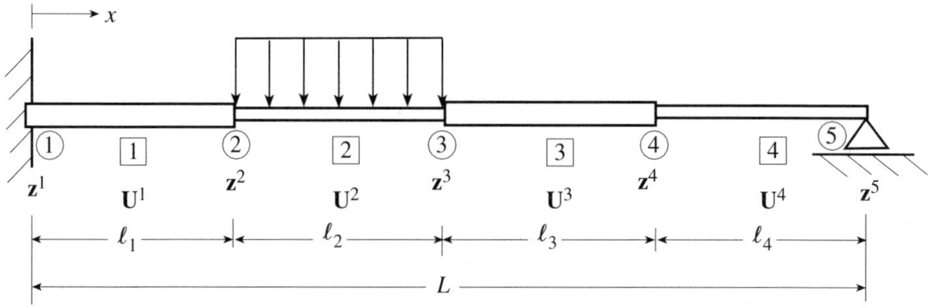

Figure 3.17 State vectors and transfer matrices for the four elements of a beam.

where $\mathbf{z}_1$ is the extended state vector at the left end, $\mathbf{U}^1$ is the extended transfer matrix of the first element of length ℓ_1, and $\mathbf{z}_2$ is the state vector at node $2(x = \ell_1)$. The remaining quantities shown in Fig. 3.17 are defined similarly.

The state vector at any of the nodes in Fig. 3.17 can be expressed in terms of the initial state vector at node 1. For example, in the case of $\mathbf{z}_3$ of Eq. (3.30) $\mathbf{z}_3 = \mathbf{U}^2\mathbf{z}_2$, introduce $\mathbf{z}_2 = \mathbf{U}^1\mathbf{z}_1$ giving $\mathbf{z}_3 = \mathbf{U}^2\mathbf{U}^1\mathbf{z}_1$. It follows that the state vectors at nodes 2, 3, 4, and 5 can be written as

$$\begin{aligned}
\mathbf{z}_2 &= \mathbf{U}^1\mathbf{z}_1 \\
\mathbf{z}_3 &= \mathbf{U}^2\mathbf{U}^1\mathbf{z}_1 \\
\mathbf{z}_4 &= \mathbf{U}^3\mathbf{U}^2\mathbf{U}^1\mathbf{z}_1 \\
\mathbf{z}_5 &= \mathbf{U}^4\mathbf{U}^3\mathbf{U}^2\mathbf{U}^1\mathbf{z}_1
\end{aligned} \tag{3.31}$$

Thus, the state vector at any location along the beam is found by progressive multiplication of the transfer matrices for all elements to the left of the location. That is, the state variables $\mathbf{z}_j$ for node j are expressed as

$$\mathbf{z}_j = \mathbf{U}^{j-1}\mathbf{U}^{j-2}\cdots\mathbf{U}^2\mathbf{U}^1\mathbf{z}_1 \tag{3.32}$$

For a beam with M elements, the state variables at the right end of the beam are given by

$$\mathbf{z}_{x=L} = \mathbf{z}_L = \mathbf{U}^M\mathbf{U}^{M-1}\cdots\mathbf{U}^2\mathbf{U}^1\mathbf{z}_1 = \mathbf{U}\mathbf{z}_1 \tag{3.33}$$

where $\mathbf{U}$ is an assembled *global* or *overall transfer matrix* that extends from left to right along the beam. In contrast to the displacement method, where assembly of the global stiffness matrix is an additive process, the global transfer matrix is obtained by multiplication of the element matrices.

The unknowns in Eqs. (3.32) and (3.33) are the initial state variables w_1, θ_1, V_1, and M_1 of $\mathbf{z}_1$, which are determined by applying the four boundary conditions, two of which occur at each of the two ends of the beam. This is referred to as an *initial value solution* of a boundary value problem. A transfer matrix solution involves two "sweeps" along the beam. Initially, the overall or global transfer matrix $\mathbf{U}$ (Eq. 3.33) is constructed, normally by a computer program that calls up stored transfer matrices as needed to perform the matrix multiplications. The four boundary conditions are applied to Eq. (3.33), leading to four linear equations that can be solved for the four unknown state variables w_1, θ_1, V_1, and M_1. Now that $\mathbf{z}_1$ has been determined, a second "sweep" along the beam using Eq. (3.32) is utilized to calculate the state variables w, θ, V, and M along the beam. In the case of the fixed, simply supported beam of Fig. 3.17, the four boundary conditions are $w_1 = 0$, $\theta_1 = 0$, $w_{x=L} = 0$, $M_{x=L} = 0$.

The transfer matrix technique is simple and systematic. It involves the product of small matrices and leads to a system matrix of the same dimensions as the element matrices. This is in contrast to the displacement method for which the size of the system matrix depends on the complexity of the structure. The shortcomings of the transfer matrix method, which applies only to structures with a chainlike topology, is that it tends to be numerically unstable when there are large variations in the value of some parameters or if the boundaries of the structure are so far apart that they have little influence on each other. Details of the use of the transfer matrix method can be found in books on structural mechanics.

3.4 DYNAMIC RESPONSES

The dynamic response of frameworks is discussed briefly here. See references on structural dynamics for more thorough coverage. The governing equations for the dynamic motion of a linear undamped system can be expressed as

$$\mathbf{M}\ddot{\mathbf{V}} + \mathbf{K}\mathbf{V} = \overline{\mathbf{P}}(t) \tag{3.34}$$

where $\mathbf{M}$ is the system mass matrix assembled from the element mass matrices of Chapter 2. In this section, matrices $\mathbf{K}$, $\mathbf{V}$, $\overline{\mathbf{P}}$, and $\mathbf{M}$ are treated often as being the reduced matrices $\tilde{\mathbf{K}}$, $\tilde{\mathbf{V}}$, $\tilde{\overline{\mathbf{P}}}$, and $\tilde{\mathbf{M}}$ discussed earlier in this chapter. Care must be taken in the use of particular mass matrices, especially for transient loading $\overline{\mathbf{P}}(t)$. For example, for nonsinusoidal vibrations the errors associated with modeling by consistent mass matrices may need to be reduced by utilizing a finer mesh of elements.

Two dynamic response problems are considered. In the first problem, the natural frequencies and corresponding mode shapes are sought. The second problem is for the computation of the motion of the system for prescribed time-dependent loading.

3.4.1 Free Vibration Analysis

Free motion corresponds to the relationship $\mathbf{M}\ddot{\mathbf{V}} + \mathbf{K}\mathbf{V} = \overline{\mathbf{P}}$ of Eq. (3.34) with the applied loading set equal to zero, that is,

$$\mathbf{M}\ddot{\mathbf{V}} + \mathbf{K}\mathbf{V} = \mathbf{0} \tag{3.35}$$

The *natural frequencies* ω (scalars) and the corresponding *mode shapes* $\boldsymbol{\phi}$ (vectors) are obtained from this relationship.

For *harmonic motion*

$$\mathbf{V} = \boldsymbol{\phi} \sin \omega t \tag{3.36}$$

The structure responds in a *mode* corresponding to a value of the frequency ω. Substitute Eq. (3.36) into Eq. (3.35), giving

$$\omega^2 \mathbf{M}\boldsymbol{\phi} = \mathbf{K}\boldsymbol{\phi}$$

or

$$(\mathbf{K} - \omega^2 \mathbf{M})\boldsymbol{\phi} = \mathbf{0} \tag{3.37}$$

For nontrivial $\boldsymbol{\phi}$, this implies that

$$\left|\mathbf{K} - \omega^2\mathbf{M}\right| = 0 \quad \text{or} \quad |\,\mathbf{K} - \lambda\mathbf{M}\,| = 0 \tag{3.38}$$

with $\omega^2 = \lambda$. This is the *characteristic equation* with *eigenvalue* λ, which can be solved for a set of values of λ.

Several efficient algorithms, with corresponding computer software, are available for solving eigenvalue problems. These computer programs often solve a problem in the form of the *standard eigenvalue problem*

$$(\mathbf{A} - \lambda\mathbf{I})\boldsymbol{\phi} = 0 \tag{3.39}$$

where $\mathbf{I}$ is the unit diagonal matrix, $\mathbf{A}$ is symmetric, λ are the eigenvalues of $\mathbf{A}$, and $\boldsymbol{\phi}$ is the *eigenvector*. Such techniques as Cholesky decomposition (Pilkey and Wunderlich, 1994) can be employed to obtain a standard form. It would appear that $\mathbf{A}$ is equal to $\mathbf{M}^{-1}\mathbf{K}$, or if λ is defined to be equal to $1/\omega^2$, $\mathbf{A}$ is equal to $\mathbf{K}^{-1}\mathbf{M}$. Use of these expressions for $\mathbf{A}$ does not necessarily effectively lead to the standard form. For example, $\mathbf{M}^{-1}\mathbf{K}$ is in general not symmetric, so that additional manipulations would be necessary to reach the standard eigenvalue problem.

For each possible rigid-body motion of a structure, the eigenvalues λ will be zero. The mode shapes $\boldsymbol{\phi}$ are "shapes" of the deformed structure and give relative magnitudes of the degrees of freedom, not the absolute values.

Orthogonality is an important property of the mode shapes. Suppose that ω_r^2, $\boldsymbol{\phi}_r$ and ω_s^2, $\boldsymbol{\phi}_s$ are two distinct solutions of the eigenvalue problem, in the sense that they satisfy

$$\mathbf{K}\boldsymbol{\phi}_r = \omega_r^2\mathbf{M}\boldsymbol{\phi}_r \tag{3.40a}$$

$$\mathbf{K}\boldsymbol{\phi}_s = \omega_s^2\mathbf{M}\boldsymbol{\phi}_s \tag{3.40b}$$

Premultiply Eq. (3.40a) by $\boldsymbol{\phi}_s^{\mathrm{T}}$ and Eq. (3.40b) by $\boldsymbol{\phi}_r^{\mathrm{T}}$:

$$\boldsymbol{\phi}_s^{\mathrm{T}}\mathbf{K}\boldsymbol{\phi}_r = \omega_r^2\boldsymbol{\phi}_s^{\mathrm{T}}\mathbf{M}\boldsymbol{\phi}_r \tag{3.41a}$$

$$\boldsymbol{\phi}_r^{\mathrm{T}}\mathbf{K}\boldsymbol{\phi}_s = \omega_s^2\boldsymbol{\phi}_r^{\mathrm{T}}\mathbf{M}\boldsymbol{\phi}_s \tag{3.41b}$$

Transpose the second relationship, remember that matrices $\mathbf{M}$ and $\mathbf{K}$ are symmetric, and subtract from Eq. (3.41a), giving

$$(\omega_r^2 - \omega_s^2)\boldsymbol{\phi}_s^{\mathrm{T}}\mathbf{M}\boldsymbol{\phi}_r = 0 \tag{3.42}$$

In general, the natural frequencies are distinct, $\omega_r \neq \omega_s$, so that

$$\boldsymbol{\phi}_s^{\mathrm{T}}\mathbf{M}\boldsymbol{\phi}_r = 0 \qquad r \neq s \tag{3.43}$$

This is the orthogonality condition for the mode shapes. From Eqs. (3.43) and (3.41a),

$$\boldsymbol{\phi}_s^{\mathrm{T}} \mathbf{K} \boldsymbol{\phi}_r = 0 \qquad r \neq s \tag{3.44}$$

which is the orthogonality condition with respect to the stiffness matrix **K**. Define, for $r = s$,

$$\boldsymbol{\phi}_s^{\mathrm{T}} \mathbf{M} \boldsymbol{\phi}_s = M_s \tag{3.45a}$$

and

$$\boldsymbol{\phi}_s^{\mathrm{T}} \mathbf{K} \boldsymbol{\phi}_s = K_s = \omega_s^2 M_s \tag{3.45b}$$

where M_s, the *generalized mass* and K_s are scalars. Assemble the vector mode shapes $\boldsymbol{\phi}_i$ as the matrix

$$\boldsymbol{\Phi} = \begin{bmatrix} \boldsymbol{\phi}_1 & \boldsymbol{\phi}_2 & \cdots & \boldsymbol{\phi}_{n_d} \end{bmatrix} \tag{3.46}$$

where n_d is the number of degrees of freedom. Then Eq. (3.45) can be expressed as

$$\boldsymbol{\Phi}^{\mathrm{T}} \mathbf{M} \boldsymbol{\Phi} = \mathbf{M}_{n_d} \qquad \boldsymbol{\Phi}^{\mathrm{T}} \mathbf{K} \boldsymbol{\Phi} = \mathbf{K}_{n_d} \tag{3.47}$$

where

$$\mathbf{M}_{n_d} = \begin{bmatrix} M_1 & & & \mathbf{0} \\ & M_2 & & \\ & & \ddots & \\ \mathbf{0} & & & M_{n_d} \end{bmatrix} \qquad \mathbf{K}_{n_d} = \begin{bmatrix} K_1 & & & \mathbf{0} \\ & K_2 & & \\ & & \ddots & \\ \mathbf{0} & & & K_{n_d} \end{bmatrix}$$

Details about this eigenvalue problem are available in numerous sources. Useful information includes the number of eigenvalues, repeated eigenvalues, and the relative accuracy of the different types of mass matrices.

3.4.2 Forced Response

The governing equations for the dynamic linear response of a structure are given by

$$\mathbf{M}\ddot{\mathbf{V}} + \mathbf{K}\mathbf{V} = \overline{\mathbf{P}} \tag{3.48}$$

where $\overline{\mathbf{P}} = \overline{\mathbf{P}}(t)$ is a prescribed time-dependent applied load. These are ordinary differential equations that can be solved using numerical integration techniques (see, e.g., Pilkey and Wunderlich, 1994). Often, especially for linear equations such as those of Eq. (3.48), a modal superposition solution is employed using the natural frequencies and mode shapes. This method is described briefly here.

For an undamped system the governing equations (Eq. 3.48) are normally coupled, that is, at least one of the equations contains more than one of the displacement components in $\mathbf{V}$. Transformation to an appropriate set of coordinates can lead to equations that are uncoupled. These new coordinates are referred to as *principal, normal, modal*, or *natural coordinates*.

It is always possible to determine principal coordinates for linear systems. The vector $\mathbf{V}$ contains the coordinates (DOF) in which the equations of motion are coupled. Let

$$\mathbf{q} = \begin{bmatrix} q_1 & q_2 & \cdots & q_{n_d} \end{bmatrix}^{\mathrm{T}} \tag{3.49}$$

where n_d is the number of DOF, be the coordinates (principal coordinates) in which the equations of motion are uncoupled.

For the relationship between the coupled $\mathbf{V}$ and uncoupled coordinates, choose

$$\mathbf{V} = \boldsymbol{\Phi}\mathbf{q} \tag{3.50}$$

where $\boldsymbol{\Phi}$ contains the mode shape vectors $\boldsymbol{\phi}_i$ of Eq. (3.46). Substitute Eq. (3.50) into Eq. (3.48) and premultiply by $\boldsymbol{\Phi}^{\mathrm{T}}$:

$$\boldsymbol{\Phi}^{\mathrm{T}}\mathbf{M}\boldsymbol{\Phi}\ddot{\mathbf{q}} + \boldsymbol{\Phi}^{\mathrm{T}}\mathbf{K}\boldsymbol{\Phi}\mathbf{q} = \boldsymbol{\Phi}^{\mathrm{T}}\overline{\mathbf{P}} \tag{3.51}$$

The orthogonality conditions of Eq. (3.47) reduce Eq. (3.51) to the set of uncoupled differential equations

$$\mathbf{M}_{n_d}\ddot{\mathbf{q}} + \mathbf{K}_{n_d}\mathbf{q} = \mathbf{P} \quad \text{or} \quad M_i\ddot{q}_i + K_i q_i = P_i \qquad i = 1, 2, \ldots, n_d \tag{3.52}$$

where $\mathbf{P} = \boldsymbol{\Phi}^{\mathrm{T}}\overline{\mathbf{P}}$, $M_i = \boldsymbol{\phi}_i^{\mathrm{T}}\mathbf{M}\boldsymbol{\phi}_i$ (Eq. 3.45a), $K_i = \boldsymbol{\phi}_i^{\mathrm{T}}\mathbf{K}\boldsymbol{\phi}_i$ (Eq. 3.45b), and $P_i = \boldsymbol{\phi}_i^{\mathrm{T}}\overline{\mathbf{P}}$. These relationships are uncoupled in the sense that only one of the q_i, $i = 1, 2, \ldots, n_d$ appears in each equation. Thus, the transformation of coordinates of Eq. (3.50), with $\boldsymbol{\Phi}$ defined by Eq. (3.46), uncouples Eq. (3.48). The coupled governing equations for the n_d-DOF system have been replaced by n_d uncoupled equations. Each of the uncoupled equations is in the form of an equation of motion for a single-DOF system.

The uncoupled equations of Eq. (3.52) have familiar solutions. In matrix form,

$$\mathbf{q} = \mathbf{A}\mathbf{q}(0) + \mathbf{B}\dot{\mathbf{q}}(0) + \mathbf{F} \tag{3.53a}$$

with

$$\mathbf{q}(0) = \begin{bmatrix} q_1(0) \\ q_2(0) \\ \vdots \\ q_{n_d}(0) \end{bmatrix} = \boldsymbol{\Phi}^{-1}\begin{bmatrix} V_{10} \\ V_{20} \\ \vdots \\ V_{n_d 0} \end{bmatrix} = \boldsymbol{\Phi}^{-1}\mathbf{V}(0)$$

$$\dot{\mathbf{q}}(0) = \begin{bmatrix} \dot{q}_1(0) \\ \dot{q}_2(0) \\ \vdots \\ \dot{q}_{n_d}(0) \end{bmatrix} = \boldsymbol{\Phi}^{-1}\begin{bmatrix} \dot{V}_{10} \\ \dot{V}_{20} \\ \vdots \\ \dot{V}_{n_d 0} \end{bmatrix} = \boldsymbol{\Phi}^{-1}\dot{\mathbf{V}}(0) \tag{3.53b}$$

$$\mathbf{A} = \begin{bmatrix} \cos\omega_1 t & & & \mathbf{0} \\ & \cos\omega_2 t & & \\ & & \ddots & \\ \mathbf{0} & & & \cos\omega_{n_d} t \end{bmatrix}$$

$$\mathbf{B} = \begin{bmatrix} \dfrac{\sin\omega_1 t}{\omega_1} & & & \mathbf{0} \\ & \dfrac{\sin\omega_2 t}{\omega_2} & & \\ & & \ddots & \\ \mathbf{0} & & & \dfrac{\sin\omega_{n_d} t}{\omega_{n_d}} \end{bmatrix}$$

and

$$\mathbf{F} = \begin{bmatrix} F_1 & & & \mathbf{0} \\ & F_2 & & \\ & & \ddots & \\ \mathbf{0} & & & F_{n_d} \end{bmatrix}$$

where

$$F_i = \frac{1}{M_i\ \omega_i} \int_0^t P_i(\tau)\sin\omega_i(t-\tau)\,d\tau$$

and $\mathbf{V}(0) = \begin{bmatrix} V_{10} & V_{20} & \cdots & V_{n_d0} \end{bmatrix}^{\mathrm{T}}$ and $\dot{\mathbf{V}}(0) = \begin{bmatrix} \dot{V}_{10} & \dot{V}_{20} & \cdots & \dot{V}_{n_d0} \end{bmatrix}^{\mathrm{T}}$ are the prescribed initial conditions (displacements and velocities) of the system.

In scalar form,

$$\begin{aligned} q_i &= q_i(0)\cos\omega_i t + \dot{q}_i(0)\frac{\sin\omega_i t}{\omega_i} \\ &\quad + \frac{1}{M_i\omega_i}\int_0^t P_i(\tau)\sin\omega_i(t-\tau)\,d\tau \qquad i = 1, 2, \ldots n_d \end{aligned} \tag{3.54a}$$

$$q_i(0) = \boldsymbol{\phi}_i^{\mathrm{T}}\frac{\mathbf{M}\mathbf{V}(0)}{M_i} = \left(\sum_j^{n_d}\sum_k^{n_d} m_{jk}\phi_{ij}V_{k0}\right)\Big/ M_i \tag{3.54b}$$

$$\dot{q}_i(0) = \boldsymbol{\phi}_i^{\mathrm{T}}\frac{\mathbf{M}\dot{\mathbf{V}}(0)}{M_i} = \left(\sum_j^{n_d}\sum_k^{n_d} m_{jk}\phi_{ij}\dot{V}_{k0}\right)\Big/ M_i$$

where ϕ_{ij} is the jth element in ϕ_i and m_{jk} is the element in the jth row and the kth column of $\mathbf{M}$.

Substitution of $\mathbf{q}$ of Eq. (3.53a) or q_i of Eq. (3.54a) into Eq. (3.50),

$$\mathbf{V} = \mathbf{\Phi q} = \sum_{i=1}^{n_d} q_i(t)\boldsymbol{\phi}_i$$

provides the responses of an n_d-DOF system with arbitrary loading.

If a viscous damping model is included, the governing equations can be of the form

$$\mathbf{M\ddot{V}} + \mathbf{C\dot{V}} + \mathbf{KV} = \overline{\mathbf{P}} \tag{3.55}$$

where $\mathbf{C}$ is a damping matrix. The principal coordinates for $\mathbf{M}$ and $\mathbf{K}$ will not, in general, decouple Eq. (3.55), which includes $\mathbf{C}$. This is due to the term $\mathbf{\Phi}^{\mathrm{T}}\mathbf{C\Phi}$, which is not in general a diagonal matrix. If $\mathbf{C}$ is *proportional* to $\mathbf{M}$ or $\mathbf{K}$ (i.e., $\mathbf{C} = \alpha\mathbf{M} + \beta\mathbf{K}$, where α and β are constants), the normal nodes $\mathbf{\Phi}$ for the undamped system will make $\mathbf{\Phi}^{\mathrm{T}}\mathbf{C\Phi}$ a diagonal matrix and will result in equations that are not coupled.

Several other damping models are discussed in the literature. In the case of modal damping, a damping term is added to each modal equation. Thus, if a viscous damping term is added to Eq. (3.52),

$$M_i\ddot{q}_i + C_i\dot{q}_i + K_i q_i = P_i \tag{3.56}$$

where C_i is the damping assigned to the ith mode. In practice, usually a *damping ratio* $\zeta_i = C_i/2M_i\omega_i) = C_i/C_{\mathrm{cr}}$, with the *critical damping* $C_{\mathrm{cr}} = 2\sqrt{M_i K_i}$, is assigned to each mode, so that Eq. (3.56) becomes

$$\ddot{q}_i + 2\zeta_i\,\omega_i\dot{q}_i + \omega_i^2\,q_i = \frac{P_i}{M_i} \tag{3.57}$$

Sometimes, M_i is scaled such that $M_i = 1$. The expression for C_{cr} follows by replacing ω_i by its single-degree-of-freedom equivalent $\omega_i = \sqrt{K_i/M_i}$.

If $P_i = 0$, the solution to Eq. (3.57) is

$$q_i = A_1 e^{\alpha_1 t} + A_2 e^{\alpha_2 t} \tag{3.58}$$

where

$$\alpha_{1,2} = \tfrac{1}{2}\left(-2\zeta_i\,\omega_i \pm \sqrt{4\zeta_i^2\omega_i^2 - 4\omega_i^2}\right) = \omega_i\left(-\zeta_i \pm \sqrt{\zeta_i^2 - 1}\right) \tag{3.59}$$

Three separate cases can be identified.

$$\begin{aligned} &\zeta_i < 1 \qquad q_i \text{ is underdamped and oscillates} \\ &\zeta_i > 1 \qquad q_i \text{ is overdamped and exponentially decays} \\ &\zeta_i = 1 \qquad q_i \text{ is critically damped, } C_i = C_{\mathrm{cr}} \end{aligned} \tag{3.60}$$

The solution to the ordinary differential equation of Eq. (3.57) for $\zeta_i < 1$ gives the modal coordinate

$$q_i = e^{-\zeta_i \omega_i t}\left[\frac{\dot{q}_i(0) + q_i(0)\zeta_i\omega_i}{\omega_i\sqrt{1-\zeta_i^2}} \sin \omega_i\sqrt{1-\zeta_i^2}t + q_i(0)\cos\omega_i\sqrt{1-\zeta_i^2}t\right]$$
$$+ \frac{1}{M_i\omega_i\sqrt{1-\zeta_i^2}}\int_0^t P_i(\tau)e^{-\zeta_i\omega_i(t-\tau)}\sin\omega_i\sqrt{1-\zeta_i^2}(t-\tau)\,d\tau \tag{3.61}$$

where the modal initial conditions are computed using $q_i(0) = \boldsymbol{\phi}_i^{\mathrm{T}}\mathbf{M}\mathbf{V}(0)/M_i$ and $\dot{q}_i(0) = \boldsymbol{\phi}_i^{\mathrm{T}}\mathbf{M}\dot{\mathbf{V}}(0)/M_i$.

For undamped responses, set $\zeta_i = 0$ in Eq. (3.61). For the harmonic excitation $P_i = \overline{P}_0 \sin \omega t$, the final term of Eq. (3.61) becomes

$$\frac{(\overline{P}_0/M_i)\omega_i^2}{\left[\left(1-(\omega/\omega_i)^2\right)^2 + 4\zeta_i^2(\omega/\omega_i)^2\right]^{1/2}}\sin(\omega t - \theta_i) \tag{3.62}$$

with $\theta_i = \tan^{-1}\left[2\zeta_i\omega_i\omega/\left(\omega_i^2-\omega^2\right)\right]$.

3.5 STABILITY ANALYSIS

The analysis for instability of a structural system is quite similar to the analysis for natural frequencies, which is outlined in Section 3.4. The critical load, or a related factor, is the eigenvalue in a stability analysis. The element geometric stiffness matrices $\mathbf{k}_\sigma^e$ are assembled into the system geometric stiffness matrix $\mathbf{K}_\sigma$.

For a multimember structural system, it is often necessary to define the axial forces that lead to instability in terms of each other. Frequently, it is assumed that the internal axial forces in the various members remain in constant proportion to each other, with the proportionality constants being found by a static analysis of the system for some nominal value of the applied loading. If $\mathbf{K}_\sigma$ is the global geometric stiffness matrix for a reference level of axial forces, $\lambda\mathbf{K}_\sigma$ would be the geometric stiffness matrix of another level of axial forces, where λ is a scalar multiplier called the *load factor*. The principle of virtual work gives the system equilibrium equations in the form

$$(\mathbf{K} - \lambda\mathbf{K}_\sigma)\,\mathbf{V} = 0 \tag{3.63}$$

where $\mathbf{V}$ is a vector containing the nodal displacements. This relationship can be solved reliably and efficiently for λ with standard, readily available eigenvalue problem software.

3.6 ANALYSES USING EXACT STIFFNESS MATRICES

Typically, stiffness or mass matrices are derived based on approximate polynomial shape functions. In the case of beams, as discussed in Chapter 2, it is often possible to derive exact stiffness and mass matrices since the governing differential equations can be solved exactly. For a general beam element, such as given in Table 2.2, the effect of translational and rotary inertias, shear deformation, axial force, and translational and rotary elastic foundations are included. The corresponding matrix, which is referred to as a dynamic stiffness matrix, is a function of frequency ω and axial force N.

The exact stiffness matrix $\mathbf{k}^e$ for the eth element can be derived from the dynamic stiffness matrix $\mathbf{k}^e_{\text{dyn}}$ using

$$\mathbf{k}^e = \mathbf{k}^e_{\text{dyn}}(\omega = 0) \tag{3.64}$$

As detailed in Chapter 2, the exact mass matrix for the eth element can be obtained from the dynamic stiffness matrix $\mathbf{k}^e_{\text{dyn}}$ using (Eq. 2.112)

$$\mathbf{m}^e = -\frac{\partial \mathbf{k}^e_{\text{dyn}}}{\partial \omega^2} \tag{3.65}$$

where ω^2 is the frequency squared. Similarly, the exact element geometric stiffness matrix $\mathbf{k}^e_\sigma$ can be obtained from (Eq. 2.122)

$$\mathbf{k}^e_\sigma = \frac{\partial \mathbf{k}^e_{\text{dyn}}}{\partial N} \tag{3.66}$$

where N is the axial force.

System analyses for the static response, buckling loads, or natural frequencies involve, of course, the assembled systems matrices. In the case of the dynamic stiffness matrix, suppose that the system matrix is $\mathbf{K}_{\text{dyn}}(\omega, N)$. For buckling problems, the natural frequency in this system matrix is set equal to zero, and the resulting dynamic stiffness matrix is a function of axial forces. To calculate the buckling load or natural frequencies, a determinant search can be conducted to find the roots of the equation

$$\left| \mathbf{K}_{\text{dyn}}(\omega, N) \right| = 0 \tag{3.67}$$

Since the curve of the value of the determinant $|\, \mathbf{K}_{\text{dyn}}(\omega, N)\,|$ versus the axial force or natural frequency can be poorly behaved, it is often necessary to calculate the value of the determinant at very fine intervals. Sometimes eigenvalues are missed. Studies of this problem have led to computational methods to avoid missing eigenvalues and to accelerate the search procedure (e.g., Kennedy and Williams, 1991). However, these determinant search procedures still tend to be cumbersome and time consuming.

An alternative method that avoids the determinant search for finding eigenvalues is to use the global mass matrix $\mathbf{M}(\omega, N)$ or the geometric stiffness matrix $\mathbf{K}_\sigma(\omega, N)$ and set up the eigenvalue problems

$$\left| \mathbf{K}_{\text{dyn}}(\omega, N) - \omega^2 \mathbf{M}(\omega, N) \right| = 0 \tag{3.68}$$

$$\left| \mathbf{K}_{\text{dyn}}(\omega, N) - N\mathbf{K}_\sigma(\omega, N) \right| = 0 \tag{3.69}$$

An iterative solution of this problem can converge rapidly to the solution of the eigenvalue problem. For example, consider the problem of finding the natural frequencies when the critical N is not of interest. Then, write Eq. (3.68) as

$$\left| \mathbf{K}_{\text{dyn}}(\omega) - \omega^2 \mathbf{M}(\omega) \right| = 0 \tag{3.70}$$

The iterative solution can take the form

$$\begin{aligned}
\left| \mathbf{K}_{\text{dyn}}(0) - \omega^2 \mathbf{M}(0) \right| &= 0 \Rightarrow \omega = \omega_1^0, \omega_2^0, \omega_3^0, \ldots \\
\left| \mathbf{K}_{\text{dyn}}(\omega_1^0) - \omega^2 \mathbf{M}(\omega_1^0) \right| &= 0 \Rightarrow \omega = \omega_1^1, \omega_2^1, \omega_3^1, \ldots \\
&\vdots \\
\left| \mathbf{K}_{\text{dyn}}(\omega_1^{j-1}) - \omega^2 \mathbf{M}(\omega_1^{j-1}) \right| &= 0 \Rightarrow \omega = \omega_1^j, \omega_2^j, \omega_3^j, \ldots
\end{aligned} \tag{3.71}$$

where the superscript j refers to the eigensolution for the jth iteration. The frequencies $\omega_1^0, \omega_2^0, \omega_3^0, \ldots$ correspond to those that are found using Eq. (3.70) with a consistent mass matrix for $\mathbf{M}$ and a static stiffness matrix $\mathbf{K}$ rather than $\mathbf{K}_{\text{dyn}}$. On occasion, more accurate higher eigenvalues are computed from Eq. (3.71) if $\mathbf{K}_{\text{dyn}}$ and $\mathbf{M}$ are evaluated at ω_n^{j-1}, $n > 1$, instead of at the lowest frequency ω_1^{j-1}.

REFERENCES

Kennedy, D., and Williams, F. W. (1991). More efficient use of determinants to solve transcendental structural eigenvalue problems reliably, *Comput. Struct.*, Vol. 41, pp. 973–979.

Pilkey, W. D. (1994). *Formulas for Stress, Strain, and Structural Matrices*, Wiley, New York.

Pilkey, W. D., and Wunderlich, W. (1994). *Mechanics of Structures: Variational and Computational Methods*, CRC Press, Boca Raton, Fla.

CHAPTER 4

FINITE ELEMENTS FOR CROSS-SECTIONAL ANALYSIS

In this chapter we present the essential finite element analysis concepts used in the cross-sectional analysis of beams and describe how the warping-independent section properties, such as areas or moments of inertia, are computed by integration. The description of the finite element method is in terms of a nine-node Lagrangian element in two dimensions, which is the only kind of element used for meshing cross sections in the rest of this book. Detailed accounts of the finite element method are available in a number of textbooks (e.g., Dhatt and Touzot, 1984; Pilkey and Wunderlich, 1994; and Zienkiewicz, 1977).

4.1 SHAPE FUNCTIONS

As indicated in Chapter 2, reference elements can be used to simplify calculations when dealing with elements of complex shape. The geometry of a reference element Ω_r with its simple geometry, is mapped to the geometry of the real element Ω by a transformation. This transformation defines the coordinates of each point of the real domain Ω in terms of the coordinates of the corresponding point in the reference domain Ω_r. The geometrical transformation has the properties defined in Section 2.2.5.

An example of the geometrical transformation from a reference element to a real element is shown in Fig. 4.1. The reference element, on the left, is a square with node 1 at the point $(\eta, \zeta) = (-1, -1)$ and node 9 at $(\eta, \zeta) = (1, 1)$ of the reference domain. The part of the boundary of the reference element between any two nodes is a straight line. The corresponding part of the boundary of the real element may be curved, as shown in Fig. 4.1b.

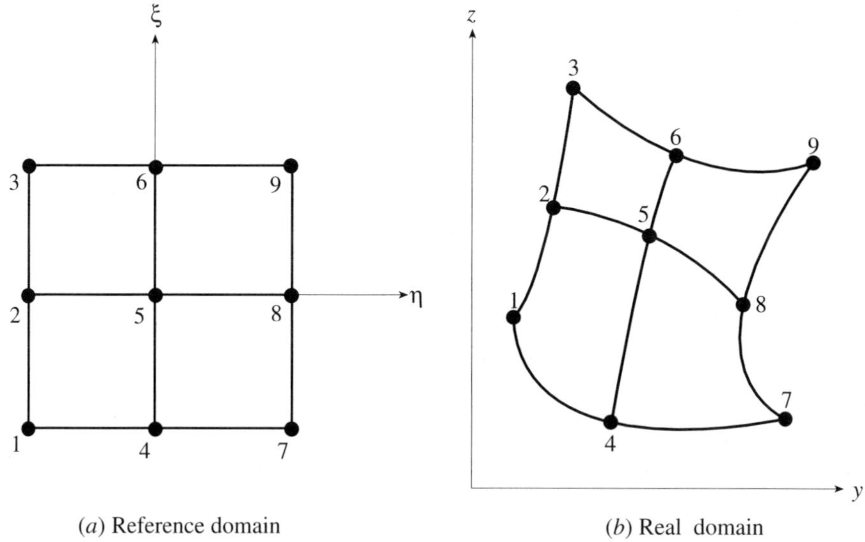

Figure 4.1 Reference and real domains for the nine-node Lagrangian element.

An interpolation formula for a function $f(\eta, \zeta)$ defined on the reference domain is written as a linear combination of a set of known basis functions $\mathbf{N}_u(\eta, \zeta)$:

$$f(\eta, \zeta) = \mathbf{N}_u(\eta, \zeta)\widehat{\mathbf{w}} \tag{4.1}$$

where $\mathbf{N}_u(\eta, \zeta)$ is composed of n_b basis functions and the coefficients (generalized displacements) $\widehat{\mathbf{w}}$ are determined by n_b values of the function $f(\eta, \zeta)$. The basis functions are usually chosen as independent monomials. For the nine-node element of Fig. 4.1, they are given by

$$\mathbf{N}_u(\eta, \zeta) = \begin{bmatrix} 1 & \eta & \zeta & \eta^2 & \eta\zeta & \zeta^2 & \eta^2\zeta & \eta\zeta^2 & \eta^2\zeta^2 \end{bmatrix} \tag{4.2}$$

Also, $\widehat{\mathbf{w}} = \begin{bmatrix} \widehat{w}_1 & \widehat{w}_2 & \widehat{w}_3 & \cdots & \widehat{w}_{n_b} \end{bmatrix}^T$, where n_b is equal to 9 for the nine-node element.

The geometric transformation for the elements of Fig. 4.1 are represented by the geometrical mapping functions $y(\eta, \zeta)$ and $z(\eta, \zeta)$. It is possible to define the geometric transformation for the nine-node quadrilateral based on the four nodes 1, 7, 3, and 9 alone. The real element sides are constrained to be straight lines. Another possibility is to utilize for the element geometrical mapping functions $y(\eta, \zeta)$ and $z(\eta, \zeta)$, the function $f(\eta, \zeta)$ interpolation formulas based on Eq. (4.1) and nine function values for the nine-node element. Elements whose geometric transformation equations and function interpolation formulas have the same form are said to be *isoparametric*. For the isoparametric nine-node element, then, geometrical mapping functions $y(\eta, \zeta)$ and $z(\eta, \zeta)$ utilize the interpolation formula for the function

$f(\eta, \zeta)$. For example, the geometrical mapping function $y(\eta, \zeta)$ expresses the nodal y coordinates of the real element as

$$y_i = y(\eta_i, \zeta_i) = \mathbf{N}_u(\eta_i, \zeta_i)\widehat{\mathbf{w}}_y \qquad 1 \le i \le 9 \tag{4.3}$$

where $\widehat{\mathbf{w}}_y = \begin{bmatrix} \widehat{w}_{y1} & \widehat{w}_{y2} & \widehat{w}_{y3} & \cdots & \widehat{w}_{y9} \end{bmatrix}^{\mathrm{T}}$. In matrix form, with

$$\begin{aligned} \mathbf{y} &= [y(\eta_1, \zeta_1) \quad y(\eta_2, \zeta_2) \quad \cdots \quad y(\eta_9, \zeta_9)]^{\mathrm{T}} \\ &= [y_1 \quad y_2 \quad \cdots \quad y_9]^{\mathrm{T}} \end{aligned} \tag{4.4}$$

these nine equations become

$$\mathbf{y} = \widehat{\mathbf{N}}_u \widehat{\mathbf{w}}_y \tag{4.5}$$

where $\widehat{\mathbf{N}}_u$ is the 9×9 matrix whose ith row is $\mathbf{N}_u(\eta_i, \zeta_i)$

$$\widehat{\mathbf{N}}_u = \begin{bmatrix} 1 & -1 & -1 & 1 & 1 & 1 & -1 & -1 & 1 \\ 1 & -1 & 0 & 1 & 0 & 0 & 0 & 0 & 0 \\ 1 & -1 & 1 & 1 & -1 & 1 & 1 & -1 & 1 \\ 1 & 0 & -1 & 0 & 0 & 1 & 0 & 0 & 0 \\ 1 & 0 & 0 & 0 & 0 & 0 & 0 & 0 & 0 \\ 1 & 0 & 1 & 0 & 0 & 1 & 0 & 0 & 0 \\ 1 & 1 & -1 & 1 & -1 & 1 & -1 & 1 & 1 \\ 1 & 1 & 0 & 1 & 0 & 0 & 0 & 0 & 0 \\ 1 & 1 & 1 & 1 & 1 & 1 & 1 & 1 & 1 \end{bmatrix} \tag{4.6}$$

In forming this matrix, use $(\eta_1, \zeta_1) = (-1, -1), (\eta_2, \zeta_2) = (0, -1), (\eta_3, \zeta_3) = (-1, 1), \ldots (\eta_9, \zeta_9) = (1, 1)$. Since the matrix $\widehat{\mathbf{N}}_u$ is invertible, the coefficient vector $\widehat{\mathbf{w}}_y$ is found from

$$\widehat{\mathbf{w}}_y = \widehat{\mathbf{N}}_u^{-1}\mathbf{y} \tag{4.7}$$

and this determines the geometrical transformation for y coordinates

$$y(\eta, \zeta) = \mathbf{N}_u(\eta, \zeta)\widehat{\mathbf{w}}_y = \mathbf{N}_u(\eta, \zeta)\widehat{\mathbf{N}}_u^{-1}\mathbf{y} \tag{4.8}$$

A similar computation shows that the z coordinates are transformed according to

$$z(\eta, \zeta) = \mathbf{N}_u(\eta, \zeta)\widehat{\mathbf{N}}_u^{-1}\mathbf{z} \tag{4.9}$$

The term $\mathbf{N}_u(\eta, \zeta)\widehat{\mathbf{N}}_u^{-1}$ defines the *shape functions* or *interpolation functions*, that is,

$$\mathbf{N}(\eta, \zeta) = \mathbf{N}_u(\eta, \zeta)\widehat{\mathbf{N}}_u^{-1} \tag{4.10}$$

Lagrangian elements are characterized by the nodal unknowns being values of the dependent function at the nodes, e.g., $f(\eta_i, \zeta_i)$ or for geometrical coordinates, $y(\eta_i, \zeta_i)$ and $z(\eta_i, \zeta_i)$. With the exception of some lower-order members of the Lagrange family, these elements have internal nodes (see Fig. 4.1). The geometrical transformation for the nine-node Lagrangian element of Fig. 4.1 is defined in terms of the shape functions by

$$y(\eta, \zeta) = \mathbf{N}(\eta, \zeta)\mathbf{y} \qquad z(\eta, \zeta) = \mathbf{N}(\eta, \zeta)\mathbf{z} \tag{4.11}$$

where $\mathbf{N}$ is a row vector of length 9 whose entries are the shape functions and $\mathbf{y}$, $\mathbf{z}$ (Eq. 4.4) are the y, z coordinates of the nodes of the real element. In expanded form, the geometrical transformation equations are

$$y(\eta, \zeta) = \sum_{i=1}^{9} N_i(\eta, \zeta) y_i \qquad z(\eta, \zeta) = \sum_{i=1}^{9} N_i(\eta, \zeta) z_i \tag{4.12}$$

The shape functions for the nine-node element are explicitly determined by substituting the inverse of the matrix $\widehat{\mathbf{N}}_u$ into Eq. (4.10). The inverse of $\widehat{\mathbf{N}}_u$ is

$$\widehat{\mathbf{N}}_u^{-1} = \frac{1}{4}\begin{bmatrix} 0 & 0 & 0 & 0 & 4 & 0 & 0 & 0 & 0 \\ 0 & -2 & 0 & 0 & 0 & 0 & 0 & 2 & 0 \\ 0 & 0 & 0 & -2 & 0 & 2 & 0 & 0 & 0 \\ 0 & 2 & 0 & 0 & -4 & 0 & 0 & 2 & 0 \\ 1 & 0 & -1 & 0 & 0 & 0 & -1 & 0 & 1 \\ 0 & 0 & 0 & 2 & -4 & 2 & 0 & 0 & 0 \\ -1 & 0 & 1 & 2 & 0 & -2 & -1 & 0 & 1 \\ -1 & 2 & -1 & 0 & 0 & 0 & 1 & -2 & 1 \\ 1 & -2 & 1 & -2 & 4 & -2 & 1 & -2 & 1 \end{bmatrix} \tag{4.13}$$

Table 4.1 lists the shape functions and their derivatives with respect to η and ζ.

TABLE 4.1 Shape Functions and Derivatives for the Nine-Node Element

i	$4N_i$	$4\dfrac{\partial N_i}{\partial \eta}$	$4\dfrac{\partial N_i}{\partial \zeta}$
1	$\eta\zeta(1-\eta)(1-\zeta)$	$\zeta(1-2\eta)(1-\zeta)$	$\eta(1-\eta)(1-2\zeta)$
2	$-2\eta(1-\eta)(1-\zeta^2)$	$-2(1-2\eta)(1-\zeta^2)$	$4\eta\zeta(1-\eta)$
3	$-\eta\zeta(1-\eta)(1+\zeta)$	$-\zeta(1-2\eta)(1+\zeta)$	$-\eta(1-\eta)(1+2\zeta)$
4	$-2\zeta(1-\eta^2)(1-\zeta)$	$4\eta\zeta(1-\zeta)$	$-2(1-\eta^2)(1-2\zeta)$
5	$4(1-\eta^2)(1-\zeta^2)$	$-8\eta(1-\zeta^2)$	$-8\zeta(1-\eta^2)$
6	$2\zeta(1-\eta^2)(1+\zeta)$	$-4\eta\zeta(1+\zeta)$	$2(1-\eta^2)(1+2\zeta)$
7	$-\eta\zeta(1+\eta)(1-\zeta)$	$-\zeta(1+2\eta)(1-\zeta)$	$-\eta(1+\eta)(1-2\zeta)$
8	$2\eta(1+\eta)(1-\zeta^2)$	$2(1+2\eta)(1-\zeta^2)$	$-4\eta\zeta(1+\eta)$
9	$\eta\zeta(1+\eta)(1+\zeta)$	$\zeta(1+2\eta)(1+\zeta)$	$\eta(1+\eta)(1+2\zeta)$

4.2 TRANSFORMATION OF DERIVATIVES AND INTEGRALS

The partial derivatives of a function $f(\eta, \zeta)$ with respect to η and ζ in the reference domain Ω_r are related to its derivatives with respect to y and z in the physical domain Ω by

$$\begin{bmatrix} \dfrac{\partial f}{\partial \eta} \\ \dfrac{\partial f}{\partial \zeta} \end{bmatrix} = \begin{bmatrix} \dfrac{\partial y}{\partial \eta} & \dfrac{\partial z}{\partial \eta} \\ \dfrac{\partial y}{\partial \zeta} & \dfrac{\partial z}{\partial \zeta} \end{bmatrix} \begin{bmatrix} \dfrac{\partial f}{\partial y} \\ \dfrac{\partial f}{\partial z} \end{bmatrix} \tag{4.14}$$

The 2×2 matrix in this equation is the *Jacobian matrix* $\mathbf{J}$ of the geometrical transformation. Substitution of Eq. (4.11) into the expression for $\mathbf{J}$ leads to

$$\mathbf{J} = \begin{bmatrix} \dfrac{\partial y}{\partial \eta} & \dfrac{\partial z}{\partial \eta} \\ \dfrac{\partial y}{\partial \zeta} & \dfrac{\partial z}{\partial \zeta} \end{bmatrix} = \begin{bmatrix} \dfrac{\partial \mathbf{N}}{\partial \eta}\mathbf{y} & \dfrac{\partial \mathbf{N}}{\partial \eta}\mathbf{z} \\ \dfrac{\partial \mathbf{N}}{\partial \zeta}\mathbf{y} & \dfrac{\partial \mathbf{N}}{\partial \zeta}\mathbf{z} \end{bmatrix} = \begin{bmatrix} \dfrac{\partial \mathbf{N}}{\partial \eta} \\ \dfrac{\partial \mathbf{N}}{\partial \zeta} \end{bmatrix} \begin{bmatrix} \mathbf{y} & \mathbf{z} \end{bmatrix} = \begin{bmatrix} J_{11} & J_{12} \\ J_{21} & J_{22} \end{bmatrix} \tag{4.15}$$

The entries of the Jacobian matrix $\mathbf{J}$ depend on the point (η, ζ), where J is evaluated, and on the nodal coordinates $\mathbf{y}$, $\mathbf{z}$ of the element.

The partial derivatives of the function $f(y, z)$ with respect to y and z in the physical domain Ω are of interest. These can be related to the derivatives of $f(y, z)$ with respect to η and ζ in the reference domain Ω_r by

$$\begin{bmatrix} \dfrac{\partial f}{\partial y} \\ \dfrac{\partial f}{\partial z} \end{bmatrix} = \begin{bmatrix} \dfrac{\partial \eta}{\partial y} & \dfrac{\partial \zeta}{\partial y} \\ \dfrac{\partial \eta}{\partial z} & \dfrac{\partial \zeta}{\partial z} \end{bmatrix} \begin{bmatrix} \dfrac{\partial f}{\partial \eta} \\ \dfrac{\partial f}{\partial \zeta} \end{bmatrix} \tag{4.16}$$

The 2×2 matrix in this equation is the inverse of the Jacobian matrix of the geometrical transformation

$$\mathbf{J}^{-1} = \begin{bmatrix} \dfrac{\partial \eta}{\partial y} & \dfrac{\partial \zeta}{\partial y} \\ \dfrac{\partial \eta}{\partial z} & \dfrac{\partial \zeta}{\partial z} \end{bmatrix} = \frac{1}{|\mathbf{J}|} \begin{bmatrix} J_{22} & -J_{12} \\ -J_{21} & J_{11} \end{bmatrix} \tag{4.17}$$

where $|\mathbf{J}|$ denotes the determinant of $\mathbf{J}$:

$$|\mathbf{J}| = J_{11}J_{22} - J_{12}J_{21} \tag{4.18}$$

The determinant $|\mathbf{J}|$ can be expressed in terms of the shape function derivatives and the element nodal coordinate vectors as

$$|\,\mathbf{J}\,| = \frac{\partial \mathbf{N}}{\partial \eta}\mathbf{y}\frac{\partial \mathbf{N}}{\partial \zeta}\mathbf{z} - \frac{\partial \mathbf{N}}{\partial \eta}\mathbf{z}\frac{\partial \mathbf{N}}{\partial \zeta}\mathbf{y} = \mathbf{y}^{\mathrm{T}}\left(\frac{\partial \mathbf{N}^{\mathrm{T}}}{\partial \eta}\frac{\partial \mathbf{N}}{\partial \zeta} - \frac{\partial \mathbf{N}^{\mathrm{T}}}{\partial \zeta}\frac{\partial \mathbf{N}}{\partial \eta}\right)\mathbf{z} \tag{4.19}$$

where Eq. (4.15) and the identity $(\mathbf{AB})^{\mathrm{T}} = \mathbf{B}^{\mathrm{T}}\mathbf{A}^{\mathrm{T}}$ have been introduced. The square matrix in the brackets, which is 9×9 for the nine-node element, depends only on η, ζ, and $|\,\mathbf{J}\,|$ depends on the nodal coordinates $\mathbf{y}$, $\mathbf{z}$ as well as on η, ζ. The determinant of the Jacobian is required to be strictly positive:

$$|\,\mathbf{J}\,(\eta, \zeta)\,| > 0 \tag{4.20}$$

at every point of the reference domain. A violation of this condition indicates that the geometrical transformation is not bijective. That is, if the determinant of the Jacobian is not strictly positive, there is no guarantee that for each point of Ω_r there is exactly one point of Ω and that any two distinct points of Ω_r are mapped to two distinct points of Ω. The sign of $|\,\mathbf{J}\,|$ should be checked whenever it occurs in a computation, and preferably also at the corners, where it is more likely to become negative.

An example with the problems that can arise if $|\,\mathbf{J}\,|$ is not strictly positive can be seen from the transformation matrix of Eq. (4.16), which, according to Eq. (4.17), is inversely proportional to $|\,\mathbf{J}\,|$. If $|\,\mathbf{J}\,|$ is zero somewhere (e.g., along a curve), the Jacobian matrix is singular and cannot be inverted.

4.3 INTEGRALS

Integrating a function $f(\eta, \zeta)$ over the reference element is simpler than integrating the corresponding function $f(y, z)$ over the real element. From the calculus, the transformation formula for integrals is

$$\int_{\Omega} f(y, z)\, dy\, dz = \int_{\Omega_r} f(\eta, \zeta)|\,\mathbf{J}\,|\, d\eta\, d\zeta \tag{4.21}$$

For an *isoparametric element*, wherein the function interpolation formulas are based on the same set of shape functions as the geometrical transformation equation (Eq. 4.11), the function $f(\eta, \zeta)$ in Eq. (4.21) is not only a function of η and ζ but will also be dependent on its nodal values $\mathbf{f}$, where

$$\mathbf{f} = [f(\eta_1, \zeta_1) \quad f(\eta_2, \zeta_2) \quad \cdots \quad f(\eta_9, \zeta_9)]^{\mathrm{T}}.$$

It follows from Eq. (4.11) that

$$f(\eta, \zeta) = \mathbf{N}(\eta, \zeta)\mathbf{f} \tag{4.22}$$

where $\mathbf{N}$ is the row vector of Eq. (4.10).

The area A_e of the real element e can be found by computing

$$A_e = \int_{\Omega} dy\, dz = \int_{\Omega_r} |\,\mathbf{J}_e\,|\, d\eta\, d\zeta \tag{4.23}$$

where $\mathbf{J}_e$ is the Jaçobian matrix for the eth element. Introduction of Eq. (4.19) into Eq. (4.23) leads to

$$A_e = \mathbf{y}^{\mathrm{T}}\mathbf{A}^*\mathbf{z} \tag{4.24}$$

where $\mathbf{A}^*$ is defined by

$$\mathbf{A}^* = \int_{-1}^{1}\int_{-1}^{1}\left(\frac{\partial \mathbf{N}^{\mathrm{T}}}{\partial \eta}\frac{\partial \mathbf{N}}{\partial \zeta} - \frac{\partial \mathbf{N}^{\mathrm{T}}}{\partial \zeta}\frac{\partial \mathbf{N}}{\partial \eta}\right) d\eta\, d\zeta \tag{4.25}$$

For the isoparametric nine-node element, the matrix $\mathbf{A}^*$ using the derivatives of the shape functions from Table 4.1 is calculated to be

$$\mathbf{A}^* = \frac{1}{6}\begin{bmatrix} 0 & -4 & 1 & 4 & 0 & 0 & -1 & 0 & 0 \\ 4 & 0 & -4 & 0 & 0 & 0 & 0 & 0 & 0 \\ -1 & 4 & 0 & 0 & 0 & -4 & 0 & 0 & 1 \\ -4 & 0 & 0 & 0 & 0 & 0 & 4 & 0 & 0 \\ 0 & 0 & 0 & 0 & 0 & 0 & 0 & 0 & 0 \\ 0 & 0 & 4 & 0 & 0 & 0 & 0 & 0 & -4 \\ 1 & 0 & 0 & -4 & 0 & 0 & 0 & 4 & -1 \\ 0 & 0 & 0 & 0 & 0 & 0 & -4 & 0 & 4 \\ 0 & 0 & -1 & 0 & 0 & 4 & 1 & -4 & 0 \end{bmatrix} \tag{4.26}$$

Given a collection of M elements in the finite element mesh, the integral over the domain made up of these elements is found by summing the contributions of all elements:

$$\int f(y, z)\, dy\, dz = \sum_{e=1}^{M}\int_{-1}^{1}\int_{-1}^{1} f(\eta, \zeta)|\,\mathbf{J}_e(\eta, \zeta)\,|\, d\eta\, d\zeta \tag{4.27}$$

In practice, integration over reference elements is performed numerically, most often by Gaussian quadrature formulas. See a reference on finite elements (e.g., Pilkey and Wunderlich, 1994) for details on numerical integration procedures. The bidirectional method of integration by Gaussian quadrature, for instance, is a direct extension of one-dimensional quadrature formulas to two-dimensional elements. A function is integrated over the region $-1 \le \eta \le 1, -1 \le \zeta \le 1$ by choosing m_1 and m_2 Gaussian integration points η_i, ζ_j in the η and ζ directions, respectively, and evaluating the integral holding ζ constant and integrating over η and then holding η constant and integrating over ζ. For quadrilateral elements, the bidirectional formula for a function $\phi(\eta, \zeta)$ is

$$\int_{-1}^{1}\int_{-1}^{1}\phi(\eta, \zeta)\, d\eta\, d\zeta = \sum_{i=1}^{m_1}\sum_{j=1}^{m_2} W_i W_j\, \phi(\eta_i, \zeta_j) \tag{4.28}$$

where W_i, W_j are the weighting coefficients. This method uses $m_1 \times m_2$ integration points and, in theory, integrates all monomials:

$$\eta^k \zeta^n \quad \text{such that} \quad 0 \leq k \leq 2m_1 - 1 \quad \text{and} \quad 0 \leq n \leq 2m_2 - 1 \tag{4.29}$$

exactly.

Table 4.2 shows the weights and integration points that can be used for various values of m_1 and m_2. Only cases for which $m_1 = m_2$ are given in this table. The use of $m_1 \times m_2$ points is referred to as "$m_1 \times m_2$ Gaussian quadrature."

TABLE 4.2 Gauss Quadrature Formulas for Quadrilateral Elements, $m_1 = m_2$[a]

$m_1 = m_2$	Number of Gauss Points $m_1 \times m_2$	Gauss Points η_i, ζ_j	Weights $W_i \times W_j$
1	1 (1×1)	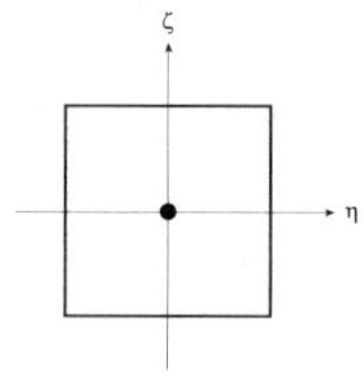	$4(= 2 \times 2)$ at center
2	4 (2×2)	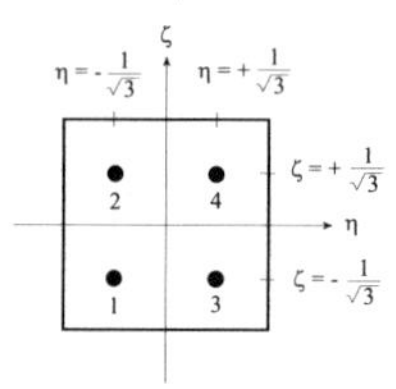	$1(= 1 \times 1)$ at points 1, 2, 3, 4
3	9 (3×3)	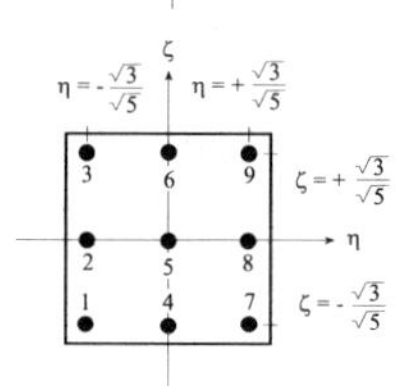	$\frac{25}{81}\left(= \frac{5}{9} \times \frac{5}{9}\right)$ at points 1, 3, 7, 9 $\frac{40}{81}\left(= \frac{5}{9} \times \frac{8}{9}\right)$ at points 2, 4, 6, 8 $\frac{64}{81}\left(= \frac{8}{9} \times \frac{8}{9}\right)$ at point 5

[a]

$$\int_{-1}^{+1} \int_{-1}^{+1} \phi(\eta, \zeta) d\eta d\zeta = \sum_{i=1}^{m_1} \sum_{j=1}^{m_2} W_i W_j \, \phi(\eta_i, \zeta_j)$$

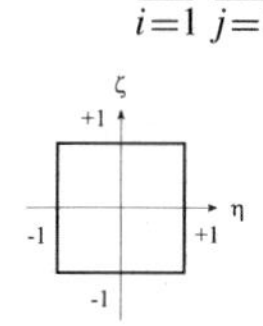

m_1 and m_2 are the number of integration (Gauss) points in η and ζ directions, respectively.

4.4 CROSS-SECTIONAL PROPERTIES

Several cross-sectional properties of a beam that are independent of warping can be found by integration using the finite-element mesh for the section. The area A of the cross section is

$$A = \int_A dy\, dz = \sum_e A_e = \sum_e \int_{-1}^{1} \int_{-1}^{1} |\, \mathbf{J}_e \,|\, d\eta\, d\zeta = \sum_e \mathbf{y}_e^{\mathrm{T}} \mathbf{A}^* \mathbf{z}_e \tag{4.30}$$

where the sum is over all elements e in the mesh, and the matrix $\mathbf{A}^*$ of Eq. (4.25) is the same for each element. For the nine-node element, $\mathbf{A}^*$ is the constant matrix given in Eq. (4.26).

The expression of Eq. (4.30) can be evaluated in closed form. Since $\mathbf{A}^*$ is a sparse matrix, the expanded form of the expression for the element area A_e is compact. With Eq. (4.26) and

$$\mathbf{z}_e = [z(\eta_1, \zeta_1) \quad z(\eta_2, \zeta_2) \quad \cdots \quad z(\eta_9, \zeta_9)]^{\mathrm{T}} = [z_1 \quad z_2 \quad \cdots \quad z_9]^{\mathrm{T}} \tag{4.31}$$

we find that

$$A_e = \tfrac{2}{3} \mathbf{y}_e^{\mathrm{T}} \mathbf{Z} \tag{4.32}$$

where

$$\mathbf{Z} = [Z_1 \quad Z_2 \quad Z_3 \quad Z_4 \quad Z_5 \quad Z_6 \quad Z_7 \quad Z_8 \quad Z_9]^{\mathrm{T}} \tag{4.33}$$

with

$$\begin{aligned}
Z_1 &= z_4 - z_2 + \frac{z_3 - z_7}{4} \\
Z_2 &= z_1 - z_3 \\
Z_3 &= z_2 - z_6 + \frac{z_9 - z_1}{4} \\
Z_4 &= z_7 - z_1 \\
Z_5 &= 0 \\
Z_6 &= z_3 - z_9 \\
Z_7 &= z_8 - z_4 + \frac{z_1 - z_9}{4} \\
Z_8 &= z_9 - z_7 \\
Z_9 &= z_6 - z_8 + \frac{z_7 - z_3}{4}
\end{aligned} \tag{4.34}$$

The cross-sectional area is obtained as $\sum A_e$. It may, however, be preferable to compute the area by Gaussian quadrature to maintain computational uniformity, because it is not plausible to write such closed-form algebraic expressions for all integrals.

The *first moments of area* are defined by

$$Q_y = \int z \, dA \qquad Q_z = \int y \, dA \tag{4.35}$$

and are calculated as

$$\begin{aligned} Q_y &= \sum_e \int_{-1}^{1} \int_{-1}^{1} \mathbf{Nz}_e |\, \mathbf{J}_e \,| \, d\eta \, d\zeta \\ Q_z &= \sum_e \int_{-1}^{1} \int_{-1}^{1} \mathbf{Ny}_e |\, \mathbf{J}_e \,| \, d\eta \, d\zeta \end{aligned} \tag{4.36}$$

The coordinates of the *centroid C* of the cross section are then found from

$$y_C = \frac{Q_z}{A} \qquad z_C = \frac{Q_y}{A} \tag{4.37}$$

The *area moments of inertia* defined by

$$I_y = \int z^2 \, dA \qquad I_z = \int y^2 \, dA \tag{4.38}$$

are computed as

$$\begin{aligned} I_y &= \sum_e \int_{-1}^{1} \int_{-1}^{1} (\mathbf{Nz}_e)^2 \,|\, \mathbf{J}_e \,| \, d\eta \, d\zeta \\ I_z &= \sum_e \int_{-1}^{1} \int_{-1}^{1} (\mathbf{Ny}_e)^2 \,|\, \mathbf{J}_e \,| \, d\eta \, d\zeta \end{aligned} \tag{4.39}$$

Similarly, the *product of inertia* defined by

$$I_{yz} = \int yz \, dA \tag{4.40}$$

is calculated as

$$I_{yz} = \sum_e \int_{-1}^{1} \int_{-1}^{1} \mathbf{Ny}_e \mathbf{Nz}_e |\, \mathbf{J}_e \,| \, d\eta \, d\zeta \tag{4.41}$$

The preceding equations for I_y, I_z, and I_{yz} express these properties in the global coordinate system in which the nodal coordinates of the elements are measured. The origin of this coordinate system is chosen arbitrarily. The area moments of inertia are often needed in the coordinate system with its origin at the centroid and axes parallel to the global coordinate axes. These properties are found by the parallel axis theorem of Eq. (1.60) with the transformation equations

$$
\begin{aligned}
I_{\overline{y}} &= I_y - z_C^2 A = I_y - \frac{Q_y^2}{A} \\
I_{\overline{z}} &= I_z - y_C^2 A = I_z - \frac{Q_z^2}{A} \\
I_{\overline{yz}} &= I_{yz} - y_C z_C A = I_{yz} - \frac{Q_y Q_z}{A}
\end{aligned}
\tag{4.42}
$$

where $I_{\overline{y}}$, $I_{\overline{z}}$, and $I_{\overline{yz}}$ are the moments of inertia about $\overline{y}$, $\overline{z}$ centroidal axes.

Example 4.1 Properties of an Asymmetric Cross Section. Return to Examples 1.1 and 1.2 in which the properties of an asymmetric cross section were found using the formulas of Chapter 1. We now continue by solving this problem using the finite-element mesh definitions of this chapter. The properties will first be solved for a convenient coordinate system. They will then be calculated with respect to the centroidal coordinate system using the parallel axis theorem.

SOLUTION. Start by dividing the cross section into two elements as shown in Fig. 4.2. Let $a = 1.0$ in. and $t = 0.1$ in. Number the nodes for each element as in Fig. 4.1a. The nodal y and z coordinates of these elements, $\mathbf{z}_e$ and $\mathbf{y}_e$, with respect to the coordinate axes of Fig. 4.2, are

$$
\mathbf{y}_1 = \begin{bmatrix} -0.05 \\ -0.05 \\ -0.05 \\ 0 \\ 0 \\ 0 \\ 0.05 \\ 0.05 \\ 0.05 \end{bmatrix} \quad
\mathbf{z}_1 = \begin{bmatrix} -2 \\ -1.025 \\ -.05 \\ -2 \\ -1.025 \\ -0.05 \\ -2 \\ -1.025 \\ -0.05 \end{bmatrix} \quad
\mathbf{y}_2 = \begin{bmatrix} -0.05 \\ -0.05 \\ -0.05 \\ 0.475 \\ 0.475 \\ 0.475 \\ 1 \\ 1 \\ 1 \end{bmatrix} \quad
\mathbf{z}_2 = \begin{bmatrix} -0.05 \\ 0 \\ 0.05 \\ -0.05 \\ 0 \\ 0.05 \\ -0.05 \\ 0 \\ 0.05 \end{bmatrix}
\tag{1}
$$

The Jacobian matrix, $\mathbf{J}_e$, for each of the elements can be calculated using Eq. (4.15) and the shape functions and their derivatives from Table 4.1. For element 1,

$$
\begin{aligned}
J_{11} &= \frac{\partial \mathbf{N}}{\partial \eta} \mathbf{y}_1 = 0.05 \\
J_{12} &= \frac{\partial \mathbf{N}}{\partial \eta} \mathbf{z}_1 = 0 \\
J_{21} &= \frac{\partial \mathbf{N}}{\partial \varsigma} \mathbf{y}_1 = 0 \\
J_{22} &= \frac{\partial \mathbf{N}}{\partial \varsigma} \mathbf{z}_1 = 0.975
\end{aligned}
$$

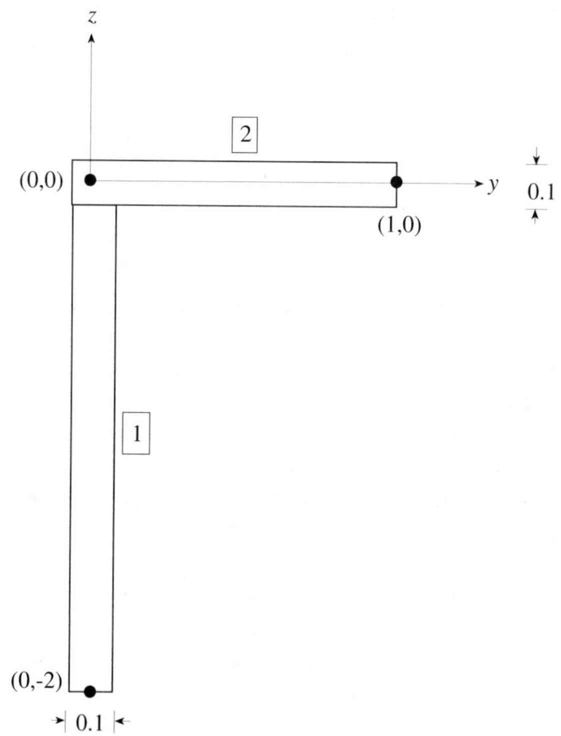

Figure 4.2 Asymmetrical cross section.

and (Eq. 4.18)

$$|\mathbf{J}_1| = 0.04875 \tag{2}$$

For element 2,

$$J_{11} = \frac{\partial \mathbf{N}}{\partial \eta}\mathbf{y}_2 = 0.525$$

$$J_{12} = \frac{\partial \mathbf{N}}{\partial \eta}\mathbf{z}_2 = 0$$

$$J_{21} = \frac{\partial \mathbf{N}}{\partial \varsigma}\mathbf{y}_2 = 0$$

$$J_{22} = \frac{\partial \mathbf{N}}{\partial \varsigma}\mathbf{z}_2 = 0.5$$

and

$$|\mathbf{J}_2| = 0.0265 \tag{3}$$

The integrals of Eq. (4.36) can be solved, using the shape functions of Table 4.1, to find the first moments of area.

$$
\begin{aligned}
Q_y &= \sum_{e=1}^{2} \int_{-1}^{1} \int_{-1}^{1} \mathbf{N}\mathbf{z}_e |\, \mathbf{J}_e \,|\, d\eta\, d\varsigma = -0.199875 \\
Q_z &= \sum_{e=1}^{2} \int_{-1}^{1} \int_{-1}^{1} \mathbf{N}\mathbf{y}_e |\, \mathbf{J}_e \,|\, d\eta\, d\varsigma = 0.049875
\end{aligned} \tag{4}
$$

Similarly, the integrals of Eqs. (4.39) and (4.41) give the moments of inertia and product of inertia

$$
\begin{aligned}
I_y &= \sum_{e=1}^{2} \int_{-1}^{1} \int_{-1}^{1} (\mathbf{N}\mathbf{z}_e)^2 |\, \mathbf{J}_e \,|\, d\eta\, d\varsigma = 0.26675 \\
I_z &= \sum_{e=1}^{2} \int_{-1}^{1} \int_{-1}^{1} (\mathbf{N}\mathbf{y}_e)^2 |\, \mathbf{J}_e \,|\, d\eta\, d\varsigma = 0.0335 \\
I_{yz} &= \sum_{e=1}^{2} \int_{-1}^{1} \int_{-1}^{1} \mathbf{N}\mathbf{y}_e \mathbf{N}\mathbf{z}_e |\, \mathbf{J}_e \,|\, d\eta\, d\varsigma = 0.0
\end{aligned} \tag{5}
$$

For the simple geometry of this problem, it can be seen from inspection that the area of the cross section is 0.3. Alternatively, the area of the cross section could be calculated using the expression (Eq. 4.30)

$$
A = \sum_{e} \mathbf{y}_e^{\mathrm{T}} \mathbf{A}^* \mathbf{z}_e \tag{6}
$$

where the matrix A^* is given by (Eq. 4.26)

$$
\mathbf{A}^* = \frac{1}{6} \begin{bmatrix}
0 & -4 & 1 & 4 & 0 & 0 & -1 & 0 & 0 \\
4 & 0 & -4 & 0 & 0 & 0 & 0 & 0 & 0 \\
-1 & 4 & 0 & 0 & 0 & -4 & 0 & 0 & 1 \\
-4 & 0 & 0 & 0 & 0 & 0 & 4 & 0 & 0 \\
0 & 0 & 0 & 0 & 0 & 0 & 0 & 0 & 0 \\
0 & 0 & 4 & 0 & 0 & 0 & 0 & 0 & -4 \\
1 & 0 & 0 & -4 & 0 & 0 & 0 & 4 & -1 \\
0 & 0 & 0 & 0 & 0 & 0 & -4 & 0 & 4 \\
0 & 0 & -1 & 0 & 0 & 4 & 1 & -4 & 0
\end{bmatrix} \tag{7}
$$

Then

$$
A = \mathbf{y}_1^{\mathrm{T}} \mathbf{A}^* \mathbf{z}_1 + \mathbf{y}_2^{\mathrm{T}} \mathbf{A}^* \mathbf{z}_2 = 0.3 \tag{8}
$$

The centroid of the cross section is obtained from Eq. (4.37):

$$y_C = \frac{Q_z}{A} = 0.166 \qquad z_C = \frac{Q_y}{A} = -0.666 \tag{9}$$

To find the moments of inertia about a coordinate system with its origin at the centroid, apply the parallel axis theorem of Eq. (4.42):

$$\begin{aligned} I_{\overline{y}} &= I_y - z_C^2 A = 0.13358 \\ I_{\overline{z}} &= I_z - y_C^2 A = 0.025208 \\ I_{\overline{y}\,\overline{z}} &= I_{yz} - y_C z_C A = 0.03323 \end{aligned} \tag{10}$$

It can readily be shown that the values calculated here are the same as those of Example 1.2 (Table 1.1).

4.5 MODULUS-WEIGHTED PROPERTIES

As described in Chapter 1, modulus-weighted section properties are used for non-homogeneous beams. All the previous formulas are then modified by replacing the differential element of area by its modulus-weighted counterpart. That is, as in Eq. (1.64),

$$d\widetilde{A} = \frac{E}{E_r}\, dA \tag{4.43}$$

where E_r is a reference elastic modulus. Each element e may be assigned a different elastic modulus E_e, and with an arbitrarily selected reference modulus E_r, the area integrals become

$$\int f(y,z)\, d\widetilde{A} = \sum_{e=1}^{M} \int_{-1}^{1} \int_{-1}^{1} f(\eta,\zeta) |\mathbf{J}_e(\eta,\zeta)| \frac{E_e}{E_r}\, d\eta\, d\zeta \tag{4.44}$$

REFERENCES

Dhatt, G., and Touzot, G. (1984). *The Finite Element Method Displayed*, Wiley, New York.

Pilkey, W. D., and Wunderlich, W. (1994). *Mechanics of Structures: Variational and Computational Methods*, CRC Press, Boca Raton, Fla.

Zienkiewicz, O. C. (1977). *The Finite Element Method*, 3rd ed., McGraw-Hill, New York.

CHAPTER 5

SAINT-VENANT TORSION

The elasticity problem of torsion as formulated by Saint-Venant is the subject of this chapter. A finite element formulation, which defines a numerical algorithm for calculating the torsional constant and the Saint-Venant torsional stresses, follows the classical statement of the problem. The determination of element stiffness and load matrices and the subsequent finite element assembly procedure are described in detail. Books such as Boresi and Chong (1987), Love (1944), Rivello (1969), and Sokolnikoff (1956) discuss the Saint-Venant problem in a classical setting. In addition, Muskhelishvili (1953) is an excellent reference, especially for nonhomogeneous cross sections. General descriptions of finite element procedures for solving boundary value problems are found in Dhatt and Touzot (1984) and Zienkiewicz (1977).

5.1 FUNDAMENTALS OF SAINT-VENANT TORSION

Coulomb formulated a solution for the torsion of thin wires in which it is assumed that cross sections remain plane after deformations and that they simply rotate. This formulation can be shown to be accurate for bars of circular cross sections. In a further step, Saint-Venant made the assumption that the cross sections of a beam subjected to pure torsion rotate about the *axis of twist* such that although the cross sections warp out of their original plane, their projections onto this plane retain their original shape. The axis of twist passes through the *center of twist*, which is the location on the cross section for which there are no y and z displacements due to a torque.

The displacements u_y and u_z of a point A of the cross section are shown in Fig. 5.1, in which the x axis is the axis of twist. The angle of twist of a section along

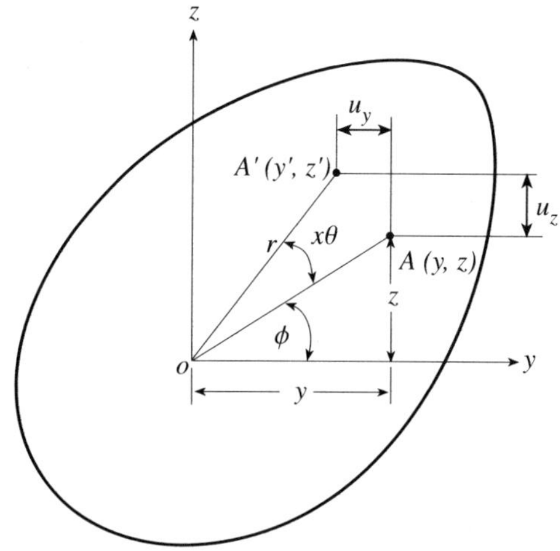

Figure 5.1 Displacements u_y and u_z of point A under twisting about the x axis.

the x axis is ϕ, so that the cylindrical coordinates of point A are (r, ϕ). As a result of further deformation, the rotation of a section of a shaft of length x is $x\theta$, where θ is the angle of twist per unit length, which is assumed to be constant. As a result of the twisting deformations, A moves to A', whose position is given by the cylindrical coordinates $(r, \phi + x\theta)$. In Fig. 5.1, $OA = OA' = r$, $y = r\cos\phi$, $z = r\sin\phi$. For small deformations, $x\theta$ is small so that $\sin(x\theta) \approx x\theta$ and $\cos(x\theta) \approx 1$. Then

$$\begin{aligned} u_y = y' - y &= r[\cos(\phi + x\theta) - \cos\phi] \\ &= r(\cos\phi\cos x\theta - \sin\phi\sin x\theta) - r\cos\phi \\ &= y(\cos x\theta - 1) - z\sin(x\theta) \approx -zx\theta \\ u_z = z' - z &= r[\sin(\phi + x\theta) - \sin\phi] \\ &= r(\cos\phi\sin x\theta + \sin\phi\cos x\theta) - r\sin\phi \\ &= y\sin x\theta + z(\cos x\theta - 1) \approx yx\theta \end{aligned} \tag{5.1}$$

As indicated above, for bars of circular cross section, the axial displacement u_x, which is also called the *warping displacement*, is zero. For cross sections of arbitrary shape, experimental evidence indicates that the axial displacement of each cross section along the bar is about the same. Thus, $u_x = f(y, z)$ where f is a function. Saint-Venant accepted this result as a fundamental assumption. It is useful to introduce θ into this expression so that $f(y, z) = \theta\omega(y, z)$. Then the axial displacement u_x is proportional to the angle of twist per unit length, and the displacement field is written in the form

$$u_x = \theta\omega(y, z) \tag{5.2}$$

The unknown function $\omega(y, z)$ in this equation is called the *warping function*.

The strains for this assumed displacement field are found from the strain–displacement relations of Eq. (1.3) as

$$\begin{aligned}
\epsilon_x &= \frac{\partial u_x}{\partial x} = \theta' \omega(y, z) = 0 \qquad \text{since } \theta \text{ is constant} \\
\epsilon_y &= \frac{\partial u_y}{\partial y} = 0 \qquad \epsilon_z = \frac{\partial u_z}{\partial z} = 0 \\
\gamma_{xy} &= \frac{\partial u_y}{\partial x} + \frac{\partial u_x}{\partial y} = \theta \left(-z + \frac{\partial \omega}{\partial y} \right) \\
\gamma_{xz} &= \frac{\partial u_z}{\partial x} + \frac{\partial u_x}{\partial z} = \theta \left(y + \frac{\partial \omega}{\partial z} \right) \\
\gamma_{yz} &= \frac{\partial u_z}{\partial y} + \frac{\partial u_y}{\partial z} = 0
\end{aligned} \tag{5.3}$$

For these strains, the constitutive relations of Eq. (1.12) give the stresses

$$\begin{aligned}
\tau_{xy} &= G\gamma_{xy} = G\theta \left(\frac{\partial \omega}{\partial y} - z \right) \\
\tau_{xz} &= G\gamma_{xz} = G\theta \left(\frac{\partial \omega}{\partial z} + y \right) \\
\sigma_x &= 0, \quad \sigma_y = 0, \quad \sigma_z = 0, \quad \tau_{yz} = 0
\end{aligned} \tag{5.4}$$

With this assumed displacement field, the dilatation, that is, the change in volume per unit volume, is identically zero (Eq. 1.41):

$$e = \nabla \cdot \mathbf{u} = \frac{\partial u_x}{\partial x} + \frac{\partial u_y}{\partial y} + \frac{\partial u_z}{\partial z} = 0 \tag{5.5}$$

where

$$\mathbf{u} = u_x \mathbf{i} + u_y \mathbf{j} + u_z \mathbf{k} \quad \text{and} \quad \nabla = \mathbf{i}\frac{\partial}{\partial x} + \mathbf{j}\frac{\partial}{\partial y} + \mathbf{k}\frac{\partial}{\partial z}.$$

Also, the Laplacians of u_y and u_z are zero:

$$\nabla^2 u_y = \frac{\partial^2 u_y}{\partial x^2} + \frac{\partial^2 u_y}{\partial y^2} + \frac{\partial^2 u_y}{\partial z^2} = 0 \qquad \nabla^2 u_z = \frac{\partial^2 u_z}{\partial x^2} + \frac{\partial^2 u_z}{\partial y^2} + \frac{\partial^2 u_z}{\partial z^2} = 0 \tag{5.6}$$

With zero body forces, the equilibrium equations of Eq. (1.39) in the y and z directions are identically satisfied; that is,

$$\begin{aligned}
(\lambda + G) \frac{\partial e}{\partial y} + G\nabla^2 u_y = 0 \\
(\lambda + G) \frac{\partial e}{\partial z} + G\nabla^2 u_z = 0
\end{aligned} \tag{5.7}$$

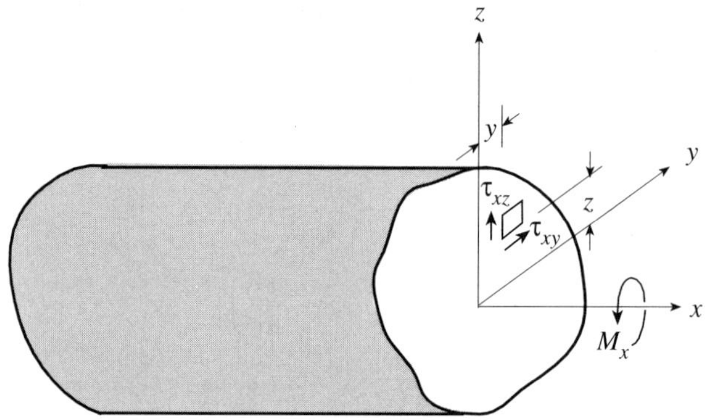

Figure 5.2 Notation employed in finding the resultant moment on a cross section.

The equilibrium equation in the x direction of Eq. (1.39), $(\lambda + G)\partial e/\partial x + G\nabla^2 u_x = 0$, with Eq. (5.2) leads to

$$\nabla^2 \omega = \frac{\partial^2 \omega}{\partial y^2} + \frac{\partial^2 \omega}{\partial z^2} = 0 \tag{5.8}$$

where $\nabla^2 = \partial^2/\partial y^2 + \partial^2/\partial z^2$. A partial differential of this form is called *Laplace's equation*. A solution to Laplace's equation is called a *harmonic function*.

On the cylindrical surface of the bar (Fig. 5.2), at the boundary of the cross section, the x component of the unit normal vector $\mathbf{n} = n_x\mathbf{i} + n_y\mathbf{j} + n_z\mathbf{k}$ is zero. If this surface is free of applied forces p_x, p_y, and p_z, the second boundary condition of Eq. (1.43), with Eq. (5.5) and $n_x = 0$, becomes

$$\mathbf{n} \cdot \boldsymbol{\nabla} u_y + \mathbf{n} \cdot \frac{\partial \mathbf{u}}{\partial y} = n_y \frac{\partial u_y}{\partial y} + n_z \frac{\partial u_y}{\partial z} + n_y \frac{\partial u_y}{\partial y} + n_z \frac{\partial u_z}{\partial y} = -n_z x\theta + n_z x\theta = 0 \tag{5.9}$$

Similarly, the condition of the third relationship of Eq. (1.43) is identically satisfied. The remaining condition, given by the first of Eq. (1.43),

$$\mathbf{n} \cdot \boldsymbol{\nabla} u_x + \mathbf{n} \cdot \frac{\partial \mathbf{u}}{\partial x} = 0 = n_x \frac{\partial u_x}{\partial x} + n_y \frac{\partial u_x}{\partial y} + n_z \frac{\partial u_x}{\partial z} + n_x \frac{\partial u_x}{\partial x} + n_y \frac{\partial u_y}{\partial x} + n_z \frac{\partial u_z}{\partial x} \tag{5.10}$$

shows that the relationship that must be satisfied on the cylindrical surface is

$$\left(\frac{\partial \omega}{\partial y} - z\right) n_y + \left(\frac{\partial \omega}{\partial z} + y\right) n_z = 0 \tag{5.11}$$

The boundaries at the ends of the bar coincide with cross sections of the bar having unit normals $n_x = \pm 1$, $n_y = 0$, $n_z = 0$. The surface conditions of Eq. (1.27) for these ends are then

$$p_x = 0 \qquad p_y = \pm\tau_{xy} \qquad p_z = \pm\tau_{xz} \tag{5.12}$$

The + sign corresponds to the end of the bar with an external normal in the direction of the positive x axis. From Eq. (5.4), for the + sign of Eq. (5.12),

$$p_x = 0 \qquad p_y = G\theta\left(\frac{\partial\omega}{\partial y} - z\right) \qquad p_z = G\theta\left(\frac{\partial\omega}{\partial z} + y\right) \tag{5.13}$$

The resultant of the stresses of Eq. (5.12) on the end of a bar is a torque. The y and z direction resultant forces on an end of area A,

$$\iint_A \tau_{xy}\, dy\, dz \qquad \iint_A \tau_{xz}\, dy\, dz \tag{5.14}$$

vanish (Pilkey and Wunderlich, 1994). The resultant moment M_x from moment equilibrium requirements on a cross section (Fig. 5.2) is the area integral

$$M_x = \iint_A \left(\tau_{xz} y - \tau_{xy} z\right)\, dy\, dz \tag{5.15}$$

or in vector notation,

$$M_x\mathbf{i} = \int_A (y\mathbf{j} + z\mathbf{k}) \times \left(p_y\mathbf{j} + p_z\mathbf{k}\right)\, dA \tag{5.16}$$

From Eq. (5.13), the torque is

$$M_x = G\theta \int_A \left[\left(\frac{\partial\omega}{\partial z} + y\right) y - \left(\frac{\partial\omega}{\partial y} - z\right) z\right] dA \tag{5.17}$$

Suppose that the torque M_x can be expressed as a function of the shear modulus of elasticity G, a material constant; the angle of twist per unit length θ; and a constant J that is based on the geometry of the cross section. Thus, let there be a constant J such that

$$M_x = GJ\theta \tag{5.18}$$

This is the same relationship found for bars of circular cross section when J is the polar moment of inertia of the cross section. In Eq. (5.18), J is the *torsional constant* and GJ is the *torsional stiffness* of the bar. By comparison of Eqs. (5.17) and (5.18), J can be determined from the warping function

$$J = \int_A \left[\left(\frac{\partial\omega}{\partial z} + y\right) y - \left(\frac{\partial\omega}{\partial y} - z\right) z\right] dA \tag{5.19}$$

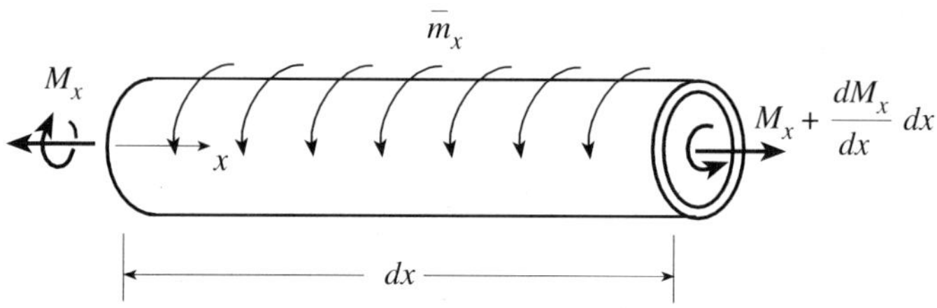

Figure 5.3 Element subject to distributed torque $\overline{m}_x$.

The problem formulation is now complete. The warping function ω is obtained by solving Eq. (5.8) subject to the surface and boundary conditions. The torsional constant J is found from Eq. (5.19). For a particular torque M_x, the stresses as given by Eqs. (5.4) and (5.18) are

$$\begin{aligned} \tau_{xy} &= \frac{M_x}{J}\left(\frac{\partial \omega}{\partial y} - z\right) \\ \tau_{xz} &= \frac{M_x}{J}\left(\frac{\partial \omega}{\partial z} + y\right) \end{aligned} \tag{5.20}$$

The torque M_x and angle of twist ϕ are related through Eq. (5.18)

$$GJ\frac{d\phi}{dx} = M_x \tag{5.21}$$

since $\theta = d\phi/dx$. The conditions of equilibrium applied to a segment of bar (Fig. 5.3) shows that the applied distributed axial moment $\overline{m}_x$ (force · length/length) is related to the axial moment M_x by

$$\frac{dM_x}{dx} = -\overline{m}_x \tag{5.22}$$

These governing equations of motion (Eqs. 5.21 and 5.22) can be combined to form the single higher-order governing equation for the angle of twist along the bar

$$\frac{d}{dx}GJ\frac{d\phi}{dx} = -\overline{m}_x \tag{5.23}$$

In terms of the state variable notation of Eq. (2.32a), Eqs. (5.21) and (5.22) appear as

$$\frac{d\mathbf{z}}{dx} = \mathbf{Az} + \overline{\mathbf{P}} \tag{5.24}$$

where

$$
\mathbf{z} = \begin{bmatrix} \phi \\ M_x \end{bmatrix} \qquad \mathbf{A} = \begin{bmatrix} 0 & \dfrac{1}{GJ} \\ 0 & 0 \end{bmatrix} \qquad \overline{\mathbf{P}} = \begin{bmatrix} 0 \\ -\overline{m}_x \end{bmatrix}
$$

If axial extension and bending in two planes are included, the governing equations for the displacements and forces along the bar are defined by

$$
\mathbf{z} = \begin{bmatrix} u & v & w & \theta_x & \theta_y & \theta_z & N & V_y & V_z & M_x & M_y & M_z \end{bmatrix}^{\mathrm{T}}
$$

$$
\mathbf{A} = \begin{bmatrix}
0 & 0 & 0 & 0 & 0 & 0 & \dfrac{1}{EA} & 0 & 0 & \dfrac{1}{GJ} & 0 & 0 \\
0 & 0 & 0 & 0 & 0 & 1 & 0 & 0 & 0 & 0 & 0 & 0 \\
0 & 0 & 0 & 0 & -1 & 0 & 0 & 0 & 0 & 0 & 0 & 0 \\
0 & 0 & 0 & 0 & 0 & 0 & 0 & 0 & 0 & \dfrac{1}{GJ} & 0 & 0 \\
0 & 0 & 0 & 0 & 0 & 0 & 0 & 0 & 0 & 0 & \dfrac{I_z}{K} & \dfrac{I_{yz}}{K} \\
0 & 0 & 0 & 0 & 0 & 0 & 0 & 0 & 0 & 0 & \dfrac{I_{yz}}{K} & \dfrac{I_y}{K} \\
0 & 0 & 0 & 0 & 0 & 0 & 0 & 0 & 0 & 0 & 0 & 0 \\
0 & 0 & 0 & 0 & 0 & 0 & 0 & 0 & 0 & 0 & 0 & 0 \\
0 & 0 & 0 & 0 & 0 & 0 & 0 & 0 & 0 & 0 & 0 & 0 \\
0 & 0 & 0 & 0 & 0 & 0 & 0 & 0 & 0 & 0 & 0 & 0 \\
0 & 0 & 0 & 0 & 0 & 0 & 0 & 0 & 1 & 0 & 0 & 0 \\
0 & 0 & 0 & 0 & 0 & 0 & 0 & -1 & 0 & 0 & 0 & 0
\end{bmatrix}
\qquad
\overline{\mathbf{P}} = \begin{bmatrix} 0 \\ 0 \\ 0 \\ 0 \\ 0 \\ 0 \\ -\overline{p}_x \\ -\overline{p}_y \\ -\overline{p}_z \\ -\overline{m}_x \\ 0 \\ 0 \end{bmatrix}
\tag{5.25}
$$

where $\theta_x = \phi$ and $K = E(I_y I_z - I_{yz}^2)$. If these equations are transformed (rotated) to the principal bending planes, they lead to the stiffness matrix of Example 2.17.

The warping-function approach (for the torsional constant and the torsional stresses on the cross section) is often referred to as a *displacement formulation* of the torsion problem. It is applicable to multiply connected cross sections. For such sections it is necessary to apply the surface conditions of Eq. (5.11) to both the internal and external boundaries. The Saint-Venant solution is exact within the limits of linear elasticity if the following conditions are satisfied:

1. The beam has a uniform cross section.
2. There is no warping restraint at any of the cross sections.
3. The torque is applied at the ends by surface forces whose distribution matches that of the calculated shear stresses.

If the warping of the beam is restrained, normal longitudinal stresses are developed even if the only loading is a torque. These normal stresses, also called *warping stresses*, are important in the analysis of thin-walled beams and are treated later in this book.

Example 5.1 Bar of Circular Cross Section. Consider a straight bar of circular cross section subject to torque M_x. The x-axis deformation of a cross section must be the same when viewed from either end of the bar. Hence, $u_x = 0$. From Eq. (5.1), the deformations in the y and z directions are

$$u_y = -zx\theta \qquad u_z = yx\theta \tag{1}$$

Equation (5.3) shows that

$$\epsilon_x = 0, \quad \epsilon_y = 0, \quad \epsilon_z = 0, \quad \gamma_{xy} = -\theta z, \quad \gamma_{xz} = \theta y, \quad \gamma_{yz} = 0 \tag{2}$$

The constitutive relations of Eq. (5.4) reduce to

$$\sigma_x = 0, \quad \sigma_y = 0, \quad \sigma_z = 0, \quad \tau_{yz} = 0, \quad \tau_{xy} = G\gamma_{xy} = -G\theta z,$$
$$\tau_{xz} = G\gamma_{xz} = G\theta y \tag{3}$$

If the body forces are equal to zero, these stress components satisfy the equations of equilibrium of Eq. (1.25).

The condition that must be satisfied on the cylindrical surface is Eq. (5.11):

$$-zn_y + yn_z = 0 \tag{4}$$

since $u_x = \theta\omega = 0$. It follows from Fig. 5.4 that

$$n_y = n\cos\beta = n\frac{y}{a} \qquad n_z = n\sin\beta = n\frac{z}{a} \tag{5}$$

so that (4) is satisfied.

On the ends of the bar, it can be shown that the resultant shear forces of Eq. (5.14) are zero, as required. The resultant moment of Eq. (5.15) is

$$M_x = \iint_A (\tau_{xz}y - \tau_{xy}z)\, dy\, dz = G\theta\int_A (y^2 + z^2)\, dA = G\theta\int_A r^2\, dA \tag{6}$$

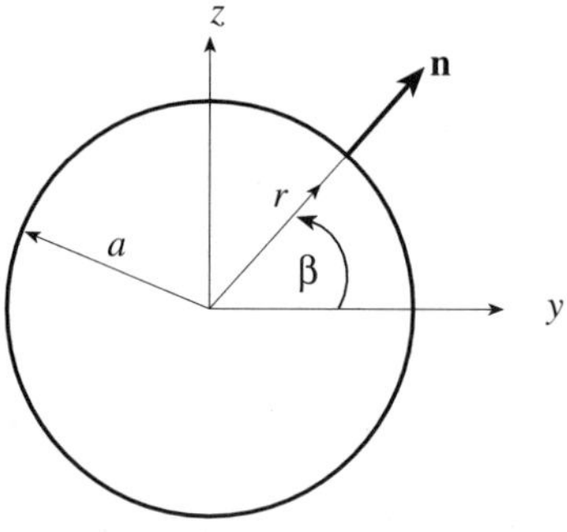

Figure 5.4 Notation for a circular cross section.

where the stresses of (3) have been employed. Also, with $y^2 + z^2 = r^2$, it is observed that for circular cross sections the torsional constant J is the polar moment of inertia, I_x.

The relationship between the angle of twist and the applied torque is (6)

$$\theta = \frac{d\phi}{dx} = \frac{M_x}{GJ} \tag{7}$$

where $J = \int_A r^2\, dA = I_x = I_p = I_z + I_y$. The solution developed here is referred to as the Coulomb solution for a bar of circular cross section subject to torsion.

The shear stresses of (3) should be reasonably accurate at a distance of several bar diameters from the ends where the torques are applied (Saint-Venant's principle). From (3) and (7)

$$\tau_{xy} = -\frac{M_x z}{J} \qquad \tau_{xz} = \frac{M_x y}{J} \tag{8}$$

The resultant shear stress is given by

$$\tau = \sqrt{\tau_{xy}^2 + \tau_{xz}^2} = \frac{M_x}{J}\sqrt{z^2 + y^2} = \frac{M_x r}{J} \tag{9}$$

which is the torsion formula given in elementary strength-of-materials textbooks. The resultant stress τ is perpendicular to the radius r and lies in the plane of the cross section.

The polar moment of inertia for a hollow shaft of inner radius r_i (diameter d_i) and outer radius r_o (diameter d_o) is (Fig. 5.5a)

$$J = \frac{\pi}{2}(r_o^4 - r_i^4) = \frac{\pi}{32}(d_o^4 - d_i^4) \tag{10}$$

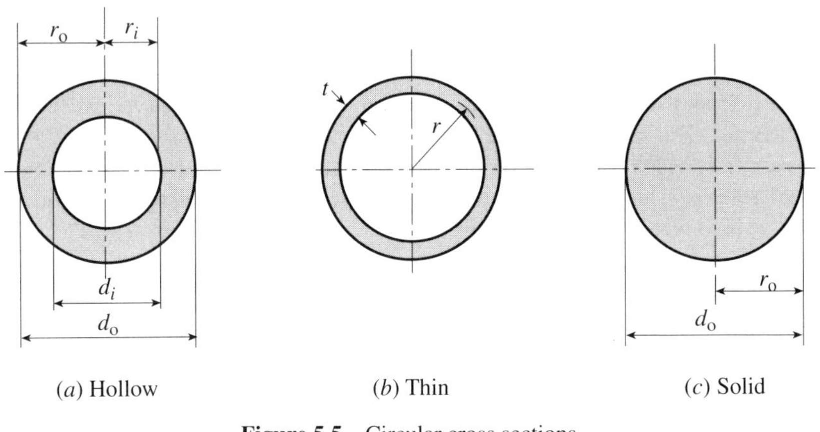

Figure 5.5 Circular cross sections.

For a very thin shaft (Fig. 5.5b) of thickness t, this reduces to the approximate expression

$$J = 2\pi r^3 t \tag{11}$$

For solid shafts (Fig. 5.5c), (10) becomes

$$J = \frac{\pi r_o^4}{2} = \frac{\pi d_o^4}{32} \tag{12}$$

Example 5.2 Torsion When the Warping Function Is Constant. Let the warping function $\omega(y, z)$ be equal to the constant c. Then $\nabla^2 \omega = 0$ of Eq. (5.8) is satisfied by $\omega = c$. The cylindrical surface condition of Eq. (5.11) becomes

$$-zn_y + yn_z = 0 \tag{1}$$

From Eq. (1.36), (1) is

$$-z\frac{dz}{ds} - y\frac{dy}{ds} = 0 \tag{2}$$

so that

$$\frac{d}{ds}\left(\frac{y^2 + z^2}{2}\right) = 0 \tag{3}$$

This implies that $y^2 + z^2 = r^2$ is constant on the surface of the bar. It is concluded that a constant warping function corresponds to the torsion of a bar of circular cross section.

Recall that the axial displacement u_x is proportional to the warping function ω (Eq. 5.2). If $u_x = 0$ at an end of the bar, c is equal to zero and there is no warping along the bar. Thus, the solution here corresponds to the Coulomb solution of Example 5.1 for the torsion of a bar of circular cross section.

It can be shown that the warping function $\omega(y, z)$ can be determined only up to a constant. This is proven by uniqueness studies in such books as Little (1973). However, the shear stresses of Eq. (5.20) are found to be unique since they involve derivatives of ω. This nonuniqueness property of $\omega(y, z)$ must be taken into account if $\omega(y, z)$ is to be employed in a study.

Example 5.3 Elliptical Cross Section—Displacement Formulation. Use the displacement formulation of the torsion equations to find the torsional constant and stresses for a uniform bar of elliptical cross section (Fig. 5.6). Let the warping function be of the form $\omega = Cyz$, where C is a constant. It is shown in Example 5.5 that this is the correct form of the warping function.

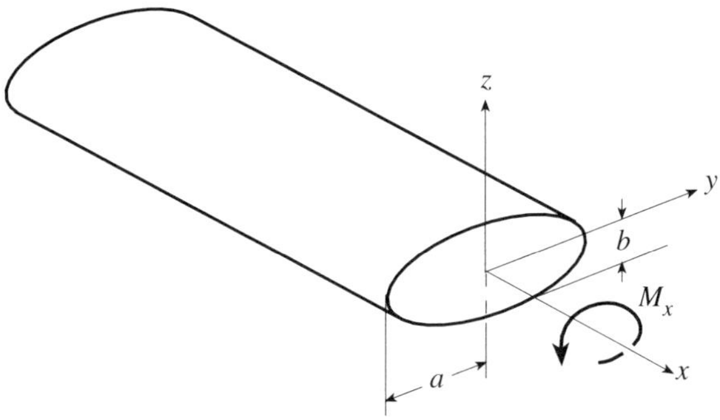

Figure 5.6 Bar of elliptical cross section.

SOLUTION. Note that the assumed ω satisfies Laplace's equation. On the surface of the bar, Eq. (5.11) must be satisfied. Introduce the direction cosines of Eq. (1.36), giving

$$\left(\frac{\partial \omega}{\partial y} - z\right)\frac{dz}{ds} - \left(\frac{\partial \omega}{\partial z} + y\right)\frac{dy}{ds} = 0 \tag{1}$$

From the equation for the ellipse of Fig. 5.6, $y^2/a^2 + z^2/b^2 = 1$, the ratio $\dfrac{dy}{ds}\Big/\dfrac{dz}{ds}$ of direction cosines can be calculated. Equation (1) becomes

$$\left(\frac{\partial \omega}{\partial y} - z\right) yb^2 + \left(\frac{\partial \omega}{\partial z} + y\right) za^2 = 0 \tag{2}$$

Substitution of the assumed warping function $\omega = Cyz$ into this expression gives

$$C = -\frac{a^2 - b^2}{a^2 + b^2} \tag{3}$$

An alternative approach for finding C is to substitute $\omega = Cyz$ into (1) to find

$$(C - 1)z\frac{dz}{ds} - (C + 1)y\frac{dy}{ds} = 0 \tag{4}$$

Integrate, giving

$$y^2\frac{C + 1}{1 - C} + z^2 = \text{constant} \tag{5}$$

Comparison of this to the ellipse equation $y^2b^2/a^2 + z^2 = b^2$ gives (3) again.

The torsional constant J and the shear stresses can be found using Eqs. (5.19) and (5.20):

$$J = \int_A \left[\left(\frac{\partial \omega}{\partial z} + y \right) y - \left(\frac{\partial \omega}{\partial y} - z \right) z \right] dA$$

$$= \frac{2}{a^2 + b^2} \int_A (b^2 y^2 + a^2 z^2)\, dy\, dz = \frac{\pi a^3 b^3}{a^2 + b^2} \tag{6}$$

$$\tau_{xy} = \frac{M_x}{J} \left(\frac{\partial \omega}{\partial y} - z \right) = -\frac{2}{ab^3 \pi} M_x z$$

$$\tau_{xz} = \frac{M_x}{J} \left(\frac{\partial \omega}{\partial z} + y \right) = \frac{2}{a^3 b \pi} M_x y \tag{7}$$

Example 5.4 Rectangular Cross Section. Calculate the displacements at the end of a rod of rectangular cross section as shown in Fig. 5.7a.

SOLUTION. Begin with a warping function of the form

$$\omega = yz + C \sin kz \sinh ky \tag{1}$$

where C and k are constants to be selected such that the boundary conditions are satisfied. Application of the boundary conditions of Eq. (5.11) leads to the warping function

$$\omega = yz + \sum_{n=1,3,5,\ldots}^{\infty} c_n \sin\left(\frac{n\pi}{2b} z \right) \sinh\left(\frac{n\pi}{2b} y \right) \tag{2}$$

where

$$c_n = (-1)^{(n+1)/2} \frac{32b^2}{n^3 \pi^3} \frac{1}{\cosh(n\pi a/2b)}$$

The torsional constant J is obtained from Eq. (5.19):

$$J = \int_A \left(y^2 + z^2 + y \frac{\partial \omega}{\partial z} - z \frac{\partial \omega}{\partial y} \right) dy\, dz \tag{3}$$

Introduce the numerical values $M_x = 100,000$ in.-lb, $a = 6$ in., $b = 4$ in., and assume that the bar is made of aluminum. Then, from Eq. (5.18), $\theta = 2.045 \times 10^{-4}$ rad/in. The displacements u_y and u_z of Eq. (5.1) and u_x of Eq. (5.2) can now be computed. The warping function is plotted in Fig. 5.7b (Reagan, 2002).

5.1.1 Force Formulation

The formulation for Saint-Venant torsion that was just presented is referred to as a *displacement formulation*. This approach resulted in a Laplace's equation expressed in terms of the warping function ω.

A *force formulation* of the Saint-Venant torsion problem can be derived as an alternative to the displacement approach. For the force form, introduce a stress function

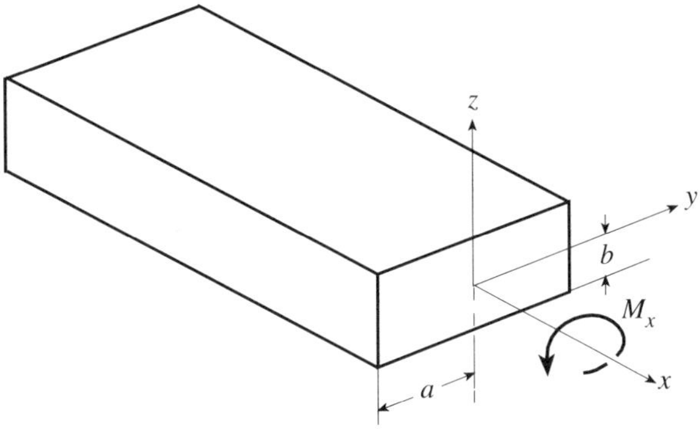

(*a*) Dimensions

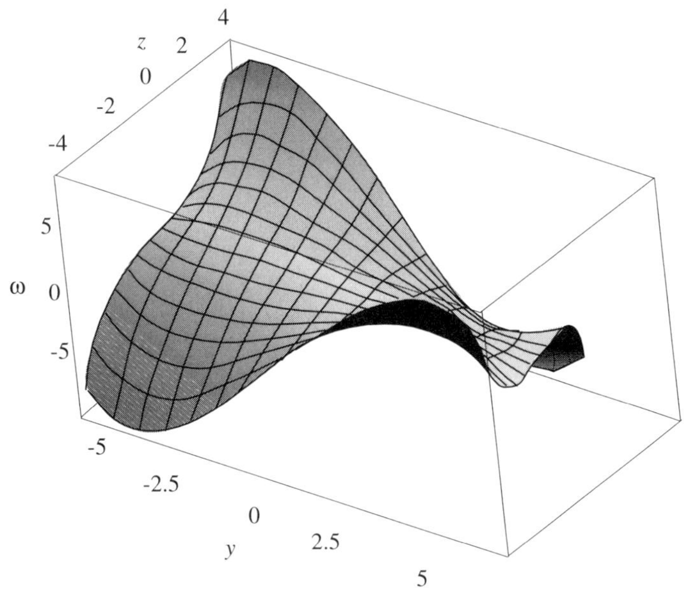

(*b*) Warping function

Figure 5.7 Bar of rectangular cross section.

$\psi(y, z)$ defined as

$$\tau_{xy} = \frac{\partial \psi}{\partial z} \qquad \tau_{xz} = -\frac{\partial \psi}{\partial y} \tag{5.26}$$

The conditions of equilibrium of Eq. (1.25) reduce to

$$\frac{\partial \tau_{xy}}{\partial y} + \frac{\partial \tau_{xz}}{\partial z} = 0 \qquad \frac{\partial \tau_{yx}}{\partial x} = \frac{\partial \tau_{zx}}{\partial x} = 0 \tag{5.27}$$

by setting the body forces equal to zero and noting that (Eq. 5.4) $\sigma_x = 0, \sigma_y = 0, \sigma_z = 0, \tau_{yz} = 0$. The stresses of Eq. (5.26) satisfy these equilibrium conditions. The torsion problem is to find the *Prandtl stress function* ψ. For the strains of Eq. (5.3), the compatibility conditions of Eq. (1.9) reduce to

$$-\frac{\partial^2 \gamma_{xz}}{\partial y^2} + \frac{\partial^2 \gamma_{xy}}{\partial y \partial z} = 0 \qquad \frac{\partial^2 \gamma_{xz}}{\partial z \partial y} - \frac{\partial^2 \gamma_{xy}}{\partial z^2} = 0 \tag{5.28}$$

Since

$$\gamma_{xy} = \frac{1}{G}\frac{\partial \psi}{\partial z} \quad \text{and} \quad \gamma_{xz} = -\frac{1}{G}\frac{\partial \psi}{\partial y}$$

Eq. (5.28) can be expressed as

$$\frac{\partial}{\partial y}\nabla^2 \psi = 0 \qquad \frac{\partial}{\partial z}\nabla^2 \psi = 0 \tag{5.29}$$

where

$$\nabla^2 = \frac{\partial^2}{\partial y^2} + \frac{\partial^2}{\partial z^2}$$

For both expressions of Eq. (5.29) to hold, $\nabla^2 \psi$ must be equal to a constant. To evaluate the constant, note from Eq. (5.26) that

$$\begin{aligned} \frac{\partial^2 \psi}{\partial y^2} &= -\frac{\partial \tau_{xz}}{\partial y} = -G\theta\left(\frac{\partial^2 \omega}{\partial y \partial z} + 1\right) \\ \frac{\partial^2 \psi}{\partial z^2} &= \frac{\partial \tau_{xy}}{\partial z} = G\theta\left(\frac{\partial^2 \omega}{\partial z \partial y} - 1\right) \end{aligned} \tag{5.30}$$

The sum of these expressions shows that

$$\nabla^2 \psi = -2G\theta \tag{5.31}$$

which is a *Poisson's equation*. This is the compatibility equation for torsion of a bar.

From Eq. (5.11), with Eqs. (5.4) and (5.26), the boundary condition becomes

$$\frac{\partial\psi}{\partial y}\frac{dy}{ds} + \frac{\partial\psi}{\partial z}\frac{dz}{ds} = 0 \quad \text{or} \quad \frac{d\psi}{ds} = 0 \tag{5.32}$$

since (Eq. 1.36) $n_y = dz/ds$ and $n_z = -dy/ds$. It follows that ψ is constant along the boundary of the cross section. The magnitude of the constant is arbitrary because the stresses are derivatives of ψ rather than being defined in terms of ψ itself. Without loss of generality, ψ is often taken to be zero along the boundary. That is, the boundary condition that accompanies Eq. (5.31) is that $\psi = 0$ along the cross-section boundary.

From Eq. (5.15), the resultant moment is given by

$$M_x = \iint_A \left(-\frac{\partial\psi}{\partial y}y - \frac{\partial\psi}{\partial z}z\right) dy\, dz \tag{5.33}$$

and, upon integration by parts and the introduction of $\psi = 0$ on the boundary, this relationship reduces to

$$M_x = 2\iint_A \psi\, dy\, dz \tag{5.34}$$

The torsional constant becomes

$$J = \frac{M_x}{G\theta} = \frac{2}{G\theta}\iint_A \psi\, dy\, dz \tag{5.35}$$

It is possible to solve analytically a few problems with simple cross-sectional shapes using these force method relationships. Begin by judiciously assuming the form of the stress function, perhaps a polynomial. Then, choose the coefficients in the assumed stress function so that Poisson's equation (Eq. 5.31) and the boundary condition of $\psi = 0$ is satisfied. This stress function can then be employed to compute the stresses, twisting moment, and the torsional constant from Eqs. (5.26), (5.34), and (5.35). Numerical methods such as those described in this book can be utilized to solve very general problems, with arbitrary cross-sectional shapes.

Example 5.5 Elliptical Cross Section—Force Formulation. Torsional stresses can be calculated for several simple cross-sectional shapes using the force method with the stress functions. The stress functions are established based on the equations that define the boundary of the section. To illustrate this procedure, examine the uniform bar of Example 5.3 that has a cross section of elliptical shape. The equation for the ellipse of Fig. 5.6 is

$$\frac{y^2}{a^2} + \frac{z^2}{b^2} = 1 \tag{1}$$

Suppose that the stress function ψ on the boundary, which must be constant, is chosen to be zero. This requirement is satisfied by the stress function

$$\psi = C\left(\frac{y^2}{a^2} + \frac{z^2}{b^2} - 1\right) \tag{2}$$

where C is a constant to be determined. Substitute ψ of (2) into Poisson's equation of Eq. (5.31), giving

$$2C\left(\frac{1}{a^2} + \frac{1}{b^2}\right) = -2G\theta \tag{3}$$

so that C is equal to

$$C = -\frac{a^2b^2}{a^2 + b^2}G\theta \tag{4}$$

The stress function of (2) is now fully defined.

The shear stress components are obtained by substituting the stress functions in Eq. (5.26):

$$\tau_{xy} = \frac{\partial\psi}{\partial z} = -\frac{2a^2z}{a^2 + b^2}G\theta \qquad \tau_{xz} = -\frac{\partial\psi}{\partial y} = \frac{2b^2y}{a^2 + b^2}G\theta \tag{5}$$

For $b < a$, the maximum shear stress occurs at $z = b$, the boundary closest to the centroid of the ellipse:

$$\tau_{\max} = \tau_{xy}\,|_{z=b} = -\frac{2a^2b}{a^2 + b^2}G\theta \tag{6}$$

Equation (5.34) gives the torque

$$\begin{aligned} M_x &= 2\iint_A \psi\, dy\, dz \\ &= -\frac{2a^2b^2}{a^2 + b^2}G\theta\left(\frac{1}{a^2}\iint_A y^2\, dy\, dz + \frac{1}{b^2}\iint_A z^2\, dy\, dz - \iint_A dy\, dz\right) \\ &= -\frac{2a^2b^2}{a^2 + b^2}G\theta\left(\frac{I_z}{a^2} + \frac{I_y}{b^2} - A\right) \end{aligned} \tag{7}$$

where

$$I_z = \iint_A y^2\, dy\, dz \qquad I_y = \iint_A z^2\, dy\, dz, \tag{8}$$

and $A = \iint_A dy\, dz$. For an ellipse, these formulas lead to $I_z = \pi a^3 b/4$, $I_y = \pi a b^3/4$, and $A = \pi ab$. Place these values in (7):

$$M_x = \frac{\pi a^3 b^3}{a^2 + b^2} G\theta \tag{9}$$

From (5) and (9), the stresses in terms of the twisting moment are

$$\tau_{xy} = -\frac{2}{ab^3\pi} M_x z \qquad \tau_{xz} = \frac{2}{a^3 b\pi} M_x y \tag{10}$$

The resultant shear force per unit area at point y, z is

$$\left(\tau_{xy}^2 + \tau_{xz}^2\right)^{1/2} = \frac{2M_x}{ab\pi}\left(\frac{z^2}{b^4} + \frac{y^2}{a^4}\right)^{1/2} \tag{11}$$

Observe that constant stress contours will be ellipses. From Eq. (5.35)

$$J = \frac{\pi a^3 b^3}{a^2 + b^2} \tag{12}$$

The quantity GJ is referred to as the *torsional rigidity* of the cross section. The same results were found in Example 5.3 using the displacement formulation with the warping function $\omega = Cyz$, where C is a constant. For $a = b$, the expression for the shear stress on an elliptical cross section reduces to the well-known stress values for a circle.

The warping displacement $u_x = \theta\omega$ can be evaluated using Eq. (5.4). From the first relationship,

$$\frac{\partial u_x}{\partial y} = \frac{\tau_{xy}}{G} + \theta z = \frac{M_x}{G}\left(\frac{-2}{ab^3\pi} + \frac{1}{J}\right) z = \frac{M_x}{G\pi a^3 b^3}(b^2 - a^2) z \tag{13}$$

where (10), Eq. (5.18), and (12) have been introduced. Integration of (13) gives

$$u_x = \frac{M_x}{G\pi a^3 b^3}(b^2 - a^2) zy + h(z) \tag{14}$$

From the second relationship of Eq. (5.4),

$$u_x = \frac{M_x}{G\pi a^3 b^3}(b^2 - a^2) zy + g(y) \tag{15}$$

These two expressions are equal only if $h(z)$ and $g(y)$ are equal to the same constant. If there is no warping displacement at the origin $[(y, z) = (0, 0)]$, this constant will be zero. The deformed shape will be asymmetrical with respect to the y and z axes and will exhibit a hyperbolic paraboloid pattern. This warping displacement corresponds to the warping assumed in Example 5.3, where ω was set equal to Cyz, where C is a constant.

Saint-Venant proposed an approximate formula for torsion of a solid section of any cross-sectional shape. He noticed that (9) can be written as

$$\frac{M_x}{\theta} = \frac{\pi a^3 b^3 G}{a^2 + b^2} = \frac{GA^4}{\pi^3 ab(a^2 + b^2)} = \frac{GA^4}{\pi^2 A(a^2 + b^2)} = \frac{GA^4}{4\pi^2 I_p} \approx \frac{GA^4}{40 I_p} \tag{16}$$

where $I_p = I_x = I_y + I_z = (a^2 + b^2)\pi ab/4 = (a^2 + b^2)A/4$. Observe in this relationship, that J of Eq. (5.35) is given by

$$J = \frac{A^4}{40 I_p} \tag{17}$$

Saint-Venant suggested that (16) can be applied to any shape if A and I_p are assigned the values of the area and centroidal polar moment of inertia for the cross section in question. This relationship sometimes gives questionable results if one cross-sectional dimension is much greater than the others.

Example 5.6 Equilateral Triangle Cross Section. The solution procedure of Example 5.5 can be employed for several other simple cross sections. For example, for the equilateral triangular cross section of Fig. 5.8, the solution can begin with the stress function

$$\psi = \frac{G\theta}{2h}\left(y - \sqrt{3}z - \frac{2h}{3}\right)\left(y + \sqrt{3}z - \frac{2h}{3}\right)\left(y + \frac{h}{3}\right) \tag{1}$$

To find M_x or J, the integral

$$\iint_A \psi \, dy \, dz = G\theta \frac{h^4}{30\sqrt{3}} \tag{2}$$

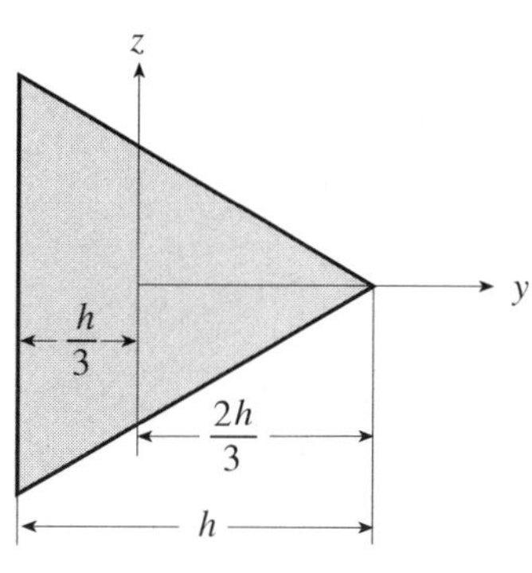

Figure 5.8 Equilateral triangle of height h.

is needed. Then, from Eq. (5.34),

$$M_x = G\theta \frac{h^4}{15\sqrt{3}} \tag{3}$$

and from Eq. (5.35),

$$J = \frac{h^4}{15\sqrt{3}} \tag{4}$$

The shear stress components are found by substituting the stress function into Eq. (5.26):

$$\tau_{xy} = \frac{\partial \psi}{\partial z} = -\frac{M_x}{2hJ}(6yz + 2hz) \qquad \tau_{xz} = -\frac{\partial \psi}{\partial y} = -\frac{M_x}{2hJ}(3y^2 - 2hy - 3z^2) \tag{5}$$

The maximum shear stress occurs at $(-h/3, 0)$, the boundary closest to the centroid of the triangle

$$\tau_{\max} = \tau_{xz} \mid_{(-h/3,0)} = -\frac{M_x h}{2J} \tag{6}$$

5.1.2 Membrane Analogy

Prandtl recognized the resemblance of the governing equations for Saint-Venant torsion of a bar and the equilibrium equations for a flat membrane stretched across an opening of the same shape as the yz plane cross section of the bar. Lateral pressure p causes deflection u in the membrane. If N is the uniform tension per unit length in the membrane, the differential equation of equilibrium for transverse motion is

$$\frac{\partial^2 u}{\partial y^2} + \frac{\partial^2 u}{\partial z^2} = -\frac{p}{N} \tag{5.36}$$

This relationship is the same as Eq. (5.31) if u is substituted for ψ and p/N for $2G\theta$. This analogy is helpful in visualizing the distribution of shear stress components. The volume between the deformed membrane and the yz plane is $\iint_A u \, dy \, dz$, so that by comparison with Eq. (5.34) the twisting moment is equal to twice the volume of the deformed membrane. Contour lines of constant u correspond to lines of constant stress function ψ. It can be shown that the slope of the membrane equals the value of the shear stress. The similarities are referred to as the *membrane analogies*. One use of the membrane analogy is in the comparison of the torques and shear stresses for two different shapes of cross section. Suppose that two membranes are stretched over the two cross sections and the same pressure p and same tension N are applied to each cross section. A comparison of the two stretched membranes permits a comparison of the shear stesses and torques for the two cross sections, provided that the two bars have the same shear modulus G and the same rate of twist θ.

5.2 CLASSICAL FORMULAS FOR THIN-WALLED CROSS SECTIONS

The torsion characteristics of most bars of thin- and thick-walled cross sections are determined quite differently. The basic tenet of being thin-walled is that the thickness of a component section should be small relative to the other cross-sectional dimensions. There is, however, no clearly defined distinction between sections that should be treated as thin and thick. Sometimes the rule of thumb

$$\frac{t_{\max}}{b} \leq 0.1 \tag{5.37}$$

is imposed, where $t_{\max}$ is the maximum thickness of a cross-section element and b is some other cross-sectional dimension. The section is considered to be suitable for analysis by thin-walled analysis methods if the foregoing condition is satisfied. If $t_{\max}/b > 0.1$, the accuracy of a thin-walled analysis theory may be questionable.

Thin-walled sections are *open* if the centerlines of the walls are not closed curves. Angle, I beam, wide-flange, and channel sections, which are formed of narrow rectangular elements, are open sections. Sections are referred to as being *closed* if they have at least one closed curve.

The cross-sectional shear stresses for a bar subjected to a pure torque form a system of self-equilibrating forces, since otherwise there would be a resultant shear force. For closed sections the stresses form a closed-loop continuous pattern around the cross section (Fig. 5.9a). In an open section the shear stress loop occurs within the individual thin sections as shown in Fig. 5.9b. The restriction of the closed-loop

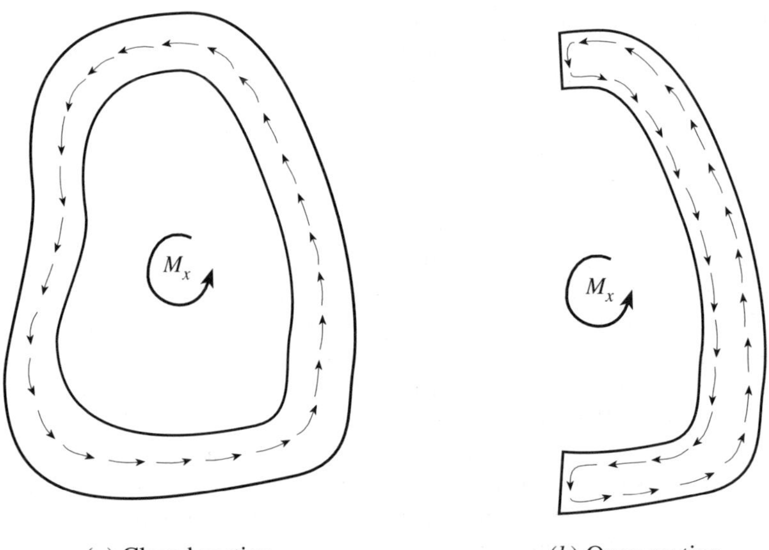

(*a*) Closed section (*b*) Open section

Figure 5.9 Shear stress patterns in closed and open cross sections.

system of shear stresses by the thinness of the walls is the basis of the comparatively low torsional strength of open sections.

5.2.1 Open Sections

We begin the study of the torsion of a bar with a thin-walled open cross section by analyzing a single thin rectangular strip as shown in Fig. 5.10. For this case, assume that the Prandtl stress function ψ does not vary with z, except perhaps at the z edges of the cross section, and is a function of y only. Ignore the decrease in stress at the z edges. Poisson's equation (Eq. 5.31) reduces to

$$\frac{d^2\psi}{dy^2} = -2G\theta \tag{5.38}$$

which is satisfied by the stress function

$$\psi = -G\theta\left(y^2 - \frac{t^2}{4}\right) \tag{5.39}$$

As required by the force formulation of Section 5.1.1, ψ is zero on the boundaries of $y = \pm t/2$. Then (Eq. 5.34)

$$M_x = 2\iint \psi\, dy\, dz = \tfrac{1}{3}bt^3G\frac{d\phi}{dx} \tag{5.40}$$

giving

$$J = \tfrac{1}{3}bt^3 \tag{5.41}$$

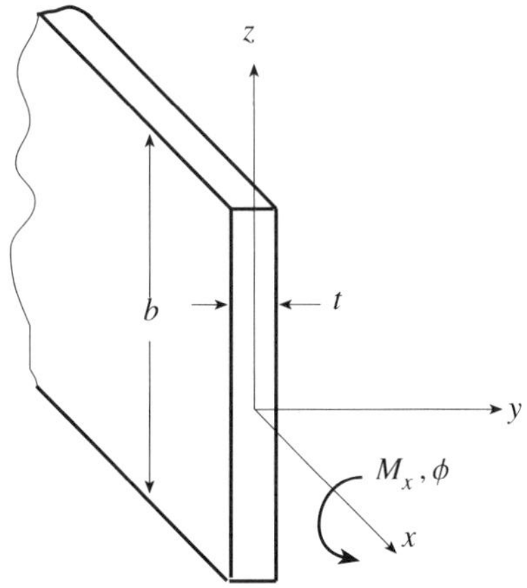

Figure 5.10 Torsion of a thin rectangular strip.

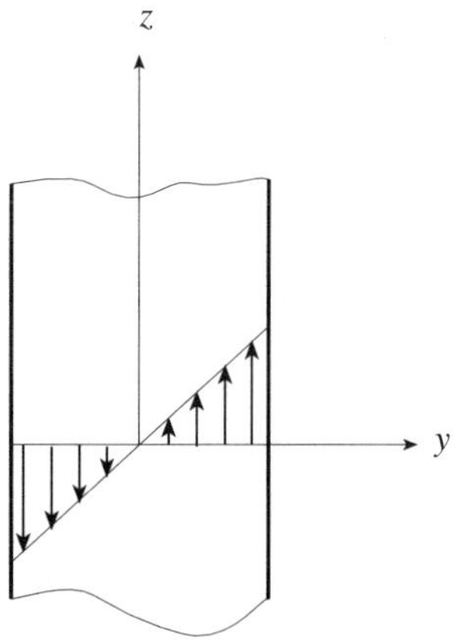

Figure 5.11 Shear stress distribution for an open cross section.

These formulas apply only when t is much smaller than b. The shear stresses of Eq. (5.26) become $\tau_{xy} = 0$, $\tau_{xz} = 2M_x y/J$, so that τ_{xz} is distributed linearly across the thickness as shown in Fig. 5.11. That is, the torsional stress varies linearly over the cross section from zero at the center to a maximum at the outer edge:

$$\tau_{\max} = \tau_{xz}\Big|_{y=\pm t/2} = -\frac{\partial\psi}{\partial y}\Big|_{y=\pm t/2} = \pm tG\frac{d\phi}{dx} = \pm\frac{tM_x}{J} \tag{5.42}$$

In this solution, it is not possible to set the value of the stress function to zero on the shorter edges of the rectangle. Consequently, the stress distribution $\tau_{xz} = 2M_x y/J$ is not valid near the shorter edges, where the boundary conditions require that the stress function be zero. In addition, the torque due to τ_{xz} is one-half the actual torque M_x. This is, in part, because the neglected shear stresses τ_{xy} are concentrated near the shorter edges and have longer moment arms than the stresses τ_{xz}.

For a complex thin-walled open cross section, J is usually taken to be

$$J = \tfrac{1}{3}\sum_{i=1}^{m} b_i t_i^3 \tag{5.43}$$

where m is the number of straight or curved segments of thickness t_i and height b_i that make up the cross section. This formula, which is sometimes referred to as *Saint-Venant's approximation*, follows from Eq. (5.41), being extended to bars for which the cross sections are formed of multisegments. Similar to Eq. (5.42), the maximum shear is given by

$$\tau_{\max} = \frac{t_{\max} M_x}{J} \tag{5.44}$$

Figure 5.12 Bars with open thin-walled cross sections of the same thickness and same total length.

The angle of twist is calculated using

$$\frac{d\phi}{dx} = \frac{M_x}{GJ} \tag{5.45}$$

According to these rather simple formulas, all the cross sections shown in Fig. 5.12 will experience the same angles of twist and stresses for bars undergoing the same twisting moment.

Example 5.7 Comparison of a Thin-Walled Cylinder with and without a Slit. To illustrate the difference in structural characteristics of open and closed sections, compare the stiffness and stresses of a steel tube of circular cross section, 4 in. outside radius and $\frac{1}{4}$ in. thickness, with the stiffness and stresses of the same tube with a longitudinal slit as shown in Fig. 5.13.

SOLUTION. For the torsion of a bar, define stiffness as the ratio of the torque to the angle of twist or to the rate of angle of twist. Thus, the stiffness is given by

$$\frac{M_x}{d\phi/dx} = GJ \tag{1}$$

so that for bars of the same material a comparison of stiffness is a problem of comparing J for each case.

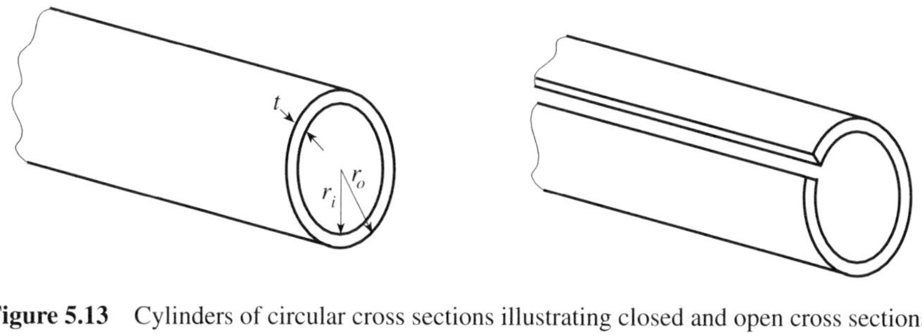

Figure 5.13 Cylinders of circular cross sections illustrating closed and open cross sections.

For a circular cross section, J is the polar moment of inertia I_p (Example 5.1). For a thin-walled tube

$$J = \frac{\pi}{2}(r_o^4 - r_i^4) = \frac{\pi}{2}(4^4 - 3.75^4) = 91.49 \text{ in}^4 \tag{2}$$

For the cylinder with the slit, J is given by Eq. (5.43):

$$J = \tfrac{1}{3}\sum_{i=1}^{m} b_i t_i^3 = \tfrac{1}{3}bt^3 = \tfrac{1}{3}(24.35)\left(\tfrac{1}{4}\right)^3 = 0.13 \text{ in}^4 \tag{3}$$

where the midline circumference (b_i) is $\pi(8 - \frac{1}{4}) = 24.35$ in. The use of the midline, inner circumference, or outer circumference is usually not numerically significant in this calculation, as they are so close to each other in value for a thin-walled section. It follows that the tube is

$$\frac{J_{\text{tube}}}{J_{\text{slit}}} = \frac{91.49}{0.13} = 721.5 \tag{4}$$

times as stiff as the slit cylinder.

The peak stress in the tube is (Example 5.1)

$$\tau_{\max} = \frac{M_x r_o}{J} = M_x \frac{4}{91.49} = 0.0437 M_x \tag{5}$$

and for the cylinder with a slit (Eq. 5.44),

$$\tau_{\max} = \frac{t_{\max} M_x}{J} = \frac{M_x}{4 \cdot 0.13} = 1.92 M_x \tag{6}$$

It can be concluded that the stresses in a cylinder with a slit are substantially higher than those in a tube.

5.2.2 Closed Sections, Hollow Shafts

Simple and reasonably accurate formulas can be obtained for hollow, cylindrical shafts of noncircular cross sections with wall thicknesses that are much smaller than the overall dimensions of the cross section. It is necessary to assume that the torsional stress τ is uniformly distributed across the thickness of these closed sections. Experiments and comparisons with more exact analyses show that this assumption is reasonable for most hollow thin-walled cross sections in the elastic range. It is convenient to replace stress (force per length^2) by force per unit length along the wall. Multiply the stress, which is assumed not to vary through the thickness, by the thickness to obtain a quantity q referred to as the *shear flow* (Fig. 5.14); that is,

$$q = \tau t \tag{5.46}$$

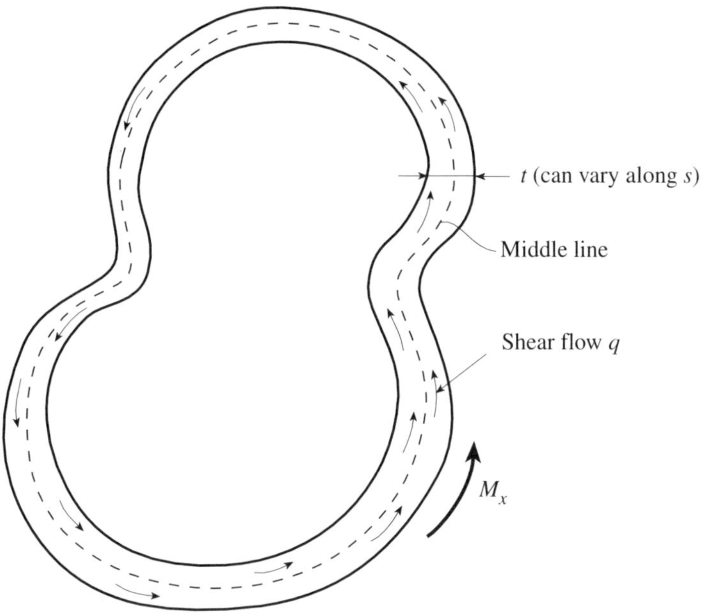

Figure 5.14 Shear flow in a thin-walled hollow shaft.

Formula Derivation Refer to the thin-walled tube of Fig. 5.15. Consider first the condition of equilibrium of the forces in the x direction for the element shown in Fig. 5.15a, $\sum F_x = 0$, $\tau_1 t_1\ \Delta x - \tau_2 t_2\ \Delta x = 0$, or $\tau_1 t_1 = \tau_2 t_2$. The product of shear stress and wall thickness (i.e., the shear flow q, is constant on any such planes). Equilibrium of moments shows that the shear stresses on the adjoining corners of an element (Fig. 5.15b) are equal (i.e., $\tau_1 = \tau_4$ and $\tau_2 = \tau_3$). Then it follows that $\tau_4 t_1 = \tau_4 t_2 = q$. At the other two corners, $\tau_1 = \tau_3$ and $\tau_2 = \tau_4$ or $\tau_3 t_1 = \tau_3 t_2 = q$. It can be concluded that q is constant around the cross section of a section perpendicular to the x axis. The words *shear flow* for q are employed because of the analogy between q and the flow of a constant quantity of water around a closed channel.

The shear flow can be shown to be related to the applied torque. Let r of Fig. 5.15c be the lever arm of the force $q\ \Delta s$ with respect to some convenient point O. Note that the element of length Δs contributes $rq\ \Delta s$ to the total moment. If Δs is small, $r\ \Delta s \approx 2\ \Delta A^*$. It follows from the configuration in Fig. 5.15c that $\Delta M_x = qr\ \Delta s = q2\ \Delta A^*$. Sum the torque contributions of all the elements to find the total torque. Thus, $M_x = 2qA^*$, where A^* is the total enclosed area. In terms of the stress $\tau = q/t$, this relationship becomes

$$\tau = \frac{M_x}{2A^* t} \tag{5.47}$$

It follows that the maximum stress occurs at the location of the thinnest section.

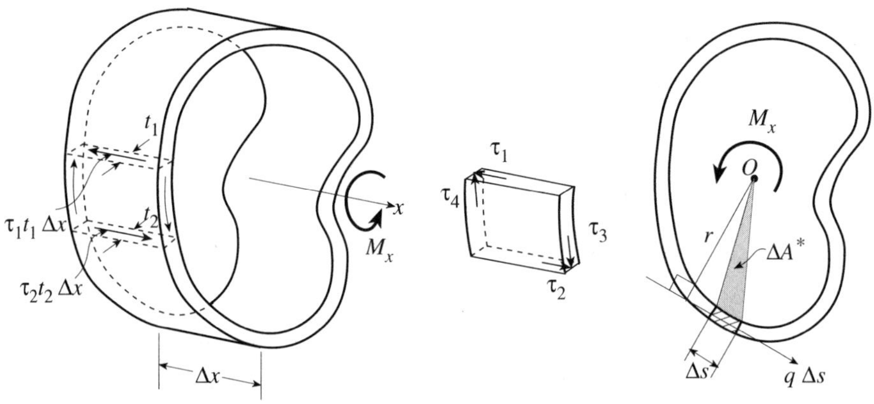

Figure 5.15 Torsion of a thin-walled hollow cylinder.

To find the angle of twist of a thin-walled hollow section, set the work done by an externally applied torque $(M_x/2)\,(d\phi/dx)$ equal to the internal (strain) energy stored in the bar (per unit length) $[\int_V (\tau\gamma/2)\,dV]$:

$$\frac{M_x}{2}\frac{d\phi}{dx} = \int_V \frac{\tau\gamma}{2}\,dV \tag{5.48}$$

where V is the volume per unit length, τ the torsional stress, and γ the corresponding strain. Expressions for work and strain energy are given in many solid mechanics books (e.g., in Pilkey and Wunderlich, 1994). Substitute into Eq. (5.48) Hooke's law $(\gamma = \tau/G)$, $\Delta V = t\,\Delta s$, and Eq. (5.47):

$$\frac{d\phi}{dx} = \frac{M_x}{4A^{*2}G}\oint \frac{ds}{t} \tag{5.49}$$

where the integration is taken completely around the profile of the hollow cross section.

In summary, the torsional stress and displacement equations for a thin-walled hollow bar are:

Stress or shear flow:

$$q = \frac{M_x}{2A^*} \qquad \tau = \frac{M_x}{2A^*t} \qquad q = \tau t \tag{5.50}$$

where A^* is the area enclosed by the middle line of the wall.

Angle of twist:

$$\frac{d\phi}{dx} = \frac{M_x}{GJ} \qquad J = \frac{4A^{*2}}{\oint (1/t)\,ds} = \frac{4A^{*2}}{\int_0^S (1/t)\,ds} \tag{5.51}$$

where s is the perimeter coordinate and S is the length of the middle line of the wall.

The expression for J of Eq. (5.51) is referred to as *Bredt's formula* (Bredt, 1896), as is the expression for τ of Eq. (5.50).

In the case of a bar of length L,

$$\phi = \frac{M_x L}{GJ} \tag{5.52}$$

If t is constant,

$$J = \frac{4A^{*2}t}{S} \tag{5.53}$$

and

$$\frac{d\phi}{dx} = \frac{M_x S}{4A^{*2}Gt} = \frac{\tau S}{2A^*G} \tag{5.54}$$

Variable Wall Thickness For a particular torque M_x, $q = M_x/2A^*$ represents the shear flow, even if the wall thickness varies. The maximum shear stress occurs where the wall thickness is thinnest, that is, $\tau_{\text{max}} = q/t_{\text{min}}$. Suppose that a hollow section has a perimeter formed of lengths $S_i (i = 1, 2, \ldots)$ with constant thicknesses t_i $(i = 1, 2, \ldots)$ and moduli G_i $(i = 1, 2, \ldots)$, respectively. Superimpose Eq. (5.54) for each segment:

$$\frac{d\phi}{dx} = \frac{M_x}{4A^{*2}}\left(\frac{S_1}{t_1G_1} + \frac{S_2}{t_2G_2} + \cdots\right) \tag{5.55}$$

Special Observations It follows from Eq. (5.50) that although the shear flow q is constant around the wall, the shear stress $\tau = q/t$ can vary because t can vary. The highest shear stress occurs where the wall is thinnest, that is, where t is a minimum. Note that there is no distinction between different section shapes, in the sense that according to Eq. (5.50), for the same torque M_x the shear flow q will be the same for all cross-sectional geometries with the same enclosed area A^*. Because the wall thickness is very small relative to the other dimensions of the cross section, normally there is no harm in using the outer or inner dimensions of the cross section in calculating the enclosed area if it is difficult to compute the area enclosed by the middle line of the cross section. There is the possibility of buckling when dealing with thin-walled structures. Thus, a hollow cylinder for which the stresses are below allowable stress levels may not be safe from the standpoint of stability. In the case of a circular cylinder of mild steel, buckling can occur at a normal allowable stress level for a thickness of about 1.5% of the radius.

Example 5.8 Accuracy of Bredt's Formula for Hollow Thin-Walled Shafts. Compare the stresses predicted by the exact theory for the torsion of circular shafts (Example 5.1) with those of the approximate Bredt's theory of thin-walled hollow tubes (Eq. 5.50).

SOLUTION. Begin with a thin-walled cylinder of circular cross section with inside radius r_i and outside radius r_o. For Bredt's hollow section formula (Eq. 5.50), $\tau = M_x/2A^*t$, with $t = r_o - r_i$, use an enclosed area of $A^* = \pi\,[(r_o + r_i)/2]^2$, since the midwall radius is $(r_o + r_i)/2$. This gives

$$\tau_{\text{approx}} = \frac{M_x}{2A^*t} = \frac{M_x}{2\pi\,[(r_o + r_i)/2]^2\,(r_o - r_i)} = \frac{M_x}{(\pi/2)(r_o + r_i)^2(r_o - r_i)} \tag{1}$$

For the exact theory, use Eq. (9) of Example 5.1 with $r = r_o$ and $J = I_p$, the polar moment of inertia.

$$\tau_{\max} = \frac{M_x r_o}{J} = \frac{M_x r_o}{(\pi/2)(r_o^4 - r_i^4)} = \frac{M_x r_o}{(\pi/2)(r_o^2 + r_i^2)(r_o - r_i)(r_o + r_i)} \tag{2}$$

To compare the two theories, study the ratio

$$\frac{\tau_{\text{approx}} - \tau_{\max}}{\tau_{\max}} = -\left(\frac{r_i}{r_o}\right)\frac{1 - r_i/r_o}{1 + r_i/r_o} \tag{3}$$

For $r_i/r_o = 0.95$, (3) shows that the difference between the maximum stresses of the two theories is $2\frac{1}{2}\%$. As expected, with an increase in wall thickness of the tube, the percent difference increases. Thus, the difference increases to 11% for $r_i/r_o = 0.75$.

For the tube of Example 5.7, with $r_o = 4$ in. and $t = \frac{1}{4}$ in., the difference is 3%.

Example 5.9 Thin-Walled Cylinder of Circular Cross Section. Apply Eq. (5.50) to find the torsional stresses in a hollow thin-walled circular section. Show that this gives the same formula obtained by equilibrium applied to a thin-walled cylinder of circular cross section with axial torque.

SOLUTION. Consider a thin-walled cylinder of circular cross section, of constant thickness t and midline radius r with axial torque M_x. Assume that the cylinder is

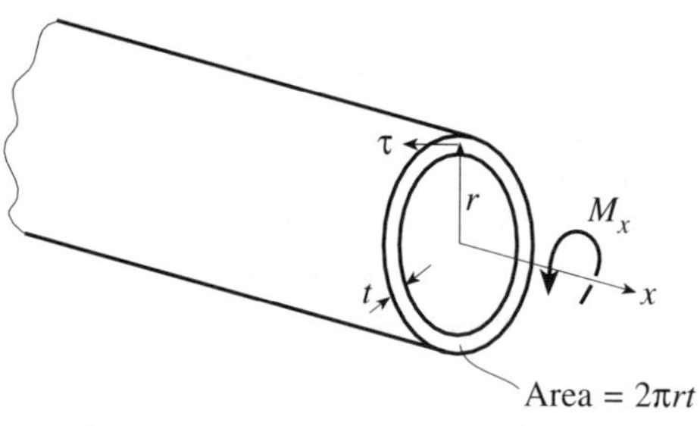

Figure 5.16 Thin-walled hollow cylinder of circular cross section.

so thin that the difference between r and the inner and outer radii does not have a significant effect on the solution. Also assume that the torsional stress is uniformly distributed throughout the wall thickness. Summation of moments about the x axis (Fig. 5.16) gives $\tau 2\pi rt \cdot r = M_x$ or $\tau = M_x/2\pi r^2 t$. From Eq. (5.50), $\tau = M_x/2A^*t$. For the circular cylinder, $A^* = \pi r^2$ so that $\tau = M_x/2\pi r^2 t$. The two formulas are the same.

Example 5.10 Torque Capacity of a Thin-Walled Tube. Find the allowable torque of the thin-walled tube of Fig. 5.17 if the average shear stress is not to exceed τ_{allow}. Also, find the angle of twist.

SOLUTION. The stress τ is given by Bredt's formula of Eq. (5.50). The area enclosed by the midthickness line is

$$A^* = \pi(r)^2 + 2(r)2r = (\pi + 4)r^2 = 7.14r^2 \tag{1}$$

Give τ its design value of τ_{allow} and calculate the allowable torque (Eq. 5.50):

$$M_x = 2A^*q = 2A^*t\tau_{\text{allow}} \tag{2}$$

With $S = 2\pi(r) + 2(2r) = (2\pi + 4)r = 10.28r$, the angle of twist per unit length for this torque is (Eq. 5.54)

$$\frac{d\phi}{dx} = \frac{M_x S}{4A^{*2}Gt} = \frac{M_x(0.05)}{Gtr^3} \tag{3}$$

Example 5.11 Stiffness of Open and Closed Cross Sections. Suppose that the open and closed cross sections shown in Fig. 5.18 have the same cross-sectional areas. Which one is stiffer if b is much greater than t, say $b \geq 10t$?

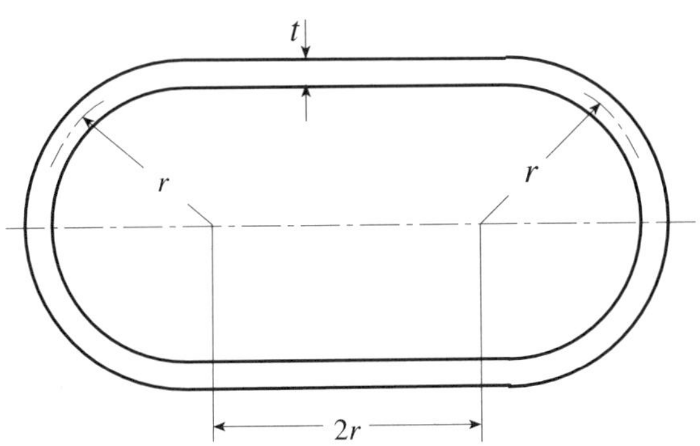

Figure 5.17 Tube cross section with semicircular ends.

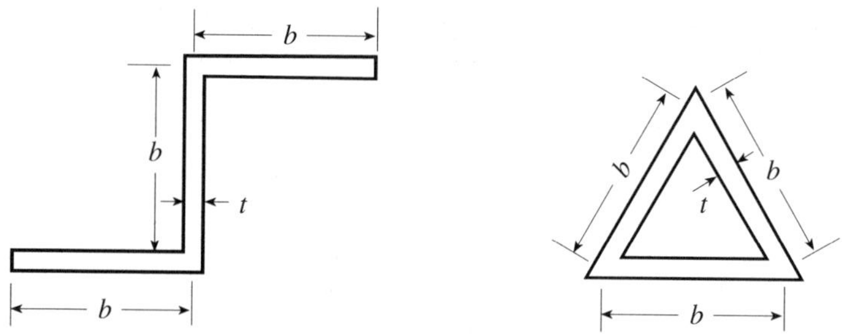

Figure 5.18 Open and closed sections of the same cross-sectional area.

SOLUTION. For the thin-walled hollow triangle section, the stiffness (ratio of torque to rate of angle of twist) is

$$\frac{M_x}{d\phi/dx} = GJ \tag{1}$$

with Eq. (5.53)

$$J = \frac{4A^{*2}t}{S} \qquad S = 3b \qquad A^* = \frac{1}{2}\left(\frac{b}{2}\right)\left(\frac{\sqrt{3}}{2}b\right) = \frac{\sqrt{3}}{8}b^2 \tag{2}$$

This gives

$$J = \tfrac{1}{16}b^3 t \tag{3}$$

so that the stiffness becomes

$$\frac{M_x}{d\phi/dx} = \frac{Gb^3t}{16} \tag{4}$$

For the open-angle section (Eq. 5.43)

$$J = \tfrac{1}{3}\sum_{i=1}^{3} b_i t_i^3 = bt^3 \tag{5}$$

giving the stiffness

$$\frac{M_x}{d\phi/dx} = Gbt^3 \tag{6}$$

If $b \geq 10t$, the closed-section bar is more than six times stiffer than the open-section bar.

The formulas for open and closed sections can be combined so that cross sections such as that of Fig. 5.19 can be analyzed. The torsional constant for this hollow tube

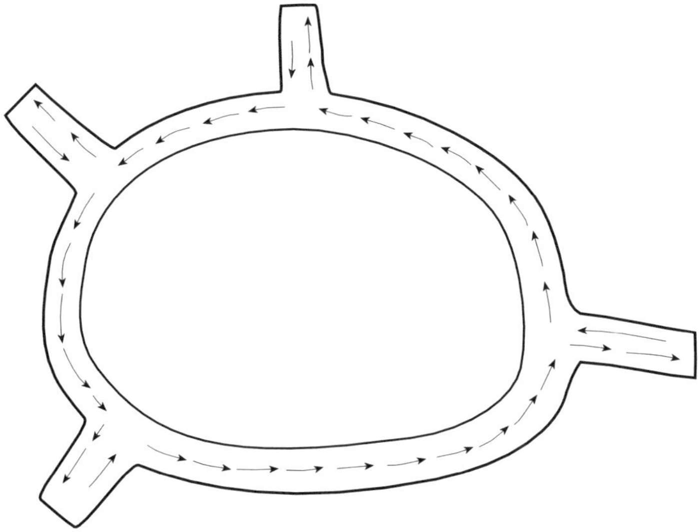

Figure 5.19 Hollow section with fins.

with fins is the sum of Eqs. (5.43) and (5.51):

$$J = \frac{1}{3}\sum_{i=1}^{m} b_i t_i^3 + \frac{4A^{*2}}{\oint (1/t)\, ds} \tag{7}$$

The stresses in the fins and hollow section are given by Eqs. (5.42) and (5.50), respectively.

Multicell Sections The formulas of this section can be extended to apply to an n-celled tube of the sort shown in Fig. 5.20. As shown in the figure, suppose that the

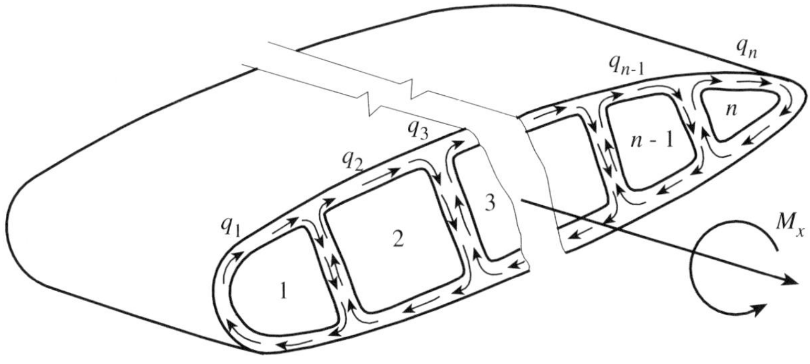

Figure 5.20 Multicell section.

section is formed of cells with independent shear flows. The equation of equilibrium for this multicell section is obtained by summation of the formulas for single cells of Eq. (5.50):

$$M_x = 2\sum_{i=1}^{n} q_i A_i^* \tag{5.56}$$

where q_i is the shear flow in the ith cell and A_i^* is the enclosed area of the ith cell. Since the equilibrium condition of Eq. (5.56) is not sufficient to determine the n unknown shear flows, this is a statically indeterminate problem. To solve this problem, turn to the deformation relation of Eq. (5.51).

Suppose that there are three cells, say 1, 2, 3. For cell 2 acting by itself, Eq. (5.51) would be

$$\frac{d\phi}{dx} = \frac{M_x}{4A_2^{*2}G}\oint_{S_2}\frac{ds}{t} = \frac{q_2}{2A_2^*G}\oint_{S_2}\frac{ds}{t} \tag{5.57}$$

where S_2 indicates integration around cell 2. However, the shear flow q_1 and q_3 that occur in adjoining cells 1 and 3 must be taken into account. Segments that belong to two adjoining cells are referred to as *webs*. The shear flow in the first web should be reduced to $q_2 - q_1$ and that in the second web to $q_2 - q_3$. Then Eq. (5.57) can be expressed as

$$\frac{d\phi}{dx} = \frac{1}{2A_2^*G}\left(q_2\oint_{S_2}\frac{ds}{t} - q_1\int_{S_{12}}\frac{ds}{t} - q_3\int_{S_{23}}\frac{ds}{t}\right) \tag{5.58}$$

where the integration in the latter two integrals is along the first and second webs of lengths S_{12} and S_{23}, respectively. Equation (5.58) can be extended to represent the ith cell:

$$\frac{d\phi}{dx} = \frac{1}{2A_i^*G}\left[q_i\oint_{S_i}\frac{ds}{t} - \sum_{\substack{j=1\\(j\neq i)}}^{m}\left(q_j\int_{S_{ij}}\frac{ds}{t}\right)\right] \qquad i = 1, 2, \ldots, n \tag{5.59}$$

where the ith cell is assumed to be bounded by m cells rather than by two. Equations (5.59) and (5.56) are $n+1$ equations for solving for the $n+1$ unknowns, $d\phi/dx$ and $q_1, q_2, \ldots, q_n$.

Example 5.12 ***Wing Section.*** Calculate the shear flows and the rate of angle of twist for the wing section of Fig. 5.21 if the torque of 200,000 in-lb is applied. Let $G = 3.8 \cdot 10^6$ psi.

SOLUTION. Assume that cell 1 is half of an ellipse with the semimajor axis = 8 in. and semiminor axis = 4 in. so that $A_1^* = \pi(8)(4)/2 = 50.3$ in^2. The areas of the other cells are $A_2^* = A_3^* = 128$ in^2. To approximate the length of the half ellipse, use

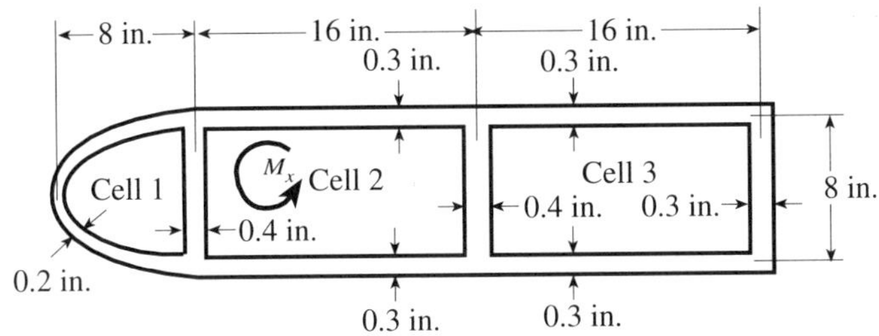

Figure 5.21 Multicell section.

$$S_1 = \pi\sqrt{\frac{a^2 + b^2}{2}} \tag{1}$$

With $a = 8$ and $b = 4$, (1) gives $S_1 = 19.9$ in. From Eqs. (5.56) and (5.59),

$$\begin{aligned}
M_x &= 2\left[q_1(50.3) + q_2(128) + q_3(128)\right] \\
\frac{d\phi}{dx} &= \frac{1}{2GA_1^*}\left[q_1\left(\frac{19.9}{0.2} + \frac{8}{0.4}\right) - q_2\frac{8}{0.4}\right] \\
\frac{d\phi}{dx} &= \frac{1}{2GA_2^*}\left[q_2\left(\frac{2\cdot 8}{0.4} + \frac{2\cdot 16}{0.3}\right) - q_1\frac{8}{0.4} - q_3\frac{8}{0.4}\right] \\
\frac{d\phi}{dx} &= \frac{1}{2GA_3^*}\left[q_3\left(\frac{2\cdot 8}{0.4} + \frac{2\cdot 16}{0.3}\right) - q_2\frac{8}{0.4}\right]
\end{aligned} \tag{2}$$

Solve these four equations for the four unknowns $d\phi/dx$, q_1, q_2, q_3 with $M_x = 200,000$ in-lb. This gives

$$\begin{aligned}
\tfrac{d\phi}{dx} &= 4.367\cdot 10^{-5} \text{ rad/in.} \\
q_1 &= 200.50 \text{ lb/in.} \\
q_2 &= 363.26 \text{ lb/in.} \\
q_3 &= 339.20 \text{ lb/in.}
\end{aligned} \tag{3}$$

5.3 COMPOSITE CROSS SECTIONS

The use of modulus-weighted properties is described in Chapter 1 for the flexural analysis of composite, or nonhomogeneous, cross sections. The elastic modulus was assumed to take on different values for different parts of the cross section, making E a function of position, $E = E(y, z)$. A reference elastic modulus E_r is chosen and modulus-weighted differential area defined (Eq. 1.64), $d\widetilde{A} = (E/E_r)\, dA$. For a finite element mesh, with E_e as the elastic modulus for element e, assign modulus weights

$$\lambda_e = \frac{E_e}{E_r} \tag{5.60}$$

The composite cross section with elementary areas dA is transformed to an equivalent cross section with the elementary areas

$$d\widetilde{A} = \lambda_e \, dA = \frac{E_e}{E_r} \, dA \tag{5.61}$$

The analysis of the transformed cross section gives stresses that are multiplied by the factor λ_e of the corresponding material to determine the actual normal stresses. The formula for these stresses, which are due to an axial load N_x and bending moments M_y, M_z, is given in Eq. (1.101).

When calculating the Saint-Venant torsional stresses for a nonhomogeneous cross section, the shear modulus is considered to be a function $G(y, z)$ of the coordinates y, z of the material points belonging to the section. The expression of Eq. (5.17) for the section torque becomes

$$M_x = \theta \int G \left[\left(\frac{\partial \omega}{\partial z} + y \right) y - \left(\frac{\partial \omega}{\partial y} - z \right) z \right] dA \tag{5.62}$$

Define a modulus-weighted differential element of area dA^* by

$$dA^* = \frac{G}{G_r} \, dA \tag{5.63}$$

where G_r is the shear modulus of the reference material. Substitute this expression into Eq. (5.62), so that the torque M_x may be rewritten as

$$M_x = \theta G_r J^* \tag{5.64}$$

where the modulus-weighted torsional constant is defined by (Eq. 5.19)

$$J^* = \int \left[\left(\frac{\partial \omega}{\partial z} + y \right) y - \left(\frac{\partial \omega}{\partial y} - z \right) z \right] dA^* \tag{5.65}$$

The nonzero shear stresses due to torsion are given by (Eqs. 5.4 and 5.64)

$$\begin{aligned} \tau_{xy} &= G\theta \left(\frac{\partial \omega}{\partial y} - z \right) = \frac{G}{G_r} \frac{M_x}{J^*} \left(\frac{\partial \omega}{\partial y} - z \right) \\ \tau_{xz} &= G\theta \left(\frac{\partial \omega}{\partial z} + y \right) = \frac{G}{G_r} \frac{M_x}{J^*} \left(\frac{\partial \omega}{\partial z} + y \right) \end{aligned} \tag{5.66}$$

The modulus weight ratio for G in terms of Poisson's ratios is (Eq. 1.13)

$$\frac{G}{G_r} = \frac{1+\nu_r}{1+\nu}\frac{E}{E_r} \tag{5.67}$$

where ν_r is the Poisson's ratio for the reference material. From Eqs. (5.63), (5.67), and (5.60) for element e in a finite element mesh, the transformed area element is

$$dA^* = \frac{G_e}{G_r}\,dA = \frac{1+\nu_r}{1+\nu_e}\frac{E_e}{E_r}\,dA = \frac{1+\nu_r}{1+\nu_e}\lambda_e\,dA \tag{5.68}$$

Then the shear stresses of Eq. (5.66) are

$$\begin{aligned} \tau_{xy} &= \lambda_e \frac{1+\nu_r}{1+\nu_e}\frac{M_x}{J^*}\left(\frac{\partial\omega}{\partial y} - z\right) \\ \tau_{xz} &= \lambda_e \frac{1+\nu_r}{1+\nu_e}\frac{M_x}{J^*}\left(\frac{\partial\omega}{\partial z} + y\right) \end{aligned} \tag{5.69}$$

In conjunction with the preceding development, it should be noted that if the Poisson's ratio is variable, the basic contention that

$$\sigma_y = \sigma_z = \tau_{yz} = 0 \tag{5.70}$$

ceases to be applicable. For two materials with very different elastic moduli, it is not unusual to find that the difference in the Poisson's ratios is very small, as in the case of copper and aluminum. If it is assumed that the different materials making up the section all have the same Poisson's ratio, the modulus weighted elastic and shear moduli have the same value:

$$\frac{G_e}{G_r} = \frac{E_e}{E_r} = \lambda_e \tag{5.71}$$

for all elements e. In addition, the validity of the assumption that σ_y, σ_z, and τ_{yz} are negligible is preserved. The modulus-weighted torsional constant is calculated with the same differential area (Eq. 5.61) as the section properties $\widetilde{I}_y$, $\widetilde{I}_z$, $\widetilde{I}_{yz}$ defined in Eq. (1.66):

$$J^* = \widetilde{J} = \int \left[\left(\frac{\partial\omega}{\partial z} + y\right)y - \left(\frac{\partial\omega}{\partial y} - z\right)z\right] d\tilde{A} \tag{5.72}$$

The formulas for the nonzero shear stresses simplify to

$$\begin{aligned} \tau_{xy} &= \lambda_e \frac{M_x}{\widetilde{J}}\left(\frac{\partial\omega}{\partial y} - z\right) \\ \tau_{xz} &= \lambda_e \frac{M_x}{\widetilde{J}}\left(\frac{\partial\omega}{\partial z} + y\right) \end{aligned} \tag{5.73}$$

5.4 STIFFNESS MATRICES

Stiffness matrices for the cross-sectional characteristics of bars subject to torsion will be studied using the principle of virtual work. As an alternative formulation, Galerkin's weighted residual method is introduced as a means of finding the stiffness matrix.

5.4.1 Principle of Virtual Work

From Section 2.1.7 the principle of virtual work appears as

$$\delta W = \delta W_{\text{int}} + \delta W_{\text{ext}} = 0 \tag{5.74}$$

For torsion this can be expressed as

$$\delta W = \delta W_{\text{int}} + \delta W_{\text{ext}} = -\int_V \delta \boldsymbol{\epsilon}^{\mathrm{T}} \boldsymbol{\sigma}\, dV + \int_L \overline{m}_x \,\delta\phi\, dx \tag{5.75}$$

or

$$-\delta W_{\text{int}} - \delta W_{\text{ext}} = \int_V \delta \boldsymbol{\epsilon}^{\mathrm{T}} \boldsymbol{\sigma}\, dV - \int_L \overline{m}_x \,\delta\phi\, dx \tag{5.76}$$

where ϕ is the angle of twist and $\overline{m}_x$ is the applied moment per unit length.

The warping function ω and the angle of twist ϕ are the essential displacements. The stresses in Eq. (5.76) can be expressed as (Eq. 5.4)

$$\begin{bmatrix} \tau_{xy} \\ \tau_{xz} \end{bmatrix} = \begin{bmatrix} G & 0 \\ 0 & G \end{bmatrix} \begin{bmatrix} \gamma_{xy} \\ \gamma_{xz} \end{bmatrix} \qquad \boldsymbol{\sigma} = \mathbf{E}\ \boldsymbol{\epsilon} \tag{5.77}$$

The strains are given by (Eq. 5.3)

$$\boldsymbol{\epsilon} = \begin{bmatrix} \gamma_{xy} \\ \gamma_{xz} \end{bmatrix} = \theta \begin{bmatrix} -z + \dfrac{\partial \omega}{\partial y} \\ y + \dfrac{\partial \omega}{\partial z} \end{bmatrix} = \frac{\partial \phi}{\partial x} \begin{bmatrix} -z + \dfrac{\partial \omega}{\partial y} \\ y + \dfrac{\partial \omega}{\partial z} \end{bmatrix} \tag{5.78}$$

Combine these two expressions to form

$$\boldsymbol{\sigma} = G\phi' \begin{bmatrix} -z + \dfrac{\partial \omega}{\partial y} \\ y + \dfrac{\partial \omega}{\partial z} \end{bmatrix} \tag{5.79}$$

there $\phi' = \partial\phi/\partial x$.

From Eq. (5.78), the variation of the strains would be

$$\delta \boldsymbol{\epsilon} = G\,\delta\phi' \begin{bmatrix} -z + \dfrac{\partial \omega}{\partial y} \\ y + \dfrac{\partial \omega}{\partial z} \end{bmatrix} + G\phi' \begin{bmatrix} \delta \dfrac{\partial \omega}{\partial y} \\ \delta \dfrac{\partial \omega}{\partial z} \end{bmatrix} \tag{5.80}$$

or

$$\delta \boldsymbol{\epsilon} = \begin{bmatrix} \left(\dfrac{\partial \omega}{\partial y} - z\right) \dfrac{\partial}{\partial x}\delta\phi + \phi' \dfrac{\partial}{\partial y}\delta\omega \\ \left(\dfrac{\partial \omega}{\partial z} + y\right) \dfrac{\partial}{\partial x}\delta\phi + \phi' \dfrac{\partial}{\partial z}\delta\omega \end{bmatrix} \tag{5.81}$$

and

$$\delta \boldsymbol{\epsilon}^{\mathrm{T}} = \left[\left(\frac{\partial \omega}{\partial y} - z\right) \frac{\partial}{\partial x}\delta\phi + \phi' \frac{\partial}{\partial y}\delta\omega \quad \left(\frac{\partial \omega}{\partial z} + y\right) \frac{\partial}{\partial x}\delta\phi + \phi' \frac{\partial}{\partial z}\delta\omega \right] \tag{5.82}$$

With Eqs. (5.82) and (5.79), the principle of virtual work expression of Eq. (5.76) becomes

$$\begin{aligned} &\int \delta \boldsymbol{\epsilon}^{\mathrm{T}} \boldsymbol{\sigma}\, dV - \int_L \overline{m}_x\, \delta\phi\, dx \\ &\quad = \int_V G\Big[\Big\{\Big(\frac{\partial \omega}{\partial y} - z\Big)\frac{\partial}{\partial x}\,\delta\phi + \phi' \frac{\partial}{\partial y}\,\delta\omega\Big\}\phi'\Big(-z + \frac{\partial \omega}{\partial y}\Big) \\ &\qquad + \Big\{\Big(\frac{\partial \omega}{\partial x} + y\Big)\frac{\partial}{\partial x}\,\delta\phi + \phi' \frac{\partial}{\partial z}\,\delta\omega\Big\}\phi'\Big(y + \frac{\partial \omega}{\partial z}\Big)\Big]\, dV - \int_L \overline{m}_x\, \delta\phi\, dx \\ &\quad = 0 \end{aligned} \tag{5.83}$$

Separate this into two expressions equal to zero, one with the $\delta\phi$ terms and the other $\delta\omega$ terms. In the case of the $\delta\phi$ terms,

$$\begin{aligned} &\int \frac{\partial}{\partial x}\,\delta\phi \left[\left(\frac{\partial \omega}{\partial y} - z\right)\left(-z + \frac{\partial \omega}{\partial y}\right) + \left(\frac{\partial \omega}{\partial z} + y\right)\left(y + \frac{\partial \omega}{\partial z}\right)\right] G\phi'\, dV \\ &\qquad - \int_L \overline{m}_x\, \delta\phi\, dx \end{aligned} \tag{5.84}$$

Simplify this expression to

$$\begin{aligned} &\int_L \frac{\partial}{\partial x}\delta\phi \int_A \left\{\left[\left(\frac{\partial \omega}{\partial y}\right)^2 + \left(\frac{\partial \omega}{\partial z}\right)^2 - z\frac{\partial \omega}{\partial y} + y\frac{\partial \omega}{\partial z}\right]\right. \\ &\qquad \left. + \left(-z\frac{\partial \omega}{\partial y} + y\frac{\partial \omega}{\partial z} + y^2 + z^2\right)\right\}\, dA\, G\phi'\, dx \\ &\qquad - \int_L \overline{m}_x\, \delta\phi\, dx = 0 \end{aligned} \tag{5.85}$$

Reorganize some of the terms on the left-hand side of Eq. (5.85):

$$
\begin{aligned}
&\int_A \left[\left(\frac{\partial \omega}{\partial y}\right)^2 + \left(\frac{\partial \omega}{\partial z}\right)^2 - z\frac{\partial \omega}{\partial y} + y\frac{\partial \omega}{\partial z} \right] dA \\
&\quad = \int_A \left[\frac{\partial}{\partial y}\left(\omega \frac{\partial \omega}{\partial y}\right) + \frac{\partial}{\partial z}\left(\omega \frac{\partial \omega}{\partial z}\right) - \omega\left(\frac{\partial^2 \omega}{\partial y^2} + \frac{\partial^2 \omega}{\partial z^2}\right) - z\frac{\partial \omega}{\partial y} + y\frac{\partial \omega}{\partial z} \right] dA \\
&\quad = \int_A \left[\frac{\partial}{\partial y}\left(\omega \frac{\partial \omega}{\partial y} - z\omega\right) + \frac{\partial}{\partial z}\left(\omega \frac{\partial \omega}{\partial z} + y\omega\right) \right] dA \\
&\quad = \oint \omega \left[\left(\frac{\partial \omega}{\partial y} - z\right) n_y + \left(\frac{\partial \omega}{\partial z} + y\right) n_z \right] ds
\end{aligned}
\tag{5.86}
$$

Here $\partial^2\omega/\partial y^2 + \partial^2\omega/\partial z^2 = 0$ from Eq. (5.8) was introduced and the final integral was obtained with the assistance of Green's integral theorem (Rubinstein and Rubinstein, 1993):

$$
\begin{aligned}
\int \left(-\frac{\partial P}{\partial z} + \frac{\partial Q}{\partial y}\right) dA &= \oint \left(-P n_z + Q n_y\right) ds \\
&= \oint \left(P\frac{dy}{ds} + Q\frac{dz}{ds}\right) ds = \oint (P\,dy + Q\,dz)
\end{aligned}
\tag{5.87}
$$

From the cylindrical surface condition of Eq. (5.11), the final integral of Eq. (5.86) vanishes. Substitute J of Eq. (5.19) into Eq. (5.85), showing that Eq. (5.84) reduces to

$$
\int_L \frac{d}{dx}\,\delta\phi\, GJ\frac{d\phi}{dx}\,dx - \int_L \delta\phi\,\overline{m}_x\,dx = 0 \tag{5.88}
$$

Integration by parts, using appropriate boundary conditions, leads to

$$
\int_L \delta\phi \left(\frac{d}{dx}GJ\frac{d\phi}{dx} + \overline{m}_x\right) dx = 0 \tag{5.89}
$$

or

$$
\frac{d}{dx}GJ\frac{d\phi}{dx} = -\overline{m}_x \tag{5.90}
$$

which is the same as Eq. (5.23). This is the governing equation for torsional motion along the longitudinal axis (x) of the bar. As can be observed in Eq. (5.25), Eq. (5.90) is the same as the governing equation for a bar in extension with a change in the definitions of variables

$$
(G \to E \qquad J \to A \qquad \phi \to u \qquad \overline{m}_x \to \overline{p}_x \qquad M_x \to N) \tag{5.91}
$$

With the adjustments in notation, the stiffness matrix for extension of Example 2.16 can be used for a bar subject to Saint-Venant torsion. The stiffness matrix for torsion is given in Example 2.17.

It was just shown that the $\delta\phi$ terms of the principle of virtual work expression of Eq. (5.83) lead to differential equations for the axial response of a bar in Saint-Venant torsion. The $\delta\omega$ terms of Eq. (5.83) will lead to an equation for the cross-sectional problem of finding ω and then the cross-sectional properties and stresses for a bar in torsion. Assemble the $\delta\omega$ terms of Eq. (5.83):

$$\begin{aligned}\int_V G\phi'^2 \left(\frac{\partial}{\partial y}\delta\omega\, \frac{\partial \omega}{\partial y} + \frac{\partial}{\partial z}\delta\omega\, \frac{\partial \omega}{\partial z} - \frac{\partial}{\partial y}\delta\omega\, z + \frac{\partial}{\partial z}\delta\omega\, y \right) dV \\ = G\int_L \phi'^2\, dx \int_A \left[\left(\frac{\partial}{\partial y}\delta\omega \frac{\partial \omega}{\partial y} + \frac{\partial}{\partial z}\delta\omega \frac{\partial \omega}{\partial z} \right) - \left(\frac{\partial}{\partial y}\delta\omega\, z - \frac{\partial}{\partial z}\delta\omega\, y \right) \right] dA \\ = 0 \end{aligned} \tag{5.92}$$

This implies that

$$\int_A \left[\left(\frac{\partial}{\partial y}\delta\omega \frac{\partial \omega}{\partial y} + \frac{\partial}{\partial z}\delta\omega \frac{\partial \omega}{\partial z} \right) - \left(\frac{\partial}{\partial y}\delta\omega\, z - \frac{\partial}{\partial z}\delta\omega\, y \right) \right] dA = 0 \tag{5.93}$$

This is the expression that will be solved for $\omega(y, z)$, which will then be used to compute cross-sectional characteristics.

Finite Element Formulation Model the cross section with finite elements and approximate the warping function ω over each element (e) by

$$\omega(y, z) = \sum N_i \omega_i = \mathbf{N}\boldsymbol{\omega}^e = \boldsymbol{\omega}^{e\mathrm{T}}\mathbf{N}^\mathrm{T} \tag{5.94}$$

where N_i are the shape functions, $\mathbf{N}$ is a vector of the shape functions defined by Eq. (4.10) for a nine-node element, ω_i are the nodal values of ω, and $\boldsymbol{\omega}^e = [\omega_1 \quad \omega_2 \quad \omega_3 \quad \cdots \quad \omega_{n_b}]^\mathrm{T}$ is a vector of nodal values of the warping function ω with n_b equal to 9 for the nine-node element. The derivatives of ω can be expressed as

$$\frac{\partial \omega}{\partial y} = \frac{\partial \mathbf{N}}{\partial y}\boldsymbol{\omega}^e \qquad \frac{\partial \omega}{\partial z} = \frac{\partial \mathbf{N}}{\partial z}\boldsymbol{\omega}^e \tag{5.95}$$

Also,

$$\delta\omega = \mathbf{N}\,\delta\boldsymbol{\omega}^e = \delta\boldsymbol{\omega}^{e\mathrm{T}}\mathbf{N}^\mathrm{T} \qquad \frac{\partial}{\partial y}\delta\omega = \delta\boldsymbol{\omega}^{e\mathrm{T}}\frac{\partial \mathbf{N}^\mathrm{T}}{\partial y} \qquad \frac{\partial}{\partial z}\delta\omega = \delta\boldsymbol{\omega}^{e\mathrm{T}}\frac{\partial \mathbf{N}^\mathrm{T}}{\partial z} \tag{5.96}$$

Insertion of Eqs. (5.95) and (5.96) into Eq. (5.93) gives

$$\begin{aligned}\int_{A_e} \delta\boldsymbol{\omega}^{e\mathrm{T}} \left[\left(\frac{\partial \mathbf{N}^\mathrm{T}}{\partial y}\frac{\partial \mathbf{N}}{\partial y} + \frac{\partial \mathbf{N}^\mathrm{T}}{\partial z}\frac{\partial \mathbf{N}}{\partial z} \right)\boldsymbol{\omega}^e - \left(z\frac{\partial \mathbf{N}^\mathrm{T}}{\partial y} - y\frac{\partial \mathbf{N}^\mathrm{T}}{\partial z} \right) \right] dA \\ = \delta\boldsymbol{\omega}^{e\mathrm{T}}\left(\mathbf{k}^e\boldsymbol{\omega}^e - \mathbf{p}^e\right) = 0 \end{aligned} \tag{5.97}$$

where the element stiffness matrix $\mathbf{k}^e$ and element loading vector $\mathbf{p}^e$ for element e are

$$\mathbf{k}^e = \int_{A_e} \left(\frac{\partial \mathbf{N}^T}{\partial y} \frac{\partial \mathbf{N}}{\partial y} + \frac{\partial \mathbf{N}^T}{\partial z} \frac{\partial \mathbf{N}}{\partial z} \right) dA \qquad \mathbf{p}^e = \int_{A_e} \left(z \frac{\partial \mathbf{N}^T}{\partial y} - y \frac{\partial \mathbf{N}^T}{\partial z} \right) dA \quad (5.98)$$

5.4.2 Weighted Residual Methods

The displacement formulation for Saint-Venant torsion has been reduced in Section 5.1 to the partial differential equation (Eq. 5.8)

$$\nabla^2 \omega = 0 \tag{5.99}$$

to be solved for the warping function ω with the boundary condition (Eq. 5.11)

$$\frac{\partial \omega}{\partial y} n_y + \frac{\partial \omega}{\partial z} n_z = \boldsymbol{\nabla} \omega \cdot \mathbf{n} = \frac{\partial \omega}{\partial n} = z n_y - y n_z = \mathbf{n} \cdot \mathbf{g} \tag{5.100}$$

where

$$\mathbf{n} = n_y \mathbf{j} + n_z \mathbf{k} \qquad \boldsymbol{\nabla} \omega = \frac{\partial \omega}{\partial y} \mathbf{j} + \frac{\partial \omega}{\partial z} \mathbf{k} \qquad \mathbf{g} = z\mathbf{j} - y\mathbf{k}$$

This section formulates this boundary value problem for solution by the weighted residual method, in particular by Galerkin's approach.

In general terms, a continuum mechanics problem requires the solution of a set of differential equations

$$\mathbf{A}(\mathbf{u}) = 0 \tag{5.101}$$

for the functions $\mathbf{u}$ in a domain Ω, which is a line, an area, or a volume, such that the functions $\mathbf{u}$ satisfy certain conditions

$$\mathbf{B}(\mathbf{u}) = 0 \tag{5.102}$$

on the boundaries Γ of the domain. In the torsion problem, the vector of operators $\mathbf{A}$ has a single nonzero entry, and the vector $\mathbf{u}$ of unknown functions has the single component ω:

$$\mathbf{A}(\omega) = \nabla^2 \omega \qquad \mathbf{B}(\omega) = \mathbf{n} \cdot \mathbf{g} - \frac{\partial \omega}{\partial n} \tag{5.103}$$

The domain Ω is the cross-sectional area of a beam subjected to torsion, and the boundaries Γ are the closed boundary curves of the cross section.

The integral statement, in weighted residual form, of the general problem is

$$\int \mathbf{v}^T \mathbf{A}(\mathbf{u})\, d\Omega + \int \overline{\mathbf{v}}^T \mathbf{B}(\mathbf{u})\, d\Gamma = 0 \tag{5.104}$$

for all weighting functions **v** and $\overline{\mathbf{v}}$, where these functions have as many components as the unknown vector of functions **u**. This integral form of this continuum mechanics problem is sometimes referred to as a *weak formulation*. Note that the residual expression for both the differential equations in the domain and boundary conditions are combined in a single expression. For the torsion problem defined by the functions of Eq. (5.103), Eq. (5.104) gives the integral formulation

$$\int v\nabla^2\omega \, dA + \int \overline{v}\left(\mathbf{n}\cdot\mathbf{g} - \frac{\partial \omega}{\partial n}\right) ds = 0 \tag{5.105}$$

where s is the arc length.

The integral form of the torsion problem of Eq. (5.105) will now be recast into another form by choosing the weighting functions v and $\overline{v}$ to be identical and integrating by parts. The latter operation is simplified by invoking Green's theorem and Green's first identity. This identity (Rubinstein and Rubinstein, 1993), which replaces the second derivatives of ω with its first derivatives, takes the form

$$\int v\nabla^2\omega \, dA = \int v\frac{\partial \omega}{\partial n} ds - \int \boldsymbol{\nabla} v \cdot \boldsymbol{\nabla}\omega \, dA \tag{5.106}$$

Green's theorem will be applied in its usual form,

$$\int \left(\frac{\partial Q}{\partial y} - \frac{\partial P}{\partial z}\right) dA = \int (P\, dy + Q\, dz) \tag{5.107}$$

For our torsion problem, take P and Q of Eq. (5.107) to be

$$P = vy \qquad Q = vz \tag{5.108}$$

Then, with $\mathbf{g} = z\mathbf{j} - y\mathbf{k}$,

$$\begin{aligned} v\mathbf{g}\cdot\mathbf{n} &= Q\mathbf{j}\cdot\mathbf{n} - P\mathbf{k}\cdot\mathbf{n} = Qn_y - Pn_z = Qt_z + Pt_y \\ &= (P\mathbf{j} + Q\mathbf{k})\cdot\mathbf{t} \end{aligned} \tag{5.109}$$

where t_y, t_z are components of the unit tangent vector ($\mathbf{t} = t_y\mathbf{j} + t_z\mathbf{k}$) and, from Eq. (1.33), $n_y = t_z$ and $n_z = -t_y$. Since, from Eq. (1.37), $d\mathbf{r} = dy\mathbf{j} + dz\mathbf{k} = \mathbf{t}\, ds$, it follows that

$$\int v\mathbf{g}\cdot\mathbf{n}\, ds = \int (P\mathbf{j} + Q\mathbf{k})\cdot\mathbf{t}\, ds = \int (P\, dy + Q\, dz) \tag{5.110}$$

Introduction of the Green's theorem of Eq. (5.107) leads to

$$\int v\mathbf{g}\cdot\mathbf{n}\, ds = \int \left(\frac{\partial vz}{\partial y} - \frac{\partial vy}{\partial z}\right) dA = \int \boldsymbol{\nabla} v\cdot\mathbf{g}\, dA \tag{5.111}$$

where

$$\nabla v = \frac{\partial v}{\partial y}\mathbf{j} + \frac{\partial v}{\partial z}\mathbf{k} \quad \text{and} \quad \mathbf{g} = z\mathbf{j} - y\mathbf{k}$$

Substitution of Eqs. (5.106) and (5.111) into the integral form of the torsion problem (Eq. 5.105) gives

$$\int \nabla v \cdot \nabla \omega \, dA - \int \nabla v \cdot \mathbf{g} \, dA = 0 \tag{5.112}$$

or

$$\int_A \left[\left(\frac{\partial v}{\partial y}\frac{\partial \omega}{\partial y} + \frac{\partial v}{\partial z}\frac{\partial \omega}{\partial z} \right) - \left(\frac{\partial v}{\partial y} z - \frac{\partial v}{\partial z} y \right) \right] dA = 0 \tag{5.113}$$

Introduce the trial function of Eqs. (5.94) and (5.95) for element e:

$$\int_{A_e} \left[\left(\frac{\partial v}{\partial y}\frac{\partial \mathbf{N}}{\partial y} + \frac{\partial v}{\partial z}\frac{\partial \mathbf{N}}{\partial z} \right) \boldsymbol{\omega}^e - \left(\frac{\partial v}{\partial y} z - \frac{\partial v}{\partial z} y \right) \right] dA = 0 \tag{5.114}$$

Galerkin's Method For *Galerkin's method* the functions v in element e are replaced by the shape functions. Thus, let

$$v = \mathbf{N}\mathbf{v}^e = \mathbf{v}^{e\mathrm{T}}\mathbf{N}^{\mathrm{T}} \tag{5.115}$$

where $\mathbf{v}^e$ contains the nodal values of v. Then Eq. (5.114) becomes

$$\begin{aligned} \int_A \mathbf{v}^{e\mathrm{T}} \left[\left(\frac{\partial \mathbf{N}^{\mathrm{T}}}{\partial y}\frac{\partial \mathbf{N}}{\partial y} + \frac{\partial \mathbf{N}^{\mathrm{T}}}{\partial z}\frac{\partial \mathbf{N}}{\partial z} \right) \boldsymbol{\omega}^e - \left(z\frac{\partial \mathbf{N}^{\mathrm{T}}}{\partial y} - y\frac{\partial \mathbf{N}^{\mathrm{T}}}{\partial z} \right) \right] dA \\ = \mathbf{v}^{e\mathrm{T}} \left(\mathbf{k}^e \boldsymbol{\omega}^e - \mathbf{p}^e \right) = 0 \end{aligned} \tag{5.116}$$

where $\mathbf{k}^e$ and $\mathbf{p}^e$ are given by Eq. (5.98). The development of the interpolation expressions of Eq. (5.115) is discussed in Chapter 4. It follows from Eqs. (5.97) and (5.116) that the principle of virtual work and Galerkin's method lead to the same element stiffness relations.

5.4.3 Isoparametric Elements

To develop a nine-node isoparametric element, make use of the transformation equations described in Chapter 4. The warping function is approximated over each element e by

$$\omega(\eta, \zeta) = \sum_{i=1}^{9} N_i \omega_i = \mathbf{N}(\eta, \zeta)\boldsymbol{\omega}^e \tag{5.117}$$

where $\boldsymbol{\omega}^e$ is the vector of nodal values ω_i of ω for the element and $\mathbf{N}$ is a vector of the shape functions N_i. The element stiffness matrix $\mathbf{k}^e$ of Eq. (5.98) can be written as

$$\mathbf{k}^e = \int_{A_e} \left(\frac{\partial \mathbf{N}^{\mathrm{T}}}{\partial y} \frac{\partial \mathbf{N}}{\partial y} + \frac{\partial \mathbf{N}^{\mathrm{T}}}{\partial z} \frac{\partial \mathbf{N}}{\partial z} \right) dy\, dz = \int_{A_e} \begin{bmatrix} \dfrac{\partial \mathbf{N}^{\mathrm{T}}}{\partial y} & \dfrac{\partial \mathbf{N}^{\mathrm{T}}}{\partial z} \end{bmatrix} \begin{bmatrix} \dfrac{\partial \mathbf{N}}{\partial y} \\ \dfrac{\partial \mathbf{N}}{\partial z} \end{bmatrix} dy\, dz \tag{5.118}$$

From Eq. (4.23), $A_e = \int_\Omega dy\, dz = \int_{\Omega_r} | \mathbf{J}_e |\, d\eta\, d\zeta$. As noted in Eq. (4.16),

$$\begin{bmatrix} \dfrac{\partial \mathbf{N}}{\partial y} \\ \dfrac{\partial \mathbf{N}}{\partial z} \end{bmatrix} = \begin{bmatrix} \dfrac{\partial \mathbf{N}}{\partial \eta} \dfrac{\partial \eta}{\partial y} + \dfrac{\partial \mathbf{N}}{\partial \zeta} \dfrac{\partial \zeta}{\partial y} \\ \dfrac{\partial \mathbf{N}}{\partial \eta} \dfrac{\partial \eta}{\partial z} + \dfrac{\partial \mathbf{N}}{\partial \zeta} \dfrac{\partial \zeta}{\partial z} \end{bmatrix} = \begin{bmatrix} \dfrac{\partial \eta}{\partial y} & \dfrac{\partial \zeta}{\partial y} \\ \dfrac{\partial \eta}{\partial z} & \dfrac{\partial \zeta}{\partial z} \end{bmatrix} \begin{bmatrix} \dfrac{\partial \mathbf{N}}{\partial \eta} \\ \dfrac{\partial \mathbf{N}}{\partial \zeta} \end{bmatrix} = \mathbf{J}_e^{-1} \begin{bmatrix} \dfrac{\partial \mathbf{N}}{\partial \eta} \\ \dfrac{\partial \mathbf{N}}{\partial \zeta} \end{bmatrix} = \mathbf{B}_e = \mathbf{B} \tag{5.119}$$

a 2×9 matrix. On occasion the subscript e will be dropped from $\mathbf{B}_e$, although this matrix always applies to element e. Also,

$$\begin{aligned} \begin{bmatrix} \dfrac{\partial \mathbf{N}^{\mathrm{T}}}{\partial y} & \dfrac{\partial \mathbf{N}^{\mathrm{T}}}{\partial z} \end{bmatrix} &= \begin{bmatrix} \dfrac{\partial \mathbf{N}^{\mathrm{T}}}{\partial \eta} \dfrac{\partial \eta}{\partial y} + \dfrac{\partial \mathbf{N}^{\mathrm{T}}}{\partial \zeta} \dfrac{\partial \zeta}{\partial y} & \dfrac{\partial \mathbf{N}^{\mathrm{T}}}{\partial \eta} \dfrac{\partial \eta}{\partial z} + \dfrac{\partial \mathbf{N}^{\mathrm{T}}}{\partial \zeta} \dfrac{\partial \zeta}{\partial z} \end{bmatrix} \\ &= \begin{bmatrix} \dfrac{\partial \mathbf{N}^{\mathrm{T}}}{\partial \eta} & \dfrac{\partial \mathbf{N}^{\mathrm{T}}}{\partial \zeta} \end{bmatrix} \begin{bmatrix} \dfrac{\partial \eta}{\partial y} & \dfrac{\partial \eta}{\partial z} \\ \dfrac{\partial \zeta}{\partial y} & \dfrac{\partial \zeta}{\partial z} \end{bmatrix} \\ &= \begin{bmatrix} \dfrac{\partial \mathbf{N}^{\mathrm{T}}}{\partial \eta} & \dfrac{\partial \mathbf{N}^{\mathrm{T}}}{\partial \zeta} \end{bmatrix} (\mathbf{J}_e^{-1})^{\mathrm{T}} = \mathbf{B}^{\mathrm{T}} \end{aligned} \tag{5.120}$$

Thus,

$$\mathbf{k}^e = \int_{-1}^{1} \int_{-1}^{1} \begin{bmatrix} \dfrac{\partial \mathbf{N}^{\mathrm{T}}}{\partial \eta} & \dfrac{\partial \mathbf{N}^{\mathrm{T}}}{\partial \zeta} \end{bmatrix} \left(\mathbf{J}_e^{-1} \right)^{\mathrm{T}} \mathbf{J}_e^{-1} \begin{bmatrix} \dfrac{\partial \mathbf{N}}{\partial \eta} \\ \dfrac{\partial \mathbf{N}}{\partial \zeta} \end{bmatrix} | \mathbf{J}_e |\, d\eta\, d\zeta \tag{5.121}$$

or

$$\mathbf{k}^e = \int_{-1}^{1} \int_{-1}^{1} \mathbf{B}^{\mathrm{T}} \mathbf{B} \, | \mathbf{J}_e |\, d\eta\, d\zeta \tag{5.122}$$

From Eq. (5.98) the element load vector $\mathbf{p}^e$ is defined by

$$\mathbf{p}^e = \int_{-1}^{1} \int_{-1}^{1} \mathbf{B}^{\mathrm{T}} \begin{bmatrix} \mathbf{Nz} \\ -\mathbf{Ny} \end{bmatrix} | \mathbf{J}_e |\, d\eta\, d\zeta \tag{5.123}$$

The element equations for the eth element are then

$$\mathbf{k}^e \boldsymbol{\omega}^e = \mathbf{p}^e \tag{5.124}$$

5.5 ASSEMBLY OF SYSTEM MATRICES

The element equations of Eq. (5.124),

$$\mathbf{k}^e \boldsymbol{\omega}^e = \mathbf{p}^e \qquad 1 \le e \le M$$

where M is the number of elements, should be assembled into the global set of linear equations

$$\mathbf{K}\boldsymbol{\omega} = \mathbf{P} \tag{5.125}$$

where $\mathbf{K}$ is the system stiffness matrix and $\mathbf{P}$ is the system load vector. This system equation is to be solved for $\boldsymbol{\omega}$, the nodal values of the warping function ω.

In solving for the unknown ω for the Saint-Venant torsion problem, two conditions must be satisfied: the equilibrium and compatibility of the nodal values of the unknown function ω. Recall that the element stiffness relations $\mathbf{k}^e \boldsymbol{\omega}^e = \mathbf{p}^e$ were derived from the principle of virtual work and from Galerkin's method. The principle of virtual work corresponds to the conditions of equilibrium and static boundary conditions for kinematically admissible displacements. Galerkin's method placed the differential equations of equilibrium (i.e., Laplace's equation for ω) along with all boundary conditions in integral form. Thus equilibrium is satisfied within each element.

The compatibility of nodal values is satisfied if the value of ω at a node shared by several elements is the same. Thus, the compatibility condition is satisfied by defining a global vector $\boldsymbol{\omega}$ of nodal values for ω. The nodes in the mesh are numbered sequentially, and each entry of $\boldsymbol{\omega}$ is associated with a single node. A *nodal connectivity* matrix $\mathbf{C}$ is defined to express the relationship between element and global nodes. The entry C_{ej} of the nodal connectivity matrix is the global node number corresponding to node j, $1 \le j \le 9$, of element e.

As an example, the global node numbers for a four-element mesh are shown in Fig. 5.22. Each element in the mesh has the local node numbers shown in Fig. 5.23. The nodal connectivity matrix for this four-element mesh will have four rows and, since each element has nine local nodes, nine columns. The entries of the nodal connectivity matrix are determined from a comparison of these two figures:

$$\mathbf{C} = \begin{bmatrix} C_{11} & C_{12} & \cdots & C_{19} \\ C_{21} & C_{22} & \cdots & C_{29} \\ C_{31} & C_{32} & \cdots & C_{39} \\ C_{41} & C_{42} & \cdots & C_{49} \end{bmatrix} = \begin{bmatrix} 1 & 2 & 3 & 6 & 7 & 8 & 11 & 12 & 13 \\ 3 & 4 & 5 & 8 & 9 & 10 & 13 & 14 & 15 \\ 11 & 12 & 13 & 16 & 17 & 18 & 21 & 22 & 23 \\ 13 & 14 & 15 & 18 & 19 & 20 & 23 & 24 & 25 \end{bmatrix} \tag{5.126}$$

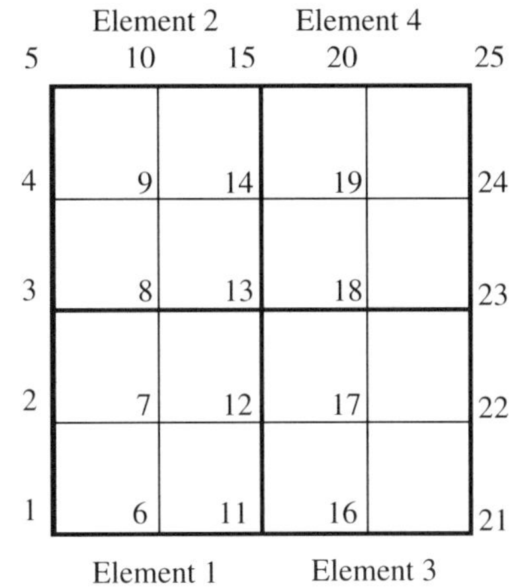

Figure 5.22 Global node and element numbers for a four-element mesh.

Each row corresponds to a particular element and each entry corresponds to a global node number arranged according to the local node numbers of Fig. 5.23.

In this formulation, element domain integrals are added together to calculate the total domain integral in Eq. (5.112). Since the algebraic equations for an element are a discretized form of the integral formulation for that element, the corresponding discretized total domain integrals are obtained by summing the element equations. Then the equilibrium conditions at the nodes are satisfied and the finite element formulation of the problem is complete, because, as mentioned earlier, the fundamental formulation here already ensures that the equilibrium conditions are met in the interior of each element.

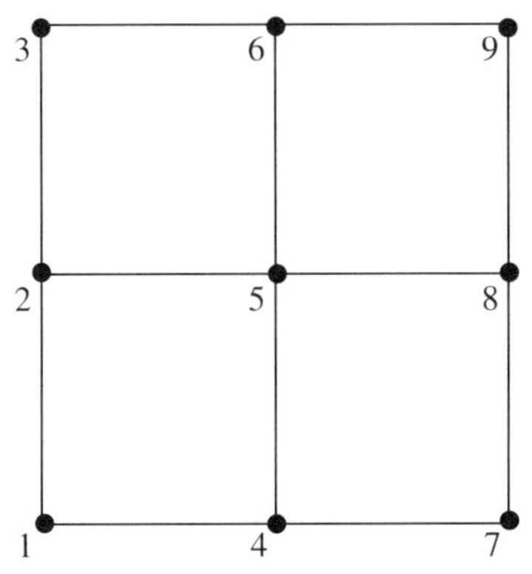

Figure 5.23 Local node numbers of a nine-node element.

The superposition of the element equations is most easily described by imagining that the 9×9 element stiffness matrices are expanded into $q \times q$ matrices and that the element load vectors are expanded to have q elements, where q is the total number of degrees of freedom for the entire mesh, or the total number of nodes when each node has a single variable assigned to it, as is the case of the torsion problem with the warping function ω as the single variable. For a torsion problem with the four-element system of Fig. 5.22, $q = 25$. Each expanded element stiffness matrix $\mathbf{k}'^e$ has zero entries corresponding to the nodal values in $\boldsymbol{\omega}$ that do not occur in the nodes of element e, and each expanded element force vector $\mathbf{p}'^e$ has zero entries corresponding to the nodes that do not belong to element e. In other words, row e of the nodal connectivity matrix $\mathbf{C}$ gives the indices of the nonzero rows and columns of the expanded $\mathbf{k}'^e$ matrix and the indices of the nonzero components of the expanded load vector $\mathbf{p}'^e$. The purpose of the expansion is to force all element vectors and matrices to have the same dimensions as the system vectors and matrices. The assembly procedure is then described by the vector equation of Eq. (5.125) ($\mathbf{K}\boldsymbol{\omega} = \mathbf{P}$):

$$\mathbf{P} = \sum_e \mathbf{p}'^e = \sum_e \mathbf{k}'^e \boldsymbol{\omega}'^e = \mathbf{K}\boldsymbol{\omega} \tag{5.127}$$

where $\boldsymbol{\omega}'^e$ is the vector expanded from the nodal warping function vector $\boldsymbol{\omega}^e$ for element e. The system stiffness matrix $\mathbf{K}$ is the sum of the expanded element stiffness matrices:

$$\mathbf{K} = \sum_e \mathbf{k}'^e \tag{5.128a}$$

and the displacement (warping function) vector is

$$\boldsymbol{\omega} = \sum_e \boldsymbol{\omega}'^e \tag{5.128b}$$

In the mesh shown in Fig. 5.22, for instance, element 2 has nonzero contributions to its expanded stiffness matrix at the rows and columns listed in the second row of the nodal connectivity matrix of Eq. (5.126):

$$\mathbf{C}_2 = [3 \quad 4 \quad 5 \quad 8 \quad 9 \quad 10 \quad 13 \quad 14 \quad 15] \tag{5.129}$$

These contributions are shown in Table 5.1, in which k_{jk} denotes the row j, column k of the 9×9 element stiffness matrix for the nine-node element of Fig. 5.23. As this example shows, the expansion of $\mathbf{k}^e$ into $\mathbf{k}'^e$ is accomplished by setting

$$k'^e_{mn} = k^e_{jk} \qquad m = C_{ej} \quad n = C_{ek} \qquad 1 \le j, \quad k \le 9 \tag{5.130}$$

with all other entries of the expanded matrix set equal to zero. Here C_{ej} from Eq. (5.126) is the element e global node number corresponding to node j, $1 \le j \le 9$. By expanding Table 5.1, the expanded stiffness matrix for element 2 becomes

TABLE 5.1 Nonzero Entries of the Expanded Stiffness Matrix for Element 2

	3	4	5	8	9	10	13	14	15
3	k_{11}	k_{12}	k_{13}	k_{14}	k_{15}	k_{16}	k_{17}	k_{18}	k_{19}
4	k_{21}	k_{22}	k_{23}	k_{24}	k_{25}	k_{26}	k_{27}	k_{28}	k_{29}
5	k_{31}	k_{32}	k_{33}	k_{34}	k_{35}	k_{36}	k_{37}	k_{38}	k_{39}
8	k_{41}	k_{42}	k_{43}	k_{44}	k_{45}	k_{46}	k_{47}	k_{48}	k_{49}
9	k_{51}	k_{52}	k_{53}	k_{54}	k_{55}	k_{56}	k_{57}	k_{58}	k_{59}
10	k_{61}	k_{62}	k_{63}	k_{64}	k_{65}	k_{66}	k_{67}	k_{68}	k_{69}
13	k_{71}	k_{72}	k_{73}	k_{74}	k_{75}	k_{76}	k_{77}	k_{78}	k_{79}
14	k_{81}	k_{82}	k_{83}	k_{84}	k_{85}	k_{86}	k_{87}	k_{88}	k_{89}
15	k_{91}	k_{92}	k_{93}	k_{94}	k_{95}	k_{96}	k_{97}	k_{98}	k_{99}

$$\mathbf{k}'^2 = \begin{bmatrix} 0 & 0 & 0 & 0 & 0 & 0 & 0 & 0 & 0 & 0 & \cdots \\ 0 & 0 & 0 & 0 & 0 & 0 & 0 & 0 & 0 & 0 & \cdots \\ 0 & 0 & k_{11} & k_{12} & k_{13} & 0 & 0 & k_{14} & k_{15} & k_{16} & \cdots \\ 0 & 0 & k_{21} & k_{22} & k_{23} & 0 & 0 & k_{24} & k_{25} & k_{26} & \cdots \\ 0 & 0 & k_{31} & k_{32} & k_{33} & 0 & 0 & k_{34} & k_{35} & k_{36} & \cdots \\ 0 & 0 & 0 & 0 & 0 & 0 & 0 & 0 & 0 & 0 & \cdots \\ 0 & 0 & 0 & 0 & 0 & 0 & 0 & 0 & 0 & 0 & \cdots \\ 0 & 0 & k_{41} & k_{42} & k_{43} & 0 & 0 & k_{44} & k_{45} & k_{46} & \cdots \\ 0 & 0 & k_{51} & k_{52} & k_{53} & 0 & 0 & k_{54} & k_{55} & k_{56} & \cdots \\ 0 & 0 & k_{61} & k_{62} & k_{63} & 0 & 0 & k_{64} & k_{65} & k_{66} & \cdots \\ \vdots & \vdots & \vdots & \vdots & \vdots & \vdots & \vdots & \vdots & \vdots & \vdots & \vdots \end{bmatrix} \tag{5.131}$$

Thus, the system stiffness matrix for the four-element mesh of Fig. 5.22 is (Eq. 5.128a)

$$\mathbf{K} = \sum_{e=1}^{4} \mathbf{k}'^e = \mathbf{k}'^1 + \mathbf{k}'^2 + \mathbf{k}'^3 + \mathbf{k}'^4$$

where, for example, $\mathbf{k}'^2$ is given by Eq. (5.131).

The expansion of $\mathbf{p}^e$ into $\mathbf{p}'^e$ is given by

$$p'^e_n = p^e_j \qquad n = C_{ej} \qquad 1 \le j \le 9 \tag{5.132}$$

with all other entries of $\mathbf{p}'^e$ set equal to zero. For the four-element mesh of Fig. 5.22, the expanded loading vector of element 2 is found by inserting the values of C given in Eq. (5.126) into Eq. (5.132):

$$\mathbf{p}'^2 = [0 \quad 0 \quad p_1 \quad p_2 \quad p_3 \quad 0 \quad 0 \quad p_4 \quad p_5 \quad p_6 \quad \cdots]^{\mathrm{T}} \tag{5.133}$$

For the four-element mesh of Fig. 5.22, the load vector is (Eq. 5.127)

$$\mathbf{P} = \sum_{e=1}^{4} \mathbf{p}'^e = \mathbf{p}'^1 + \mathbf{p}'^2 + \mathbf{p}'^3 + \mathbf{p}'^4$$

where $\mathbf{p}'^2$ is given by Eq. (5.133).

The computational algorithm for the assembly procedure is:

1. Set the system stiffness matrix **K** and the system load vector **P** equal to zero.
2. Add the contributions of the stiffness matrix of each element e to the system stiffness matrix:

$$K_{mn} = K_{mn} + k_{jk}^e \qquad m = C_{ej}, \quad n = C_{ek}, \quad 1 \leq j, k \leq 9 \tag{5.134a}$$

In practice, the assembly of the stiffness matrix is not performed as stated because **K**, a banded symmetric matrix, is not normally stored as a full square matrix. The storage scheme defines a mapping between square matrix storage locations and the banded storage locations $m \to m^{'}$, $n \to n^{'}$ so that the assembly formula is modified to

$$K_{m'n'} = K_{m'n'} + k_{jk}^e \tag{5.134b}$$

3. Assemble the load vector **P** from the element load vectors:

$$P_n = p_j^e \qquad n = C_{ej}, \quad 1 \leq j \leq 9 \tag{5.134c}$$

Assembly of System Matrices for Beam Systems The assembly of the beam elements of Chapter 2 into the system matrices of Chapter 3 can be placed in the same notation as developed in this chapter for the cross-sectional characteristics of a bar undergoing Saint-Venant torsion. That is, if distributed loads are ignored, the element matrices $\mathbf{k}^e\mathbf{v}^e = \mathbf{p}^e$ are to be assembled into system matrices $\mathbf{KV} = \overline{\mathbf{P}}$. In contrast to the planar problem of this chapter for cross-sectional properties which involves a two-dimensional mesh, a planar beam problem is modeled with one-dimensional elements. A nine-node element with a single unknown variable (DOF) per node was employed for the cross-sectional problem, whereas the beams in bending elements of Chapter 2 have a node at each end with two DOF per node. The beam degrees of freedom are end deflections and slopes. As shown in Fig. 3.6, the beam elements are numbered sequentially starting with element 1 at the left end and proceeding toward the right end. The degrees of freedom (deflection and slope) are numbered as 1 and 2 for the left end of the first element and 3 and 4 for the right end. The nodal connectivity matrix for the beam mesh is particularly simple. For a mesh of 4 elements, it is

$$\mathbf{C} = \begin{bmatrix} 1 & 2 & 3 & 4 \\ 3 & 4 & 5 & 6 \\ 5 & 6 & 7 & 8 \\ 7 & 8 & 9 & 10 \end{bmatrix} \tag{5.135a}$$

and if there are M elements the $M \times 4$ connectivity matrix is generated by continuing the pattern. In other words, for the kth beam element the row

$$C_k = [2k - 1 \quad 2k \quad 2k + 1 \quad 2(k + 1)] \tag{5.135b}$$

is appended to the connectivity matrix **C**. Thus, the global node numbers that make up the kth row C_k of the connectivity matrix corresponds to w_a, θ_a, w_b, and θ_b for beam element k, where a is the left end and b is the right end of the beam element. The assembly of the element load vectors $\mathbf{p}^e$ into the system load vector $\overline{\mathbf{P}}$ is identical to the process utilized for the cross-sectional problem. In summary, the computational algorithm for the assembly procedure is

1. Set the system stiffness matrix **K** and the system load vector **P** equal to zero:
2. Add the contributions of the stiffness matrix of each element e to the system stiffness matrix:

$$K_{mn} = K_{mn} + k_{ij}^e \quad m = C_{ej} \quad n = C_{ej} \quad 1 \leq j, \quad k \leq 4 \tag{5.136a}$$

3. Assemble the load vector $\overline{\mathbf{P}}$ from the element load vectors:

$$\overline{P}_n = p_j^e \qquad n = C_{ej} \qquad 1 \leq j \leq 4 \tag{5.136b}$$

5.6 CALCULATION OF THE TORSIONAL CONSTANT AND STRESSES

The global equations $\mathbf{K}\boldsymbol{\omega} = \mathbf{P}$, with the assembled matrices **K** and **P**, can be solved for the nodal values $\boldsymbol{\omega}$ of the warping function. The various response parameters, such as stresses, can now be calculated. Unlike the stiffness matrix for the transverse motion of beams (Eq. 3.19), for which the displacement vector **V** contains nodal translations and rotations, the stiffness matrix for cross-sectional torsional properties (Eq. 5.125), for which the displacement vector $\boldsymbol{\omega}$ contains nodal values of the warping function, is not, in general, singular.

The torsional constant J is given by Eq. (5.19):

$$\begin{aligned}
J &= \int \left[\left(\frac{\partial \omega}{\partial z} + y \right) y - \left(\frac{\partial \omega}{\partial y} - z \right) z \right] dA \\
&= \int \left[z^2 + y^2 - \left(z \frac{\partial \omega}{\partial y} - y \frac{\partial \omega}{\partial z} \right) \right] dA \\
&= I_y + I_z - \int \left(z \frac{\partial \omega}{\partial y} - y \frac{\partial \omega}{\partial z} \right) dA \\
&= \int \mathbf{g} \cdot (\mathbf{g} - \boldsymbol{\nabla} \omega) \; dA = I_y + I_z - \int \mathbf{g} \cdot \boldsymbol{\nabla} \omega \; dA
\end{aligned}$$

$$= I_y + I_z - \int [z \quad -y] \begin{bmatrix} \dfrac{\partial \omega}{\partial y} \\ \dfrac{\partial \omega}{\partial z} \end{bmatrix} dA \tag{5.137}$$

with

$$\mathbf{g} = z\mathbf{j} - y\mathbf{k} \quad \text{and} \quad \boldsymbol{\nabla}\omega = \frac{\partial \omega}{\partial y}\mathbf{j} + \frac{\partial \omega}{\partial z}\mathbf{k}$$

In terms of the isoparametric element, with Eqs. (4.11), (4.23), and (5.95),

$$\begin{aligned} J &= I_y + I_z - \sum_e \int_{-1}^{1} \int_{-1}^{1} [\mathbf{Nz} \quad -\mathbf{Ny}]\, \mathbf{B}\boldsymbol{\omega}^e \mid \mathbf{J}_e \mid d\eta\, d\zeta \\ &= I_y + I_z - \sum_e \int_{-1}^{1} \int_{-1}^{1} [\mathbf{Nz} \quad -\mathbf{Ny}]\, \mathbf{B} \mid \mathbf{J}_e \mid d\eta\, d\zeta\, \boldsymbol{\omega}^e \end{aligned} \tag{5.138}$$

If Eq. (5.123) is introduced,

$$\begin{aligned} J &= I_y + I_z - \sum_e \mathbf{p}^{e\mathrm{T}} \boldsymbol{\omega}^e = I_y + I_z - \mathbf{P}^{\mathrm{T}} \boldsymbol{\omega} \\ &= I_y + I_z - \boldsymbol{\omega}^{\mathrm{T}} \mathbf{P} = I_y + I_z - \boldsymbol{\omega}^{\mathrm{T}} \mathbf{K} \boldsymbol{\omega} \end{aligned} \tag{5.139}$$

where $\mathbf{P}$ is given in Eq. (5.127).

To compute the shear stresses for a given torque M_x, begin with Eqs. (5.4) and (5.18):

$$\begin{aligned} \tau_{xy} &= G\theta \left(\frac{\partial \omega}{\partial y} - z \right) = \frac{M_x}{J} \left(\frac{\partial \omega}{\partial y} - z \right) \\ \tau_{xz} &= G\theta \left(\frac{\partial \omega}{\partial z} + y \right) = \frac{M_x}{J} \left(\frac{\partial \omega}{\partial z} + y \right) \end{aligned} \tag{5.140}$$

or

$$\begin{bmatrix} \tau_{xy} \\ \tau_{xz} \end{bmatrix} = \frac{M_x}{J} \begin{bmatrix} \dfrac{\partial \omega}{\partial y} - z \\ \dfrac{\partial \omega}{\partial z} + y \end{bmatrix} = \frac{M_x}{J} \left(\begin{bmatrix} \dfrac{\partial \omega}{\partial y} \\ \dfrac{\partial \omega}{\partial z} \end{bmatrix} - \begin{bmatrix} z \\ -y \end{bmatrix} \right) \tag{5.141}$$

The shear stresses at any point of element e are given by

$$\boldsymbol{\tau}^e = \begin{bmatrix} \tau_{xy} \\ \tau_{xz} \end{bmatrix}^e = \frac{M_x}{J} \left(\mathbf{B}\boldsymbol{\omega}^e - \mathbf{h}^e \right) \tag{5.142}$$

where, from Eq. (4.11),

$$\mathbf{h}^e = \begin{bmatrix} z \\ -y \end{bmatrix} = \begin{bmatrix} \mathbf{N}\mathbf{z}_e \\ -\mathbf{N}\mathbf{y}_e \end{bmatrix} \tag{5.143}$$

and, from Eqs. (5.119), (4.15), and (4.17),

$$\mathbf{B}\boldsymbol{\omega}^e = \mathbf{J}_e^{-1} \begin{bmatrix} \dfrac{\partial \mathbf{N}}{\partial \eta} \\ \dfrac{\partial \mathbf{N}}{\partial \zeta} \end{bmatrix} \boldsymbol{\omega}^e = \frac{1}{|\,\mathbf{J}_e\,|} \begin{bmatrix} \dfrac{\partial \mathbf{N}}{\partial \zeta}\mathbf{z}_e & -\dfrac{\partial \mathbf{N}}{\partial \eta}\mathbf{z}_e \\ -\dfrac{\partial \mathbf{N}}{\partial \zeta}\mathbf{y}_e & \dfrac{\partial \mathbf{N}}{\partial \eta}\mathbf{y}_e \end{bmatrix} \begin{bmatrix} \dfrac{\partial \mathbf{N}}{\partial \eta}\boldsymbol{\omega}^e \\ \dfrac{\partial \mathbf{N}}{\partial \zeta}\boldsymbol{\omega}^e \end{bmatrix} \tag{5.144}$$

Extrapolation of Stress from the Gauss Points When calculating the stresses at the nodes of the element, stresses are first computed at the Gaussian integration points, which are often the optimal sampling locations for stress (Cook et al., 1989). The stresses at the Gauss points are multiplied by a *smoothing matrix* to obtain the nodal stresses. Smoothed values from adjacent elements are then averaged at the element nodes. Stress smoothing is discussed in journal articles by Hinton and Campbell (1974) and Hinton et al. (1975).

Equation (4.28) provides a formula for bidirectional integration by Gaussian quadrature. For the case of $m_1 = m_2 = 3$, the integration (Gauss) points and weighting factors are given in Table 5.2 and the Gauss points are illustrated in Fig. 5.24. Equation (4.28) appears as

$$\int_{-1}^{1}\int_{-1}^{1} \phi(\eta, \zeta)\, d\eta\, d\zeta = \sum_{i=1}^{m_1=3}\sum_{j=1}^{m_2=3} W_i(m_1)W_j(m_2)\phi(\eta_i, \zeta_j) \tag{5.145}$$

The total number of integration points in the domain is $m_1 \times m_2 = 3 \times 3 = 9$. Thus, when 3×3-order Gaussian quadrature is used for a nine-node element, the number of stresses computed at the Gauss points and the number of nodal stresses are both nine. The smoothed stresses may therefore be assumed to be given by the set of equations

$$\tilde{\sigma}\,(\eta, \zeta) = \mathbf{N}(\eta, \zeta)\mathbf{c} \tag{5.146}$$

where $\mathbf{N}$ is the row vector of Eq. (4.10) and $\tilde{\sigma}$ represents either the stress component τ_{xy} or the component τ_{xz}. Treat each component separately. The unknown coefficient vector $\mathbf{c}$ is found by assuming that the smoothed stress components at the Gauss points are the same as the stress components computed at these points by Eq. (5.142). This gives nine equations:

$$\tilde{\sigma}\,(\eta_k, \zeta_k) = \mathbf{N}(\eta_k, \zeta_k)\mathbf{c} \qquad 1 \le k \le 9 \tag{5.147}$$

where (η_k, ζ_k) is the kth Gauss point. These equations determine $\mathbf{c}$ to be

$$\mathbf{c} = \mathbf{H}^{-1}\boldsymbol{\sigma}_G \tag{5.148}$$

TABLE 5.2 Gaussian Integration Points and Weights

i	η_i	ζ_i	Weight
1	$-\sqrt{\frac{3}{5}}$	$-\sqrt{\frac{3}{5}}$	$\frac{25}{81}$
2	$-\sqrt{\frac{3}{5}}$	0	$\frac{40}{81}$
3	$-\sqrt{\frac{3}{5}}$	$\sqrt{\frac{3}{5}}$	$\frac{25}{81}$
4	0	$-\sqrt{\frac{3}{5}}$	$\frac{40}{81}$
5	0	0	$\frac{64}{81}$
6	0	$\sqrt{\frac{3}{5}}$	$\frac{40}{81}$
7	$\sqrt{\frac{3}{5}}$	$-\sqrt{\frac{3}{5}}$	$\frac{25}{81}$
8	$\sqrt{\frac{3}{5}}$	0	$\frac{40}{81}$
9	$\sqrt{\frac{3}{5}}$	$\sqrt{\frac{3}{5}}$	$\frac{25}{81}$

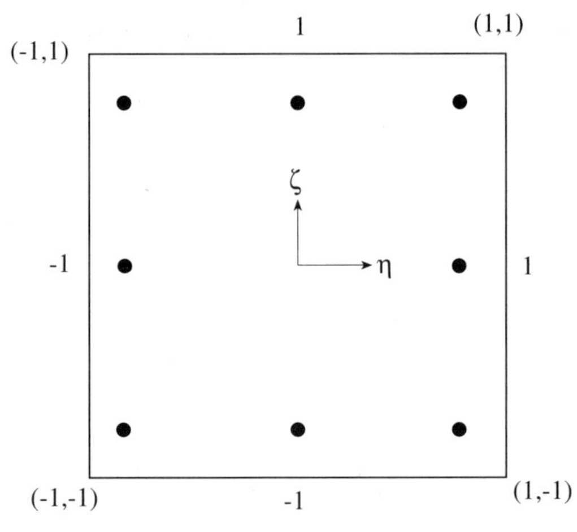

Figure 5.24 Gaussian integration points for square region with $m_1 = m_2 = 3$.

where the column vector $\boldsymbol{\sigma}_G$ contains the nine stresses at the Gauss points obtained from Eq. (5.142):

$$\boldsymbol{\sigma}_G = [\widetilde{\sigma}(\eta_1, \zeta_1) \quad \widetilde{\sigma}(\eta_2, \zeta_2) \quad \widetilde{\sigma}(\eta_3, \zeta_3) \quad \widetilde{\sigma}(\eta_4, \zeta_4) \quad \widetilde{\sigma}(\eta_5, \zeta_5)$$
$$\widetilde{\sigma}(\eta_6, \zeta_6) \quad \widetilde{\sigma}(\eta_7, \zeta_7) \quad \widetilde{\sigma}(\eta_8, \zeta_8) \quad \widetilde{\sigma}(\eta_9, \zeta_9)]^{\mathrm{T}} \tag{5.149}$$

and $\mathbf{H}$ is the 9×9 matrix

$$\mathbf{H} = \begin{bmatrix} \mathbf{N}(\eta_1, \zeta_1) \\ \mathbf{N}(\eta_2, \zeta_2) \\ \mathbf{N}(\eta_3, \zeta_3) \\ \mathbf{N}(\eta_4, \zeta_4) \\ \mathbf{N}(\eta_5, \zeta_5) \\ \mathbf{N}(\eta_6, \zeta_6) \\ \mathbf{N}(\eta_7, \zeta_7) \\ \mathbf{N}(\eta_8, \zeta_8) \\ \mathbf{N}(\eta_9, \zeta_9) \end{bmatrix} \tag{5.150}$$

The smoothed stresses are given by (Eq. 5.146)

$$\widetilde{\sigma}(\eta, \zeta) = \mathbf{N}(\eta, \zeta)\mathbf{H}^{-1}\boldsymbol{\sigma}_G \tag{5.151}$$

Since the kth shape function is 1 at node k, with all other shape functions zero, the smoothed stresses at the nodes are given by

$$\widetilde{\boldsymbol{\sigma}} = \mathbf{H}^{-1}\boldsymbol{\sigma}_G \tag{5.152}$$

where $\widetilde{\boldsymbol{\sigma}}$ is a column vector of the stresses at the nine nodes shown in Fig. 4.1.

The matrix $\mathbf{H}^{-1}$ can be written in the form

$$\mathbf{H}^{-1} = \begin{bmatrix} q & d & b & d & c & p & b & p & e \\ 0 & r & 0 & 0 & a & 0 & 0 & f & 0 \\ b & d & q & p & c & d & e & p & b \\ 0 & 0 & 0 & r & a & f & 0 & 0 & 0 \\ 0 & 0 & 0 & 0 & 1 & 0 & 0 & 0 & 0 \\ 0 & 0 & 0 & f & a & r & 0 & 0 & 0 \\ b & p & e & d & c & p & q & d & b \\ 0 & f & 0 & 0 & a & 0 & 0 & r & 0 \\ e & p & b & p & c & d & b & d & q \end{bmatrix} \tag{5.153}$$

The Gaussian integration points and corresponding weighting factors can be identified in Table 5.2. For our 3×3 Gaussian integration, these constants are listed in Table 5.2 for the nodes ordered as shown in Fig. 4.1.

For the Gaussian points shown in Table 5.2, the values of the entries in the matrix $\mathbf{H}^{-1}$ are calculated to be

$$a = -\frac{2}{3}, \quad b = \frac{5}{18}, \quad c = \frac{4}{9}, \quad d = -\frac{5+\sqrt{15}}{9}, \quad e = \frac{5(4-\sqrt{15})}{18}$$
$$f = \frac{5-\sqrt{15}}{6}, \quad p = \frac{-5+\sqrt{15}}{9}, \quad q = \frac{5(4+\sqrt{15})}{18}, \quad r = \frac{5+\sqrt{15}}{6} \tag{5.154}$$

Example 5.13 Rectangular Cross Section. Figure 5.25 shows a nonhomogeneous rectangular cross section, with the left half made of aluminum and the right half copper. The input data file is discussed in Appendix B.

The calculated cross-section area is 83.36 in^2. This is modulus weighted. The geometric area of the section is, however, 60 in^2. The reference elastic modulus E_r is chosen to be equal to the modulus of elasticity E_a of aluminum, so that the area is calculated as

$$A = A_a + A_c\frac{E_c}{E_a} = 30 + 30\left(\frac{18.5}{10.4}\right) = 83.36 \text{ in}^2 \tag{1}$$

where E_c denotes the modulus of elasticity of copper.

The torsional constant for this section is found to be $J = 106.22$ in^4. An analytical formula for the torsional constant of a rectangular strip of two materials is derived in Muskhelishvili (1953). This formula, normalized by dividing through by the shear modulus G_a, is

$$J = \frac{1}{3}(L_a + \mu L_c)t^3 - 3.361\frac{t^4}{16}\frac{1+\mu^2}{1+\mu} \tag{2}$$

where t is the thickness, L_a and L_c are the widths of the aluminum and copper parts, and μ is the ratio of the shear moduli:

$$\mu = \frac{G_c}{G_a} = \frac{E_c}{E_a} = \frac{18.5}{10.4} \tag{3}$$

The second equality holds because Poisson's ratio has been taken to be identical for the two materials. The formula gives $J = 106.12$ in^4, in close agreement with the numerically determined value.

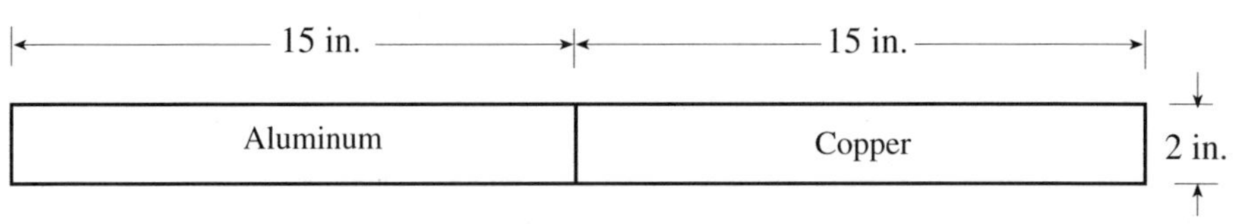

Figure 5.25 Rectangular strip made of two materials.

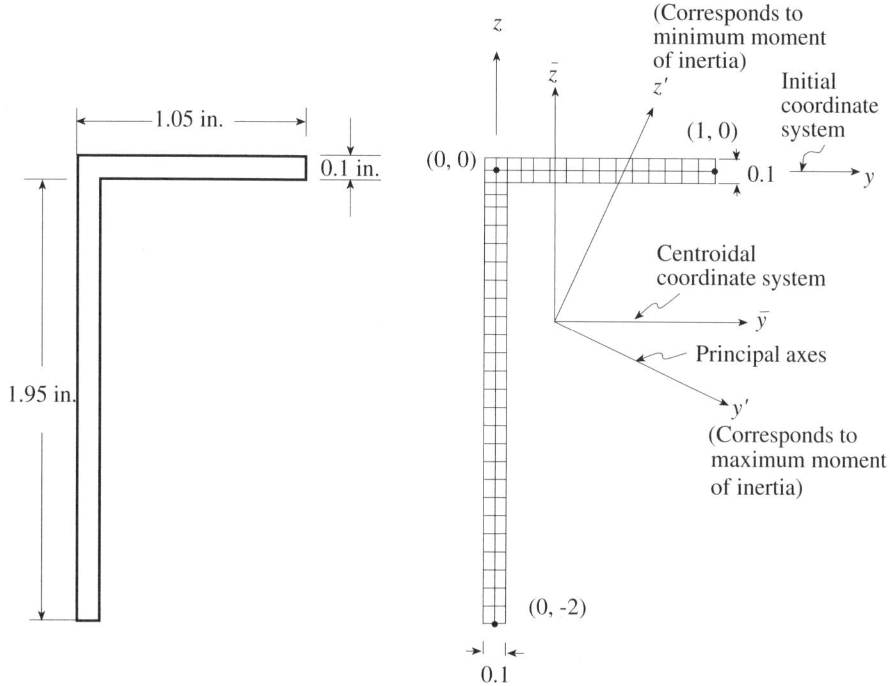

(*a*) Angle section (*b*) Mesh

Figure 5.26 Thin-walled L section.

Example 5.14 Properties of an L Section. Return again to the L section of Examples 1.1, 1.2, and 4.1 (Fig. 5.26). The thin-walled beam program of Appendix A can be used to find both cross-sectional properties as well as the stress distribution over the cross section. As in Example 4.1, let a and t of Example 1.1 be 1.0 in. and 0.1 in., respectively. For the results given in this example and in Table 1.1, the mesh of Fig. 5.26*b* was employed. The input data for a similar L section is discussed in Appendix B.

If we consider the cross section to be made up of two segments, one of length 1.95 in. and thickness 0.1 in. and the other of length 1.05 in. and thickness 0.1 in. (Fig. 5.26), the torsional constant J can be calculated using Eq. (5.43):

$$J = \tfrac{1}{3} \sum_{i=1}^{2} b_i t_i^3 = \tfrac{1}{3}[1.95(0.1)^3 + 1.05(0.1)^3] = 0.001 \text{ in}^4 \tag{1}$$

The torsional constant, J, is calculated by the computer program as $J = 0.00099$ in^4, a 1% difference.

Suppose that a torque of 1000 in.-lb, M_x, is placed on the beam. The maximum torsional stress for the multisegment thin-walled beam can be calculated using Eq. (5.44):

$$\tau_{\max} = \frac{t_{\max} M_x}{J} = \frac{0.1(1000)}{0.001} = 100{,}000 \text{ psi} \tag{2}$$

Using the thin-walled beam computer program, the maximum shear stress is approximately 98,000 psi, a 2% difference. Thus, the thin-walled open section formulas of Section 5.2.1 are in close agreement with the finite element code in this case.

Example 5.15 Comparison of a Thin-Walled Cylinder with and without a Slit. Use the thin-walled beam computer program of Appendix A to compare the torsional constants of the two thin-walled cylinders of Example 5.7 (Fig. 5.13). Both of the cylinders have a circular cross section with a 4-in. outside radius and $\frac{1}{4}$ in. thickness. However, one of the tubes has a longitudinal slit. The input data for a similar example is discussed in Appendix B.

SOLUTION. For the cylinder with no slit, the torsional constant J is the polar moment of inertia and was calculated in Example 5.7 as

$$J = \frac{\pi}{2}(r_o^4 - r_i^4) = \frac{\pi}{2}(4^4 - 3.75^4) = 91.49 \text{ in}^4 \tag{1}$$

Using the finite element computer program with a closed section, $J = 91.49$ in^4.

For the cylinder with a longitudinal slit, the torsional constant J can be found using the open-section formula of Eq. (5.43) and was calculated in Example 5.7 to be

$$J = \tfrac{1}{3}\sum_{i=1}^{m} b_i t_i^3 = \tfrac{1}{3}(24.35)\left(\tfrac{1}{4}\right)^3 = 0.13 \text{ in}^4 \tag{2}$$

Using the thin-walled beam computer program, the torsional constant of the open section was found to be $J = 0.13$ in^4. Thus, for both the tube with a slit and the tube without a slit, the finite element program gives the results obtained in Example 5.7.

5.7 ALTERNATIVE COMPUTATIONAL METHODS

The finite element method is the computational technique of choice in this book. However, other techniques, in particular boundary solution methods, are available to solve efficiently the shear stress–related problems. The boundary element and boundary integration methods are outlined in this section. These methods are more suitable for solid sections than for thin-walled members. Liu (1993) and Schramm and Pilkey (1994) discuss these methods in depth.

The boundary solution methods involve the solution of integral equations obtained from variational formulations or from governing differential equations. Initially, integral equations with unknowns on the boundary are developed. Then, various numerical methods can be utilized to solve the integral equations. Two numerical methods are introduced here: the boundary element method and the direct integration method.

Boundary integral equations are developed for the displacement formulation of the torsion problem. Similar boundary integral equations can be derived for the force formulation of Saint-Venant's torsion problem. The other shear stress–related problems in this book can also be represented in terms of boundary integral equations.

Blended Interpolation Function Elements Before considering boundary solution methods, it is of interest to consider how to couple the elements of a cross section that possesses both thin-walled and solid components. It is convenient to have *transition elements* to couple the different components. In contrast to standard isoparametric elements, the transition elements can have different numbers of intermediate nodes on the element edges. They are appropriate for the coupling of solid and thin-walled components as well as for use in local mesh refinements. Since these elements use blended interpolation for the generation of shape functions, they are often referred to as *blended interpolation function* (BIF) *elements*. A typical BIF element and its use in coupling solid and thin-walled cross sections are shown in Fig. 5.27. The development of BIF elements is discussed in such references as Cavendish et al. (1977) and Röhr (1985).

5.7.1 Boundary Integral Equations

This formulation of an integral equation begins with a weighted residual expression in which the governing differential equation and the boundary conditions are multiplied by a weighting function and integrated over the cross section and the boundary. The integrals are then transformed so that the unknowns appear only on the boundary. A thorough discussion of this procedure is provided in Pilkey and Wunderlich (1994).

This displacement form for the governing differential equations for Saint-Venant torsion are given by (Eq. 5.8)

$$\nabla^2 \omega = 0 \tag{5.155}$$

within the boundary and the condition (Eq. 5.11)

$$\frac{\partial \omega}{\partial n} = z n_y - y n_z = \overline{q} \tag{5.156}$$

on the boundary, where $q = \partial \omega / \partial n$ and the bar over $\overline{q}$ indicates a prescribed condition. An integral equation formulation can be obtained using a weighted residual method. If v and $\overline{v}$ are set equal to the weighting function ω^*, the integral formulation

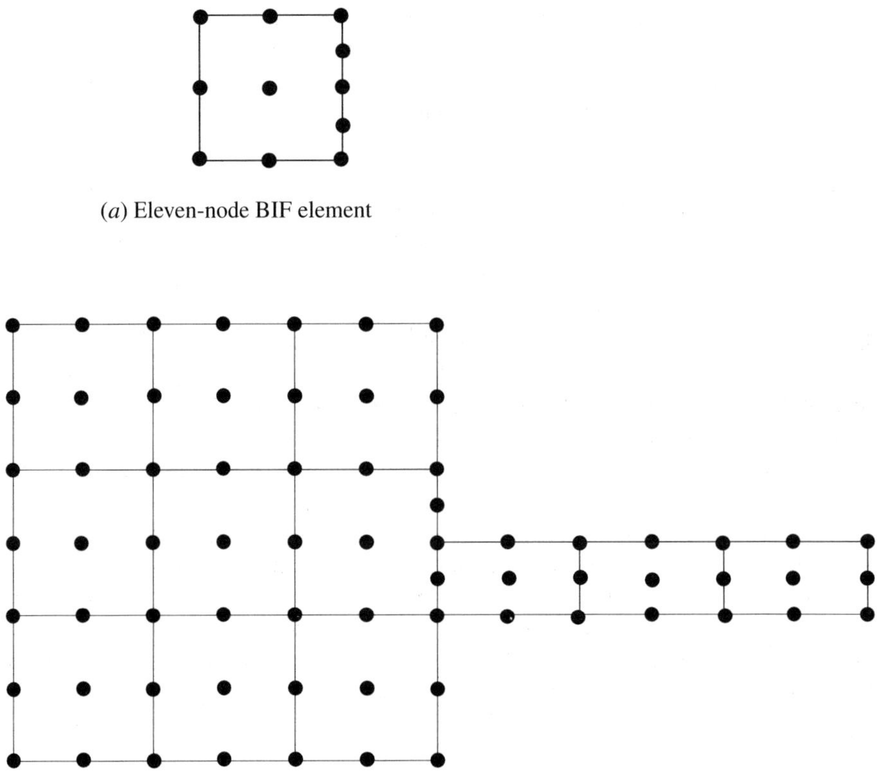

(*a*) Eleven-node BIF element

(*b*) Coupling of solid and thin-walled cross sections

Figure 5.27 BIF element and its use.

of Eq. (5.105) can be expressed as

$$\int_A \omega^* \left(\nabla^2 \omega\right) \, dA - \int_S \omega^* \left(q - \overline{q}\right) \, dS = 0 \tag{5.157}$$

where A is the area of the cross section and S is the boundary. This corresponds to an *extended Galerkin's formula*.

Apply Gauss's integral theorem to Eq. (5.157) to obtain

$$\int_A \omega \left(\nabla^2 \omega^*\right) \, dA + \int_S \left(\overline{q}\omega^* - \omega q^*\right) \, dS = 0 \tag{5.158}$$

where $q^* = \partial\omega^*/\partial n$. The quantity ω^*, which will be a singular function, is referred to as a *fundamental solution*. Introduce

$$\nabla^2 \omega^* = -2\pi\delta\left(\xi, x\right) \tag{5.159}$$

where δ is a Dirac delta function and ξ, x are two points in the cross section. The solution to Eq. (5.159) is

$$\omega^* = \ln \frac{1}{r} \tag{5.160}$$

with $r = r(\xi, x)$ as the distance from ξ, where the delta function is applied to x a point under consideration. The derivative $\partial\omega^*/\partial n$ can be shown to be

$$\frac{\partial \omega^*}{\partial n} = -\frac{1}{r^2}\left(r_y n_y + r_z n_z\right) \tag{5.161}$$

where r_y and r_z are the components of r in the y and z directions, respectively, and n_y and n_z are the direction cosines of the normal with respect to the y and z axes, respectively. Substitute Eqs. (5.159), (5.160), and (5.161) into Eq. (5.158) and introduce $\ln(1/r) = -\ln r$

$$2\pi\omega(\xi) + \int_S \overline{q} \ln r \, dS - \int_S \omega \frac{r_y n_y + r_z n_z}{r^2} \, dS = 0 \tag{5.162}$$

where the point ξ is inside the boundary and integrands of the line integrals are referred to point x on the boundary.

The intention for boundary integral equation formulations is to establish an integral with unknowns occurring on the boundary only. This can be accomplished by moving point ξ to the boundary. Then, since r is the distance between ξ and x, r is zero and the integrands of the boundary integral of Eq. (5.162) become singular. These singularities can be investigated by letting the boundary S be represented as

$$S = (S - S_\epsilon) + S_\epsilon \tag{5.163}$$

for which S_ϵ is a semicircle of radius ϵ and going to the limit of the integrals as $\epsilon \to 0$. Equation (5.162) becomes

$$c\omega(\xi) + \int_S \overline{q} \ln r \, dS - \int_S \omega \frac{r_y n_y + r_z n_z}{r^2} \, dS = 0 \tag{5.164}$$

with

$$c = \begin{cases} 2\pi & \text{when } \xi \text{ is inside the boundary of the cross section} \\ \pi & \text{when } \xi \text{ is on the boundary that is smooth at } \xi \\ \text{constant} & \text{when } \xi \text{ is on the boundary that is not smooth at } \xi \end{cases}$$

This boundary integral equation, which contains the unknown variable $\omega(\xi)$ both inside and outside the integral and is referred to as *Fredholm's integral of the second kind*, can be solved for $\omega(\xi)$.

An analytical solution of the integral equation of Eq. (5.164) is difficult to obtain; hence, numerical solutions are sought. The most common approach is the *bound-*

ary element method. Alternatively, the *direct boundary integration method* can be employed.

5.7.2 Boundary Element Method

The boundary element method is similar to the finite element in the sense that approximate trial solutions are employed for a region that is discretized into elements. The basic steps involved in the boundary element method are:

1. Discretize the boundary S into elements.
2. Assign approximate shape functions for the unknowns for each element.
3. Introduce the elements with shape functions into Eq. (5.164), leading to a system of linear equations in the unknowns at the nodes on the boundary.
4. Impose the boundary conditions and solve the equations for the unknowns on the boundary.
5. Find the unknowns as desired inside the boundary and calculate the cross-sectional properties and stresses.

The boundary (perimeter) S of the cross section is discretized into elements; typically, the elements are straight lines (constant or linear elements) or curves (quadratic or higher order), as shown in Fig. 5.28. The trial function for element e can be expressed as

$$\omega = \mathbf{N}\boldsymbol{\omega}^e \tag{5.165}$$

which is similar to the finite element expression. The element shape functions are contained in $\mathbf{N}$. In the case of a constant element with one node at the centroid, the unknown warping function ω along the element is assumed to be equal to the warping function at the nodal point. This then is a single-degree-of-freedom element.

The discrete version of Eq. (5.164) appears as

$$c\omega\left(\xi_j\right) - \sum_{e=1}^{M}\left(\int_{S_e} \frac{r_y n_y + r_z n_z}{r^2}\mathbf{N}^e\, dS\right)\omega^e = \sum_{e=1}^{M}\left(\int_{S_e} \ln r\ \mathbf{N}^e\, dS\right)\mathbf{q}^e \tag{5.166}$$

where c is defined in Eq. (5.164), $\omega\left(\xi_j\right)$ is the value of the warping function at ξ_j, ξ_j is the point where node j is located, S_e is the length of the eth element, r is the distance from point ξ_j to a point on S_e, and M is the number of boundary elements. These equations can be assembled into a system of linear equations, which are then solved for the warping function at the nodes.

Finally, other variables of interest, such as the torsional constant and stresses, can be computed.

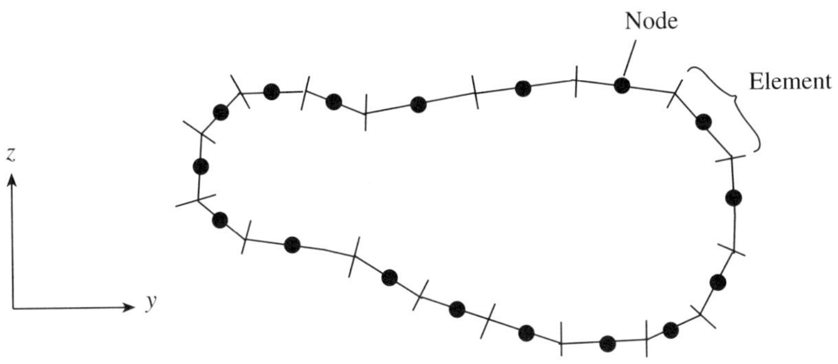

(*a*) Constant elements

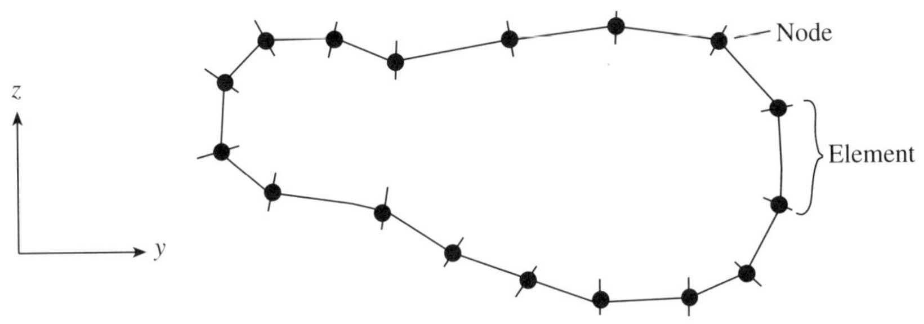

(*b*) Linear elements

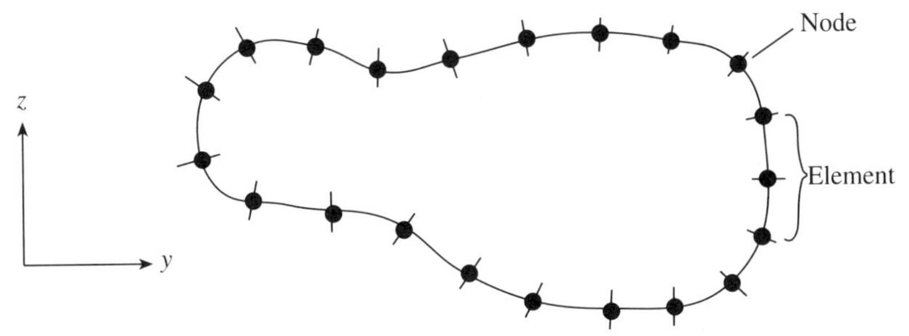

(*c*) Quadratic elements

Figure 5.28 Boundary element discretization for two-dimensional cross-sectional problems.

5.7.3 Direct Integration of the Integral Equations

Direct integration of the boundary integral equations provides an alternative, especially for two-dimensional problems such as the cross-sectional calculations for beams, to the finite element and boundary element methods. Basically, with the direct integration method, the boundary of the domain is divided into m segments, and then a numerical integration scheme such as Gauss quadrature (Section 4.3) is applied to each segment. The integral for the complete boundary is obtained by summing the integrals for the segments. Define ξ_i to be a Gauss integration point on the boundary and S_ℓ to be the length of the ℓth segment. Then Eq. (5.164) can be expressed as

$$c\omega(\xi_i) - \sum_{\ell=1}^{m} \int_{S_\ell} \omega \frac{r_y n_y + r_z n_z}{r^2} \, dS = -\sum_{\ell=1}^{m} \int_{S_\ell} \overline{q} \ln r \, dS \tag{5.167}$$

where the unknowns are the warping function values at the Gauss integration points. Equation (5.167) leads to a set of linear equations that can be solved to find the distribution of the warping function. These results can be utilized to compute other characteristics, such as the torsional constant and stresses. Numerical experimentation shows (Pilkey and Wunderlich, 1994) that, typically, an accurate torsional constant is calculated with significantly fewer integration points for direct integration than nodes for the boundary element method.

REFERENCES

Becker, K., and Braun, M. (1987). Numerical solution of the torsion problem for axisymmetric bodies using different boundary integral equations, in *Boundary Elements IX*, Brebbia, C., ed., Springer-Verlag, Berlin.

Boresi, A. P., and Chong, K. P. (1987). *Elasticity in Engineering Mechanics*, Elsevier, Amsterdam, The Netherlands.

Bredt, R. (1896). "Kritische Bemerkungen zur Drehungselastisität," *Zeitschrift des Vereines deutscher Ingenieure*, Vol. 40, p. 785.

Cavendish, J. C., Gordon, W. J., and Hall, C. A. (1977). Substructured macro elements based on locally blended interpolation, *Int. J. Numer. Methods Eng.*, Vol. 11, pp. 1405–1421.

Christiansen, S. A. (1978). 'Review of some integral equations for solving the Saint-Venant torsion problem, *J. Elasticity*, Vol. 8, pp. 1–20.

Cook, R. D., Malkus, D. S., and Plesha, M. E. (1989). *Concepts and Applications of Finite Element Analysis*, 3rd ed., Wiley, New York.

Dhatt, G., and Touzot, G. (1984). *The Finite Element Method Displayed*, Wiley, New York.

Heise, U. (1973). Combined application of finite element methods and Richardson extrapolation to the torsion problem, in *The Mathematics of Finite Elements and Applications*, J. R. Whitman, ed., Academic Press, San Diego, Calif.

Hinton, E., and Campbell, J. S. (1974). Local and global smoothing of discontinuous finite element functions using a least squares method, *Int. J. Numer. Methods Eng.*, Vol. 8, pp. 461–480.

Hinton, E., Scott, F. C., and Ricketts, R. E. (1975). Local least squares smoothing for parabolic isoparametric elements, *Int. J. Numer. Methods Eng.*, Vol. 9, pp. 235–256.

Hromadka, T. V., II, and Pardoen, G. C. (1985). Application of the CVBM to non-uniform St. Venant torsion, *Comput. Methods Appl. Mech. Eng.*, Vol. 53, pp. 149–161.

Katsikadelis, J. T., and Sapountzakis, E. J. (1985). Torsion of composite bars by boundary element method, *J. Eng. Mech.*, Vol. 111, pp. 1197–1210.

Little, R. W. (1973). *Elasticity*. Prentice Hall, Upper Saddle River, N.J.

Liu, Yongquan (1993). Computation of beam cross-sectional properties and stresses, Ph.D. dissertation, University of Virginia, Charlottesville, Va.

Love, A. E. (1944). *A Treatise on the Mathematical Theory of Elasticity*, 4th ed., Dover, New York.

Menken, C. M. (1989). A hybrid formulation for determining torsion and warping constants, *J. Mech. Sci.*, Vol. 31, pp. 25–35.

Moan, T. (1973). Finite element stress field solution of the problem of Saint Venant torsion, *Int. J. Numer. Methods Eng.*, Vol. 5, pp. 455–458.

Muskhelishvili, N. I. (1953). *Some Basic Problems of the Mathematical Theory of Elasticity*, Wolters-Noordhoff, Groningen, The Netherlands.

Nguyen, S. H. (1992). An accurate finite element formulation for linear elastic torsion calculations, *Comput. Struct.*, Vol. 42, pp. 707–711.

Noor, A. K., and Andersen, C. M. (1975). Mixed isoparametric elements for Saint-Venant torsion, *Comput. Methods Appl. Mech. Eng.*, Vol. 6, pp. 195–218.

Pilkey, W. D. (1994). *Formulas for Stress, Strain, and Structural Matrices*, Wiley, New York.

Pilkey, W. D., and Wunderlich, W. (1994). *Mechanics of Structures, Variational and Computational Methods*, CRC Press, Boca Raton, Fla.

Pilkey, W. D., Schramm, U., Liu, Y. Q., and Antes, H. (1994). Boundary integration as a means of solving two-dimensional problems, *Finite Elements Anal. Des.*, Vol. 18, pp. 17–29.

Reagan, S. (2002). Constrained torsion of prismatic bars, Ph.D. dissertation, University of Virginia, Charlottesville, Va.

Rivello, R. M. (1969). *Theory and Analysis of Flight Structures*, McGraw-Hill, New York.

Röhr, U. (1985). Lokale finite Elementnetzverfeinerungen bei Plätten und Scheibenaufgaben mittels gemischter Interpolation, *Schiffbauforschung*, Vol. 24, pp. 39–50.

Rubinstein, I., and Rubinstein, L. (1993). *Partial Differential Equations in Classical Mathematical Physics*, Cambridge University Press, Cambridge.

Schramm, U., and Pilkey, W. D. (1993). Structural shape optimization for the torsion problem using direct integration and B-splines. *Comput. Methods Appl. Mech. Eng.*, Vol. 107, pp. 251–268.

Schramm, U., and Pilkey, W. D. (1994). Higher order boundary elements for shape optimization using rational B-spline, *Eng. Anal. Boundary Elements*, Vol. 14, pp. 255–266.

Sokolnikoff, I. S. (1956). *Mathematical Theory of Elasticity*, McGraw-Hill, New York.

Surana, K. S. (1979). Isoparametric elements for cross-sectional properties and stress analysis of beams, *Int. J. Numer. Methods Eng.*, Vol. 14, pp. 475–497.

Yoo, C. H. (1986). Cross-sectional properties of thin-walled multi-cellular section, *Comput. Struct.*, Vol. 22, pp. 53–61.

Zienkiewicz, O. C. (1977). *The Finite Element Method*, 3rd ed., McGraw-Hill, New York.

CHAPTER 6

BEAMS UNDER TRANSVERSE SHEAR LOADS

This chapter deals with the shear stresses in a beam with transverse applied loading. Initially, approximate engineering beam theory formulas are developed. Then, more accurate shear stresses due to transverse loading are obtained by using the theory of elasticity. In particular, for a cantilevered, prismatic beam a relationship between the cross-sectional shear force and shear stress is derived. It is assumed that this cross-sectional relationship applies for any loading, any cross-sectional shape, and any beam boundary conditions. A finite element analysis is used to compute the shear stresses.

The location of the shear center is discussed in this chapter. Also, shear deformation coefficients, which are useful in computing shear deformation effects on deflections, are determined here. Finally, the deflection of beams, using stiffness matrices, including the influence of shear deformation, is studied.

6.1 TRANSVERSE SHEAR STRESSES IN A PRISMATIC BEAM

6.1.1 Approximate Shear Stress Formulas Based on Engineering Beam Theory

In this section we deal with a beam theory approximation for the shear stresses due to transverse loading of a beam. To find the shear stress corresponding to the transverse shear force, consider the element and slice of Fig. 6.1. The cross-sectional area in the yz plane of the slice (Fig. 6.1b) is A' and the width is b. If τ is the average shear stress on the area $b\ dx$, the total horizontal force developed is $\tau b\ dx$. From Fig. 6.1d

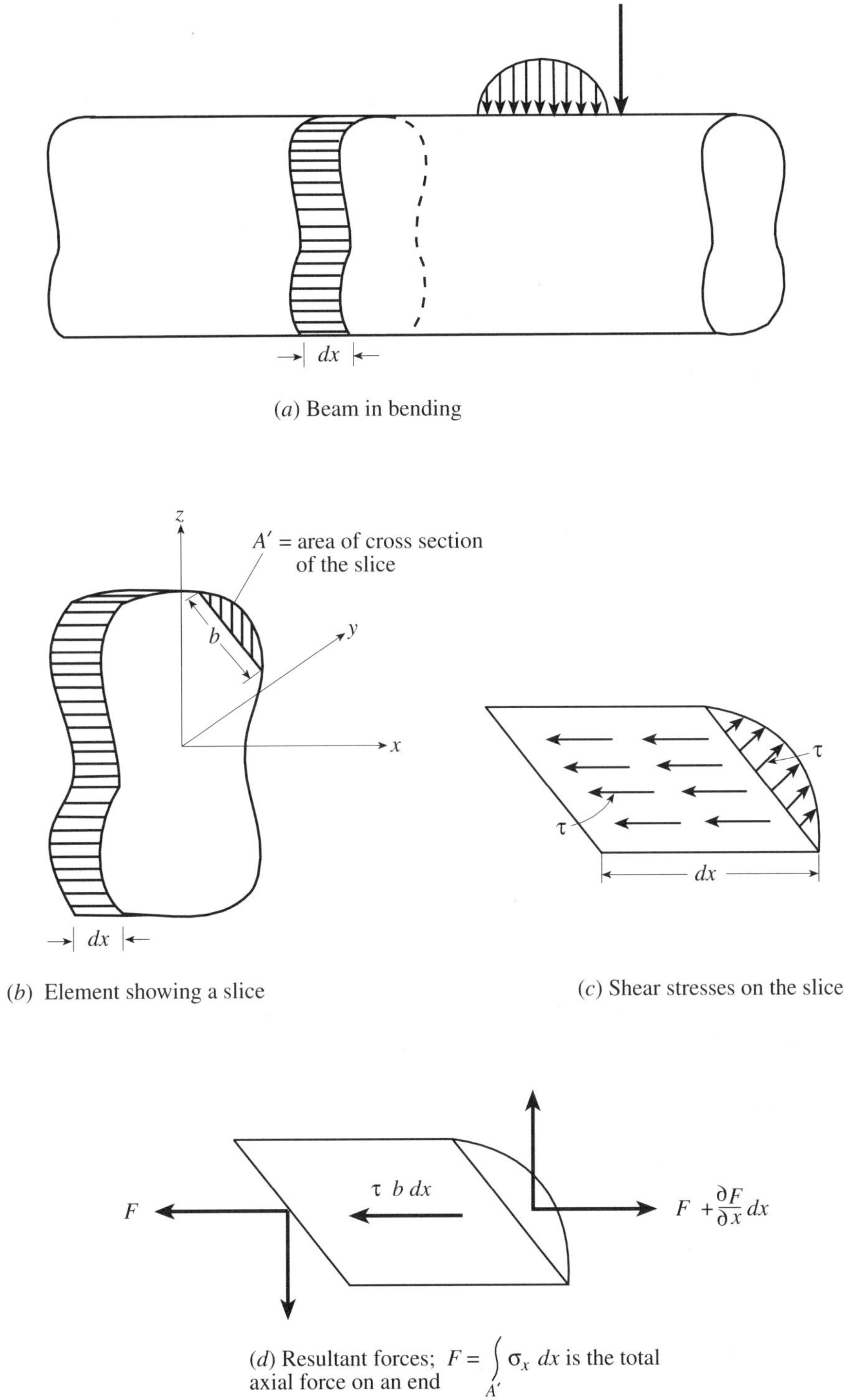

Figure 6.1 Forces on a slice of an element of a beam.

the axial equilibrium requirement is

$$\tau b\, dx = \frac{\partial F}{\partial x}\, dx \tag{6.1}$$

or

$$\tau = \frac{1}{b}\frac{\partial F}{\partial x} = \frac{1}{b}\int_{A'} \frac{\partial \sigma_x}{\partial x}\, dA \tag{6.2}$$

Substitution of σ_x of Eq. (1.57) into this expression gives

$$\tau = \frac{1}{b}\left(-\frac{I_{\overline{yz}}(\partial M_y/\partial x) + I_{\overline{y}}(\partial M_z/\partial x)}{I_{\overline{y}}I_{\overline{z}} - I_{\overline{yz}}^2} Q_z' + \frac{I_{\overline{z}}(\partial M_y/\partial x) + I_{\overline{yz}}(\partial M_z/\partial x)}{I_{\overline{y}}I_{\overline{z}} - I_{\overline{yz}}^2} Q_y' \right) \tag{6.3}$$

where

$$Q_y' = \int_{A'} \overline{z}\, dA \quad \text{and} \quad Q_z' = \int_{A'} \overline{y}\, dA \tag{6.4}$$

are the first moments of the cross-sectional area of the slice about the centroidal axes. That is, Q_y' and Q_z' are the moments of the area beyond b about the centroidal axes. From Eq. (2.31),

$$\frac{\partial M_y}{\partial x} = V_z \quad \text{and} \quad \frac{\partial M_z}{\partial x} = -V_y \tag{6.5}$$

so that Eq. (6.3) can be written as

$$\tau = \frac{Q_z' I_{\overline{y}} - Q_y' I_{\overline{yz}}}{b(I_{\overline{y}}I_{\overline{z}} - I_{\overline{yz}}^2)} V_y + \frac{Q_y' I_{\overline{z}} - Q_z' I_{\overline{yz}}}{b(I_{\overline{y}}I_{\overline{z}} - I_{\overline{yz}}^2)} V_z \tag{6.6}$$

This expression, with an appropriate selection of b, provides the average shearing stress on a particular cut of width b. If the width b is parallel to the y axis, Eq. (6.6) gives τ_{xz}, which is equal to τ_{zx}. Similarly, if b is parallel to the z axis, τ of Eq. (6.6) represents $\tau_{yx} = \tau_{xy}$.

The familiar formula from elementary mechanics is obtained from Eq. (6.6) by considering the net forces M_z and V_y in the xz plane to be zero and $\overline{y}, \overline{z}$ to be the principal axes of bending, so that $I_{\overline{yz}}$ is zero. Then

$$\tau_{xz} = \frac{V_z Q_y'}{I_y b} \tag{6.7}$$

In all of the formulas above it is important to remember that the shear stress is assumed to be constant along the dimension b. This assumption is not made for the theory of elasticity solution of the next section. For thin cross sections, it is customary to use the product τb rather than τ. This product is q, the shear flow due to transverse shear or bending.

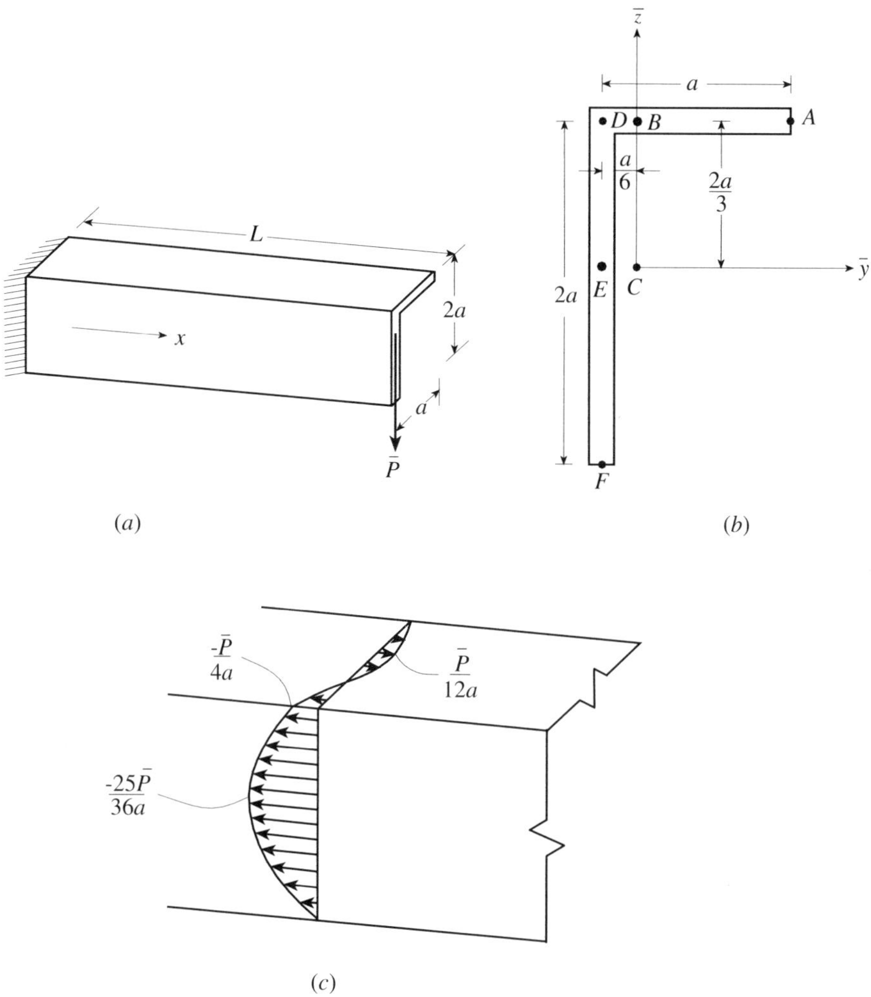

Figure 6.2 Shear flow in an angle section.

Example 6.1 Shear Stress in an Angle Section. Consider the angle section of Fig. 6.2a with a concentrated force $\overline{P}$ that acts in the negative z direction. From statics the shear force V_z in any section is $-\overline{P}$ and V_y is zero. The moments and product of inertia about the centroid for this cross section were found in Example 1.1 (Fig. 1.9) to be

$$I_{\bar{y}} = \tfrac{4}{3}a^3 t \qquad I_{\bar{z}} = \tfrac{1}{4}a^3 t \qquad I_{\bar{y}\bar{z}} = \tfrac{1}{3}a^3 t \tag{1}$$

Inserting these values into Eq. (6.6) and multiplying by $b = t$, the thickness of the thin-walled section, gives an equation for the shear flow along the section:

$$\tau t = q = -\frac{\frac{9}{8}Q'_y - \frac{3}{2}Q'_z}{a^3 t}\overline{P} \tag{2}$$

To complete the problem, we must find expressions for the first moments of cross-sectional area, Q'_y and Q'_z, on the section. Recall that Q'_y is defined as the moment of the area beyond a particular point of interest about the $\overline{y}$ axis. To facilitate this process it should also be noted that since the coordinate axes are located at the centroid, the first moment of the area from a point, p, to one end of the cross section is the same in magnitude as the first moment from p to the other end of the section. However, the two moments will differ in sign. Thus, we can define the areas with respect to either the end at point A or the end at point F, keeping in mind that the moments that are defined with respect to the end at point F must be negated before they can be substituted into (2). The calculation of Q'_y and Q'_z, especially the determination of their signs, can appear to be rather complicated. This calculation is discussed in-depth in Example 7.2.

We will calculate the moments of areas along the four segments BA, BD, ED, and EF. On BA we define the area A' as the area from the point that is distance y from the centroid to the end of segment BA at point A. The quantity $Q_{z_{BA}}$ is equal to $\overline{y}A'$, where $\overline{y}$ is the distance in the y direction from the $\overline{z}$ axis to the centroid of area A'. Since the length of segment BA is $5a/6$, A' is equal to $(5a/6 - y)t$ and $\overline{y}$ is equal to $y + (5a/6 - y)/2 = (5a/6 + y)/2$. Thus,

$$Q'_{z_{BA}} = \overline{y}A' = \left(\frac{5a}{6} - y\right)t\left(\frac{5a}{6} + y\right)\frac{1}{2} = \frac{25a^2t}{72} - \frac{t}{2}y^2 \tag{3}$$

Similar reasoning gives

$$Q'_{y_{BA}} = \overline{z}A' = \left(\frac{5a}{6} - y\right)t\left(\frac{2a}{3}\right) = \frac{5a^2t}{9} - \frac{2at}{3}y$$

For section BD, first consider the moment of the area about the $\overline{z}$ axis, $Q'_{z_{BD}}$. In this case we define the area A' to be the area from points between B and D to the end at point F. Note, however, that since we are defining the area with respect to point F instead of A, the quantity must be negated in order to follow our sign convention.

$$Q'_{z_{BD}} = -\left[-\frac{t}{2}\left(\frac{a}{6} + y\right)\left(\frac{a}{6} - y\right) + 2at\left(-\frac{a}{6}\right)\right] = \frac{25a^2t}{72} - \frac{t}{2}y^2 \tag{4}$$

For $Q'_{y_{BD}}$ the area will again be defined with respect to the end at point A. Thus, it is not necessary to negate the quantity

$$Q'_{y_{BD}} = -yt\frac{2a}{3} + \left(\frac{5a}{6}\right)t\left(\frac{2a}{3}\right) = -\frac{2at}{3}y + \frac{5a^2t}{9} \tag{5}$$

Using the same approach, the moments of the cross-sectional areas along segments ED and EF are

$$\begin{aligned} Q'_{zED} &= \frac{at}{6}z + \frac{2a^2t}{9} \\ Q'_{yED} &= \frac{8a^2t}{9} - \frac{t}{2}z^2 \\ Q'_{zEF} &= \frac{2a^2t}{9} + \frac{at}{6}z \\ Q'_{yEF} &= \frac{8a^2t}{9} - \frac{t}{2}z^2 \end{aligned} \tag{6}$$

Substituting these quantities into (2) leads to expressions for the shear flow on the four segments:

$$\begin{aligned} q_{BA} = q_{BD} &= \frac{\overline{P}}{a^3t}\left(-\frac{3t}{4}y^2 + \frac{3at}{4}y - \frac{5a^2t}{48}\right) \\ q_{ED} = q_{EF} &= \frac{\overline{P}}{a^3t}\left(\frac{9t}{16}z^2 + \frac{at}{4}z - \frac{2a^2t}{3}\right) \end{aligned} \tag{7}$$

The final distribution is shown in Fig. 6.2c. The stress on section DF is negative since it results in a downward shearing force on beam sections.

This engineering beam theory can be extended to calculate the normal stress σ_z, which is usually negligible (see, e.g., Oden and Ripperger, 1981).

6.1.2 Theory of Elasticity Solution

In this section we present Saint-Venant's elasticity solution for a homogeneous prismatic beam subjected to transverse shear loads. A relationship between the cross-sectional shear force and shear stress is to be derived for a particular set of boundary conditions. Since analytical solutions are available in the literature for a cantilevered beam (see, e.g., Sokolnikoff, 1956), the beam shown in Fig. 6.3 is chosen for study. It is to be assumed that the cross-sectional relationships derived for a cantilevered beam subjected to a transverse shear load applies for a beam with any loading, any cross-sectional shape, and any beam boundary conditions. Finally, a finite element solution is used to compute the shear stresses.

Shear Stresses due to a Transverse Load in the z Direction Suppose that a transverse shear force V_z is applied at the *shear center* S_1 (see Section 6.2), so that no twisting moment is exerted by it. The total length of the beam is L, and it may be assumed that the end with the centroid at point O is fixed. The beam axis x is the line joining the centroids of the cross sections. At the cross section with centroid C,

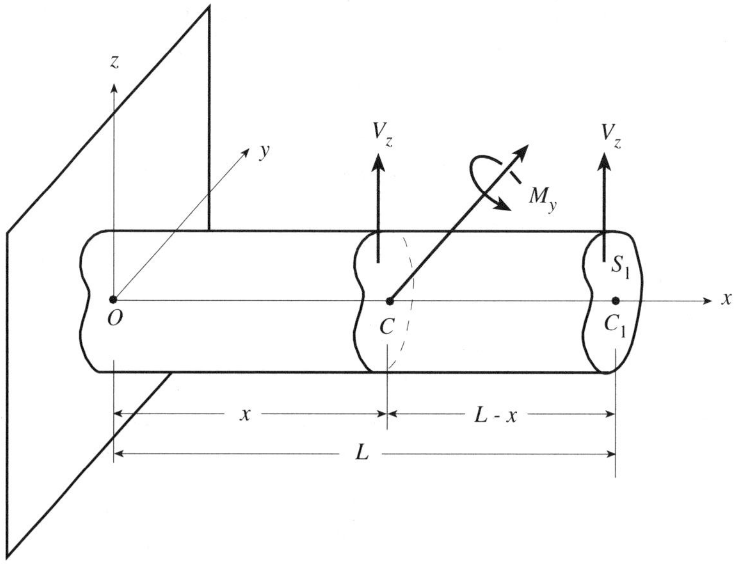

Figure 6.3 Beam subjected to a shear force.

at a distance x from the fixed end, the internal forces are the shear force V_z and the bending moment M_y:

$$M_y = -V_z(L - x) \tag{6.8}$$

Saint-Venant assumed that the stresses σ_y, σ_z, and τ_{yz} are negligibly small, leaving σ_x, τ_{xy}, and τ_{xz} as unknown stresses. In addition, he assumed that the distribution of the normal stress σ_x is given by Eq. (1.57) for a beam in pure bending:

$$\begin{aligned}\sigma_x &= \frac{I_z M_y z}{I_y I_z - I_{yz}^2} - \frac{I_{yz} M_y y}{I_y I_z - I_{yz}^2} \\ &= B(L - x)\left(I_{yz} y - I_z z\right)\end{aligned} \tag{6.9}$$

where for convenience the superscript bars over the coordinates have been dropped and the constant B is defined as

$$B = \frac{V_z}{I_y I_z - I_{yz}^2} \tag{6.10}$$

The state of stress is to be determined by solving for the two stresses τ_{xy} and τ_{xz}, called *transverse* or *direct* shear stresses, over the cross section.

With the assumption that the body forces are zero, the equations of equilibrium (Eq. 1.25) for this problem become

$$
\begin{aligned}
\frac{\partial \sigma_x}{\partial x} + \frac{\partial \tau_{xy}}{\partial y} + \frac{\partial \tau_{xz}}{\partial z} &= 0 \\
\frac{\partial \tau_{xy}}{\partial x} &= 0 \\
\frac{\partial \tau_{xz}}{\partial x} &= 0
\end{aligned}
\tag{6.11}
$$

Substitution of Eq. (6.9) into the first relationship of Eq. (6.11) leads to

$$
\frac{\partial \tau_{xy}}{\partial y} + \frac{\partial \tau_{xz}}{\partial z} = B\left(I_{yz} y - I_z z\right) \tag{6.12}
$$

The last two relationships of Eq. (6.11) show that the shear stresses are independent of x, which means that the shear stress distributions, functions of y and z, are the same over any cross section.

The strains are given by Hooke's stress–strain equations (Eq. 1.12) as

$$
\begin{aligned}
\epsilon_x &= \frac{\sigma_x}{E} & \epsilon_y &= -\frac{\nu}{E}\sigma_x & \epsilon_z &= -\frac{\nu}{E}\sigma_x \\
\gamma_{xy} &= \frac{2(1+\nu)}{E}\tau_{xy} & \gamma_{xz} &= \frac{2(1+\nu)}{E}\tau_{xz} & \gamma_{yz} &= 0
\end{aligned}
\tag{6.13}
$$

These strains satisfy four of the compatibility conditions of Eq. (1.9) identically. The remaining two conditions are

$$
\begin{aligned}
2\frac{\partial^2 \epsilon_y}{\partial x \partial z} &= \frac{\partial}{\partial y}\left(-\frac{\partial \gamma_{xz}}{\partial y} + \frac{\partial \gamma_{xy}}{\partial z} + \frac{\partial \gamma_{yz}}{\partial x}\right) \\
2\frac{\partial^2 \epsilon_z}{\partial x \partial y} &= \frac{\partial}{\partial z}\left(-\frac{\partial \gamma_{xy}}{\partial z} + \frac{\partial \gamma_{yz}}{\partial x} + \frac{\partial \gamma_{xz}}{\partial y}\right)
\end{aligned}
\tag{6.14}
$$

Substitution of the strains of Eq. (6.13) and then the stress σ_x of Eq. (6.9) into Eq. (6.14) yields

$$
\begin{aligned}
\frac{\partial}{\partial y}\left(\frac{\partial \tau_{xz}}{\partial y} - \frac{\partial \tau_{xy}}{\partial z}\right) &= \frac{\nu}{1+\nu} I_z B \\
\frac{\partial}{\partial z}\left(\frac{\partial \tau_{xz}}{\partial y} - \frac{\partial \tau_{xy}}{\partial z}\right) &= \frac{\nu}{1+\nu} I_{yz} B
\end{aligned}
\tag{6.15}
$$

The stresses τ_{xy} and τ_{xz} are next expressed in terms of a function $\Phi(y, z)$, which is assumed to have continuous partial derivatives of third order and is chosen such that the two compatibility conditions of Eq. (6.15) are satisfied identically:

$$\begin{aligned}\tau_{xy} &= \frac{B}{2(1+\nu)}\left[\frac{\partial\Phi}{\partial y} + \nu\left(I_{yz}\frac{y^2-z^2}{2} - I_z yz\right)\right] \\ \tau_{xz} &= \frac{B}{2(1+\nu)}\left[\frac{\partial\Phi}{\partial z} + \nu\left(I_{yz} yz + I_z\frac{y^2-z^2}{2}\right)\right]\end{aligned} \tag{6.16}$$

These expressions for the shear stresses can be written in the form

$$\begin{aligned}\tau_{xy} &= \frac{V_z}{\Delta}\left(\frac{\partial\Phi}{\partial y} - h_y\right) \\ \tau_{xz} &= \frac{V_z}{\Delta}\left(\frac{\partial\Phi}{\partial z} - h_z\right)\end{aligned} \tag{6.17}$$

where Δ depends on Poisson's ratio and the cross-sectional geometry

$$\Delta = 2(1+\nu)\left(I_y I_z - I_{yz}^2\right) \tag{6.18}$$

and

$$\begin{aligned}h_y &= \nu\left(I_z yz - I_{yz}\frac{y^2-z^2}{2}\right) \\ h_z &= -\nu\left(I_{yz} yz + I_z\frac{y^2-z^2}{2}\right)\end{aligned} \tag{6.19}$$

When the expressions of Eq. (6.17) are substituted into Eq. (6.12), the partial differential equation governing the function Φ is obtained:

$$\nabla^2\Phi = 2\left(I_{yz} y - I_z z\right) \tag{6.20}$$

where

$$\nabla^2 = \frac{\partial^2}{\partial y^2} + \frac{\partial^2}{\partial z^2}$$

A condition to be satisfied is that the cylindrical surface of the beam be free of surface forces. On the cylindrical surface the x component of the outward unit normal vector **n** is zero. The stress normal to the boundary curve must be zero:

$$\tau_{xy} n_y + \tau_{xz} n_z = 0 \tag{6.21}$$

where n_y, n_z are the y, z components of **n**. Note that Eq. (6.21) is the same as Eq. (5.11) if the stress–strain relationships of Eq. (5.4) are introduced. Substitute Eq. (6.16) into Eq. (6.21):

$$\left[\frac{\partial \Phi}{\partial y} + \nu\left(I_{yz}\frac{y^2 - z^2}{2} - I_z yz\right)\right] n_y + \left[\frac{\partial \Phi}{\partial z} + \nu\left(I_{yz} yz + I_z\frac{y^2 - z^2}{2}\right)\right] n_z = 0 \tag{6.22}$$

or

$$\mathbf{n} \cdot \boldsymbol{\nabla}\Phi + \mathbf{n} \cdot \left[\nu\left(I_{yz}\frac{y^2 - z^2}{2} - I_z yz\right)\mathbf{j} + \nu\left(I_{yz} yz + I_z\frac{y^2 - z^2}{2}\right)\mathbf{k}\right] = 0 \tag{6.23}$$

where

$$\boldsymbol{\nabla} = \mathbf{j}\frac{\partial}{\partial y} + \mathbf{k}\frac{\partial}{\partial z} \quad \text{and} \quad \mathbf{n} = n_y\mathbf{j} + n_z\mathbf{k}$$

Equation (6.23) can be expressed as

$$\mathbf{n} \cdot \boldsymbol{\nabla}\Phi - \mathbf{n} \cdot \mathbf{h} = 0 \tag{6.24}$$

where

$$\mathbf{h} = h_y\mathbf{j} + h_z\mathbf{k} = \nu\left(I_z yz - I_{yz}\frac{y^2 - z^2}{2}\right)\mathbf{j} - \nu\left(I_{yz} yz + I_z\frac{y^2 - z^2}{2}\right)\mathbf{k} \tag{6.25}$$

Since the directional derivative of a point of a surface along the normal is

$$\frac{\partial \Phi}{\partial n} = n_y\frac{\partial \Phi}{\partial y} + n_z\frac{\partial \Phi}{\partial z} = \mathbf{n} \cdot \boldsymbol{\nabla}\Phi$$

the boundary condition of Eq. (6.21) can be represented as

$$\frac{\partial \Phi}{\partial n} = \mathbf{n} \cdot \boldsymbol{\nabla}\Phi = \mathbf{n} \cdot \mathbf{h} \tag{6.26}$$

This boundary value problem of solving Eq. (6.20), $\nabla^2\Phi = 2\left(I_{yz}y - I_z z\right)$, subject to the boundary condition of Eq. (6.26), $\mathbf{n} \cdot \boldsymbol{\nabla}\Phi = \mathbf{n} \cdot \mathbf{h}$, provides a relationship between the shear force V_z and the cross-sectional shear stresses τ_{xy} and τ_{xz}. Although this relationship was derived for a particular beam (a cantilevered beam with end loading V_z), it is assumed to apply to a beam of any boundary condition and any loading.

Shear Stresses due to a Transverse Load in the y Direction Consider the cantilevered beam of Fig. 6.3 with the addition of a shear force V_y at $x = L$. Also, set $V_z = 0$. In this case, the bending moment is

$$M_z = V_y(L - x) \tag{6.27}$$

and the normal stress distribution is given by

$$\sigma_x = \frac{I_{yz} M_z z - I_y M_z y}{I_y I_z - I_{yz}^2} = H(L - x)\left(I_{yz} z - I_y y\right) \tag{6.28}$$

where

$$H = \frac{V_y}{I_y I_z - I_{yz}^2} \tag{6.29}$$

Substitute Eq. (6.28) into the first equilibrium equation of Eq. (6.11). This gives the equilibrium relation

$$\frac{\partial \tau_{xy}}{\partial y} + \frac{\partial \tau_{xz}}{\partial z} = H\left(I_{yz} z - I_y y\right) \tag{6.30}$$

Substitute the strains of Eq. (6.13) and then the stress σ_x of Eq. (6.28) into Eq. (6.14), giving the two compatibility conditions

$$\begin{aligned} \frac{\partial}{\partial y}\left(\frac{\partial \tau_{xz}}{\partial y} - \frac{\partial \tau_{xy}}{\partial z}\right) &= -\frac{\nu}{1+\nu} I_{yz} H \\ \frac{\partial}{\partial z}\left(\frac{\partial \tau_{xz}}{\partial y} - \frac{\partial \tau_{xy}}{\partial z}\right) &= -\frac{\nu}{1+\nu} I_y H \end{aligned} \tag{6.31}$$

The stresses are expressed in terms of a function $\Psi(y, z)$, which is assumed to have continuous partial derivatives of third order and such that the two compatibility conditions of Eq. (6.31) are identically satisfied:

$$\begin{aligned} \tau_{xy} &= \frac{V_y}{\Delta}\left(\frac{\partial \Psi}{\partial y} - d_y\right) \\ \tau_{xz} &= \frac{V_y}{\Delta}\left(\frac{\partial \Psi}{\partial z} - d_z\right) \end{aligned} \tag{6.32}$$

where d_y, d_z are the y, z components of the vector $\mathbf{d}$ defined by

$$\mathbf{d} = d_y \mathbf{j} + d_z \mathbf{k} = \nu\left(I_y \frac{y^2 - z^2}{2} - I_{yz} yz\right)\mathbf{j} + \nu\left(I_y yz + I_{yz}\frac{y^2 - z^2}{2}\right)\mathbf{k} \tag{6.33}$$

The partial differential equation for Ψ is obtained by substituting Eq. (6.32) into the equilibrium equation (Eq. 6.30)

$$\nabla^2 \Psi = 2\left(I_{yz} z - I_y y\right) \tag{6.34}$$

The boundary condition for the cylindrical surface of the beam to be free of surface forces requires that

$$\frac{\partial \Psi}{\partial n} = \mathbf{n} \cdot \boldsymbol{\nabla} \Psi = \mathbf{n} \cdot \mathbf{d} \tag{6.35}$$

6.1.3 Composite Cross Section

For nonhomogeneous cross sections the elastic modulus E is a function of position [i.e., $E = E(y, z)$]. Let the reference modulus be given by E_r. For the analysis of shear stresses due to transverse shear loads, it will be assumed that the different materials making up the section all have the same Poisson's ratio ν. The boundary value problems for transverse shear have a strong dependence on ν, and a formulation that allows different materials to have different Poisson's ratios is considerably more complicated than the analysis given below for a nonhomogeneous section with a single ν. As discussed in Chapter 5, for two materials with very different moduli of elasticity, it is not uncommon for the difference in Poisson's ratios to be very small.

To derive stress formulas for transverse shear loads for a composite cross section with the elastic modulus function $E(y, z)$, the development in Section 6.1.2 for a transverse shear force V_z applied at the shear center of a homogeneous section will be retraced, with the assumption of homogeneity dropped at each step. Saint-Venant's assumption that the distribution of the normal stress σ_x is given by Eq. (1.57) or (6.9) for a beam in pure bending is replaced by

$$\sigma_x = \frac{E}{E_r}\left(\frac{\widetilde{I}_z M_y z}{\widetilde{I}_y \widetilde{I}_z - \widetilde{I}_{yz}^2} - \frac{\widetilde{I}_{yz} M_y y}{\widetilde{I}_y \widetilde{I}_z - \widetilde{I}_{yz}^2}\right) = \widetilde{B}\frac{E}{E_r}(L - x)\left(\widetilde{I}_{yz} y - \widetilde{I}_z z\right) \tag{6.36}$$

for a composite cross section where the constant $\widetilde{B}$ is

$$\widetilde{B} = \frac{V_z}{\widetilde{I}_y \widetilde{I}_z - \widetilde{I}_{yz}^2} \tag{6.37}$$

The equations of equilibrium retain the same form as those of Eq. (6.11) for a homogeneous section. The first equilibrium equation becomes (Eq. 6.12)

$$\frac{\partial \tau_{xy}}{\partial y} + \frac{\partial \tau_{xz}}{\partial z} = \widetilde{B}\frac{E}{E_r}\left(\widetilde{I}_{yz} y - \widetilde{I}_z z\right) \tag{6.38}$$

The strains satisfy Hooke's stress–strain equations (Eq. 6.13). As in the case of homogeneous cross sections, these strains satisfy four of the compatibility conditions identically, and the remaining two conditions are given by Eq. (6.14). Since E is a function of y and z, it will be convenient to rewrite these Hooke's stress–strain equations in the form

$$\begin{aligned} &\epsilon_x = \overline{\sigma}_x && \epsilon_y = -\nu\overline{\sigma}_x && \epsilon_z = -\nu\overline{\sigma}_x \\ &\gamma_{xy} = 2(1+\nu)\overline{\tau}_{xy} && \gamma_{xz} = 2(1+\nu)\overline{\tau}_{xz} && \gamma_{yz} = 0 \end{aligned} \tag{6.39}$$

where a bar over a symbol indicates that the stress corresponding to that symbol is divided by E. The substitution of these stress–strain relations into the two compatibility conditions of Eq. (6.14) gives

$$\frac{\partial}{\partial y}\left(\frac{\partial \overline{\tau}_{xz}}{\partial y} - \frac{\partial \overline{\tau}_{xy}}{\partial z}\right) = \frac{\nu}{E_r(1+\nu)}\widetilde{I}_z\widetilde{B}$$
$$\frac{\partial}{\partial z}\left(\frac{\partial \overline{\tau}_{xz}}{\partial y} - \frac{\partial \overline{\tau}_{xy}}{\partial z}\right) = \frac{\nu}{E_r(1+\nu)}\widetilde{I}_{yz}\widetilde{B} \qquad (6.40)$$

The cross section is assumed to be made up of a finite number of regions of different materials. Over each of these regions, the elastic modulus E has a constant value, so that the equilibrium relationship of Eq. (6.38) rewritten in the form

$$\frac{\partial \overline{\tau}_{xy}}{\partial y} + \frac{\partial \overline{\tau}_{xz}}{\partial z} = \frac{\widetilde{B}}{E_r}\left(\widetilde{I}_{yz}y - \widetilde{I}_z z\right) \qquad (6.41)$$

is valid over each homogeneous region. The stresses are next expressed in terms of a function $\Phi(y, z)$, which is assumed to have continuous partial derivatives of third order, and such that the two compatibility conditions of Eq. (6.40) are identically satisfied:

$$\overline{\tau}_{xy} = \frac{\widetilde{B}}{2E_r(1+\nu)}\left[\frac{\partial \Phi}{\partial y} + \nu\left(\widetilde{I}_{yz}\frac{y^2 - z^2}{2} - \widetilde{I}_z yz\right)\right]$$
$$\overline{\tau}_{xz} = \frac{\widetilde{B}}{2E_r(1+\nu)}\left[\frac{\partial \Phi}{\partial z} + \nu\left(\widetilde{I}_{yz}yz + \widetilde{I}_z\frac{y^2 - z^2}{2}\right)\right] \qquad (6.42)$$

When these expressions are substituted into the equilibrium equation of Eq. (6.41), the partial differential equation governing Φ is obtained:

$$\nabla^2\Phi = 2\left(\widetilde{I}_{yz}y - \widetilde{I}_z z\right) \qquad (6.43)$$

The boundary condition that the cylindrical surface of the beam be free of surface forces is expressed in terms of the function Φ by (Eq. 6.26)

$$\frac{\partial \Phi}{\partial n} = \mathbf{n}\cdot\nabla\Phi = \mathbf{n}\cdot\widetilde{\mathbf{h}} \qquad (6.44)$$

where the vector $\widetilde{\mathbf{h}}$ is defined as (Eq. 6.19)

$$\widetilde{\mathbf{h}} = \widetilde{h}_y\mathbf{j} + \widetilde{h}_z\mathbf{k} = \nu\left(\widetilde{I}_z yz - \widetilde{I}_{yz}\frac{y^2 - z^2}{2}\right)\mathbf{j} - \nu\left(\widetilde{I}_z\frac{y^2 - z^2}{2} + \widetilde{I}_{yz}yz\right)\mathbf{k} \qquad (6.45)$$

The expressions for the shear stresses become

$$\tau_{xy} = \frac{E}{E_r}\frac{V_z}{\widetilde{\Delta}}\left(\frac{\partial \Phi}{\partial y} - \widetilde{h}_y\right) \qquad \tau_{xz} = \frac{E}{E_r}\frac{V_z}{\widetilde{\Delta}}\left(\frac{\partial \Phi}{\partial z} - \widetilde{h}_z\right) \qquad (6.46)$$

where $\widetilde{\Delta}$ depends on Poisson's ratio and the cross-sectional geometry

$$\widetilde{\Delta} = 2(1+\nu)\left(\widetilde{I}_y\widetilde{I}_z - \widetilde{I}_{yz}^2\right) \tag{6.47}$$

The preceding paragraph shows that the shear stresses due to transverse shear loads for a nonhomogeneous cross section are obtained by using modulus-weighted cross-sectional properties. The stresses due to the forces V_y and V_z are given by the formulas

$$\begin{aligned} \tau_{xy} &= \frac{E}{E_r}\left[\frac{V_y}{\widetilde{\Delta}}\left(\frac{\partial\Psi}{\partial y} - \widetilde{d}_y\right) + \frac{V_z}{\widetilde{\Delta}}\left(\frac{\partial\Phi}{\partial y} - \widetilde{h}_y\right)\right] \\ \tau_{xz} &= \frac{E}{E_r}\left[\frac{V_y}{\widetilde{\Delta}}\left(\frac{\partial\Psi}{\partial z} - \widetilde{d}_z\right) + \frac{V_z}{\widetilde{\Delta}}\left(\frac{\partial\Phi}{\partial z} - \widetilde{h}_z\right)\right] \end{aligned} \tag{6.48}$$

where

$$\widetilde{\mathbf{d}} = \tilde{d}_y\mathbf{j} + \tilde{d}_z\mathbf{k} = \nu\left(\widetilde{I}_y\frac{y^2 - z^2}{2} - \widetilde{I}_{yz}yz\right)\mathbf{j} + \nu\left(\widetilde{I}_y yz + \widetilde{I}_{yz}\frac{y^2 - z^2}{2}\right)\mathbf{k} \tag{6.49}$$

and Ψ is the shear function for V_y, which satisfies the partial differential equation

$$\nabla^2\Psi = 2\left(\widetilde{I}_{yz}z - \widetilde{I}_y y\right) \tag{6.50}$$

and the boundary condition

$$\frac{\partial\Psi}{\partial n} = \mathbf{n}\cdot\nabla\Psi = \mathbf{n}\cdot\widetilde{\mathbf{d}} \tag{6.51}$$

The analysis of stresses given in this section is not completely rigorous because the boundary conditions at the interfaces between different material regions have not been considered.

6.1.4 Finite Element Solution Formulation

V_y Problem In Section 6.1.2, two boundary value problems have been formulated, one involving a transverse shear force V_y and the other the force V_z. The first problem is to solve (Eq. 6.34)

$$\nabla^2\Psi = 2\left(I_{yz}z - I_y y\right) \tag{6.52}$$

for the function Ψ subject to the boundary condition (Eq. 6.35)

$$\mathbf{n}\cdot\nabla\Psi = \mathbf{n}\cdot\mathbf{d} \tag{6.53}$$

The second problem is to solve for Φ (Eq. 6.20):

$$\nabla^2\Phi = 2\left(I_{yz}y - I_z z\right) \tag{6.54}$$

with (Eq. 6.26)

$$\mathbf{n} \cdot \nabla \Phi = \mathbf{n} \cdot \mathbf{h} \tag{6.55}$$

The definitions of the vectors **h** and **d** are (Eqs. 6.25 and 6.33)

$$\begin{aligned} \mathbf{h} &= \nu \left(I_z yz - I_{yz} \frac{y^2 - z^2}{2} \right) \mathbf{j} - \nu \left(I_z \frac{y^2 - z^2}{2} + I_{yz} yz \right) \mathbf{k} \\ \mathbf{d} &= \nu \left(I_y \frac{y^2 - z^2}{2} - I_{yz} yz \right) \mathbf{j} + \nu \left(I_y yz + I_{yz} \frac{y^2 - z^2}{2} \right) \mathbf{k} \end{aligned} \tag{6.56}$$

As in Chapter 5, Galerkin's method will be used to arrive at a finite element formulation of the boundary value problems. Applied to the first problem, the integral formulation is

$$\int v \left[\nabla^2 \Psi - 2 \left(I_{yz} z - I_y y \right) \right] dA + \int \overline{v} \left(\mathbf{n} \cdot \mathbf{d} - \mathbf{n} \cdot \nabla \Psi \right) ds = 0 \tag{6.57}$$

This equation is transformed by choosing the functions v and $\overline{v}$ to be identical, and using Green's first identity (Rubinstein and Rubinstein, 1993)

$$\int v \nabla^2 \Psi \, dA = \int v \mathbf{n} \cdot \nabla \Psi \, ds - \int \nabla v \cdot \nabla \Psi \, dA \tag{6.58}$$

This transformation gives

$$\int \left[-\nabla v \cdot \nabla \Psi - 2v \left(I_{yz} z - I_y y \right) \right] dA + \int v \, \mathbf{d} \cdot \mathbf{n} \, ds = 0 \tag{6.59}$$

The path integral in this equation is written as an area integral using the following form of the divergence theorem in two dimensions:

$$\int v \, \mathbf{d} \cdot \mathbf{n} \, ds = \int \left[v \nabla \cdot \mathbf{d} + \mathbf{d} \cdot \nabla v \right] dA \tag{6.60}$$

The integral formulation of the problem for Ψ then becomes

$$\int \nabla v \cdot \nabla \Psi \, dA + \int 2v \left(I_{yz} z - I_y y \right) dA - \int v \nabla \cdot \mathbf{d} \, dA - \int \mathbf{d} \cdot \nabla v \, dA = 0 \tag{6.61}$$

The first integral of Eq. (6.61) will lead to the element stiffness matrix. The function Ψ is approximated over each element e, as in Chapter 5, by

$$\Psi(\eta, \zeta) = \mathbf{N}(\eta, \zeta) \boldsymbol{\Psi}^e \tag{6.62}$$

where $\mathbf{\Psi}^e$ are the values of Ψ at the nodes of element e and $\mathbf{N}$ is the row vector of shape functions for the nine-node element as defined in Chapter 4. In the Galerkin formulation, the trial functions v in the integral form are replaced by the shape functions in arriving at the discretized approximation of Eq. (6.61). Follow the procedure used in Chapter 5 for the Saint-Venant torsion problem. This leads to an expression of the form (Eq. 5.116) $\mathbf{v}^{e\mathrm{T}}(\mathbf{k}^e\mathbf{\Psi}^e - \mathbf{p}^e) = 0$, so that the element domain equations are given by

$$\mathbf{k}^e\mathbf{\Psi}^e = \mathbf{p}^e \qquad 1 \le e \le M \tag{6.63}$$

The number of elements is M. The element stiffness matrix is the one developed in Chapter 5 (Eq. 5.122) and is given by

$$\mathbf{k}^e = \int_{-1}^{1}\int_{-1}^{1} \mathbf{B}^\mathrm{T}\mathbf{B} \mid \mathbf{J}_e \mid d\eta\, d\zeta \tag{6.64}$$

where $\mathbf{B}$ is the 2×9 matrix (Eq. 5.119)

$$\mathbf{B} = \mathbf{B}_e = \mathbf{J}_e^{-1}\begin{bmatrix} \dfrac{\partial \mathbf{N}}{\partial \eta} \\ \dfrac{\partial \mathbf{N}}{\partial \zeta} \end{bmatrix}$$

To define the element load vector $\mathbf{p}^e$ for this problem, it is necessary to transform the last three integrals in Eq. (6.61) to their discrete nodal form for each element e. Since

$$\boldsymbol{\nabla}\cdot\mathbf{d} = \nu\left(I_y y - I_{yz}z\right) + \nu\left(I_y y - I_{yz}z\right) = 2\nu\left(I_y y - I_{yz}z\right) \tag{6.65}$$

we find that

$$2(I_{yz}z - I_y y) - \boldsymbol{\nabla}\cdot\mathbf{d} = 2(1+\nu)\left(I_{yz}z - I_y y\right) \tag{6.66}$$

Then the middle two integrals in Eq. (6.61) combine, giving with Eqs. (4.11) and (5.115),

$$2(1+\nu)\int_{-1}^{1}\int_{-1}^{1} \mathbf{N}^\mathrm{T}\left(I_{yz}\mathbf{N}\mathbf{z} - I_y\mathbf{N}\mathbf{y}\right) \mid \mathbf{J}_e \mid d\eta\, d\zeta \tag{6.67}$$

for element e. The term $\mathbf{v}^{e\mathrm{T}}$ has been factored out of Eq. (6.67) and ignored since we are dealing with the components of $\mathbf{k}^e\mathbf{\Psi}^e - \mathbf{p}^e = 0$ and not $\delta\mathbf{v}^{e\mathrm{T}}\,(\mathbf{k}^e\mathbf{\Psi}^e - \mathbf{p}^e) = 0$. Using Eq. (5.115), the quantity $d^e \cdot \boldsymbol{\nabla} v = \mathbf{d}\cdot\boldsymbol{\nabla} v$ can be written as

$$\mathbf{d}\cdot\boldsymbol{\nabla} v = \mathbf{v}^{e\mathrm{T}}\begin{bmatrix} \dfrac{\partial \mathbf{N}^T}{\partial y} & \dfrac{\partial \mathbf{N}^\mathrm{T}}{\partial z} \end{bmatrix}\begin{bmatrix} d_y \\ d_z \end{bmatrix}$$

Thus, the contribution of the last integral in Eq. (6.61) to element e is (Eq. 5.120)

$$\frac{\nu}{2}\int_{-1}^{1}\int_{-1}^{1}\mathbf{B}^{\mathrm{T}}\begin{bmatrix}d_1\\ d_2\end{bmatrix} \mid \mathbf{J}_e \mid d\eta\, d\zeta \tag{6.68}$$

where, with the abbreviations,

$$r^e = r = (\mathbf{Ny}_e)^2 - (\mathbf{Nz}_e)^2 = (\mathbf{Ny})^2 - (\mathbf{Nz})^2 \qquad q^e = q = 2\mathbf{Ny}_e\mathbf{Nz}_e = 2\mathbf{NyNz} \tag{6.69}$$

the quantities d_1, d_2 are expressed as

$$d_1 = I_y r - I_{yz} q \qquad d_2 = I_{yz} r + I_y q \tag{6.70}$$

Again the term $\mathbf{v}^{e\mathrm{T}}$ has been factored out of Eq. (6.68) and ignored. The load vector is the sum of Eqs. (6.67) and (6.68):

$$\mathbf{p}^e = \int_{-1}^{1}\int_{-1}^{1}\left[\frac{\nu}{2}\mathbf{B}^{\mathrm{T}}\begin{bmatrix}d_1\\ d_2\end{bmatrix} + 2(1+\nu)\mathbf{N}^{\mathrm{T}}\left(I_y\mathbf{Ny} - I_{yz}\mathbf{Nz}\right)\right] \mid \mathbf{J}_e \mid d\eta\, d\zeta \tag{6.71}$$

The assembly of the system stiffness matrix $\mathbf{K}$ from the element stiffness matrices $\mathbf{k}^e$ and the assembly of the system load vector $\mathbf{P}$ from the element load vectors $\mathbf{p}^e$ are performed as described in Chapter 5. The system stiffness matrix $\mathbf{K}$ is the one derived in Chapter 5 for the torsion problem. The final global stiffness equation appears as

$$\mathbf{K\Psi} = \mathbf{P}_y \tag{6.72}$$

which can be solved for the nodal values $\mathbf{\Psi}$ of the function Ψ. The subscript y indicates the problem corresponds to the transverse shear load V_y.

***V_z* Problem** The integral formulation of the boundary value problem for Φ is obtained from Eq. (6.61) by substituting $\mathbf{h}$ for $\mathbf{d}$ and changing the second integrand to the expression on the right side of the Poisson equation (Eq. 6.20) for Φ. Then

$$\int \nabla v \cdot \nabla\Phi\, dA + \int 2v\left(I_{yz}y - I_z z\right) dA - \int v\nabla\cdot\mathbf{h}\, dA - \int \mathbf{h}\cdot\nabla v\, dA = 0 \tag{6.73}$$

is the z-direction equivalent of Eq. (6.61). The element stiffness matrices and the system stiffness matrix for this problem are the same as those for Ψ. The element load vector is computed by noting that

$$2(I_{yz}y - I_z z) - \nabla\cdot\mathbf{h} = 2(1+\nu)\left(I_{yz}y - I_z z\right) \tag{6.74}$$

and defining

$$h_1 = -I_{yz} r + I_z q \qquad h_2 = -I_z r - I_{yz} q \tag{6.75}$$

The element load vector for Φ is

$$\mathbf{p}^e = \int_{-1}^{1}\int_{-1}^{1}\left[\frac{\nu}{2}\mathbf{B}^{\mathrm{T}}\begin{bmatrix} h_1 \\ h_2 \end{bmatrix} + 2(1+\nu)\mathbf{N}^{\mathrm{T}}\left(I_z\mathbf{Nz} - I_{yz}\mathbf{Ny}\right)\right] \mid \mathbf{J}_e \mid d\eta\, d\zeta \qquad (6.76)$$

The assembled stiffness equations appear as

$$\mathbf{K\Phi} = \mathbf{P}_z \qquad (6.77)$$

where the subscript z is included to indicate that Eq. (6.77) corresponds to shear force V_z.

Shear Stresses It was shown above that the discretized forms of the boundary value problems for Ψ and Φ are written as

$$\mathbf{K\Psi} = \mathbf{P}_y \qquad \mathbf{K\Phi} = \mathbf{P}_z \qquad (6.78)$$

The total transverse shear stresses in element e are given by (Eqs. 6.32 and 6.17)

$$\boldsymbol{\tau}^e = \begin{bmatrix} \tau_{xy} \\ \tau_{xz} \end{bmatrix}^e = \frac{V_y}{\Delta}\left(\boldsymbol{\nabla}\Psi^e - \mathbf{d}^e\right) + \frac{V_z}{\Delta}\left(\boldsymbol{\nabla}\Phi^e - \mathbf{h}^e\right) \qquad (6.79)$$

where

$$\Delta = 2(1+\nu)\left(I_y I_z - I_{yz}^2\right)$$

$$\mathbf{d}^e = \begin{bmatrix} d_1 \\ d_2 \end{bmatrix}^e \qquad \mathbf{h}^e = \begin{bmatrix} h_1 \\ h_2 \end{bmatrix}^e$$

$$\boldsymbol{\nabla}\Psi^e = \begin{bmatrix} \dfrac{\partial \Psi}{\partial y} \\[2mm] \dfrac{\partial \Psi}{\partial z} \end{bmatrix}^e \qquad \boldsymbol{\nabla}\Phi^e = \begin{bmatrix} \dfrac{\partial \Phi}{\partial y} \\[2mm] \dfrac{\partial \Phi}{\partial z} \end{bmatrix}^e \qquad (6.80)$$

Approximate the functions Ψ^e and Φ^e over each element e as

$$\Psi^e(\eta, \zeta) = \mathbf{N}(\eta, \zeta)\boldsymbol{\Psi}^e \qquad \Phi^e(\eta, \zeta) = \mathbf{N}(\eta, \zeta)\boldsymbol{\Phi}^e \qquad (6.81)$$

so that with Eq. (5.119) the gradients of the shear functions Ψ^e, Φ^e are

$$\boldsymbol{\nabla}\Psi^e = \mathbf{B}\boldsymbol{\Psi}^e \qquad \boldsymbol{\nabla}\Phi^e = \mathbf{B}\boldsymbol{\Phi}^e \qquad (6.82)$$

The finite element expression for the shear stresses becomes

$$\begin{bmatrix} \tau_{xy} \\ \tau_{xz} \end{bmatrix}^e = \frac{V_y}{\Delta}\left(\mathbf{B}\boldsymbol{\Psi}^e - \frac{\nu}{2}\begin{bmatrix} d_1 \\ d_2 \end{bmatrix}^e\right) + \frac{V_z}{\Delta}\left(\mathbf{B}\boldsymbol{\Phi}^e - \frac{\nu}{2}\begin{bmatrix} h_1 \\ h_2 \end{bmatrix}^e\right) \qquad (6.83)$$

6.2 SHEAR CENTER

If the resultant shear force on a cross section passes through a particular point no torsion will occur. This point is called the *shear center*. Various definitions of the shear center are discussed in Fung (1969). Initially, theory of elasticity based formulas for the location of the shear center that depend on the material properties will be derived. Later, other formulas will be obtained that are purely geometric properties of the cross section.

6.2.1 *y* Coordinate of the Shear Center

Suppose that the internal shear force V_z at the cross section with centroid C is applied at the shear center S of the cross section. The *bending axis* or *flexural axis* of a beam can now be clarified. The *centroidal axis* of a beam is defined as a line that passes through the centroids of the cross sections along the beam. The bending axis, which is parallel to the centroidal axis, passes through the shear centers of the cross sections. Thus, the axis through which the resultants of the reactions and transverse loading act leads to "pure bending." As used in Chapter 1, pure bending corresponds to zero torsion, which contrasts with a frequently used definition of a beam with zero internal shear forces.

The force–couple equivalent of the force V_z at the centroid C is the force V_z and the torsional couple whose moment is equal to the moment of V_z about C. Since the shear stress distribution is statically equivalent to this force–couple system, the moment of V_z about C can be calculated from the shear stresses (Fig. 6.4)

$$M_x = y_S V_z = \int (\tau_{xz} y - \tau_{xy} z)\, dA \tag{6.84}$$

This equation determines the y coordinate (y_S) of the shear center S.

Substitute Eq. (6.17) into Eq. (6.84) and observe that the coordinate y_S is given by

$$y_S = \frac{1}{\Delta} \int \left(y \frac{\partial \Phi}{\partial z} - y h_z - z \frac{\partial \Phi}{\partial y} + z h_y \right) dA \tag{6.85}$$

or, from Eq. (6.19),

$$y_S = \frac{1}{\Delta} \left[\frac{\nu}{2} \int (I_z y + I_{yz} z)(y^2 + z^2)\, dA - \int \mathbf{g} \cdot \boldsymbol{\nabla} \Phi \, dA \right] \tag{6.86}$$

where

$$\Delta = 2(1+\nu)(I_y I_z - I_{yz}^2), \quad \mathbf{g} = z\mathbf{j} - y\mathbf{k}, \quad \text{and} \quad \boldsymbol{\nabla} = \mathbf{j}\frac{\partial}{\partial y} + \mathbf{k}\frac{\partial}{\partial z}$$

Note that Eq. (6.86) depends on Poisson's ratio, a material constant.

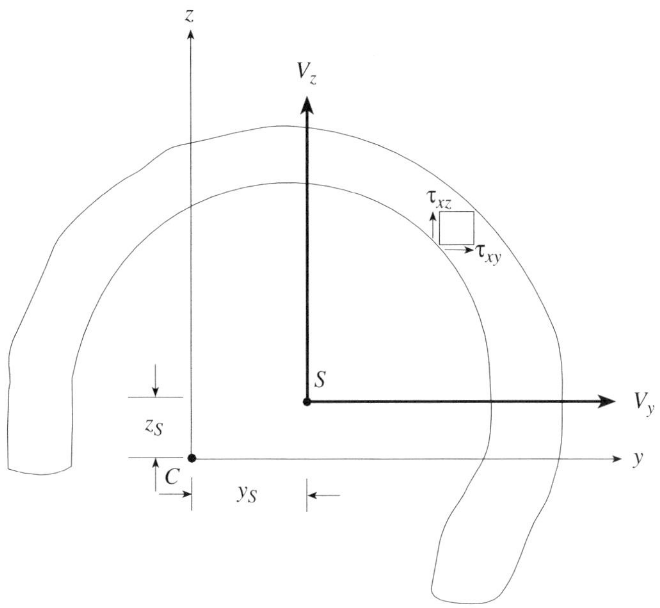

Figure 6.4 Centroid C and shear center S.

6.2.2 Axis of Symmetry

If the cross section is symmetric about the z axis, $I_{yz} = 0$ (Fig. 6.5). As a consequence, the differential equation for Φ of Eq. (6.20) reduces to

$$\nabla^2 \Phi = -2I_z z \tag{6.87}$$

Consider any point (y_1, z_1) on the cross section. At (y_1, z_1), $\nabla^2\Phi(y_1, z_1) = -2I_z z_1$. Next consider the mirror-image point $(-y_1, z_1)$. Then $\nabla^2\Phi(-y_1, z_1) = -2I_z z_1$. It follows that $\nabla^2\Phi(-y_1, z_1) = \nabla^2\Phi(y_1, z_1)$, showing that $\nabla^2\Phi$ is an even function of y. Since $\nabla^2\Phi(-y, z) = \nabla^2\Phi(y, z) = -2I_z z$, the differential equation of Eq. (6.87) exhibits symmetry about the z axis.

Consider now the boundary condition of Eq. (6.26) for this cross section, which is symmetric about the z axis:

$$\frac{\partial \Phi}{\partial n} = \mathbf{n} \cdot \mathbf{h} = n_y \nu I_z yz - n_z \nu I_z \frac{y^2 - z^2}{2} \tag{6.88}$$

On the boundary at (y_0, z_0),

$$\frac{\partial \Phi}{\partial n}(y_0, z_0) = \nu I_z \left(n_{y_0} y_0 z_0 - n_{z_0} \frac{y_0^2 - z_0^2}{2} \right) \tag{6.89}$$

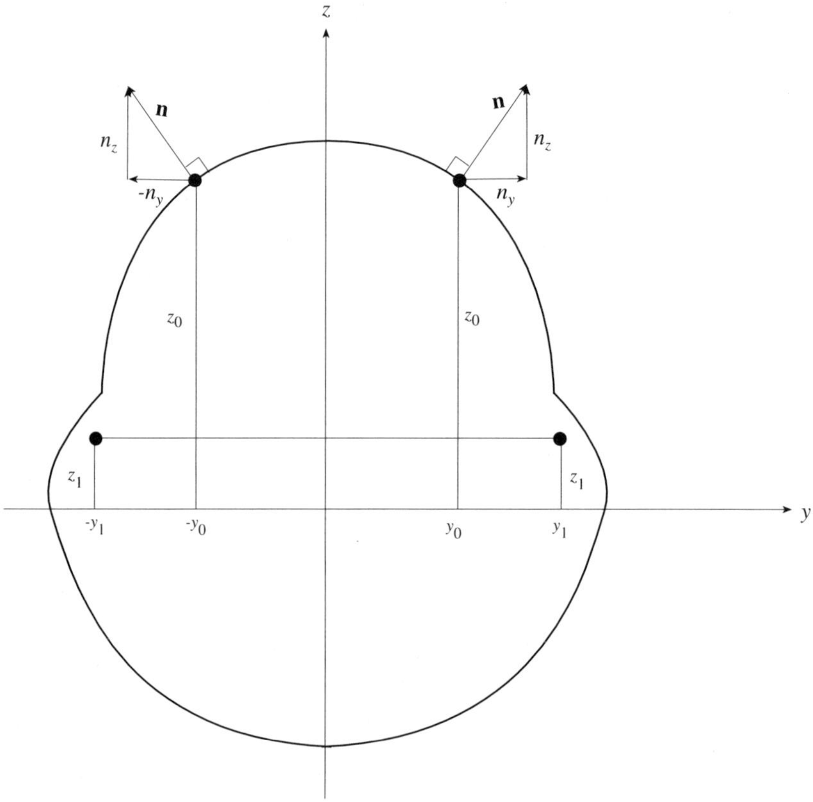

Figure 6.5 Cross section symmetric with respect to the z axis.

For the mirror-image point $(-y_0, z_0)$,

$$\frac{\partial \Phi}{\partial n}(-y_0, z_0) = \nu I_z \left[n_{-y_0}(-y_0)z_0 - n_{z_0}\frac{y_0^2 - z_0^2}{2} \right] \tag{6.90}$$

From Fig. 6.5, $n_{-y_0} = -n_{y_0}$, so that at any boundary point,

$$\frac{\partial \Phi}{\partial n}(-y_0, z_0) = \nu I_z \left(n_{y_0} y_0 z_0 - n_{z_0}\frac{y_0^2 - z_0^2}{2} \right) = \frac{\partial \Phi}{\partial n}(y_0, z_0) \tag{6.91}$$

Thus, it is seen that the boundary condition exhibits symmetry with respect to the z axis.

It has been shown that the functions $\Phi(y, z)$ and $\Phi(-y, z)$ satisfy the same differential equation and the same boundary condition. It follows that

$$\Phi(y, z) = \Phi(-y, z) \tag{6.92}$$

and that Φ is an even function of y.

Turn now to the shear center formula of Eq. (6.86) for this cross section (Fig. 6.5), which is symmetric about the z axis.

$$y_S = \frac{1}{\Delta}\frac{\nu}{2}\int I_z y(y^2 + z^2)\, dA - \frac{1}{\Delta}\int \mathbf{g}\cdot\boldsymbol{\nabla}\Phi\, dA \tag{6.93}$$

The function $y(y^2 + z^2)$ is odd in y, so that the integral

$$\int y(y^2 + z^2)\, dA \tag{6.94}$$

is zero. Next, consider the term in the second integral on the right side of Eq. (6.93)

$$\mathbf{g}\cdot\boldsymbol{\nabla}\Phi = z\frac{\partial\Phi}{\partial y} - y\frac{\partial\Phi}{\partial z} \tag{6.95}$$

Because Φ is even in y, $\partial\Phi/\partial y$ is odd in y. Hence $z\partial\Phi/\partial y$ of Eq. (6.95) is odd in y. Since Φ is even in y, $\partial\Phi/\partial z$ is also even in y. Since y is odd in y, the product $y\partial\Phi/\partial z$ of Eq. (6.95) is odd in y. It can be concluded that $\mathbf{g}\cdot\boldsymbol{\nabla}\Phi$ is odd in y and that the integral

$$\int \mathbf{g}\cdot\boldsymbol{\nabla}\Phi\, dA \tag{6.96}$$

is zero.

Finally, it is apparent from Eq. (6.93) that $y_S = 0$. That is, the shear center of a cross section that is symmetric with respect to z is on the z axis.

6.2.3 Location of Shear Centers for Common Cross Sections

There are several characteristics of shear centers that are easily verified (see Fig. 6.6).

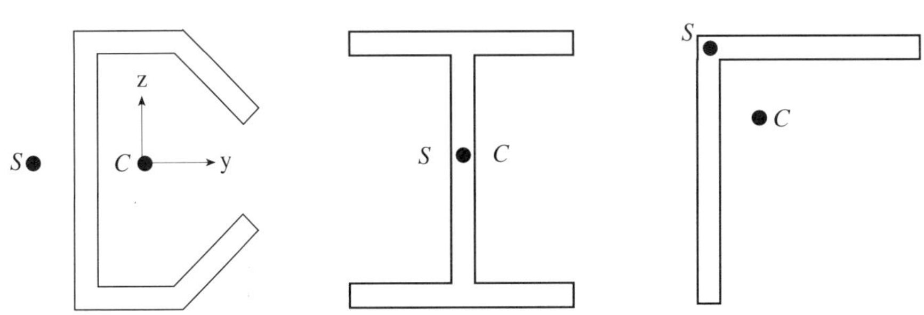

(a) One axis of symmetry (b) Two axes of symmetry (c) Angle section

Figure 6.6 Examples of the location of shear centers. The centroid and shear center are denoted by C and S, respectively.

1. For a cross section with two axes of symmetry, the shear center is at the centroid of the section.
2. For a cross section with one axis of symmetry, the shear center falls on this axis.
3. For a cross section with two intersecting flat flanges, the shear center is at the location of the intersection of the flanges.

6.2.4 *z* Coordinate of the Shear Center

The z coordinate of the shear center is found by setting the torsional couple due to the shear stress distribution equal to the torsional moment of V_y (Fig. 6.4):

$$M_x = -z_S V_y = \int (\tau_{xz} y - \tau_{xy} z)\, dA \tag{6.97}$$

The z coordinate of the shear center S is given in terms of the function Ψ by

$$z_S = \frac{1}{\Delta}\left[\frac{\nu}{2}\int (I_y z + I_{yz} y)(y^2 + z^2)\, dA + \int \mathbf{g} \cdot \boldsymbol{\nabla}\Psi\, dA\right] \tag{6.98}$$

6.2.5 Finite Element Solution Formulation

The coordinates of the shear center S are given by Eqs. (6.86) and (6.98) as

$$\begin{aligned} y_S &= \frac{1}{\Delta}\left[\frac{\nu}{2}\int \left(I_z y + I_{yz} z\right)\left(y^2 + z^2\right)\, dA - \int \mathbf{g} \cdot \boldsymbol{\nabla\Phi}\, dA\right] \\ z_S &= \frac{1}{\Delta}\left[\frac{\nu}{2}\int \left(I_y z + I_{yz} y\right)\left(y^2 + z^2\right)\, dA + \int \mathbf{g} \cdot \boldsymbol{\nabla\Psi}\, dA\right] \end{aligned} \tag{6.99}$$

The first integral in each of these equations is evaluated by quadrature. The second integrals are evaluated as in the calculation of the torsional constant J in Chapter 5, Eq. (5.138)

$$\begin{aligned} \int \mathbf{g} \cdot \boldsymbol{\nabla}\Phi\, dA &= \sum_e \int_{-1}^{1}\int_{-1}^{1} \left[\mathbf{Nz} \quad -\mathbf{Ny}\right] \mathbf{B}\boldsymbol{\Phi}^e \mid \mathbf{J}_e \mid d\eta\, d\zeta \\ &= \mathbf{P}^{\mathrm{T}}\boldsymbol{\Phi}^e \\ \int \mathbf{g} \cdot \boldsymbol{\nabla}\Psi\, dA &= \sum_e \int_{-1}^{1}\int_{-1}^{1} \left[\mathbf{Nz} \quad -\mathbf{Ny}\right] \mathbf{B}\boldsymbol{\Psi}^e \mid \mathbf{J}_e \mid d\eta\, d\zeta \\ &= \mathbf{P}^{\mathrm{T}}\boldsymbol{\Psi}^e \end{aligned} \tag{6.100}$$

where $\mathbf{P}$ is the global load vector for the torsion problem of Chapter 5.

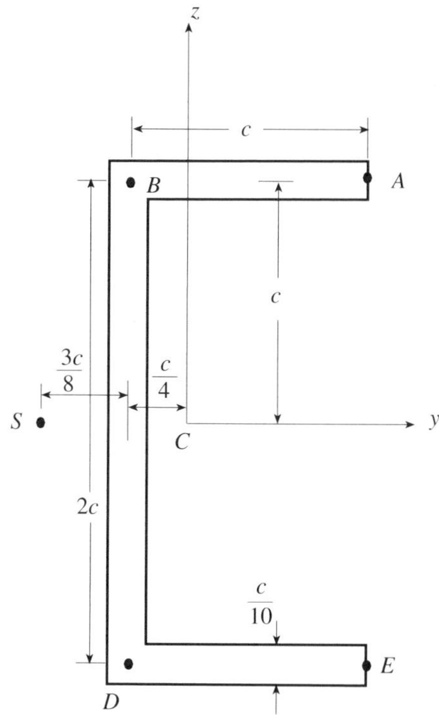

Figure 6.7 Channel section. Classical location of shear center is shown.

Example 6.2 Shear Center of a Channel Section. Consider the channel section of Fig. 6.7 with $c = 1.0$. Part of the output file for section properties as calculated by the computer program of the appendices is shown in Table 6.1. The shear center coordinates with respect to the centroid show that the shear center lies to the left (0.62056) of the cross section on the y axis. Recall from Section 6.2.3 that when a cross section has one axis of symmetry, the shear center lies on that axis.

TABLE 6.1 Part of an Output File for a Channel Section for $\nu = 0.3$

Y Centroid	0.24937
Z Centroid	1.00000
Y Shear Center wrt Centroid	−0.62056
Z Shear Center wrt Centroid	0.00000
Y Shear Coefficient	3.09621
Z Shear Coefficient	2.34102
YZ Shear Coefficient	0.00000
Torsional Constant	0.00133

6.2.6 Trefftz's Definition of the Shear Center

The shear center equations derived thus far, Eqs. (6.86) and (6.98) which are based on the theory of elasticity, show that the location of S is dependent on Poisson's ratio. Consequently, the shear center defined in this way is not a purely geometric property of the cross section. A second definition, given by Trefftz (1936), removes this dependence on Poisson's ratio. In discussing the cross-sectional properties that are dependent on the warping function ω, it will be more convenient to adopt Trefftz's definition. Shear center equations based on Trefftz's definition are viable for thin-walled beams and could have been introduced in Chapter 7 where sectorial properties are considered.

If a cantilever beam of length L is subjected to a twisting moment at its free end, the elastic strain energy stored in the beam due to pure torsion is

$$U_t = \frac{L}{2G} \int \left(t_{xy}^2 + t_{xz}^2 \right) dA \tag{6.101}$$

where t_{xy} and t_{xz} are torsional shear stresses. If the same beam is subjected to transverse loads V_y, V_z, the strain energy in the beam due to these loads is the flexural energy

$$U_f = \frac{1}{2E} \int \sigma_x^2 \, dx \, dA + \frac{L}{2G} \int \left(\tau_{xy}^2 + \tau_{xz}^2 \right) dA \tag{6.102}$$

where τ_{xy}, τ_{xz} are the transverse shear stresses. If the beam is subjected to pure torsion and then to transverse loads without removing the twisting torque, the total strain energy stored in it is given by

$$U = \frac{1}{2E} \int \sigma_x^2 \, dx \, dA + \frac{L}{2G} \int \left[\left(t_{xy} + \tau_{xy} \right)^2 + \left(t_{xz} + \tau_{xz} \right)^2 \right] dA \tag{6.103}$$

When the transverse loading is not torsion-free, the pure twisting moment does additional work on the beam because of the rotation caused by the transverse loads. When the transverse loads are applied at the shear center, they do not, by definition, cause any additional twisting of the beam. This means that the twisting moment performs no additional work, and the total strain energy may be obtained simply by adding the flexural and torsional energies of Eqs. (6.101) and (6.102):

$$U = U_t + U_f \tag{6.104}$$

Substitution of Eqs. (6.101), (6.102), and (6.103) into Eq. (6.104) shows that this condition requires that the coupling term between the transverse and torsional shear stresses be zero:

$$\int \left(t_{xy} \tau_{xy} + t_{xz} \tau_{xz} \right) dA = 0 \tag{6.105}$$

This equation is the definition of torsion-free flexure according to Trefftz.

The shear stresses due to pure torsion were found in Chapter 5 to be (Eq. 5.4)

$$t_{xy} = G\theta\left(\frac{\partial\omega}{\partial y} - z\right) \qquad t_{xz} = G\theta\left(\frac{\partial\omega}{\partial z} + y\right) \tag{6.106}$$

With these stresses, Trefftz's condition of Eq. (6.105) for torsion-free flexure becomes

$$\int\left[\left(\frac{\partial\omega}{\partial y} - z\right)\tau_{xy} + \left(\frac{\partial\omega}{\partial z} + y\right)\tau_{xz}\right] dA = 0 \tag{6.107}$$

Suppose that V_z is the transverse load applied. Then, if y_S is the distance in the y-coordinate direction from the centroid to the shear center to V_z and if y and z are measured from the centroid, a condition of equilibrium gives

$$y_S V_z = \int \left(\tau_{xz} y - \tau_{xy} z\right) dA \tag{6.108}$$

Equation (6.107) can be rewritten as

$$\int \left(\tau_{xz} y - \tau_{xy} z\right) dA = -\int\left(\frac{\partial\omega}{\partial y}\tau_{xy} + \frac{\partial\omega}{\partial z}\tau_{xz}\right) dA \tag{6.109}$$

The integral on the right can be expressed as

$$\begin{aligned}\int\left(\frac{\partial\omega}{\partial y}\tau_{xy} + \frac{\partial\omega}{\partial z}\tau_{xz}\right) dA &= \int\left(\frac{\partial\omega\tau_{xy}}{\partial y} + \frac{\partial\omega\tau_{xz}}{\partial z}\right) dA \\ &\quad - \int \omega\left(\frac{\partial\tau_{xy}}{\partial y} + \frac{\partial\tau_{xz}}{\partial y}\right) dA\end{aligned} \tag{6.110}$$

The first integral on the right of Eq. (6.110) is transformed by an application of Green's theorem of Eq. (5.107) into

$$\int\left(\frac{\partial\omega\tau_{xy}}{\partial y} + \frac{\partial\omega\tau_{xz}}{\partial z}\right) dA = \int \omega\left(-\tau_{xz}\, dy + \tau_{xy}\, dz\right) \tag{6.111}$$

The boundary condition of Eq. (5.11) for the cylindrical surface of the beam to be free of surface forces can be written as

$$\tau_{xy} n_y + \tau_{xz} n_z = 0 \tag{6.112}$$

or, with $\boldsymbol{\tau} = \tau_{xy}\mathbf{j} + \tau_{xz}\mathbf{k}$, $\boldsymbol{\tau}\cdot\mathbf{n} = 0$, which expresses that the total shear stress at any point on the boundary is tangent to the boundary. Since (Eq. 1.36) $n_y = dz/ds$ and $n_z = -dy/ds$, this boundary condition takes the form

$$\frac{\tau_{xz}}{\tau_{xy}} = \frac{dz}{dy} \tag{6.113}$$

Equation (6.113) shows that the integral on the right of Eq. (6.111) is zero.

The second integral on the right of Eq. (6.110) may be rewritten using the equilibrium equation in the x direction of Eq. (6.11):

$$\int \omega \left(\frac{\partial \tau_{xy}}{\partial y} + \frac{\partial \tau_{xz}}{\partial z} \right) dA = -\int \omega \frac{\partial \sigma_x}{\partial x} \, dA = \int \omega \frac{V_z}{I_y I_z - I_{yz}^2} \left(I_{yz} y - I_z z \right) dA \tag{6.114}$$

where Eq. (6.9) has been introduced. It follows that Eq. (6.108) can be expressed as

$$y_S V_z = \int \left(\tau_{xz} y - \tau_{xy} z \right) dA = \int \omega \frac{V_z}{I_y I_z - I_{yz}^2} \left(I_{yz} y - I_z z \right) dA \tag{6.115}$$

Similarly, with only V_y applied at the free end of the beam, the same calculations lead to the condition

$$-z_S V_y = \int \left(\tau_{xz} y - \tau_{xy} z \right) dA = \int \omega \frac{V_y}{I_y I_z - I_{yz}^2} \left(I_{yz} z - I_y y \right) dA \tag{6.116}$$

The warping-dependent section properties defined by

$$I_{y\omega} = \int y\omega(y, z) \, dA \qquad I_{z\omega} = \int z\omega(y, z) \, dA \tag{6.117}$$

are called *sectorial products of area*. Substitution of these sectorial products into Eqs. (6.115) and (6.116) shows that the shear center coordinates according to Trefftz's definition are given by

$$y_S = \frac{I_{yz} I_{y\omega} - I_z I_{z\omega}}{I_y I_z - I_{yz}^2} \tag{6.118}$$

$$z_S = \frac{I_y I_{y\omega} - I_{yz} I_{z\omega}}{I_y I_z - I_{yz}^2} \tag{6.119}$$

Since the warping function ω depends only on the cross-sectional shape, so do the sectorial products, and the coordinates of the shear center S are determined solely by the geometry of the cross section.

Example 6.3 Shear Center of a Channel Section. Consider again the channel section of Fig. 6.7 with $c = 1$. Table 6.2 gives part of the output file for section properties as calculated by the computer program of the appendixes. The program calculates the shear centers using both the Trefftz method and the theory of elasticity based method

TABLE 6.2 Shear Centers for Channel Sections of Different Poisson's Ratios

	ν		
	0.0	0.333	0.5
Y shear center with respect to centroid	−0.62055	−0.62056	−0.62057
Z Shear center with respect to centroid	0.00000	0.00000	0.00000
Y Shear center with respect to centroid (Trefftz)	−0.62055	−0.62055	−0.62055
Z Shear center with respect to centroid (Trefftz)	0.00000	0.00000	0.00000

discussed in Section 6.2.5. The shear centers have been calculated for several different values of Poisson's ratio ν. Note that the shear center calculated using the Trefftz method does not depend on ν and thus remains the same for all three cases. The shear center locations from the more accurate theory of elasticity based approach do, however, change slightly for different Poisson's ratios. Both the theory of elasticity method and the Trefftz method are implemented with finite elements in the computer program of the appendixes.

Traditional formulas for the location of shear centers for common cross sections are provided in Table 6.3. A much more extensive collection of formulas is available in Pilkey (1994).

6.3 SHEAR DEFORMATION COEFFICIENTS

Standard Bernoulli–Euler engineering beam theory with Timoshenko's (1921) shear deformation supplement was discussed in Chapter 2. If the effects of shear deformation, and rotary inertia are included, this is referred to as the *Timoshenko beam theory*. Various *shear deformation* or *shear stiffness coefficients* are reviewed in Kaneko (1975), Renton (1991), and Hutchinson (2001). Timoshenko's shear coefficients were defined as the ratio of the maximum shear stress to the average shear stress on cross sections. To improve this theory, Cowper (1966) suggested the use of an elasticity solution, based on the geometric assumption that the average cross-sectional transverse displacements of a particular cross section of a beam can be interpreted as the deflection of the longitudinal axis of the beam. The shear coefficients obtained by Cowper are only for symmetric cross sections, oriented along the principal bending axes for which I_{yz} is zero. This led to two shear coefficients, α_{yy} and α_{zz}. Mason and Herrmann (1968) attempted to expand Cowper's method for arbitrarily shaped cross sections and in an arbitrarily oriented coordinate system. This led to the additional coefficients α_{yz} and α_{zy}, which were not equal. The use of asymmetrical shear coefficients leads to asymmetrical structural matrices.

A different formulation for shear deformation coefficients (Schramm et al., 1994) is to assume that the strain energy for a beam as represented by the theory of elasticity is equal to the strain energy for a one-dimensional beam based on technical beam theory. This approach is utilized here. The shear deformation coefficients α_{ij}, $i, j = y, z$, derived using this energy method form a symmetric tensor. The principal axes

TABLE 6.3 Classical Shear Center Locations

Shape	Location of Shear Center (S) y_S or z_S
Sector of thin circle 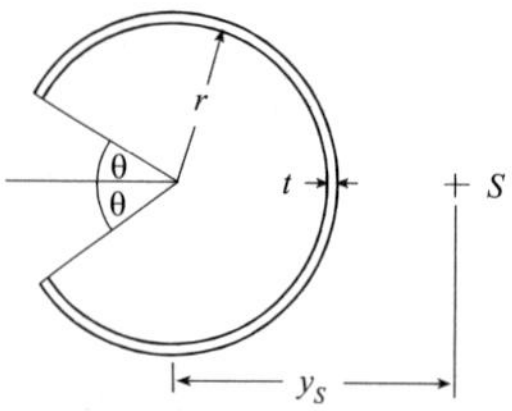	$\dfrac{2r}{(\pi-\theta)+\sin\theta\cos\theta}[(\pi-\theta)\cos\theta+\sin\theta]$ For split tube ($\theta=0$), use $y_S=2r$.
Semicircular solid section 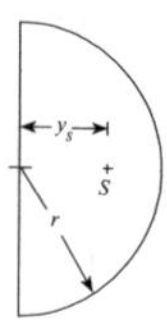	$\dfrac{8}{15\pi}\dfrac{3+4\nu}{1+\nu}r$
Channel 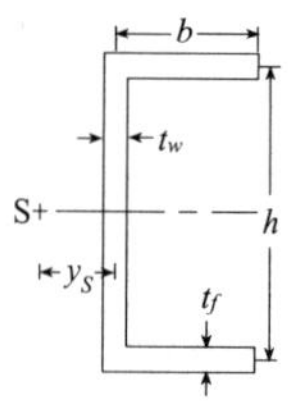	$\dfrac{3t_f b^2}{6bt_f+ht_w}$
I-beam 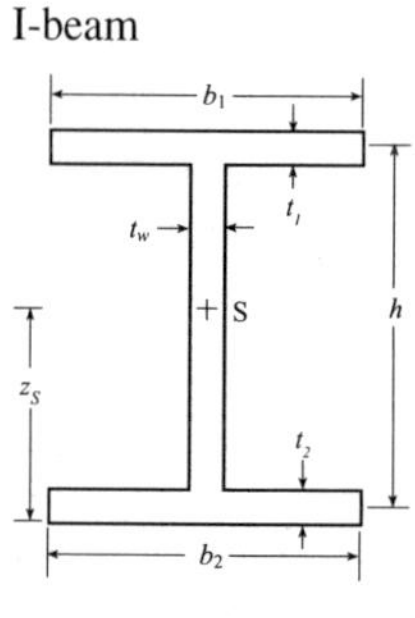	$\dfrac{t_1 b_1^3 h}{t_1 b_1^3+t_2 b_2^3}$ If $b_1=b_2$ and $t_1=t_2$, then $z_S=\frac{1}{2}h$
Thin-walled U section 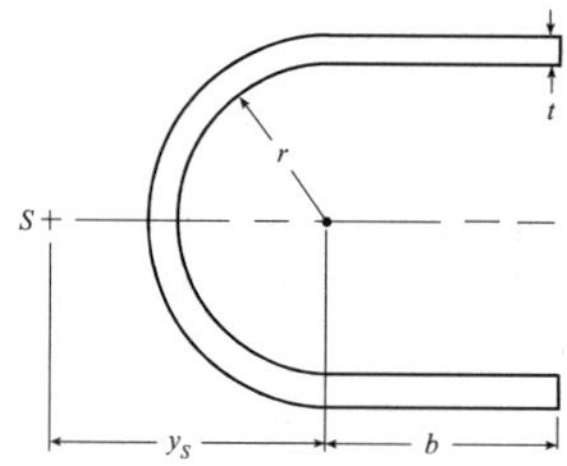	$\dfrac{4r^2+2b^2+2\pi br}{4b+\pi r}$

of this tensor are different, in general, from the principal axes of the tensor for the moments of inertia.

6.3.1 Derivation

From the theory of elasticity, the shear strain energy per unit length of the beam is

$$U_0 = \int \frac{\tau_{xy}^2 + \tau_{xz}^2}{2G}\, dA \tag{6.120}$$

In engineering beam theory, an approximate formula for the strain energy per unit length due to shear deformation is used by introducing shear factors

$$U_{\text{beam}} = \frac{1}{2}\left(\frac{\alpha_{yy}V_y^2}{GA} + \frac{\alpha_{zz}V_z^2}{GA} + \frac{\alpha_{yz}V_yV_z}{GA} + \frac{\alpha_{zy}V_zV_y}{GA}\right) \tag{6.121}$$

These factors will be determined by assuming that the shear strain energies given by the elasticity formula and the approximate beam theory formula are equal. The coefficient $\alpha_z = \alpha_{zz}$ is used in the calculation of the shear deformation in the z direction due to a shear force in the same direction. The shear form factor k_s of Eq. (2.12) and the shear deformation coefficient α_z are related as $\alpha_z = 1/k_s$. The coefficient $\alpha_y = \alpha_{yy}$ is defined similarly. The shear deformation in the y direction due to a shear force in the z direction is denoted by α_{yz}. The coefficient α_{zy} is, of course, related to the shear deformation in the z direction due to the y shear force.

To determine α_{yy}, for example, let V_z be zero, and set U_0 equal to U_{beam}. Introduce the stresses τ_{xy} and τ_{xz} from Eq. (6.32).

$$\frac{V_y^2}{\Delta^2}\int \left[\left(\frac{\partial\Psi}{\partial y} - d_y\right)^2 + \left(\frac{\partial\Psi}{\partial z} - d_z\right)^2\right] dA = \frac{\alpha_{yy}V_y^2}{A} \tag{6.122}$$

This gives

$$\alpha_{yy} = \frac{A}{\Delta^2}\int \left[\left(\frac{\partial\Psi}{\partial y} - d_y\right)^2 + \left(\frac{\partial\Psi}{\partial z} - d_z\right)^2\right] dA \tag{6.123}$$

The coefficient α_{zz} is determined in a similar fashion, setting V_y equal to zero. To determine α_{yz}, both V_y and V_z are applied. In this case, utilize the total shear stresses τ_{xy} and τ_{xz}, which are obtained by superposition of the stress formulas of Eqs. (6.17) and (6.32)

$$
\begin{aligned}
\tau_{xy} &= \frac{V_y}{\Delta}\left(\frac{\partial \Psi}{\partial y} - d_y\right) + \frac{V_z}{\Delta}\left(\frac{\partial \Phi}{\partial y} - h_y\right) \\
\tau_{xz} &= \frac{V_y}{\Delta}\left(\frac{\partial \Psi}{\partial z} - d_z\right) + \frac{V_z}{\Delta}\left(\frac{\partial \Phi}{\partial y} - h_z\right)
\end{aligned} \tag{6.124}
$$

It is readily observed that $\alpha_{yz} = \alpha_{zy}$.

The four shear deformation coefficients are given by

$$
\begin{aligned}
\alpha_{yy} &= \frac{A}{\Delta^2}\int\left[\left(\frac{\partial \Psi}{\partial y} - d_y\right)^2 + \left(\frac{\partial \Psi}{\partial z} - d_z\right)^2\right] dA \\
\alpha_{zz} &= \frac{A}{\Delta^2}\int\left[\left(\frac{\partial \Phi}{\partial y} - h_y\right)^2 + \left(\frac{\partial \Phi}{\partial z} - h_z\right)^2\right] dA \\
\alpha_{yz} = \alpha_{zy} &= \frac{A}{\Delta^2}\int\left[\left(\frac{\partial \Psi}{\partial y} - d_y\right)\left(\frac{\partial \Phi}{\partial y} - h_y\right) + \left(\frac{\partial \Psi}{\partial z} - d_z\right)\left(\frac{\partial \Phi}{\partial z} - h_z\right)\right] dA
\end{aligned} \tag{6.125}
$$

or, in vector notation,

$$
\begin{aligned}
\alpha_{yy} &= \frac{A}{\Delta^2}\int (\mathbf{\nabla}\Psi - \mathbf{d})\cdot(\mathbf{\nabla}\Psi - \mathbf{d})\ dA \\
\alpha_{zz} &= \frac{A}{\Delta^2}\int (\mathbf{\nabla}\Phi - \mathbf{h})\cdot(\mathbf{\nabla}\Phi - \mathbf{h})\ dA \\
\alpha_{yz} = \alpha_{zy} &= \frac{A}{\Delta^2}\int (\mathbf{\nabla}\Phi - \mathbf{h})\cdot(\mathbf{\nabla}\Psi - \mathbf{d})\ dA
\end{aligned} \tag{6.126}
$$

with $\mathbf{\nabla} = \mathbf{j}(\partial/\partial y) + \mathbf{k}(\partial/\partial z)$.

Schramm et al. (1997) presented a more comprehensive theory including the effect of torsion of the bar. In this case, the strain energy per unit length of Eq. (6.121) is replaced by

$$
\begin{aligned}
U_{\text{beam}} = \frac{1}{2}\Bigg(&\frac{\alpha_{xx} M_x^2}{GJ} + \frac{\alpha_{yy} V_y^2}{GA} + \frac{\alpha_{zz} V_z^2}{GA} + \frac{\alpha_{xy} M_x V_y}{GJ} + \frac{\alpha_{yx} V_y M_x}{GJ} \\
&+ \frac{\alpha_{xz} M_x V_z}{GJ} + \frac{\alpha_{zx} V_z M_x}{GJ} + \frac{\alpha_{yz} V_y V_z}{GA} + \frac{\alpha_{zy} V_z V_y}{GA}\Bigg)
\end{aligned} \tag{6.127}
$$

where α_{xx} is a *torsional coefficient* and α_{xy}, α_{yx}, α_{xz}, and α_{zx} are *torsion-shear coefficients*.

6.3.2 Principal Shear Axes

These shear deformation coefficients can be expressed in matrix form as

$$
\mathbf{A} = \begin{bmatrix} \alpha_{yy} & \alpha_{yz} \\ \alpha_{zy} & \alpha_{zz} \end{bmatrix} = \frac{A}{\Delta^2}\int \mathbf{B}^{\mathrm{T}}\mathbf{B}\ dA \tag{6.128}
$$

where

$$\mathbf{B} = \left[\boldsymbol{\nabla}\Psi - \mathbf{d} \quad \boldsymbol{\nabla}\Phi - \mathbf{h}\right] \tag{6.129}$$

The matrix **A** is a symmetric tensor. The principal values of the tensor can be determined from an eigenvalue problem (Schramm et al., 1997). The angle φ^s of the principal axes, which are the *principal shear axes*, of this tensor are obtained from

$$\tan 2\varphi^s = \frac{2\alpha_{yz}}{\alpha_{yy} - \alpha_{zz}} \tag{6.130}$$

In general, the angle φ^s differs from φ of Eq. (1.82) for the principal bending axes. For a symmetrical cross section, both angles are the same. That is, the principal bending and principal shear axes are the same for symmetrical cross sections. It is readily shown (Schramm et al., 1997) that the principal values of the shear deformation coefficients are always greater than or equal to 1.

6.3.3 Finite Element Solution Formulation

For the shear coefficient $\alpha_{yy} = \alpha_y$, define

$$\kappa_y = \frac{\alpha_{yy}\Delta^2}{A} \tag{6.131}$$

where A is the cross-sectional area and $\Delta = 2(1 + \nu)(I_y I_z - I_{yz}^2)$. Introduce the formula of Eq. (6.126) for α_{yy}:

$$\kappa_y = \int (\boldsymbol{\nabla}\Psi - \mathbf{d}) \cdot (\boldsymbol{\nabla}\Psi - \mathbf{d})\, dA \tag{6.132}$$

which in discretized form becomes

$$\kappa_y = \sum_e \int_{-1}^{1} \int_{-1}^{1} (\boldsymbol{\Psi}^{eT}\mathbf{B}^T - \mathbf{d}^T)(\mathbf{B}\boldsymbol{\Psi}^e - \mathbf{d}) \mid \mathbf{J}_e \mid d\eta\, d\zeta \tag{6.133}$$

where $\boldsymbol{\Psi}^e$ is the vector of nodal values in element e of the shear function Ψ for transverse shear loading V_y in the y direction.

The expression for κ_y may be rewritten as

$$\begin{aligned}
\kappa_y &= \sum_e \int_{-1}^{1} \int_{-1}^{1} (\boldsymbol{\Psi}^{eT}\mathbf{B}^T\mathbf{B}\boldsymbol{\Psi}^e - 2\boldsymbol{\Psi}^{eT}\mathbf{B}^T\mathbf{d} + \mathbf{d}^T\mathbf{d}) \mid \mathbf{J}_e \mid \, d\eta\, d\zeta \\
&= \sum_e \boldsymbol{\Psi}^{eT}\mathbf{k}^e\boldsymbol{\Psi}^e + \sum_e \int_{-1}^{1} \int_{-1}^{1} (-2\boldsymbol{\Psi}^{eT}\mathbf{B}^T\mathbf{d} + \mathbf{d}^T\mathbf{d}) \mid \mathbf{J}_e \mid d\eta\, d\zeta
\end{aligned}$$

$$= \boldsymbol{\Psi}^{\mathrm{T}}\mathbf{K}\boldsymbol{\Psi} + \sum_e \int_{-1}^{1}\int_{-1}^{1} (-2\boldsymbol{\Psi}^{e\mathrm{T}}\mathbf{B}^{\mathrm{T}}\mathbf{d} + \mathbf{d}^{\mathrm{T}}\mathbf{d}) \mid \mathbf{J}_e \mid d\eta\, d\zeta \tag{6.134}$$

$$= \boldsymbol{\Psi}^{\mathrm{T}}\mathbf{P}_y - 2\sum_e \int_{-1}^{1}\int_{-1}^{1} \boldsymbol{\Psi}^{e\mathrm{T}}\mathbf{B}^{\mathrm{T}}\mathbf{d} \mid \mathbf{J}_e \mid d\eta\, d\zeta$$

$$+ \frac{1}{4}\nu^2(I_y^2 + I_{yz}^2) \int (y^2 + z^2)^2\, dA$$

where

$$\mathbf{k}^e = \int_{-1}^{1}\int_{-1}^{1} \mathbf{B}^{\mathrm{T}}\mathbf{B} \mid \mathbf{J}_e \mid d\eta\, d\zeta$$

The element matrix $\mathbf{k}^e$ is assembled to form $\mathbf{K}$. Also, with

$$\mathbf{p}^e = \int_{-1}^{1}\int_{-1}^{1} \mathbf{B}^{\mathrm{T}}\mathbf{d} \mid \mathbf{J}_e \mid\, d\eta\, d\zeta \tag{6.135}$$

κ_y can be expressed as

$$\kappa_y = \boldsymbol{\Psi}^{\mathrm{T}}(\mathbf{P}_y - 2\mathbf{P}_{y2}) + \sum_e \int_{-1}^{1}\int_{-1}^{1} \mathbf{d}^{\mathrm{T}}\,\mathbf{d} \mid \mathbf{J}_e \mid\, d\eta\, d\zeta$$

where $\mathbf{K}\boldsymbol{\Psi} = \mathbf{P}_y$ and $\sum_e \boldsymbol{\Psi}^{e\mathrm{T}}\mathbf{p}^e = \boldsymbol{\Psi}^{\mathrm{T}}\mathbf{P}_{y2}$.

For the shear coefficient $\alpha_{zz} = \alpha_z$, let κ_z be defined by

$$\kappa_z = \frac{\alpha_{zz}\Delta^2}{A} \tag{6.136}$$

From Eq. (6.126) for α_{zz},

$$\kappa_z = \sum_e \int_{-1}^{1}\int_{-1}^{1} (\boldsymbol{\Phi}^{e\mathrm{T}}\mathbf{B}^{\mathrm{T}} - \mathbf{h}^{\mathrm{T}})(\mathbf{B}\boldsymbol{\Phi}^e - \mathbf{h}) \mid \mathbf{J}_e \mid d\eta\, d\zeta \tag{6.137}$$

where $\boldsymbol{\Phi}^e$ is the vector of nodal values in element e of the shear function Φ for transverse shear loading V_z in the z direction. This expression may be evaluated as

$$\begin{aligned}\kappa_z &= \boldsymbol{\Phi}^{\mathrm{T}}\mathbf{P}_z - 2\sum_e \int_{-1}^{1}\int_{-1}^{1} \boldsymbol{\Phi}^{e\mathrm{T}}\mathbf{B}^{\mathrm{T}}\mathbf{h} \mid \mathbf{J}_e \mid d\eta\, d\zeta \\ &\quad + \frac{1}{4}\nu^2(I_z^2 + I_{yz}^2) \int (y^2 + z^2)^2\, dA\end{aligned} \tag{6.138}$$

In the case of the shear coefficient α_{yz}, define κ_{yz} as

$$\kappa_{yz} = \frac{\alpha_{yz}\Delta^2}{A} \tag{6.139}$$

The formula for α_{yz} in Eq. (6.126) gives

$$\kappa_{yz} = \sum_e \int_{-1}^{1}\int_{-1}^{1} (\mathbf{\Psi}^{e\mathrm{T}}\mathbf{B}^{\mathrm{T}} - \mathbf{d}^{\mathrm{T}})(\mathbf{B}\mathbf{\Phi}^e - \mathbf{h}) \mid \mathbf{J}_e \mid d\eta\, d\zeta \tag{6.140}$$

This is evaluated as follows:

$$\begin{aligned}\kappa_{yz} &= \mathbf{P}_y^{\mathrm{T}}\mathbf{\Phi} - \sum_e \int_{-1}^{1}\int_{-1}^{1} (\mathbf{\Psi}^{e\mathrm{T}}\mathbf{B}^{\mathrm{T}}\mathbf{h} + \mathbf{\Phi}^{e\mathrm{T}}\mathbf{B}^{\mathrm{T}}\mathbf{d} - \mathbf{d}^{\mathrm{T}}\mathbf{h}) \mid \mathbf{J}_e \mid d\eta\, d\zeta \\ &= \mathbf{P}_y^{\mathrm{T}}\mathbf{\Phi} - \sum_e \int_{-1}^{1}\int_{-1}^{1} (\mathbf{\Psi}^{e\mathrm{T}}\mathbf{B}^{\mathrm{T}}\mathbf{h} + \mathbf{\Phi}^{e\mathrm{T}}\mathbf{B}^{\mathrm{T}}\mathbf{d}) \mid \mathbf{J}_e \mid d\eta\, d\zeta \\ &\quad - \frac{1}{4}\nu^2 I_{yz}(I_y + I_z)\int (y^2 + z^2)^2\, dA\end{aligned} \tag{6.141}$$

It is important to recognize that the subscripts chosen here to represent shear deformation coefficients differ from those for moments of inertia. In some of the solid mechanics literature the moments of inertia are defined as $I_{yy} = I_y = \int y^2\, dA$ and $I_{zz} = I_z = \int z^2\, dA$. More common definitions are $I_{yy} = I_y = \int z^2\, dA$ and $I_{zz} = I_z = \int y^2\, dA$. Although the latter definitions for the subscripts of the moments of inertia are employed in this book, the shear coefficient subscripts used here for α_{zz} and α_{yy} correspond to the former definition for the subscripts of the moments of inertia.

Example 6.4 Bar of Open Circular Cross Section. Figure 6.8 shows an open circular cross section with an inner radius of 8 in. and a thickness of 1.25 in. Part of the output file for section properties as calculated by the program of the appendixes

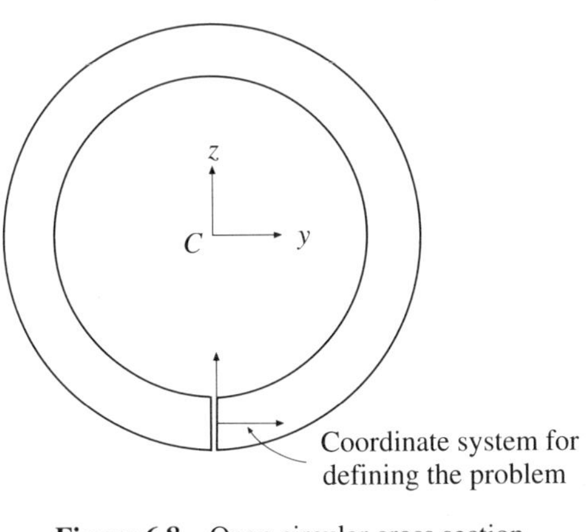

Figure 6.8 Open circular cross section.

TABLE 6.4 Part of an Output File for an Open Circular Cross Section

Y Centroid	0.00000
Z Centroid	8.00000
Y Shear Center with regard to Centroid	0.00000
Z Shear Center with regard to Centroid	15.90306
Y Shear Coefficient	5.93977
Z Shear Coefficient	1.98015
YZ Shear Coefficient	0.00000
Torsional Constant	32.23967

is shown in Table 6.4. More complete output is given in Table B.8. The shear center is approximately one diameter away from the centroid on the z axis, or one and a half diameters away from the slit. This result is often included in handbooks with structural mechanics formulas.

Example 6.5 Shear Deformation Coefficients for a Rectangular Cross Section. Consider the rectangular cross section shown in Fig. 6.9 with height (thick-

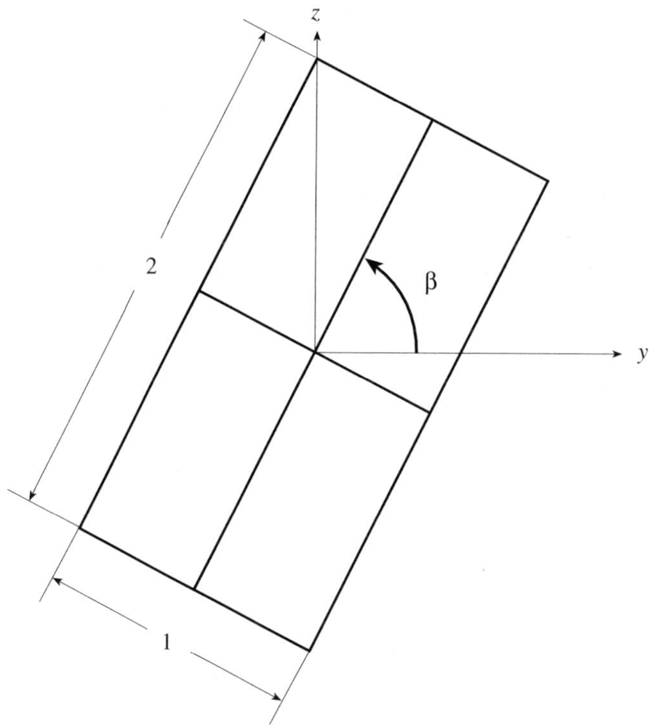

Figure 6.9 Rectangle.

TABLE 6.5 Shear Deformation Coefficients for a Rectangular Cross Section

		ν		
		0	0.3	0.5
$\alpha_{\text{classical}}$		1.2	1.1769	1.1667
Principal	α_1	1.2	1.27479	1.35605
	α_2	1.2	1.20056	1.20118
$\beta = 30°$	α_{yy}	1.2	1.21912	1.23990
	α_{yz}	0.0	−0.32142	−0.67062
	α_{zz}	1.2	1.25624	1.31733
$\beta = 60°$	α_{yy}	1.2	1.23768	1.27861
	α_{yz}	0.0	−0.03711	−0.07744
	α_{zz}	1.2	1.23768	1.27861
$\beta = 90°$	α_{yy}	1.2	1.27479	1.35605
	α_{yz}	0.0	0.00000	0.00000
	α_{zz}	1.2	1.20056	1.20118

ness) h and width b. The shear deformation coefficients are calculated for three different values of Poisson's ratio and for three angles. For this doubly symmetric section, the shear center coincides with the centroid. Some of the results are shown in Table 6.5. The principal shear coefficients are denoted by α_1 and α_2. Also shown in the Table are the "classical" values of the shear deformation coefficient calculated using

$$\alpha_{\text{classical}} = \frac{12 + 11\nu}{10(1 + \nu)}. \tag{1}$$

This rectangular cross section is useful in studying the concept of *shear locking* (Pilkey and Wunderlich, 1994). Traditionally, for rectangular cross sections of any aspect ratio (b/h in Fig. 6.10), a shear deformation coefficient of $\alpha_{yy} = \alpha_{zz} = 1.2$ is employed. This corresponds to (1) with $\nu = 0$. Often shear locking is explained by noting that quantities similar to

$$\overline{k} = \frac{1}{\alpha_{zz}} \frac{GA}{EI_y} = \frac{6}{\alpha_{zz}(1 + \nu)h^2} \tag{2}$$

for beams and

$$\zeta = \frac{6k_s(1 - \nu)}{h^2} = \frac{6(1 - \nu)}{\alpha_{zz} h^2} \qquad k_s = \frac{1}{\alpha_{zz}} \tag{3}$$

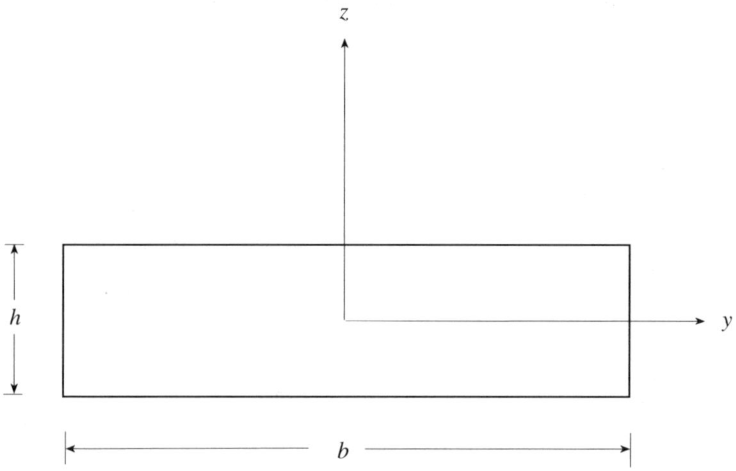

Figure 6.10 Rectangular cross section with thickness h and width b.

for rectangular plates, which are constants in the response expressions, become inordinately large as the cross section becomes very thin. Table 6.6 illustrates this behavior in these constasnts for $\alpha_{zz} = 1.2$. This leads to unrealistic results of transverse displacements approaching infinity. In addition, Table 6.6 shows for increasing aspect ratios b/h and $\nu = 0.3$, values of α_{zz}, calculated using the theory of this chapter. Also shown are the corresponding values of $\overline{k} = GA/\alpha_{zz}EI_y$ and $\zeta = 6k_s(1-\nu)/h^2$. Note that as the cross section becomes thinner and the aspect ratios b/h become larger, α_{zz} increases as do $\overline{k}$ and ζ. However, the increase in $\overline{k}$ and ζ are not as great as for the case with $\alpha_{zz} = 1.2$. Of course, for ζ to be applicable to thin plates it is necessary that b/h be quite large. It appears that if the shear deformation coefficients presented here are employed, the shear deformation of the member is described more accurately than with the traditional coefficients, and shear

TABLE 6.6 Shear Deformation Coefficients for a Rectangular Cross Section, with $\nu = 0.3$, for Various b/h Ratios

α_{zz} of Eq. (6.136)				$\alpha_{zz} = 1.2$		
b/h	α_{zz}	$\overline{k}$	ζ	b/h	$\overline{k}$	ζ
1	1.2074	3.86	3.51	1	3.85	3.5
2	1.2748	14.59	13.28	2	15.38	14.0
5	2.0920	55.25	50.28	5	96.15	87.5
10	5.5908	82.53	75.11	10	384.62	350.0
50	131.64	89.11	81.09	50	9615.38	8750.0
100	530.78	88.05	80.12	100	38461.54	35000.0

The calculations are made for a fixed b (width) and a progressively smaller h (height).

locking might be avoided. Also, as the cross section becomes thinner, $1/\alpha_{zz}$ and k_s approach zero, and the shear deformation effects disappear as expected.

The shear deformation coefficients computed here using finite elements are based on a theory of elasticity formulation. These coefficients can be obtained for cross sections of any shape. For a rectangular cross section it can be shown that the shear coefficients calculated with the finite element formulation of this chapter for a rectangular cross section coincide with coefficients found using a formula developed by Renton (1991, 1997) and an exact theory of elasticity approach by Sanchez (2001). The latter case corresponds to the exact beam theory proposed by Ladevèze and Simmonds (1996).

Example 6.6 Shear Deformation Coefficients for a Trapezoidal Cross Section. As an example of a cross section with no axis of symmetry, consider the trapezoid shown in Fig. 6.11. The results, as given in Table 6.7, show that the shear deformation coefficients $\alpha_{yz} = \alpha_{zy}$ are not zero when calculated for the principal bending axes. This implies that the inclusion of shear due to transverse loads leads, in

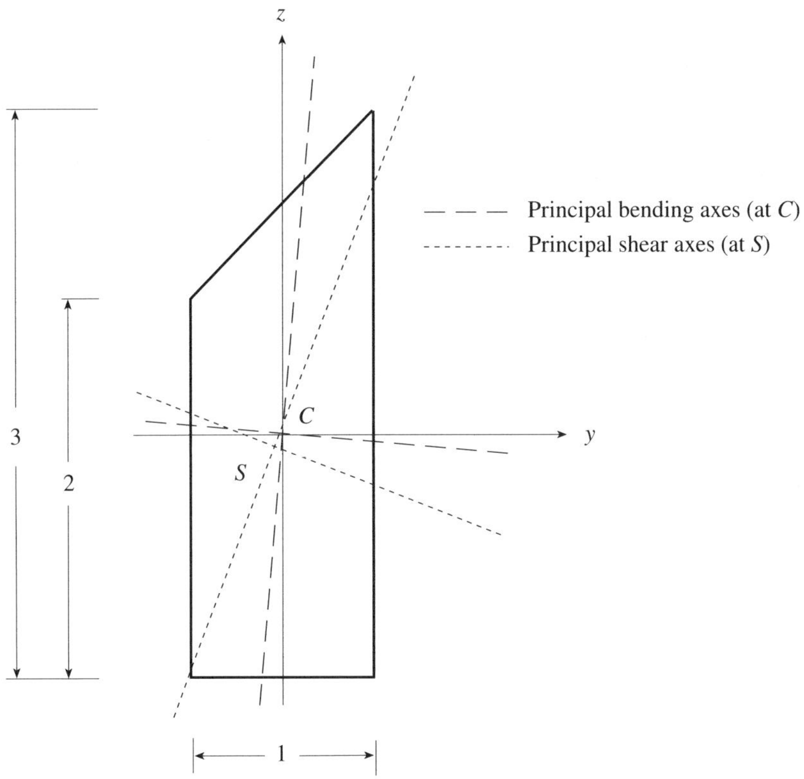

Figure 6.11 Principal axes of a trapezoid. The shear axes are centered at the shear center and correspond to $\nu = 0$.

TABLE 6.7 Shear Deformation Coefficients for a Trapezoid

		ν		
		0	0.3	0.5
y, z axes corresponding to the figure	α_{yy}	1.313417	1.465389	1.630496
	α_{yz}	−0.063785	−0.076064	−0.089406
	α_{zz}	1.169516	1.170704	1.171995
Shear center	y_S	−0.069460	−0.071690	−0.072681
	z_S	−0.072013	−0.090135	−0.098190
Principal shear axes at shear center	α_1	1.337619	1.483865	1.647313
	α_2	1.114531	1.152229	1.164731
	$\alpha_{yz\ \max}$	0.111574	0.165818	0.241291
	φ^s	−20.77868°	−13.65235°	−10.65227°
Principal bending axes at centroid	φ	−4.860108°	−4.860108°	−4.860108°
	α_{yy}	1.323148	1.476062	1.642190
	α_{yz}	−0.057145	−0.050085	−0.049400
	α_{zz}	1.159780	1.159977	1.160191

general, to coupled system equations. From Table 6.7 and Fig. 6.11 observe that the angles to the principal bending axes φ and the principal shear axes φ^s are different.

Example 6.7 Shear Deformation Coefficients for L Sections. Consider two L-shaped cross sections, the first with legs of equal length (Fig. 6.12) so that the line

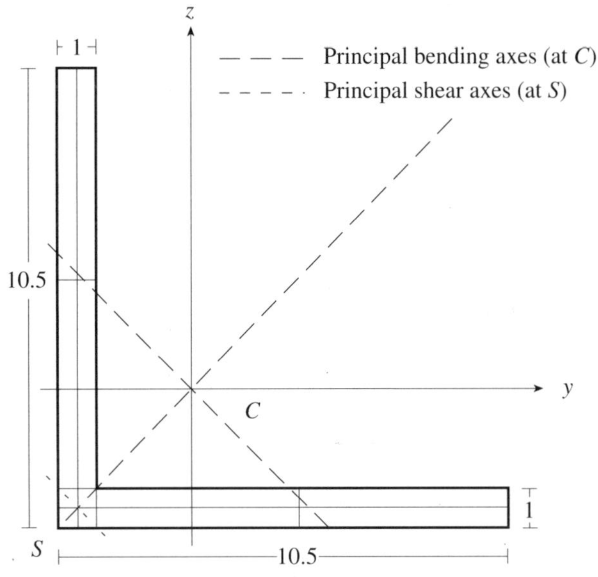

Figure 6.12 Symmetric L section, principal axes, $\nu = 0.5$.

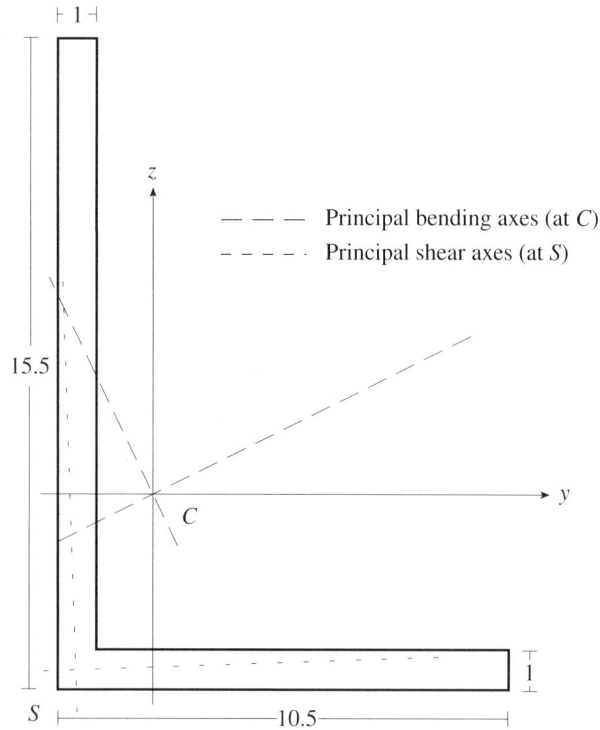

Figure 6.13 Unsymmetric L section, principal axes, $\nu = 0.5$.

evenly between the legs is an axis of symmetry, and the second with one leg shorter than the other (Fig. 6.13). The shear deformation coefficients are calculated for coordinate axes parallel to the legs of the L sections and for the principal bending axes and are given in Tables 6.8 and 6.9. It can be observed that for the asymmetric case of Fig. 6.13, the principal shear and principal bending axes are different, although this is not the case for the symmetric cross section of Fig. 6.12. The cross terms $\alpha_{yz} = \alpha_{zy}$ vanish for the principal bending axes of the symmetric cross section and are nonzero for the asymmetric case.

6.3.4 Traditional Analytical Formulas

Formulas for approximate shear deformation coefficients can be obtained by using the approximate shear stress formulas of Section 6.1.1. Equation (6.120) provides an expression for the shear strain energy per unit length of beam:

$$U_0 = \int_A \frac{\tau_{xy}^2 + \tau_{xz}^2}{2G} \, dA \tag{6.142}$$

TABLE 6.8 Symmetric L-Section

		ν		
		0	0.3	0.5
Torsional constant	J	6.539189	6.539189	6.539189
Centroidal y, z axes corresponding to the figure	α_{yy}	2.301059	2.302935	2.304973
	α_{yz}	0.014969	0.016782	0.018753
	α_{zz}	2.301059	2.302935	2.304973
Shear center	y_S	−2.464750	−2.464971	−2.465070
	z_S	−2.464750	−2.464971	−2.465070
Principal shear axes at shear center	α_1	2.316028	2.319718	2.323726
	α_2	2.286090	2.286153	2.286220
	$\alpha_{yz\ \max}$	0.014969	0.016782	0.018753
	φ^s	0.785398	0.785398	0.785398
Principal bending axes at centroid	φ	0.785398	0.785398	0.785398

TABLE 6.9 Unsymmetric L-Section

		ν		
		0	0.3	0.5
Torsional constant	J	8.211705	8.211705	8.211705
Centroidal y, z axes corresponding to the figure	α_{yy}	3.058207	3.061764	3.065628
	α_{yz}	0.039510	0.041123	0.042876
	α_{zz}	1.898375	1.899138	1.899967
Shear center	y_S	−1.997641	−1.997605	−1.997589
	z_S	−4.423925	−4.424391	−4.424592
Principal shear axes at shear center	α_1	3.059551	3.063217	3.067203
	α_2	1.897031	1.897685	1.898392
	$\alpha_{yz\ \max}$	0.581260	0.582766	0.584405
	φ^s	0.034013	0.035312	0.036717
Principal bending axes at centroid	φ	0.430248	0.430248	0.430248

Substitute in Eq. (6.142) the engineering beam theory stress τ_{xz} in Eq. (6.7) and the equivalent expression for τ_{xy}:

$$\begin{aligned} U_0 &= \int_A \left[\frac{V_y^2}{2G} \left(\frac{Q'_z}{I_z b} \right)^2 + \frac{V_z^2}{2G} \left(\frac{Q'_y}{I_y b} \right)^2 \right] dA \\ &= \frac{V_y^2}{2G} \alpha_{yy} + \frac{V_z^2}{2G} \alpha_{zz} \end{aligned} \tag{6.143}$$

where we have chosen to define the shear deformation coefficients α_{zz} and α_{yy} as

$$\alpha_{zz} = \frac{A}{I_y^2} \int_A \left(\frac{Q'_y}{b} \right)^2 dA \qquad \alpha_{yy} = \frac{A}{I_z^2} \int_A \left(\frac{Q'_z}{b} \right)^2 dA \tag{6.144}$$

Example 6.8 Rectangular Cross Section. Calculate approximate shear deformation factors for the rectangular cross section of Fig. 6.14.

SOLUTION. For the rectangular cross section of Fig. 6.14, the shaded area is $A' = (h/2 - z)b$ and

$$Q'_y = \int_{A'} \bar{z}\, dA = \bar{z} A' = \left(\frac{(h/2 - z)}{2} + z \right) \left(\frac{h}{2} - z \right) b = \frac{1}{2} b \left(\frac{1}{4} h^2 - z^2 \right) \tag{1}$$

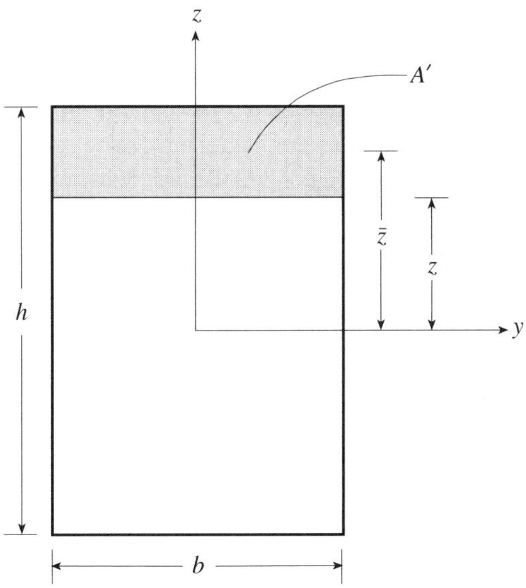

Figure 6.14 Rectangular cross section.

With $I_y = \frac{1}{12}bh^3$, Eq. (6.144) becomes

$$\begin{aligned}\alpha_{zz} &= \frac{A}{I_y^2}\int_A \left(\frac{Q'_y}{b}\right)^2 dA \\ &= \frac{bh}{(\frac{1}{12}bh^3)^2}\int_{-h/2}^{h/2}\left[\frac{1}{2}\left(\frac{1}{4}h^2 - z^2\right)\right]^2 b\,dz = \frac{6}{5}\end{aligned} \tag{2}$$

The same reasoning leads to the shear deformation coefficient $\alpha_{yy} = 6/5$.

The traditional formula (Cowper, 1966) of Eq. (1) of Example 6.5, which depends on the material properties, is

$$\alpha_{zz} = \frac{12 + 11\nu}{10(1+\nu)} \tag{3}$$

which reduces to the value $\frac{6}{5}$ of (2) for $\nu = 0$. As indicated by the more accurate figures in Table 6.6, for $\nu = 0.3$, the shear deformation coefficient of $\frac{6}{5}$ applies when b and h are of similar magnitudes. The approach which is highlighted in this chapter leads to coefficients that depend on the material properties and the dimensions of the cross section.

6.4 DEFLECTION RESPONSE OF BEAMS WITH SHEAR DEFORMATION

The displacements and forces along a beam with shear deformation taken into account are readily obtained by following the procedures outlined in Chapter 2. Principal shear axes, which for asymmetrical cross sections are different from the principal bending axes, were defined in Section 6.3. Because of this difference, deflection components in the y and z directions are, in general, coupled even if the coordinate system for y and z is chosen to correspond to the principal bending axes. If the shape of the cross section is symmetrical about some axis, the principal shear axes coincide with the principal bending axes. If the y, z axes for the symmetrical cross section are chosen to coincide with these principal axes, $\alpha_{yz} = 0$ and $I_{yz} = 0$, and the y and z components of the deflection (v and w) will not be coupled.

The coupled structural matrices for beam elements with arbitrary cross-section shapes will be derived in this section. In the case of the static response of a beam with shear deformation taken into account, an exact solution can be obtained.

6.4.1 Governing Equations

The equations of motion of Chapter 2 can be modified to take into account shear deformation using the shear coefficients α_{yy}, α_{zz}, α_{yz}. We choose for the moment to ignore the inertia terms and the terms involving the axial force N. Begin with the first-order differential equations for a beam in an arbitrarily oriented y, z coordinate system, with the x axis along the beam. The familiar beam notation is employed. The material is linearly elastic with Young's modulus E, Poisson's ratio ν, shear modulus G $(G = E/[2(1+\nu)])$, area A, radii of gyration r_y and r_z, volume density ρ^*, and mass per unit length ρ. Loading for the beam includes distributed transverse forces $\overline{p}_y(x, t)$ and $\overline{p}_z(x, t)$, temperature change $\Delta T(x, y, z)$, and axial compression force N. The transverse loads pass through the shear center so that no torsion occurs.

$$
\begin{aligned}
\frac{\partial v}{\partial x} &= \theta_z + \alpha_{yy}\frac{V_y}{GA} + \alpha_{yz}\frac{V_z}{GA} \\
\frac{\partial \theta_z}{\partial x} &= \frac{(M_z + M_{Tz})I_y}{E(I_y I_z - I_{yz}^2)} + \frac{(M_y + M_{Ty})I_{yz}}{E(I_y I_z - I_{yz}^2)} \\
\frac{\partial V_y}{\partial x} &= \rho\frac{\partial^2 v}{\partial t^2} - \overline{p}_y(x, t) \\
\frac{\partial M_z}{\partial x} &= -V_y - N\theta_z + \rho r_y^2\frac{\partial^2 \theta_z}{\partial t^2} \\
\frac{\partial w}{\partial x} &= -\theta_y + \alpha_{zy}\frac{V_y}{GA} + \alpha_{zz}\frac{V_z}{GA} \\
\frac{\partial \theta_y}{\partial x} &= \frac{(M_y + M_{Ty})I_z}{E(I_y I_z - I_{yz}^2)} + \frac{(M_z + M_{Tz})I_{yz}}{E(I_y I_z - I_{yz}^2)} \\
\frac{\partial V_z}{\partial x} &= \rho\frac{\partial^2 w}{\partial t^2} - \overline{p}_z(x, t) \\
\frac{\partial M_y}{\partial x} &= V_z - N\theta_y + \rho r_z^2\frac{\partial^2 \theta_y}{\partial t^2}
\end{aligned}
\tag{6.145}
$$

These equations are written in a y, z coordinate system corresponding to the principal bending axes if I_{yz} is set equal to zero and the principal shear axes if α_{yz} is set equal to zero.

In these equations the terms with the radii of gyration r_y and r_z in the y and z directions can be expressed as $\rho r_y^2 = \rho^* I_z$ and $\rho r_z^2 = \rho^* I_y$. The quantities M_{Ty} and M_{Tz} are thermal moments given by

$$
\begin{aligned}
M_{Ty} &= \int_A E\alpha\, \Delta T(x, y, z) z\, dA \\
M_{Tz} &= -\int_A E\alpha\, \Delta T(x, y, z) y\, dA
\end{aligned}
\tag{6.146}
$$

where α is the linear coefficient of thermal expansion, a material property.

In matrix form these equations appear as

$$\frac{d}{dx}\begin{bmatrix} v \\ \theta_z \\ V_y \\ M_z \\ w \\ \theta_y \\ V_z \\ M_y \end{bmatrix} = \begin{bmatrix} 0 & 1 & a_{yy} & 0 & 0 & 0 & a_{yz} & 0 \\ 0 & 0 & 0 & b_{yy} & 0 & 0 & 0 & b_{yz} \\ 0 & 0 & 0 & 0 & 0 & 0 & 0 & 0 \\ 0 & 0 & -1 & 0 & 0 & 0 & 0 & 0 \\ 0 & 0 & a_{yz} & 0 & 0 & -1 & a_{zz} & 0 \\ 0 & 0 & 0 & b_{yz} & 0 & 0 & 0 & b_{zz} \\ 0 & 0 & 0 & 0 & 0 & 0 & 0 & 0 \\ 0 & 0 & 0 & 0 & 0 & 0 & 1 & 0 \end{bmatrix} \begin{bmatrix} v \\ \theta_z \\ V_y \\ M_z \\ w \\ \theta_y \\ V_z \\ M_y \end{bmatrix} + \begin{bmatrix} 0 \\ b_{yy}M_{Tz} + b_{yz}M_{Ty} \\ -\overline{p}_y(x) \\ 0 \\ 0 \\ b_{yz}M_{Tz} + b_{zz}M_{Ty} \\ -\overline{p}_z(x) \\ 0 \end{bmatrix}$$

$$\frac{d}{dx}\ \mathbf{z} \quad = \quad \mathbf{A} \quad \mathbf{z} \quad + \quad \overline{\mathbf{P}} \tag{6.147}$$

or

$$\frac{d}{dx}\mathbf{z} = \mathbf{A}\mathbf{z} + \overline{\mathbf{P}} \tag{6.148}$$

where

$$a_{ij} = \frac{\alpha_{ij}}{GA} \quad \text{and} \quad b_{ij} = \frac{I_{ij}}{E(I_{yy}I_{zz} - I_{yz}^2)} \qquad i, j = y, z \tag{6.149}$$

Recall that $I_{yy} = I_y$ and $I_{zz} = I_z$.

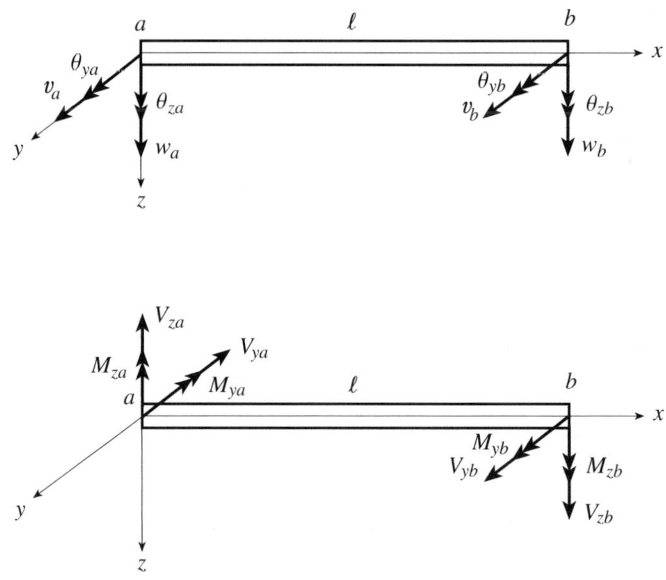

Figure 6.15 Positive displacements and forces for Sign Convention 1 for the transfer matrix solution for element *e*.

6.4.2 Transfer Matrix

The transfer matrix is determined in this work primarily so that it is available for conversion to a stiffness matrix. Integration of Eq. (6.147) leads to Eq. (2.51), that is,

$$\mathbf{z}(x) = e^{\mathbf{A}x}\mathbf{z}_a + e^{\mathbf{A}x}\int_{x=a}^{x} e^{-\mathbf{A}\tau}\overline{\mathbf{P}}(\tau)\,d\tau \tag{6.150}$$

and for element e of length ℓ extending from $x = a$ to $x = b$ (Eq. 2.52),

$$\mathbf{z}_b = \mathbf{U}^e\left[\mathbf{z}_a + \int_a^b \left(\mathbf{U}^e\right)^{-1}\overline{\mathbf{P}}\,d\tau\right] = \mathbf{U}^e\mathbf{z}_a + \overline{\mathbf{z}}^e \tag{6.151}$$

The transfer matrix $\mathbf{U}^e$ for the eth element is obtained using Eq. (2.55), where $\mathbf{A}^s$ is zero for any s greater than 3. Sign convention 1 as shown in Fig. 6.15 applies for the displacements and forces at the ends of element e, which is of length ℓ. Equation (2.55) gives the transfer matrix

$$\mathbf{U}^e = \begin{bmatrix} 1 & \ell & \dfrac{-b_{yy}\ell^3 + 6a_{yy}\ell}{6} & \dfrac{b_{yy}\ell^2}{2} & 0 & 0 & \dfrac{b_{yz}\ell^3 + 6a_{yz}\ell}{6} & \dfrac{b_{yz}\ell^2}{2} \\ 0 & 1 & -\dfrac{b_{yy}\ell^2}{2} & b_{yy}\ell & 0 & 0 & \dfrac{b_{yz}\ell^2}{2} & b_{yz}\ell \\ 0 & 0 & 1 & 0 & 0 & 0 & 0 & 0 \\ 0 & 0 & -\ell & 1 & 0 & 0 & 0 & 0 \\ 0 & 0 & \dfrac{b_{yz}\ell^3 + 6a_{yz}\ell}{6} & -\dfrac{b_{yz}\ell^2}{2} & 1 & -\ell & \dfrac{-b_{zz}\ell^3 - 6a_{zz}\ell}{6} & -\dfrac{b_{zz}\ell^2}{2} \\ 0 & 0 & -\dfrac{b_{yz}\ell^2}{2} & b_{yz}\ell & 0 & 1 & \dfrac{b_{zz}\ell^2}{2} & b_{zz}\ell \\ 0 & 0 & 0 & 0 & 0 & 0 & 1 & 0 \\ 0 & 0 & 0 & 0 & 0 & 0 & \ell & 1 \end{bmatrix} \tag{6.152}$$

The loading vector is given by Eq. (2.53) with Eq. (2.56). The matrix $\mathbf{U}^e(-x)$ is obtained from $\mathbf{U}^e$ of Eq. (6.152) by replacing ℓ with $-x$. If the applied in-span loading is constant or linearly distributed, let

$$\begin{aligned} \overline{p}_i(x) &= \overline{p}_{ia} + \frac{x}{\ell}(\overline{p}_{ib} - \overline{p}_{ia}) \\ M_{Ti}(x) &= M_{Tia} + \frac{x}{\ell}(M_{Tib} - M_{Tia}) \\ i &= y, z \end{aligned} \tag{6.153}$$

where the subscripts a and b indicate that $\overline{p}_i$ or M_{Ti} is evaluated at $x = a$ and $x = b$. The load vector $\overline{\mathbf{z}}^e$ is given by

$$
\overline{\mathbf{z}}^e = \begin{bmatrix} 8b_{yy}\ell^4 - 80a_{yy}\ell^2 \\ 30b_{yy}\ell^3 \\ -120\ell \\ 80\ell^2 \\ -8b_{yz}\ell^4 - 80a_{yz}\ell^2 \\ 30b_{yz}\ell^3 \\ 0 \\ 0 \end{bmatrix} \frac{\overline{p}_{ya}}{240} + \begin{bmatrix} 2b_{yy}\ell^4 - 40a_{yy}\ell^2 \\ 10b_{yy}\ell^3 \\ -120\ell \\ 40\ell^2 \\ -2b_{yz}\ell^4 - 40a_{yz}\ell^2 \\ 10b_{yz}\ell^3 \\ 0 \\ 0 \end{bmatrix} \frac{\overline{p}_{yb}}{240}
$$

$$
+ \begin{bmatrix} -8b_{yz}\ell^4 - 80a_{yz}\ell^2 \\ -30b_{yz}\ell^3 \\ 0 \\ 0 \\ 8b_{zz}\ell^4 - 80a_{zz}\ell^2 \\ -30b_{zz}\ell^3 \\ -120\ell \\ -80\ell^2 \end{bmatrix} \frac{\overline{p}_{za}}{240} + \begin{bmatrix} -2b_{yz}\ell^4 - 40a_{yz}\ell^2 \\ -10b_{yz}\ell^3 \\ 0 \\ 0 \\ 2b_{zz}\ell^4 - 40a_{zz}\ell^2 \\ -10b_{zz}\ell^3 \\ -120\ell \\ -40\ell^2 \end{bmatrix} \frac{\overline{p}_{zb}}{240}
$$

$$
+ \begin{bmatrix} 2b_{yy}\ell^2 \\ 3b_{yy}\ell \\ 0 \\ 0 \\ -2b_{yz}\ell^2 \\ 3b_{yz}\ell \\ 0 \\ 0 \end{bmatrix} \frac{M_{Tza}}{6} + \begin{bmatrix} b_{yy}\ell^2 \\ 3b_{yy}\ell \\ 0 \\ 0 \\ -b_{yz}\ell^2 \\ 3b_{yz}\ell \\ 0 \\ 0 \end{bmatrix} \frac{M_{Tzb}}{6}
$$

$$
+ \begin{bmatrix} 2b_{yz}\ell^2 \\ 3b_{yz}\ell \\ 0 \\ 0 \\ -2b_{zz}\ell^2 \\ 3b_{zz}\ell \\ 0 \\ 0 \end{bmatrix} \frac{M_{Tya}}{6} + \begin{bmatrix} b_{yz}\ell^2 \\ 3b_{yy}\ell \\ 0 \\ 0 \\ -b_{zz}\ell^2 \\ 3b_{zz}\ell \\ 0 \\ 0 \end{bmatrix} \frac{M_{Tyb}}{6} \tag{6.154}
$$

6.4.3 Stiffness Matrix

Although most stiffness or mass matrices are derived based on approximate polynomial shape functions, beams present a special case since often it is possible to derive exact stiffness and mass matrices because the governing differential equations can be solved exactly. This approach is followed here for beams including the effects of shear deformation.

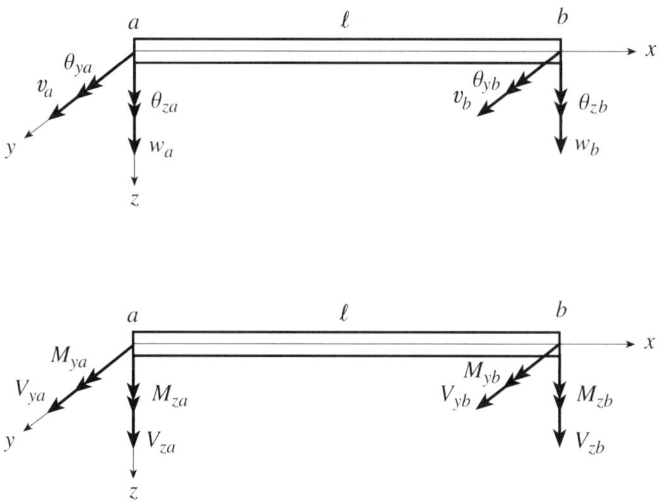

Figure 6.16 Positive displacements and forces for Sign Convention 2 for the stiffness matrix solution for element e.

The stiffness matrix for the beam element will conform to Sign Convention 2, as shown in Fig. 6.16. This convention differs from the sign convention of Fig. 6.15 in that the forces on end a are in the opposite direction. The stiffness matrix of Eq. (2.60), that is,

$$\mathbf{p}^e = \mathbf{k}^e \mathbf{v}^e \tag{6.155}$$

is obtained by reorganizing the transfer matrix of Eq. (6.152). The resulting stiffness matrix will be exact in the sense that Eq. (6.147) is solved without approximations. Follow the procedure detailed in Chapter 2. Define the force vector

$$\begin{aligned}\mathbf{p}^e &= \begin{bmatrix}\mathbf{p}_{ya} & \mathbf{p}_{za} & \mathbf{p}_{yb} & \mathbf{p}_{zb}\end{bmatrix}^{\mathrm{T}} \\ &= \begin{bmatrix}V_{ya} & M_{za} & V_{za} & M_{ya} & V_{yb} & M_{zb} & V_{zb} & M_{zb}\end{bmatrix}^{\mathrm{T}}\end{aligned} \tag{6.156}$$

and the displacement vector

$$\begin{aligned}\mathbf{v}^e &= \begin{bmatrix}\mathbf{v}_{ya} & \mathbf{v}_{za} & \mathbf{v}_{yb} & \mathbf{v}_{zb}\end{bmatrix}^{\mathrm{T}} \\ &= \begin{bmatrix}v_a & \theta_{za} & w_a & \theta_{ya} & v_b & \theta_{zb} & w_b & \theta_{yb}\end{bmatrix}^{\mathrm{T}}\end{aligned} \tag{6.157}$$

The stiffness matrix for this beam element with an arbitrarily shaped cross section and in an arbitrarily oriented coordinate system is

$$
\mathbf{k}^e = \frac{1}{D}\begin{bmatrix}
k_{11} & & & & & & & \\
\frac{\ell}{2}k_{11} & k_{22} & & & & & & \\
k_{31} & \frac{\ell}{2}k_{31} & k_{33} & & & \text{symmetric} & & \\
-\frac{\ell}{2}k_{31} & k_{42} & -\frac{\ell}{2}k_{33} & k_{44} & & & & \\
-k_{11} & -\frac{\ell}{2}k_{11} & -k_{31} & \frac{\ell}{2}k_{31} & k_{11} & & & \\
\frac{\ell}{2}k_{11} & k_{62} & \frac{\ell}{2}k_{31} & k_{82} & -\frac{\ell}{2}k_{11} & k_{22} & & \\
-k_{31} & -\frac{\ell}{2}k_{31} & -k_{33} & \frac{\ell}{2}k_{33} & k_{31} & -\frac{\ell}{2}k_{31} & k_{33} & \\
-\frac{\ell}{2}k_{31} & k_{82} & -\frac{\ell}{2}k_{33} & k_{84} & \frac{\ell}{2}k_{31} & k_{42} & \frac{\ell}{2}k_{33} & k_{44}
\end{bmatrix} \tag{6.158}
$$

where

$$
\begin{aligned}
D &= \ell\left[\left(b_{zz}b_{yy} - b_{yz}^2\right)\ell^4 + 12\left(a_{yy}b_{zz} + 2a_{yz}b_{yz} + a_{zz}b_{yy}\right)\ell^2 \right. \\
&\qquad \left. + 144\left(a_{yy}a_{zz} - a_{yz}^2\right)\right] \\
a_{ij} &= \frac{\alpha_{ij}}{GA} \quad \text{and} \quad b_{ij} = \frac{I_{ij}}{E\left(I_{yy}I_{zz} - I_{yz}^2\right)} \qquad i, j = y, z.
\end{aligned}
$$

The entries of this matrix are

$$
\begin{aligned}
k_{11} &= 12\left(b_{zz}\ell^2 + 12a_{zz}\right) \\
k_{31} &= 12\left(b_{yz}\ell^2 - 12a_{yz}\right) \\
k_{33} &= 12\left(b_{yy}\ell^2 + 12a_{yy}\right) \\
k_{22} &= 4\left[b_{zz}\left(b_{yy}b_{zz} - b_{yz}^2\right)\ell^4 \right. \\
&\qquad + 3\left(4a_{zz}b_{yy}b_{zz} - 3a_{zz}b_{yz}^2 + 2a_{yz}b_{zz}b_{yz} + a_{yy}b_{zz}^2\right)\ell^2 \\
&\qquad \left. + 36b_{zz}\left(a_{yy}a_{zz} - a_{yz}^2\right)\right] \Big/ \left(b_{yy}b_{zz} - b_{yz}^2\right)
\end{aligned}
$$

$$
\begin{aligned}
k_{42} &= -4\Big[b_{yz}\left(b_{yy}b_{zz} - b_{yz}^2\right)\ell^4 \\
&\quad - 3\left(3a_{yz}b_{yy}b_{zz} - 5a_{yz}b_{yz}^2 - a_{yy}b_{zz}b_{yz} - a_{zz}b_{yz}b_{yy}\right)\ell^2 \\
&\quad + 36b_{yz}\left(a_{yy}a_{zz} - a_{yz}^2\right)\Big] \Big/ \left(b_{yy}b_{zz} - b_{yz}^2\right) \\
k_{44} &= 4\Big[b_{yy}\left(b_{yy}b_{zz} - b_{yz}^2\right)\ell^4 \\
&\quad + 3\left(4a_{yy}b_{yy}b_{zz} - 3a_{yy}b_{yz}^2 + 3a_{yz}b_{yy}b_{yz} + a_{zz}b_{yy}^2\right)\ell^2 \\
&\quad + 36b_{yy}\left(a_{yy}a_{zz} - a_{yz}^2\right)\Big] \Big/ \left(b_{yy}b_{zz} - b_{yz}^2\right) \\
k_{62} &= 2\Big[b_{yy}\left(b_{yy}b_{zz} - b_{yz}^2\right)\ell^4 \\
&\quad + 3\left(4a_{zz}b_{yy}b_{zz} - 6a_{zz}b_{yz}^2 - 4a_{yz}b_{zz}b_{yz} - 2a_{yy}b_{zz}^2\right)\ell^2 \\
&\quad - 72b_{zz}\left(a_{yy}a_{zz} - a_{yz}^2\right)\Big] \Big/ \left(b_{yy}b_{zz} - b_{yz}^2\right) \\
k_{82} &= -2\Big[b_{yz}\left(b_{yy}b_{zz} - b_{yz}^2\right)\ell^4 \\
&\quad - 6\left(3a_{yz}b_{yy}b_{zz} - a_{yz}b_{yz}^2 + a_{yy}b_{zz}b_{yz} + a_{zz}b_{yz}b_{yy}\right)\ell^2 \\
&\quad - 72b_{yz}\left(a_{yy}a_{zz} - a_{yz}^2\right)\Big] \Big/ \left(b_{yy}b_{zz} - b_{yz}^2\right) \\
k_{84} &= 2\Big[b_{yy}\left(b_{yy}b_{zz} - b_{yz}^2\right)\ell^4 \\
&\quad + 6\left(2a_{yy}b_{yy}b_{zz} - 3a_{yy}b_{yz}^2 - 2a_{yz}b_{yy}b_{yz} - a_{zz}b_{yy}^2\right)\ell^2 \\
&\quad - 72b_{yy}\left(a_{yy}a_{zz} - a_{yz}^2\right)\Big] \Big/ \left(b_{yy}b_{zz} - b_{yz}^2\right)
\end{aligned}
\tag{6.159}
$$

Example 6.9 A Cantilever Beam with a Trapezoidal Cross Section. Consider a beam with a trapezoidal cross section, as shown in Fig. 6.17. This trapezoid, which does not have an axis of symmetry, is oriented 5.8° from a horizontal axis. The beam is cantilevered and has a force at the free end. The length of the beam is 10 in. The area of the cross section is 2.5 in^2.

The principal bending axes are found (Table 6.7) to be oriented at an angle of 4.9°, as shown in Fig. 6.17. The x, y coordinate system is chosen to be aligned with the principal bending axes. The moments of inertia for these principal axes are found to be $I_z = 0.19688956$, $I_y = 1.4144668$, $I_{yz} = 0$. For $\nu = 0.3$, the shear deformation coefficients relative to the principal bending axes are found using the formulas of Eq. (6.126) to be $\alpha_{zz} = 1.159977$, $\alpha_{yy} = 1.476062$, and $\alpha_{yz} = -0.050085$. As explained in Section 6.3.2, the principal shear axes differ from the principal bending axes, so that α_{yz} is, as expected, not equal to zero for this asymmetric cross section.

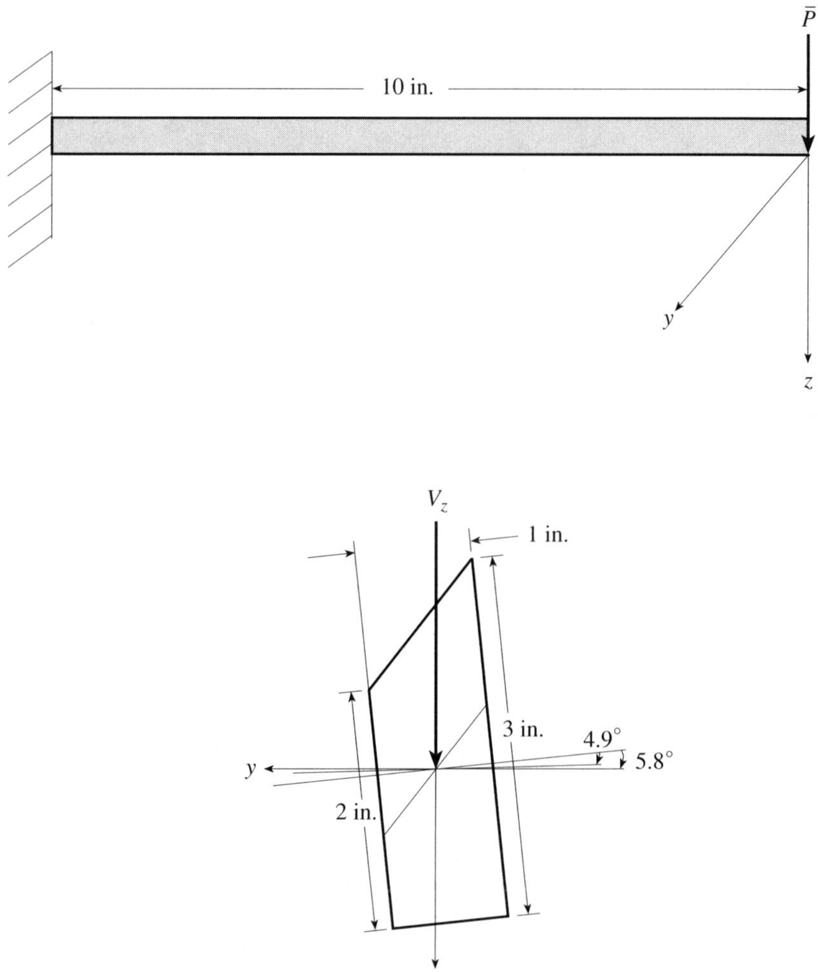

Figure 6.17 Cantilever beam with a trapezoidal cross section.

The load $\overline{\mathbf{P}}$ is applied along the z axis. The deflection of the beam is readily computed using the methods of Chapters 2 and 3. Table 6.10 portrays the beam displacements for various shear deformation coefficient scenarios. This table lists the nondimensional displacements $\overline{v} = v_{x=L} \backslash L$ and $\overline{w} = w_{x=L} \backslash L$ and the rotations $\theta_y|_{x=L}$ and $\theta_z|_{x=L}$ under the tip loading, which is chosen to be $P = 0.3EI_y/L^2$.

The first column in Table 6.10, with all shear deformation coefficients being set equal to zero, corresponds to Bernoulli–Euler beam theory. In this case, familiar formulas are used for the displacements:

TABLE 6.10 Free End Displacements of a Cantilever Beam with Various Sets of Shear Deformation Coefficients

	Bernoulli–Euler Theory	Frequently Used Default Values of α_{yy} and α_{zz}	Utilize Proper Values for α_{yy} and α_{zz}, but Set $\alpha_{yz} = 0$	Use the Correct Values for α_{yy}, α_{zz}, and α_{yz}
α_{yy}	0	1	1.476062	1.476062
α_{zz}	0	1	1.159977	1.159977
α_{yz}	0	0	0	−0.050085
$\overline{w}$	0.1	0.1044131	0.1051191	0.1051191
$\overline{v}$	0	0	0	−0.000221
θ_y	0.15	0.15	0.15	0.15
θ_z	0	0	0	0

$$\overline{w} = \frac{w_{x=L}}{L} = \frac{PL^2}{3EI_y} \qquad \theta_y \bigg|_{x=L} = \frac{PL^2}{2EI_y} \tag{1}$$

The displacements $\overline{v}$ and θ_z are zero since y and z are chosen to be the principal bending axes and $\overline{\mathbf{P}}$ is applied along the z coordinate.

Some general-purpose structural analysis programs recommend that the user input the two shear coefficients α_{yy} and α_{zz} as being equal to one. The cross coefficient α_{yz} is usually not utilized in these programs. The values in the second column in Table 6.10 correspond to these coefficients. Note that the deflection $\overline{w}$ increases somewhat.

Often, the users of general-purpose structural analysis programs are asked to input values of α_{zz} and α_{yy}, but α_{yz} is not taken into account. This is equivalent to the assumption that shear deformations are uncoupled in the principal bending planes. This is shown in the third column of Table 6.10, where α_{yy} and α_{zz} are calculated by using Eq. (6.126) and α_{yz} is set equal to zero. The deflection $\overline{w}$ continues to increase.

The final column shown in Table 6.10 corresponds to the solution for the displacements being based on the transfer matrix of Eq. (6.152) or the stiffness matrix of Eq. (6.158), in which the three shear coefficients are given their proper values. Note that a nonzero deflection occurs in the y direction even though the load is applied along z, a principal bending axis. Of course, this y-direction deflection is due to the cross coefficient α_{yz}. Also, note that this theory of shear deformation does not affect the rotations θ_y and θ_z.

6.4.4 Exact Geometric Stiffness Matrix for Beams with Axial Loading

Begin with the governing equations of Eq. (6.145) with the dynamic response terms and the transverse loading terms ignored. Also, consider the equations in a coordinate system corresponding to the principal bending axes for which $I_{yz} = 0$. These relations then appear as

$$
\begin{aligned}
\frac{dv}{dx} &= \theta_z + \alpha_{yy}\frac{V_y}{GA} + \alpha_{yz}\frac{V_z}{GA} \\
\frac{d\theta_z}{dx} &= \frac{M_z}{EI_z} \\
\frac{dV_y}{dx} &= 0 \\
\frac{dM_z}{dx} &= -V_y - N\theta_z \\
\frac{dw}{dx} &= -\theta_y + \alpha_{yz}\frac{V_y}{GA} + \alpha_{zz}\frac{V_z}{GA} \\
\frac{d\theta_y}{dx} &= \frac{M_y}{EI_y} \\
\frac{dV_z}{dx} &= 0 \\
\frac{dM_y}{dx} &= V_z - N\theta_y
\end{aligned}
\tag{6.160}
$$

The axial load N is compressive. In matrix form, with

$$
a_{ij} = \frac{\alpha_{ij}}{GA} \qquad i, j = y, z \quad \text{and} \quad b_{zz} = \frac{1}{EI_y} \qquad b_{yy} = \frac{1}{EI_z} \tag{6.161}
$$

these relations appear as

$$
\frac{d}{dx}\begin{bmatrix} v \\ \theta_z \\ V_y \\ M_z \\ w \\ \theta_y \\ V_z \\ M_y \end{bmatrix} = \begin{bmatrix} 0 & 1 & a_{yy} & 0 & 0 & 0 & a_{yz} & 0 \\ 0 & 0 & 0 & b_{yy} & 0 & 0 & 0 & 0 \\ 0 & 0 & 0 & 0 & 0 & 0 & 0 & 0 \\ 0 & -N & -1 & 0 & 0 & 0 & 0 & 0 \\ 0 & 0 & a_{yz} & 0 & 0 & -1 & a_{zz} & 0 \\ 0 & 0 & 0 & 0 & 0 & 0 & 0 & b_{zz} \\ 0 & 0 & 0 & 0 & 0 & 0 & 0 & 0 \\ 0 & 0 & 0 & 0 & 0 & -N & 1 & 0 \end{bmatrix} \begin{bmatrix} v \\ \theta_z \\ V_y \\ M_z \\ w \\ \theta_y \\ V_z \\ M_y \end{bmatrix}
$$
$$
\frac{d}{dx}\mathbf{z} = \mathbf{A}\,\mathbf{z} \tag{6.162}
$$

First find the transfer matrix, which can be used as a vehicle for determining the stiffness matrix. The transfer matrix $\mathbf{U}^e$ for element e of length ℓ is expressed as (Eq. 2.52) $\mathbf{z}_b = \mathbf{U}^e\mathbf{z}_a$, with (Eq. 2.54) $\mathbf{U}^e = e^{\mathbf{A}\ell}$. For a $n \times n$ matrix, $\mathbf{A}$, a function $f(\mathbf{A})$ can be replaced by a polynomial $P(\mathbf{A})$ in $\mathbf{A}$ of order $n - 1$ (Pilkey and Wunderlich, 1994):

$$
\begin{aligned}
\mathbf{U}^e = e^{\mathbf{A}\ell} &= c_o\mathbf{I} + c_1(\mathbf{A}\ell) + c_2(\mathbf{A}\ell)^2 + c_3(\mathbf{A}\ell)^3 \\
&\quad + c_4(\mathbf{A}\ell)^4 + c_5(\mathbf{A}\ell)^5 + c_6(\mathbf{A}\ell)^6 + c_7(\mathbf{A}\ell)^7
\end{aligned}
\tag{6.163}
$$

where $c_i, i = 0, 1, \ldots, 7$, are constants. From the Cayley–Hamilton theorem (a matrix satisfies its own characteristic equation),

$$\begin{aligned} e^{\lambda_i \ell} &= c_o + c_1(\lambda_i \ell) + c_2(\lambda_i \ell)^2 + c_3(\lambda_i \ell)^3 \\ &\quad + c_4(\lambda_i \ell)^4 + c_5(\lambda_i \ell)^5 + c_6(\lambda_i \ell)^6 + c_7(\lambda_i \ell)^7 \end{aligned} \tag{6.164}$$

where λ_i are the eigenvalues of the matrix $\mathbf{A}$. For the case that λ_j is a multiple eigenvalue (i.e., $\lambda_j = \lambda_{j+1} = \cdots = \lambda_{j+m}, m \leq 8$), m equations of the type of Eq. (6.164) are missing. These equations are obtained by recognizing that $e^{\lambda_i \ell}$ and the polynomial on the right-hand side of Eq. (6.164) have the same first m derivatives for the eigenvalue λ_j.

The eigenvalues of matrix $\mathbf{A}$ are the roots of the characteristic equation of $\mathbf{A}$:

$$| \lambda \mathbf{I} - \mathbf{A} | = 0 \tag{6.165}$$

or

$$\lambda^4 \left[\lambda^4 + (b_{zz} + b_{yy}) N \lambda^2 + b_{zz} b_{yy} N^2 \right] = 0 \tag{6.166}$$

The roots of Eq. (6.166) are found to be

$$\begin{aligned} &\lambda_1 = \lambda_2 = \lambda_3 = \lambda_4 = 0 \\ &\lambda_5 = -\lambda_6 = \xi_1 i \qquad i = \sqrt{-1} \\ &\lambda_7 = -\lambda_8 = \xi_2 i \end{aligned} \tag{6.167}$$

where $\xi_1 = \sqrt{N/EI_y}$ and $\xi_2 = \sqrt{N/EI_z}$. Substitute these roots into Eq. (6.164), taking into consideration the multiple eigenvalues $\lambda_1 = \lambda_2 = \lambda_3 = \lambda_4 = 0$:

$$\begin{aligned} &1 = c_o \qquad 1 = c_1 \qquad 1 = 2c_2 \qquad 1 = 6c_3 \\ e^{\xi_1 \ell i} &= c_o + c_1(\xi_1 \ell i) + c_2(\xi_1 \ell i)^2 + c_3(\xi_1 \ell i)^3 + c_4(\xi_1 \ell i)^4 \\ &\quad + c_5(\xi_1 \ell i)^5 + c_6(\xi_1 \ell i)^6 + c_7(\xi_1 \ell i)^7 \\ e^{-\xi_1 \ell i} &= c_o - c_1(\xi_1 \ell i) + c_2(\xi_1 \ell i)^2 - c_3(\xi_1 \ell i)^3 + c_4(\xi_1 \ell i)^4 \\ &\quad - c_5(\xi_1 \ell i)^5 + c_6(\xi_1 \ell i)^6 - c_7(\xi_1 \ell i)^7 \\ e^{\xi_2 \ell i} &= c_o + c_1(\xi_2 \ell i) + c_2(\xi_2 \ell i)^2 + c_3(\xi_2 \ell i)^3 + c_4(\xi_2 \ell i)^4 \\ &\quad + c_5(\xi_2 \ell i)^5 + c_6(\xi_2 \ell i)^6 + c_7(\xi_2 \ell i)^7 \\ e^{-\xi_2 \ell i} &= c_o - c_1(\xi_2 \ell i) + c_2(\xi_2 \ell i)^2 - c_3(\xi_2 \ell i)^3 + c_4(\xi_2 \ell i)^4 \\ &\quad - c_5(\xi_2 \ell i)^5 + c_6(\xi_2 \ell i)^6 - c_7(\xi_2 \ell i)^7 \end{aligned} \tag{6.168}$$

Solve Eq. (6.168) for $c_i, i = 0, 1, \ldots, 7$.

$$
\begin{aligned}
&c_o = 1 \qquad c_1 = 1 \qquad c_2 = \tfrac{1}{2} \qquad c_3 = \tfrac{1}{6} \\
&c_4 = \frac{\dfrac{\cos(\xi_1\ell) + (\xi_1\ell)^2/2 - 1}{(\xi_1\ell)^4}(\xi_2\ell)^2 - \dfrac{\cos(\xi_2\ell) + (\xi_2\ell)^2/2 - 1}{(\xi_2\ell)^4}(\xi_1\ell)^2}{(\xi_2\ell)^2 - (\xi_1\ell)^2} \\
&c_5 = \frac{\dfrac{\sin(\xi_1\ell)/(\xi_1\ell) + (\xi_1\ell)^2/6 - 1}{(\xi_1\ell)^4}(\xi_2\ell)^2 - \dfrac{\sin(\xi_2\ell)/(\xi_2\ell) + (\xi_2\ell)^2/6 - 1}{(\xi_2\ell)^4}(\xi_1\ell)^2}{(\xi_2\ell)^2 - (\xi_1\ell)^2} \\
&c_6 = \frac{\dfrac{\cos(\xi_1\ell) + (\xi_1\ell)^2/2 - 1}{(\xi_1\ell)^4} - \dfrac{\cos(\xi_2\ell) + (\xi_2\ell)^2/2 - 1}{(\xi_2\ell)^4}}{(\xi_2\ell)^2 - (\xi_1\ell)^2} \\
&c_7 = \frac{\dfrac{\sin(\xi_1\ell)/(\xi_1\ell) + (\xi_1\ell)^2/6 - 1}{(\xi_1\ell)^4} - \dfrac{\sin(\xi_2\ell)/(\xi_2\ell) + (\xi_2\ell)^2/6 - 1}{(\xi_2\ell)^4}}{(\xi_2\ell)^2 - (\xi_1\ell)^2}
\end{aligned}
\tag{6.169}
$$

For $\xi_1 = \xi_2 = \xi$, which implies that $I_y = I_z = I$,

$$
\begin{aligned}
&c_o = 1 \qquad c_1 = 1 \qquad c_2 = \tfrac{1}{2} \qquad c_3 = \tfrac{1}{6} \\
&c_4 = \frac{\xi\ell\sin(\xi\ell) + 6\cos(\xi\ell) + 2(\xi\ell)^2 + 6}{2(\xi\ell)^4} \\
&c_5 = \frac{21\sin(\xi\ell) - 3\xi\ell\cos(\xi\ell) + 2(\xi\ell)^3 - 18\xi\ell}{6(\xi\ell)^5} \\
&c_6 = \frac{\xi\ell\sin(\xi\ell) + 4\cos(\xi\ell) + (\xi\ell)^2 - 4}{2(\xi\ell)^6} \\
&c_7 = \frac{15\sin(\xi\ell) - 3\xi\ell\cos(\xi\ell) + (\xi\ell)^3 - 12\xi\ell}{6(\xi\ell)^7}
\end{aligned}
\tag{6.170}
$$

If $N = 0$, the values of c_i are simply

$$
\begin{aligned}
&c_o = 1 \qquad c_1 = 1 \qquad c_2 = \tfrac{1}{2} \qquad c_3 = \tfrac{1}{6} \\
&c_4 = \tfrac{1}{24} \qquad c_5 = \tfrac{1}{120} \qquad c_6 = \tfrac{1}{720} \qquad c_7 = \tfrac{1}{5040}
\end{aligned}
\tag{6.171}
$$

Note that in all cases, c_0, c_1, c_2, and c_3 remain the same. The transfer matrix of Eq. (6.163) becomes

$$
\mathbf{U}^e = \begin{bmatrix}
1 & u_{12} & u_{13} & -u_{23} & 0 & 0 & u_{53} & 0 \\
0 & u_{22} & u_{23} & u_{12}b_{yy} & 0 & 0 & 0 & 0 \\
0 & 0 & 1 & 0 & 0 & 0 & 0 & 0 \\
0 & -u_{12}N & -u_{12} & u_{22} & 0 & 0 & 0 & 0 \\
0 & 0 & u_{53} & 0 & 1 & u_{56} & u_{57} & -u_{67} \\
0 & 0 & 0 & 0 & 0 & u_{66} & u_{67} & -u_{56}b_{zz} \\
0 & 0 & 0 & 0 & 0 & 0 & 1 & 0 \\
0 & 0 & 0 & 0 & 0 & u_{56}N & -u_{56} & u_{66}
\end{bmatrix} \tag{6.172}
$$

where

$$
u_{12} = (-\lambda_{1z}\ell^2 N + 1)\ell \qquad u_{22} = -\lambda_{2z}\ell^2 N + 1 \qquad u_{13} = (-\lambda_{1z}\ell^2 + a_{yy})\ell
$$

$$
u_{23} = -\lambda_{2z}\ell^2 \qquad u_{53} = a_{yz}\ell \qquad u_{56} = (\lambda_{1y}\ell^2 N - 1)\ell
$$

$$
u_{66} = -\lambda_{2y}\ell^2 N + 1 \qquad u_{57} = (-\lambda_{1y}\ell^2 + a_{zz})\ell \qquad u_{67} = \lambda_{2y}\ell^2
$$

and

$$
\lambda_{1z} = b_{yy}^3 c_7 \ell^4 N^2 - b_{yy}^2 c_5 \ell^2 N + \tfrac{1}{6} b_{yy} \qquad \lambda_{2z} = b_{yy}^3 c_6 \ell^4 N^2 - b_{yy}^2 c_4 \ell^2 N + \tfrac{1}{2} b_{yy}
$$

$$
\lambda_{1y} = b_{zz}^3 c_7 \ell^4 N^2 - b_{zz}^2 c_5 \ell^2 N + \tfrac{1}{6} b_{zz} \qquad \lambda_{2y} = b_{zz}^3 c_6 \ell^4 N^2 - b_{zz}^2 c_4 \ell^2 N + \tfrac{1}{2} b_{zz}
$$

This transfer matrix is rearranged into the stiffness matrix $\mathbf{k}^e$ of $\mathbf{p}^e = \mathbf{k}^e \mathbf{v}^e$ with $\mathbf{p}^e$ and $\mathbf{v}^e$, defined by Eqs. (6.156 and 6.157), respectively. This stiffness matrix is exact, that is,

$$
\mathbf{p}^e = \mathbf{k}^e_{\text{exact}} \mathbf{v}^e
$$

where, for this straight beam with arbitrarily shaped cross section with the coordinate systems aligned with the principal bending axes:

$$
\mathbf{k}^e_{\text{exact}} = \frac{1}{D} \begin{bmatrix}
k_{11} & & & & & & & \\
k_{21} & k_{22} & & & & & & \\
k_{31} & k_{32} & k_{33} & & & \text{symmetric} & & \\
k_{41} & k_{42} & k_{43} & k_{44} & & & & \\
-k_{11} & -k_{21} & -k_{31} & -k_{41} & k_{11} & & & \\
k_{21} & k_{62} & k_{32} & k_{82} & -k_{21} & k_{22} & & \\
-k_{31} & -k_{32} & -k_{33} & -k_{43} & k_{31} & -k_{32} & k_{33} & \\
k_{41} & k_{82} & k_{43} & k_{84} & -k_{41} & k_{42} & -k_{43} & k_{44}
\end{bmatrix} \tag{6.173}
$$

where

$$
\begin{aligned}
D = \ell(&-a_{yz}^2 b_{zz} b_{yy} + a_{yy} a_{zz} b_{zz} b_{yy} - a_{yy} b_{zz} b_{yy} \ell^2 \lambda_{1y} + a_{yz}^2 b_{zz} b_{yy} \ell^2 N \lambda_{1y} \\
&- a_{yy} a_{zz} b_{zz} b_{yy} \ell^2 N \lambda_{1y} + a_{yy} b_{zz} b_{yy} \ell^4 N \lambda_{1y}^2 - a_{zz} b_{zz} b_{yy} \ell^2 \lambda_{1z}
\end{aligned}
$$

$$
\begin{aligned}
&+ a_{yz}^2 b_{zz} b_{yy} \ell^2 N \lambda_{1z} - a_{yy} a_{zz} b_{zz} b_{yy} \ell^2 N \lambda_{1z} + b_{zz} b_{yy} \ell^4 \lambda_{1y} \lambda_{1z} \\
&+ a_{yy} b_{zz} b_{yy} \ell^4 N \lambda_{1y} \lambda_{1z} + a_{zz} b_{zz} b_{yy} \ell^4 N \lambda_{1y} \lambda_{1z} - a_{yz}^2 b_{zz} b_{yy} \ell^4 N^2 \lambda_{1y} \lambda_{1z} \\
&+ a_{yy} a_{zz} b_{zz} b_{yy} \ell^4 N^2 \lambda_{1y} \lambda_{1z} - b_{zz} b_{yy} \ell^6 N \lambda_{1y}^2 \lambda_{1z} - a_{yy} b_{zz} b_{yy} \ell^6 N^2 \lambda_{1y}^2 \lambda_{1z} \\
&+ a_{zz} b_{zz} b_{yy} \ell^4 N \lambda_{1z}^2 - b_{zz} b_{yy} \ell^6 N \lambda_{1y} \lambda_{1z}^2 - a_{zz} b_{zz} b_{yy} \ell^6 N^2 \lambda_{1y} \lambda_{1z}^2 \\
&+ b_{zz} b_{yy} \ell^8 N^2 \lambda_{1y}^2 \lambda_{1z}^2 + a_{yy} b_{yy} \ell^2 \lambda_{2y}^2 - b_{yy} \ell^4 \lambda_{2y}^2 \lambda_{1z} - a_{yy} b_{yy} \ell^4 N \lambda_{2y}^2 \lambda_{1z} \\
&+ b_{yy} \ell^6 N \lambda_{2y}^2 \lambda_{1z}^2 + a_{zz} b_{zz} \ell^2 \lambda_{2z}^2 - b_{zz} \ell^4 \lambda_{1y} \lambda_{2z}^2 - a_{zz} b_{zz} \ell^4 N \lambda_{1y} \lambda_{2z}^2 \\
&+ b_{zz} \ell^6 N \lambda_{1y}^2 \lambda_{2z}^2 + \ell^4 \lambda_{2y}^2 \lambda_{2z}^2)
\end{aligned}
$$

$$
a_{ij} = \frac{\alpha_{ij}}{GA} \quad \text{and} \quad b_{ij} = \frac{I_{ij}}{E\left(I_{yy} I_{zz} - I_{yz}^2\right)} \qquad i, y = y, z \tag{6.174}
$$

The stiffness coefficients of $\mathbf{k}^e$ are given by

$$
\begin{aligned}
k_{11} &= b_{yy}(-1 + \ell^2 N \lambda_{1z})(-a_{zz} b_{zz} + b_{zz} \ell^2 \lambda_{1y} + a_{zz} b_{zz} \ell^2 N \lambda_{1y} \\
&\quad - b_{zz} \ell^4 N \lambda_{1y}^2 - \ell^2 \lambda_{2y}^2) \\
k_{21} &= \ell(a_{zz} b_{zz} - b_{zz} \ell^2 \lambda_{1y} - a_{zz} b_{zz} \ell^2 N \lambda_{1y} + b_{zz} \ell^4 N \lambda_{1y}^2 + \ell^2 \lambda_{2y}^2) \lambda_{2z} \\
k_{31} &= a_{yz} b_{yy} b_{zz} (-1 + \ell^2 N \lambda_{1y})(1 - \ell^2 N \lambda_{1z}) \\
k_{41} &= a_{yz} b_{yy} \ell (1 - \ell^2 N \lambda_{1z}) \lambda_{2y} \\
k_{22} &= -a_{yz}^2 b_{zz} + a_{yy} a_{zz} b_{zz} - a_{yy} b_{zz} \ell^2 \lambda_{1y} + a_{yz}^2 b_{zz} \ell^2 N \lambda_{1y} - a_{yy} a_{zz} b_{zz} \ell^2 N \lambda_{1y} \\
&\quad + a_{yy} b_{zz} \ell^4 N \lambda_{1y}^2 - a_{zz} b_{zz} \ell^2 \lambda_{1z} + b_{zz} \ell^4 \lambda_{1z} \lambda_{1y} + a_{zz} b_{zz} \ell^4 N \lambda_{1y} \lambda_{1z} \\
&\quad - b_{zz} \ell^6 N \lambda_{1y}^2 \lambda_{1z} + a_{yy} \ell^2 \lambda_{2y}^2 - \ell^4 \lambda_{1z} \lambda_{2y}^2 + a_{zz} b_{zz} \ell^2 \lambda_{2z} + a_{yz}^2 b_{zz} \ell^2 N \lambda_{2z} \\
&\quad - a_{yy} a_{zz} b_{zz} \ell^2 N \lambda_{2z} - b_{zz} \ell^4 \lambda_{1y} \lambda_{2z} + a_{yy} b_{zz} \ell^4 N \lambda_{1y} \lambda_{2z} - a_{zz} b_{zz} \ell^4 N \lambda_{1y} \lambda_{2z} \\
&\quad - a_{yz}^2 b_{zz} \ell^4 N^2 \lambda_{1y} \lambda_{2z} + a_{yy} a_{zz} b_{zz} \ell^4 N^2 \lambda_{1y} \lambda_{2z} + b_{zz} \ell^6 N \lambda_{1y}^2 \lambda_{2z} \\
&\quad - a_{yy} b_{zz} \ell^6 N^2 \lambda_{1y}^2 \lambda_{2z} + \ell^4 \lambda_{2y}^2 \lambda_{2z} - a_{yy} \ell^4 N \lambda_{2y}^2 \lambda_{2z} \\
k_{32} &= a_{yz} b_{zz} \ell (-1 + \ell^2 N \lambda_{1y}) \lambda_{2z} \\
k_{42} &= a_{yz} \ell^2 \lambda_{2y} (b_{yy} - 2 b_{yy} \ell^2 N \lambda_{1z} + b_{yy} \ell^4 N^2 \lambda_{1z}^2 - \lambda_{2z} + \ell^2 N \lambda_{2z}^2) \\
k_{62} &= a_{yz}^2 b_{zz} - a_{yy} a_{zz} b_{zz} + a_{yy} b_{zz} \ell^2 \lambda_{1y} - a_{yz}^2 b_{zz} \ell^2 N \lambda_{1y} + a_{yy} a_{zz} b_{zz} \ell^2 N \lambda_{1y} \\
&\quad - a_{yy} b_{zz} \ell^4 N \lambda_{1y}^2 + a_{zz} b_{zz} \ell^2 \lambda_{1z} - b_{zz} \ell^4 \lambda_{1y} \lambda_{1z} - a_{zz} b_{zz} \ell^4 N \lambda_{1y} \lambda_{1z} \\
&\quad + b_{zz} \ell^6 N \lambda_{1y}^2 \lambda_{1z} - a_{yy} \ell^2 \lambda_{2y}^2 + \ell^4 \lambda_{1z} \lambda_{2y}^2 \\
k_{82} &= a_{yz} \ell^2 \lambda_{2y} \lambda_{2z}
\end{aligned} \tag{6.175}
$$

$$k_{33} = b_{zz}\left(-1+\ell^2 N\lambda_{1y}\right)\left(-a_{yy}b_{yy}+b_{yy}\ell^2\lambda_{1z}+a_{yy}b_{yy}\ell^2 N\lambda_{1z}\right.$$
$$\left.-b_{yy}\ell^4 N\lambda_{1z}^2-\ell^2\lambda_{2z}^2\right)$$

$$k_{43} = \ell\lambda_{2y}\left(-a_{yy}b_{yy}+b_{yy}\ell^2\lambda_{1z}+a_{yy}b_{yy}\ell^2 N\lambda_{1z}-b_{yy}\ell^4 N\lambda_{1z}^2-\ell^2\lambda_{2z}^2\right)$$

$$\begin{aligned} k_{44} = & -a_{yz}^2 b_{yy}+a_{yy}a_{zz}b_{yy}-a_{yy}b_{yy}\ell^2\lambda_{1y}-a_{zz}b_{yy}\ell^2\lambda_{1z}+a_{yz}^2 b_{yy}\ell^2 N\lambda_{1z} \\ & -a_{yy}a_{zz}b_{yy}\ell^2 N\lambda_{1z}+b_{yy}\ell^4\lambda_{1y}\lambda_{1z}+a_{yy}b_{yy}\ell^4 N\lambda_{1y}\lambda_{1z}+a_{zz}b_{yy}\ell^4 N\lambda_{1z}^2 \\ & -b_{yy}\ell^6 N\lambda_{1y}\lambda_{1z}^2+a_{yy}b_{yy}\ell^2\lambda_{2y}+a_{yz}^2 b_{yy}\ell^2 N\lambda_{2y}-a_{yy}a_{zz}b_{yy}\ell^2 N\lambda_{2y} \\ & -b_{yy}\ell^4\lambda_{1z}\lambda_{2y}-a_{yy}b_{yy}\ell^4 N\lambda_{1z}\lambda_{2y}+a_{zz}b_{yy}\ell^4 N\lambda_{1z}\lambda_{2y} \\ & -a_{yz}^2 b_{yy}\ell^4 N^2\lambda_{1z}\lambda_{2y}+a_{zz}a_{yy}b_{yy}\ell^4 N^2\lambda_{1z}\lambda_{2y}+b_{yy}\ell^6 N\lambda_{1z}^2\lambda_{2y} \\ & -a_{zz}b_{yy}\ell^6 N^2\lambda_{1z}^2\lambda_{2y}+a_{zz}\ell^2 N^2\lambda_{2z}^2-\ell^4\lambda_{2z}^2\lambda_{1y}+\ell^4\lambda_{2z}^2\lambda_{2y} \\ & -a_{zz}\ell^4 N\lambda_{2z}^2\lambda_{2y} \end{aligned}$$

$$\begin{aligned} k_{84} = & a_{yz}^2 b_{yy}-a_{yy}a_{zz}b_{yy}+a_{yy}b_{yy}\ell^2\lambda_{1y}+a_{zz}b_{yy}\ell^2\lambda_{1z}-a_{yz}^2 b_{yy}\ell^2 N\lambda_{1z} \\ & +a_{zz}a_{yy}b_{yy}\ell^2 N\lambda_{1z}-b_{yy}\ell^4\lambda_{1z}\lambda_{1y}-a_{yy}b_{yy}\ell^4 N\lambda_{1z}\lambda_{1y}-a_{zz}b_{yy}\ell^4 N\lambda_{1z}^2 \\ & +b_{yy}\ell^6 N\lambda_{1y}\lambda_{1z}^2-a_{zz}\ell^2\lambda_{2z}^2+\ell^4\lambda_{1y}\lambda_{2z}^2 \end{aligned}$$

Example 6.10 Critical Loads for Beams with Different Cross Sections. The effects of shear deformation on the buckling loads of beams of different end restraints can be studied by imposing the appropriate boundary conditions to the transfer matrix of Eq. (6.172) or the stiffness matrix of Eq. (6.173).

Hinged–Hinged The boundary conditions of a column hinged at both ends ($x = a, x = b$) are

$$v_a = M_{za} = w_a = M_{ya} = v_b = M_{zb} = w_b = M_{yb} = 0 \tag{1}$$

Apply these to either the column transfer matrix or stiffness matrix. In the case of the transfer matrix

$$\begin{bmatrix} v_b = 0 \\ \theta_{zb} \\ V_{yb} \\ M_{zb} = 0 \\ w_b = 0 \\ \theta_{yb} \\ V_{zb} \\ M_{yb} = 0 \end{bmatrix} = \begin{bmatrix} 1 & u_{12} & u_{13} & -u_{23} & 0 & 0 & u_{53} & 0 \\ 0 & u_{22} & u_{23} & u_{12}b_{yy} & 0 & 0 & 0 & 0 \\ 0 & 0 & 1 & 0 & 0 & 0 & 0 & 0 \\ 0 & -u_{12}N & -u_{12} & u_{22} & 0 & 0 & 0 & 0 \\ 0 & 0 & u_{53} & 0 & 1 & u_{56} & u_{57} & -u_{67} \\ 0 & 0 & 0 & 0 & 0 & u_{66} & u_{67} & -u_{56}b_{zz} \\ 0 & 0 & 0 & 0 & 0 & 0 & 1 & 0 \\ 0 & 0 & 0 & 0 & 0 & u_{56}N & -u_{56} & u_{66} \end{bmatrix} \begin{bmatrix} v_a = 0 \\ \theta_{za} \\ V_{ya} \\ M_{za} = 0 \\ w_a = 0 \\ \theta_{ya} \\ V_{za} \\ M_{ya} = 0 \end{bmatrix} \tag{2}$$

or

$$\begin{bmatrix} 0 \\ 0 \\ 0 \\ 0 \end{bmatrix} = \begin{bmatrix} u_{12} & u_{13} & 0 & u_{53} \\ -u_{12}N & -u_{12} & 0 & 0 \\ 0 & u_{53} & u_{56} & u_{57} \\ 0 & 0 & u_{56}N & -u_{56} \end{bmatrix} \begin{bmatrix} \theta_{za} \\ V_{ya} \\ \theta_{ya} \\ V_{za} \end{bmatrix} \tag{3}$$

To find the buckling load, set the determinant of the homogeneous equations of (3) equal to zero.

$$\left[(a_{yy}a_{zz} - a_{yz}^2)N^2 - (a_{yy} + a_{zz})N + 1\right](\lambda_{1y}\ell^2 N - 1)(\lambda_{1z}\ell^2 N - 1) = 0 \tag{4}$$

Note from the definition of a_{ij}, $i, j = (y, z)$ that two of the factors of (4) can be expressed as

$$a_{yy}a_{zz} - a_{yz}^2 = \frac{\alpha_{yy}\alpha_{zz} - \alpha_{yz}^2}{(GA)^2} \qquad a_{yy} + a_{zz} = \frac{\alpha_{yy} + \alpha_{zz}}{GA} \tag{5}$$

Recall that the shear deformation coefficients α_{ij}, $i, j = y, z$, form a tensor. The first and second invariants of this tensor are

$$\alpha_{yy} + \alpha_{zz} = \alpha_{\overline{yy}} + \alpha_{\overline{zz}} \quad \text{and} \quad \alpha_{yy}\alpha_{zz} - \alpha_{yz}^2 = \alpha_{\overline{yy}}\alpha_{\overline{zz}} \tag{6}$$

where $\alpha_{\overline{yy}}$ and $\alpha_{\overline{zz}}$ are the shear deformation coefficients about coordinates along the principal shear axes $\overline{y}$ and $\overline{z}$. For this orientation, $\alpha_{\overline{yz}} = 0$.

Substitute (5) and (6) into (4) and solve for the roots. Two roots are found to be

$$N_1 = \frac{GA}{\alpha_{\overline{yy}}} \qquad N_2 = \frac{GA}{\alpha_{\overline{zz}}} \tag{7}$$

Equation (4) also yields the roots

$$N_3 = \frac{n^2\pi^2 EI_z}{\ell^2} \qquad N_4 = \frac{n^2\pi^2 EI_y}{\ell^2} \qquad n = 1, 2, 3, \ldots \tag{8}$$

Note that (8) are the *Euler formulas* for the critical loads of a pinned–pinned column. Since the buckling would occur at the lowest values, set $n = 1$.

Normally, the shear-related roots of N_1 and N_2 of (7) are higher than the Euler formulas (8). Note that shear deformation has no effect on the Euler formulas of (8). It is apparent from (7) that shear deformation does induce a pure shear buckling mode.

Free-Fixed For a beam free at $x = a$ and fixed at $x = b$, the boundary conditions are

$$V_{ya} = M_{za} = V_{za} = M_{ya} = v_b = \theta_{zb} = w_b = \theta_{yb} = 0 \tag{9}$$

The characteristic equation is found to be

$$(\lambda_{2y}\ell^2 N - 1)(\lambda_{2z}\ell^2 N - 1) = 0 \tag{10}$$

which gives the critical loads

$$N_1 = \frac{\pi^2 EI_z}{4\ell^2} \qquad N_2 = \frac{\pi^2 EI_y}{4\ell^2} \tag{11}$$

Shear deformation does not affect these critical loads, which, again, are the Euler buckling loads.

Guided–Fixed The buckling conditions for a guided–fixed beam are

$$V_{ya} = \theta_{za} = V_{za} = \theta_{ya} = v_b = \theta_{zb} = w_b = \theta_{yb} = 0 \tag{12}$$

These lead to the characteristic equation

$$(\lambda_{1y}\ell^2 N - 1)(\lambda_{1z}\ell^2 N - 1) = 0 \tag{13}$$

from which it is evident that shear deformation does not influence the critical loads, which are the same as those of (8).

Hinged-Fixed In the case of a hinged–fixed beam the boundary conditions are

$$v_a = M_{za} = w_a = M_{ya} = v_b = \theta_{zb} = w_b = \theta_{yb} = 0 \tag{14}$$

The characteristic equation is found to be

$$\begin{aligned}
&\left(a_{yy}a_{zz} - a_{yz}^2\right)\lambda_{2y}\lambda_{2z}\ell^4 N^2 + \Big[a_{yy}\left(\lambda_{1y} - \lambda_{2y}\right)\lambda_{2z}\ell^4 + a_{zz}\left(\lambda_{1z} - \lambda_{2z}\right)\lambda_{2y}\ell^4 \\
&\quad - \left(a_{yy}a_{zz} - a_{yz}^2\right)\left(\lambda_{2y} + \lambda_{2z}\right)\ell^2\Big] N + \left(\lambda_{1y} - \lambda_{2y}\right)\left(\lambda_{1z} - \lambda_{2z}\right)\ell^4 \\
&\quad - \left[a_{yy}\left(\lambda_{1y} - \lambda_{2y}\right) + a_{zz}\left(\lambda_{1z} - \lambda_{2z}\right)\right]\ell^2 + \left(a_{yy}a_{zz} - a_{yz}^2\right) = 0
\end{aligned} \tag{15}$$

In this case, shear deformation and bending are coupled. A numerical example of this column follows.

Example 6.11 Buckling Load of a Fixed–Hinged Beam with Shear Deformation. Consider a beam subject to an axial compressive force N with a L-shaped cross section as shown in Fig. 6.18. One end is fixed and the other hinged.

The orientation of the principal bending axes is readily found and shown in the figure. For the principal bending axes, which are designated as y, z, the moments of inertia are $I_y = 7\,235\,926.0 \text{ mm}^4$, $I_z = 1\,321\,977.7 \text{ mm}^4$, and $I_{yz} = 0$. The cross-

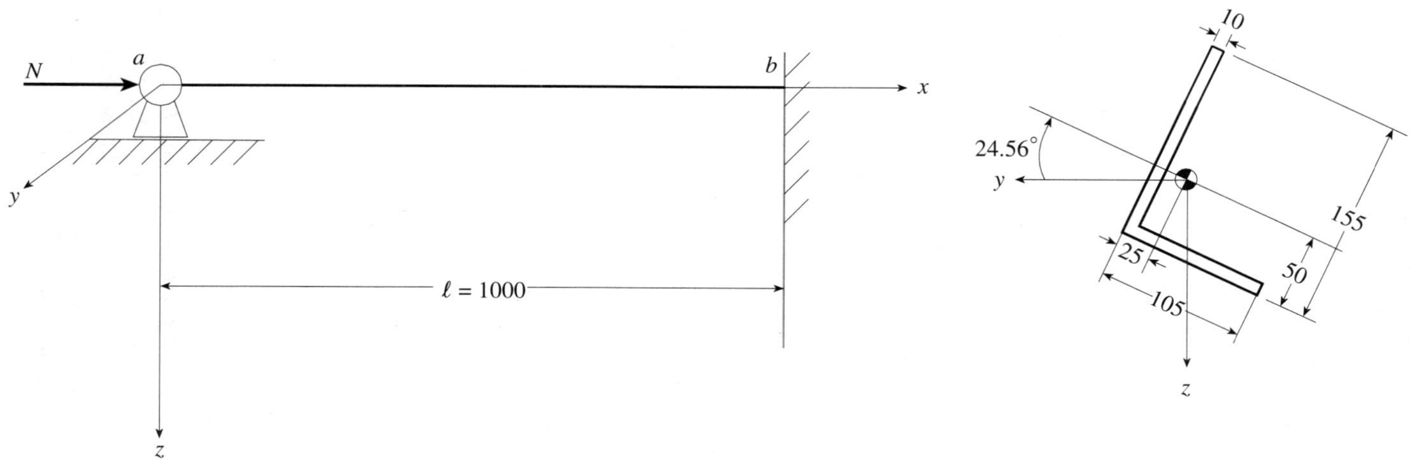

Figure 6.18 Hinged-fixed beam of L-shaped cross section with compressive axial force.

sectional area is 2500.0 mm^2. For $\nu = 0.3$, Eq. (6.126) gives $\alpha_{yy} = 2.8886346$, $\alpha_{zz} = 2.0676036$, $\alpha_{yz} = -0.4115954$ relative to the principal bending axes. The column can be modeled with a single element since the transfer and stiffness matrices are exact. The boundary conditions for this fixed–hinged bar reduce the stiffness matrix to

$$\begin{bmatrix} k_{22} & k_{42} \\ k_{42} & k_{44} \end{bmatrix} \begin{bmatrix} \theta_{za} \\ \theta_{ya} \end{bmatrix} = \begin{bmatrix} 0 \\ 0 \end{bmatrix} \tag{1}$$

A determinant search of (1) can be used to find the critical or buckling load.

The Bernoulli–Euler beam buckling load for a fixed–hinged beam is given by

$$N_{\text{cr}} = 20.190729 \frac{EI_z}{L^2} \tag{2}$$

Table 6.11 shows the critical axial loads for various sets of shear deformation coefficients. These loads are expressed in terms of λ, defined as

$$N_{\text{cr}} = \lambda (N_{\text{cr}})_{\text{Bernoulli–Euler beam}} \tag{3}$$

The meaning of the columns of Table 6.11 were discussed in Example 6.11. Note that for this case, the more precise the incorporation of shear deformation, the lower the buckling load.

6.4.5 Shape Function–Based Geometric Stiffness and Mass Matrices

The exact stiffness matrix for the static response of a beam with shear deformation was readily obtained in Section 6.4.3 by reorganizing a transfer matrix into the stiffness matrix format. The same procedure was used in Section 6.4.4 to obtain the exact geometric stiffness matrix. In theory the procedure should also lead to an exact mass matrix. Alternatively, the exact geometric stiffness and mass matrices can be derived using Eqs. (3.68) and (3.67) if the element dynamic stiffness matrix is available. In practice, however, even with computational symbolic manipulators, the

TABLE 6.11 Buckling Loads for a Fixed-Hinged Column with Various Sets of Shear Deformation Coefficients

	Bernoulli–Euler Theory	Frequently Used Default Values of α_{yy} and α_{zz}	Utilize Proper Values for α_{yy} and α_{zz}, but Set $\alpha_{yz} = 0$	Use the Correct Values for α_{yy}, α_{zz}, and α_{yz}
α_{yy}	0	1	2.8886350	2.8886350
α_{zz}	0	1	2.0676040	2.0676040
α_{yz}	0	0	0.0000000	−0.4115954
λ	1.0000000	0.9971820	0.9914641	0.9914580

operations required are inefficient. Hence, it is useful to develop approximate consistent geometric stiffness and mass matrices using familiar trial function technology that involves the development of shape functions. The eigenvalue problems for the buckling load or natural frequencies using consistent matrices leads to a much simpler computational problem than if exact matrices are employed since the matrices are not functions of the eigenvalues, and hence the cumbersome iterative procedures described in Section 3.6 are avoided. The development of the mass matrix requires at the outset that the differential equations of Eq. (6.145) be made time independent by assuming that

$$\begin{aligned} v\,(x,t) &= v(x)e^{i\omega t} \\ \theta_x\,(x,t) &= \theta_x(x)e^{i\omega t} \\ V_y\,(x,t) &= V_y(x)e^{i\omega t} \\ M_z\,(x,t) &= M_z(x)e^{i\omega t} \\ w\,(x,t) &= w(x)e^{i\omega t} \\ \theta_y\,(x,t) &= \theta_y(x)e^{i\omega t} \\ V_z\,(x,t) &= V_z(x)e^{i\omega t} \\ M_y\,(x,t) &= M_y(x)e^{i\omega t} \end{aligned} \tag{6.176}$$

The interpolation polynomials necessary to develop the consistent matrices are readily obtained from the exact transfer and stiffness matrices for a massless beam element without axial load. Begin by expressing the state vector at any point x between the ends of a beam element in the form

$$\begin{bmatrix} v(x) \\ \theta_z(x) \\ V_y(x) \\ M_z(x) \\ w(x) \\ \theta_y(x) \\ V_z(x) \\ M_y(x) \end{bmatrix} = \mathbf{U}^e(x) \begin{bmatrix} v_a \\ \theta_{za} \\ V_{ya} \\ M_{za} \\ w_a \\ \theta_{ya} \\ V_{za} \\ M_{ya} \end{bmatrix} \tag{6.177}$$

The transfer matrix $\mathbf{U}^e(x)$ relates the state variables at x in terms of the state variables at a. This transfer matrix, which should be expressed using Sign Convention

2, can be obtained from the transfer matrix of Eq. (6.152) by substituting x for ℓ and changing from Sign Convention 1 to Sign Convention 2. From the stiffness matrix of Eq. (6.158), write

$$\begin{bmatrix} V_{ya} \\ M_{za} \\ V_{za} \\ M_{ya} \end{bmatrix} = \overline{\mathbf{k}}^e \begin{bmatrix} v_a \\ \theta_{za} \\ w_a \\ \theta_{ya} \\ v_b \\ \theta_{zb} \\ w_b \\ \theta_{yb} \end{bmatrix} \tag{6.178}$$

The 4×8 matrix $\mathbf{k}^e$ is the upper half of the 8×8 matrix $\overline{\mathbf{k}}^e$. Substitute Eq. (6.178) into Eq. (6.177), giving

$$\begin{bmatrix} v(x) \\ \theta_z(x) \\ w(x) \\ \theta_y(x) \end{bmatrix} = \begin{bmatrix} \mathbf{N}_v \\ \mathbf{N}_{\theta_z} \\ \mathbf{N}_w \\ \mathbf{N}_{\theta_y} \end{bmatrix} \begin{bmatrix} v_a \\ \theta_{za} \\ w_a \\ \theta_{ya} \\ v_b \\ \theta_{zb} \\ w_b \\ \theta_{yb} \end{bmatrix} = \mathbf{N}(x)\mathbf{v}^e \tag{6.179}$$

where $\mathbf{N}(x)$ contains the shape functions. The expressions for $V_y(x)$, $M_z(x)$, $V_z(x)$, and $M_y(x)$ of Eq. (6.177) are intentionally not shown in Eq. (6.178). These interpolation polynomials are exact for a massless beam element without axial force since the transfer and stiffness matrices for the static response are exact.

Geometric Stiffness Matrix The geometric stiffness matrix is given by (Eq. 2.127)

$$\mathbf{k}_\sigma^e = \int_\ell \frac{d}{dx} \begin{bmatrix} \mathbf{N}_v \\ \mathbf{N}_w \end{bmatrix}^{\mathrm{T}} \frac{d}{dx} \begin{bmatrix} \mathbf{N}_v \\ \mathbf{N}_w \end{bmatrix} dx \tag{6.180}$$

Substitution of $\mathbf{N}_v$ and $\mathbf{N}_w$ of Eq. (6.179) into this expression and integration, with the assistance of a symbolic manipulator, leads to the consistent geometric stiffness matrix (Pilkey et al., 1995) for a straight beam element of an arbitrarily shaped cross section with an arbitrarily oriented coordinate system.

$$\mathbf{k}_\sigma^e = \frac{1}{30D^2}\begin{bmatrix} k_{\sigma 11} & & & & & & & \\ k_{\sigma 21} & k_{\sigma 22} & & & & & & \\ k_{\sigma 31} & \frac{\ell}{2}k_{\sigma 31} & k_{\sigma 33} & & & \text{symmetric} & & \\ -\frac{\ell}{2}k_{\sigma 31} & -\frac{\ell^2}{4}k_{\sigma 31} & k_{\sigma 43} & k_{\sigma 44} & & & & \\ -k_{\sigma 11} & -k_{\sigma 21} & -k_{\sigma 31} & \frac{\ell}{2}k_{\sigma 31} & k_{\sigma 11} & & & \\ k_{\sigma 21} & k_{\sigma 62} & \frac{\ell}{2}k_{\sigma 31} & -\frac{\ell^2}{4}k_{\sigma 31} & -k_{\sigma 21} & k_{\sigma 22} & & \\ -k_{\sigma 31} & -\frac{\ell}{2}k_{\sigma 31} & -k_{\sigma 33} & -k_{\sigma 43} & k_{\sigma 31} & -\frac{\ell}{2}k_{\sigma 31} & k_{\sigma 33} & \\ -\frac{\ell}{2}k_{\sigma 31} & -\frac{\ell^2}{4}k_{\sigma 31} & k_{\sigma 43} & k_{\sigma 84} & \frac{\ell}{2}k_{\sigma 31} & -\frac{\ell^2}{4}k_{\sigma 31} & -k_{\sigma 43} & k_{\sigma 44} \end{bmatrix} \tag{6.181}$$

where

$$\begin{aligned} D &= \ell\Big[\left(b_{zz}b_{yy} - b_{yz}^2\right)\ell^4 + 12\left(a_{yy}b_{zz} + 2a_{yz}b_{yz} + a_{zz}b_{yy}\right)\ell^2 \\ &\quad + 144\left(a_{yy}a_{zz} - a_{yz}^2\right)\Big] \end{aligned} \tag{6.182}$$

$$a_{ij} = \frac{\alpha_{ij}}{GA} \quad \text{and} \quad b_{ij} = \frac{I_{ij}}{E\left(I_{yy}I_{zz} - I_{yz}^2\right)} \qquad i, j = y, z$$

with $I_{yy} = I_y$, $I_{zz} = I_z$. The entries of this matrix are

$$\begin{aligned} k_{\sigma 11} = 36\Big[&\left(b_{zz}b_{yy} - b_{yz}^2\right)^2\ell^8 \\ &+ 4\left(6a_{zz}b_{yy} + 11a_{yz}b_{yz} + 5a_{yy}b_{zz}\right)\left(b_{zz}b_{yy} - b_{yz}^2\right)\ell^6 \\ &+ 24\Big(5a_{yy}^2b_{zz}^2 + 20a_{yy}a_{zz}b_{zz}b_{yy} + 20a_{yy}a_{yz}b_{zz}b_{yz} - 10a_{yz}^2b_{zz}b_{yy} \\ &- 10a_{yy}a_{zz}b_{yz}^2 + 31a_{yz}^2b_{yz}^2 + 22a_{yz}a_{zz}b_{yz}b_{yy} + 6a_{zz}^2b_{yy}^2\Big)\ell^4 \\ &+ 24\left(a_{zz}b_{yz} + a_{yz}b_{zz}\right)^2\ell^4 \\ &+ 2800\left(a_{zz}b_{yy} + 2a_{yz}b_{yz} + a_{yy}b_{zz}\right)\left(a_{yy}a_{zz} - a_{yz}^2\right)\ell^2 \\ &+ 17{,}280\left(a_{yy}a_{zz} - a_{yz}^2\right)^2\Big]\ell \end{aligned}$$

$$k_{\sigma 21} = 3\Big[\Big(b_{zz}b_{yy} - b_{yz}^2\Big)\ell^4 + 24\Big(a_{zz}b_{yy} + a_{yz}b_{yz}\Big)\Big(b_{zz}b_{yy} - b_{yz}^2\Big)\ell^2$$
$$+ 144\Big(a_{zz}b_{yy} + a_{yz}b_{yz}\Big)^2$$
$$+ 144\Big(a_{zz}b_{yz} + a_{yz}b_{zz}\Big)^2\Big]\ell^6$$

$$k_{\sigma 31} = -72\Big[\Big(a_{yz}b_{yy} + a_{zz}b_{yz} + a_{yy}b_{yz} + a_{yz}b_{zz}\Big)\Big(b_{zz}b_{yy} - b_{yz}^2\Big)\ell^2$$
$$+ 12\Big(a_{zz}b_{yz} + a_{yz}b_{zz}\Big)\Big(a_{yz}b_{yz} + a_{yy}b_{zz}\Big)$$
$$+ 12\Big(a_{zz}b_{yy} + a_{yz}b_{yz}\Big)\Big(a_{yz}b_{yy} + a_{yy}b_{yz}\Big)\Big]\ell^5$$

$$k_{\sigma_{22}} = 4\Big[\Big(b_{zz}b_{yy} - b_{yz}^2\Big)^2\ell^8$$
$$+ 3\Big(8a_{zz}b_{yy} + 13a_{yz}b_{yz} + 5a_{yy}b_{zz}\Big)\Big(b_{zz}b_{yy} - b_{yz}^2\Big)\ell^6$$
$$+ 18\Big(5a_{yy}^2b_{zz}^2 + 20a_{yy}a_{yz}b_{zz}b_{yz} + 20a_{yy}a_{zz}b_{zz}b_{yy}$$
$$- 10a_{yz}^2b_{zz}b_{yy} - 10a_{yy}a_{zz}b_{yz}^2$$
$$+ 33a_{yz}^2b_{yz}^2 + 26a_{yz}a_{zz}b_{yz}b_{yy} + 8a_{zz}^2b_{yy}^2\Big)\ell^4$$
$$+ 54\Big(a_{zz}b_{yz} + a_{yz}b_{zz}\Big)^2\ell^4$$
$$+ 2160\Big(a_{zz}b_{yy} + 2a_{yz}b_{yz} + a_{yy}b_{zz}\Big)\Big(a_{yy}a_{zz} - a_{yz}^2\Big)\ell^2$$
$$+ 12{,}960\Big(a_{yy}a_{zz} - a_{yz}^2\Big)^2\Big]\ell^3$$

$$k_{\sigma 62} = -\Big[\Big(b_{zz}b_{yy} - b_{yz}^2\Big)^2\ell^8$$
$$+ 12\Big(2a_{zz}b_{yy} + 7a_{yz}b_{yz} + 5a_{yy}b_{zz}\Big)\Big(b_{zz}b_{yy} - b_{yz}^2\Big)\ell^6$$
$$+ 72\Big(2a_{zz}^2b_{yy}^2 + 14a_{yz}a_{zz}b_{yz}b_{yy} + 20a_{yy}a_{zz}b_{zz}b_{yy}$$
$$- 10a_{yz}^2b_{zz}b_{yy} - 10a_{yy}a_{zz}b_{yz}^2$$
$$+ 27a_{yz}^2b_{yz}^2 + 20a_{yy}a_{yz}b_{zz}b_{yz} + 5a_{yy}^2b_{zz}^2\Big)\ell^4$$
$$- 216\Big(a_{zz}b_{yz} + a_{yz}b_{zz}\Big)^2\ell^4$$
$$+ 8640\Big(a_{zz}b_{yy} + 2a_{yz}b_{yz} + a_{yy}b_{zz}\Big)\Big(a_{yy}a_{zz} - a_{yz}^2\Big)\ell^2$$
$$+ 51{,}840\Big(a_{yy}a_{zz} - a_{yz}^2\Big)^2\Big]\ell^3$$

$$
\begin{aligned}
k_{\sigma 33} = 36\Big[&\left(b_{zz}b_{yy} - b_{yz}^2\right)^2 \ell^8 \\
&+ 4\left(5a_{zz}b_{yy} + 11a_{yz}b_{yz} + 6a_{yy}b_{zz}\right)\left(b_{zz}b_{yy} - b_{yz}^2\right)\ell^6 \\
&+ 24\Big(5a_{zz}^2b_{yy}^2 + 20a_{yz}a_{zz}b_{yz}b_{yy} + 20a_{yy}a_{zz}b_{zz}b_{yy} \\
&- 10a_{yz}^2b_{zz}b_{yy} - 10a_{yy}a_{zz}b_{yz}^2 \\
&+ 31a_{yz}^2b_{yz}^2 + 22a_{yy}a_{yz}b_{zz}b_{yz} + 6a_{yy}^2b_{zz}^2\Big)\ell^4 \\
&+ 24\left(a_{yz}b_{yy} + a_{yy}b_{yz}\right)^2\ell^4 \\
&+ 2880\left(a_{zz}b_{yy} + 2a_{yz}b_{yz} + a_{yy}b_{zz}\right)\left(a_{yy}a_{zz} - a_{yz}^2\right)\ell^2 \\
&+ 17{,}280\left(a_{yy}a_{zz} - a_{yz}^2\right)^2\Big]\ell
\end{aligned}
$$

$$
\begin{aligned}
k_{\sigma 43} = -3\Big[&\left(b_{zz}b_{yy} - b_{yz}^2\right)\ell^4 \\
&+ 24\left(a_{yz}b_{yz} + a_{yy}b_{zz}\right)\left(b_{zz}b_{yy} - b_{yz}^2\right)\ell^2 \\
&+ 144\left(a_{yz}b_{yy} + a_{yy}b_{yz}\right)^2 \\
&+ 144\left(a_{yz}b_{yz} + a_{yy}b_{zz}\right)^2\Big]\ell^6
\end{aligned}
$$

$$
\begin{aligned}
k_{\sigma 44} = 4\Big[&\left(b_{zz}b_{yy} - b_{yz}^2\right)^2 \ell^8 \\
&+ 3\left(5a_{zz}b_{yy} + 13a_{yz}b_{yz} + 8a_{yy}b_{zz}\right)\left(b_{zz}b_{yy} - b_{yz}^2\right)\ell^6 \\
&+ 18\Big(5a_{zz}^2b_{yy}^2 + 20a_{yz}a_{zz}b_{yz}b_{yy} \\
&+ 20a_{yy}a_{zz}b_{zz}b_{yy} - 10a_{yz}^2b_{zz}b_{yy} - 10a_{yy}a_{zz}b_{yz}^2 \\
&+ 33a_{yz}^2b_{yz}^2 + 26a_{yy}a_{yz}b_{zz}b_{yz} + 8a_{yy}^2b_{zz}^2\Big)\ell 4 \\
&+ 54\left(a_{yz}b_{yy} + a_{yy}b_{yz}\right)^2\ell^4 \\
&+ 2160\left(a_{zz}b_{yy} + 2a_{yz}b_{yz} + a_{yy}b_{zz}\right)\left(a_{yy}a_{zz} - a_{yz}^2\right)\ell^2 \\
&+ 12{,}960\left(a_{yy}a_{zz} - a_{yz}^2\right)^2\Big]\ell^3
\end{aligned}
$$

$$
\begin{aligned}
k_{\sigma 84} = -\Big[&\left(b_{zz}b_{yy} - b_{yz}^2\right)^2 \ell^8 + 12\left(5a_{zz}b_{yy}\right. \\
&\left. + 7a_{yz}b_{yz} + 2a_{yy}b_{zz}\right)\left(b_{zz}b_{yy} - b_{yz}^2\right)\ell^6 \\
&+ 72\ell^4\left(2a_{yy}^2 b_{zz}^2 + 14a_{yy}a_{yz}b_{zz}b_{yz} + 20a_{yy}a_{zz}b_{zz}b_{yy}\right. \\
&- 10a_{yz}^2 b_{zz}b_{yy} - 10a_{yy}a_{zz}b_{yz}^2 \\
&+ 27a_{yz}^2 b_{yz}^2 + 20a_{yz}a_{zz}b_{yz}b_{yy} \\
&\left. + 5a_{zz}^2 b_{yy}^2\right) - 216\ell^4\left(a_{yz}b_{yy} + a_{yy}b_{yz}\right)^2 \\
&+ 8640\left(a_{zz}b_{yy} + 2a_{yz}b_{yz} + a_{yy}b_{zz}\right)\left(a_{yy}a_{zz} - a_{yz}^2\right)\ell^2 \\
&+ 51{,}840\left(a_{yy}a_{zz} - a_{yz}^2\right)^2\Big]\ell^3
\end{aligned}
\tag{6.183}
$$

The total element stiffness matrix is given by

$$
\mathbf{k}_{\text{total}}^e = \mathbf{k}^e - N\mathbf{k}_\sigma^e \tag{6.184}
$$

Assembly of the element matrices gives the system matrices in the eigenvalue problem form

$$
\mathbf{K} - N\mathbf{K}_\sigma = 0 \tag{6.185}
$$

which can be placed in standard eigenvalue form and solved for the critical value of N, the buckling load. Note that in contrast to Eq. (3.69), $\mathbf{K}$ and $\mathbf{K}_\sigma$ are not functions of N so that the solution of the eigenvalue problem of Eq. (6.185) does not entail an iterative process, other than that used, perhaps, to solve this as a standard eigenvalue problem.

Mass Matrix It is convenient to separate the mass matrix into a contribution $\mathbf{m}_u^e$ from the transverse inertia and a contribution $\mathbf{m}_\theta^e$ from the rotary inertia. Thus,

$$
\mathbf{m}^e = \mathbf{m}_u^e + \mathbf{m}_\theta^e \tag{6.186}
$$

From Section 2.3, the two mass matrices are given by

$$
\begin{aligned}
\mathbf{m}_u^e &= \int_\ell \begin{bmatrix} \mathbf{N}_v \\ \mathbf{N}_w \end{bmatrix}^{\mathrm{T}} \begin{bmatrix} \rho & 0 \\ 0 & \rho \end{bmatrix} \begin{bmatrix} \mathbf{N}_v \\ \mathbf{N}_w \end{bmatrix} dx \\
\mathbf{m}_\theta^e &= \int_\ell \begin{bmatrix} \mathbf{N}_{\theta_z} \\ \mathbf{N}_{\theta_y} \end{bmatrix}^{\mathrm{T}} \begin{bmatrix} \rho r_z^2 & 0 \\ 0 & \rho r_y^2 \end{bmatrix} \begin{bmatrix} \mathbf{N}_{\theta_z} \\ \mathbf{N}_{\theta_y} \end{bmatrix} dx
\end{aligned}
\tag{6.187}
$$

with $\rho = \rho^* A$, $\rho r_z^2 = \rho^* I_z$, and $\rho r_y^2 = \rho^* I_y$, where ρ is the mass per unit length and ρ^* is the mass density (mass per unit volume). Substitute $\mathbf{N}_v$, $\mathbf{N}_w$, $\mathbf{N}_{\theta_z}$, and $\mathbf{N}_{\theta_y}$ from Eq. (6.179) into Eq. (6.187) and integrate. This provides the consistent mass matrices (Pilkey et al., 1995) for a straight beam element of arbitrarily shaped cross section and in an arbitrarily oriented coordinate system.

Transverse Inertia

$$\mathbf{m}_u^e = \frac{\rho}{420D^2} \times \begin{bmatrix} m_{u11} & & & & & & & \\ m_{u21} & m_{u22} & & & & \text{symmetric} & & \\ m_{u31} & m_{u32} & m_{u33} & & & & & \\ m_{u41} & m_{u42} & m_{u43} & m_{u44} & & & & \\ m_{u51} & -m_{u61} & -m_{u31} & -m_{u41} & m_{u11} & & & \\ m_{u61} & m_{u62} & m_{u32} & m_{u42} & -m_{u21} & m_{u22} & & \\ -m_{u31} & -m_{u32} & m_{u73} & m_{u83} & m_{u31} & -m_{u32} & m_{u33} & \\ m_{u41} & m_{u42} & m_{u83} & m_{u84} & -m_{u41} & m_{u42} & -m_{u43} & m_{u44} \end{bmatrix} \tag{6.188}$$

where

$$D = \ell\left[\left(b_{zz}b_{yy} - b_{yz}^2\right)\ell^4 + 12\left(a_{yy}b_{zz} + 2a_{yz}b_{yz} + a_{zz}b_{yy}\right)\ell^2 + 144\left(a_{yy}a_{zz} - a_{yz}^2\right)\right] \tag{6.189}$$

$$a_{ij} = \frac{\alpha_{ij}}{GA} \quad \text{and} \quad b_{ij} = \frac{I_{ij}}{E\left(I_{yy}I_{zz} - I_{yz}^2\right)} \qquad i, j = y, z$$

The entries of this matrix are

$$\begin{aligned} m_{u11} = 12\Big[&13\left(b_{zz}b_{yy} - b_{yz}^2\right)^2\ell^8 \\ &+ 6\left(52a_{zz}b_{yy} + 101a_{yz}b_{yz} + 49a_{yy}b_{zz}\right)\left(b_{zz}b_{yy} - b_{yz}^2\right)\ell^6 \\ &+ 24\Big(78a_{zz}^2b_{yy}^2 + 303a_{zz}a_{yz}b_{yy}b_{yz} + 294a_{zz}a_{yy}b_{zz}b_{yy} \\ &- 147a_{yz}^2b_{zz}b_{yy} - 147a_{yy}a_{zz}b_{yz}^2 \\ &+ 442a_{yz}^2b_{yz}^2 + 70a_{yy}^2b_{zz}^2 + 287a_{yy}a_{yz}b_{zz}b_{yz}\Big)\ell^4 \\ &+ 24\left(a_{zz}b_{yz} + a_{yz}b_{zz}\right)^2\ell^4 \end{aligned}$$

$$
\begin{aligned}
&+ 2016\left(21a_{zz}b_{yy} + 41a_{yz}b_{yz} + 20a_{yy}b_{zz}\right)\left(a_{yy}a_{zz} - a_{yz}^2\right)\ell^2 \\
&+ 241{,}920\left(a_{yy}a_{zz} - a_{yz}^2\right)^2\Big]\ell^3
\end{aligned}
$$

$$
\begin{aligned}
m_{u21} = 2\Big[&11\left(b_{zz}b_{yy} - b_{yz}^2\right)^2\ell^8 \\
&+ 33\left(8a_{zz}b_{yy} + 15a_{yz}b_{yz} + 7a_{yy}b_{zz}\right)\left(b_{zz}b_{yy} - b_{yz}^2\right)\ell^6 \\
&+ 36\Big(44a_{zz}^2b_{yy}^2 + 165a_{yz}a_{zz}b_{yz}b_{yy} + 154a_{yy}a_{zz}b_{zz}b_{yy} \\
&- 77a_{yz}^2b_{zz}b_{yy} - 77a_{yy}a_{zz}b_{yz}^2 \\
&+ 233a_{yz}^2b_{yz}^2 + 35a_{yy}^2b_{zz}^2 + 147a_{yy}a_{yz}b_{zz}b_{yz}\Big)\ell^4 \\
&+ 72\left(a_{yz}b_{zz} + a_{zz}b_{yz}\right)^2 \\
&+ 3024\left(11a_{zz}b_{yy} + 21a_{yz}b_{yz} + 10a_{yy}b_{zz}\right)\left(a_{yy}a_{zz} - a_{yz}^2\right)\ell^2 \\
&+ 181{,}440\left(a_{yy}a_{zz} - a_{yz}^2\right)^2\Big]\ell^4
\end{aligned}
$$

$$
\begin{aligned}
m_{u31} = -36\Big[&3\left(a_{yz}b_{yy} + a_{zz}b_{yz} + a_{yy}b_{yz} + a_{yz}b_{zz}\right)\left(b_{zz}b_{yy} - b_{yz}^2\right)\ell^4 \\
&+ 4\Big(9a_{zz}a_{yz}b_{yy}^2 + 7a_{zz}^2b_{yz}b_{yy} + 9a_{yy}a_{zz}b_{yz}b_{yy} + 16a_{yz}^2b_{yz}b_{yy} \\
&+ 7a_{yz}a_{zz}b_{zz}b_{yy} + 7a_{yy}a_{yz}b_{zz}b_{yy} + 16a_{yz}a_{zz}b_{yz}^2 + 16a_{yy}a_{yz}b_{yz}^2 \\
&+ 16a_{yz}^2b_{zz}b_{yz} + 9a_{yy}a_{zz}b_{zz}b_{yz} + 7a_{yy}^2b_{zz}b_{yz} + 9a_{yy}a_{yz}b_{zz}^2\Big)\ell^2 \\
&+ 336\left(a_{yz}b_{yy} + a_{zz}b_{yz} + a_{yy}b_{yz} + a_{yz}b_{zz}\right)\left(a_{yy}a_{zz} - a_{yz}^2\right)\Big]\ell^5
\end{aligned}
$$

$$
\begin{aligned}
m_{u41} = 6\Big[&\left(9a_{yz}b_{yy} + 9a_{yy}b_{yz} + 2a_{zz}b_{yz} + 2a_{yz}b_{zz}\right)\left(b_{zz}b_{yy} - b_{yz}^2\right)\ell^4 \\
&+ 12\Big(9a_{yz}a_{zz}b_{yy}^2 + 9a_{yy}a_{zz}b_{yz}b_{yy} + 16a_{yz}^2b_{yz}b_{yy} + 16a_{yy}a_{yz}b_{yz}^2 \\
&+ 2a_{yz}a_{zz}b_{yz}^2 + 2a_{yz}^2b_{zz}b_{yz} + 2a_{yy}a_{zz}b_{zz}b_{yz} + 2a_{yy}a_{yz}b_{zz}^2 \\
&+ 7a_{yy}a_{yz}b_{zz}b_{yy} + 7a_{yy}^2b_{zz}b_{yz}\Big)\ell^2 \\
&+ 1008\left(a_{yz}b_{yy} + a_{yy}b_{yz}\right)\left(a_{yy}a_{zz} - a_{yz}^2\right)\Big]\ell^6
\end{aligned}
$$

$$
\begin{aligned}
m_{u51} = 18\Big[&3\left(b_{zz}b_{yy} - b_{yz}^2\right)^2\ell^8 \\
&+ 12\left(6a_{zz}b_{yy} + 13a_{yz}b_{yz} + 7a_{yy}b_{zz}\right)\left(b_{zz}b_{yy} - b_{yz}^2\right)\ell^6
\end{aligned}
$$

$$
\begin{aligned}
&+ 16\Big(27a_{zz}^2 b_{yy}^2 + 117a_{yz}a_{zz}b_{yz}b_{yy} + 126a_{yy}a_{zz}b_{zz}b_{yy} \\
&- 63a_{yz}^2 b_{zz}b_{yy} - 63a_{yy}a_{zz}b_{yz}^2 \\
&+ 188a_{yz}^2 b_{yz}^2 + 35a_{yy}^2 b_{zz}^2 + 133a_{yy}a_{yz}b_{zz}b_{yz}\Big)\ell^4 \\
&- 16\Big(a_{zz}b_{yz} + a_{yz}b_{zz}\Big)^2 \ell^4 \\
&+ 1344\Big(9a_{zz}b_{yy} + 19a_{yz}b_{yz} + 10a_{yy}b_{zz}\Big)\Big(a_{yy}a_{zz} - a_{yz}^2\Big)\ell^2 \\
&+ 80{,}640\Big(a_{yy}a_{zz} - a_{yz}^2\Big)^2\Big]\ell^3
\end{aligned}
$$

$$
\begin{aligned}
m_{u61} = -\Big[&13\Big(b_{zz}b_{yy} - b_{yz}^2\Big)^2 \ell^8 \\
&+ 6\Big(52a_{zz}b_{yy} + 115a_{yz}b_{yz} + 63a_{yy}b_{zz}\Big)\Big(b_{zz}b_{yy} - b_{yz}^2\Big)\ell^6 \\
&+ 72\Big(26a_{zz}^2 b_{yy}^2 + 115a_{zz}a_{yz}b_{yz}b_{yy} \\
&+ 126a_{yy}a_{zz}b_{zz}b_{yy} - 63a_{yz}^2 b_{zz}b_{yy} - 63a_{yy}a_{zz}b_{yz}^2 \\
&+ 187a_{yz}^2 b_{yz}^2 + 35a_{yy}^2 b_{zz}^2 + 133a_{yy}a_{yz}b_{zz}b_{yz}\Big)\ell^4 - 144\Big(a_{zz}b_{yz} + a_{yz}b_{zz}\Big)^2 \ell^4 \\
&+ 6048\Big(9a_{zz}b_{yy} + 19a_{yz}b_{yz} + 10a_{yy}b_{zz}\Big)\Big(a_{yy}a_{zz} - a_{yz}^2\Big)\ell^2 \\
&- 362{,}880\Big(a_{yy}a_{zz} - a_{yz}^2\Big)^2\Big]\ell^4
\end{aligned}
$$

$$
\begin{aligned}
m_{u22} = 4\Big[&\Big(b_{zz}b_{yy} - b_{yz}^2\Big)^2 \ell^8 \\
&+ 3\Big(8a_{zz}b_{yy} + 15a_{yz}b_{yz} + 7a_{yy}b_{zz}\Big)\Big(b_{zz}b_{yy} \\
&- b_{yz}^2\Big)\ell^6 + 18\Big(8a_{zz}^2 b_{yy}^2 + 30a_{yz}a_{zz}b_{yz}b_{yy} \\
&+ 28a_{yy}a_{zz}b_{zz}b_{yy} - 14a_{yz}^2 b_{zz}b_{yy} - 14a_{yy}a_{zz}b_{yz}^2 \\
&+ 43a_{yz}^2 b_{yz}^2 + 28a_{yy}a_{yz}b_{yz}b_{zz} + 7a_{yy}^2 b_{zz}^2\Big)\ell^4 \\
&+ 18\Big(a_{zz}b_{yz} + a_{yz}b_{zz}\Big)^2 \ell^4 \\
&+ 3024\Big(a_{zz}b_{yy} + 2a_{yz}b_{yz}\Big)\Big(a_{yy}a_{zz} - a_{yz}^2\Big)\ell^2 \\
&+ 18{,}144\Big(a_{yy}a_{zz} - a_{yz}^2\Big)^2\Big]\ell^5
\end{aligned}
$$

$$
\begin{aligned}
m_{u32} = -6\Big[&\Big(2a_{yz}b_{yy} + 9a_{zz}b_{yz} + 2a_{yy}b_{yz} + 9a_{yz}b_{zz}\Big)\Big(b_{zz}b_{yy} - b_{yz}^2\Big)\ell^4 \\
&+ 12\Big(2a_{yz}a_{zz}b_{yy}^2 + 2a_{yy}a_{zz}b_{yz}b_{yy} + 7a_{zz}^2b_{yz}b_{yy} \\
&+ 7a_{yz}a_{zz}b_{zz}b_{yy} + 2a_{yz}^2b_{yz}b_{yy} + 2a_{yy}a_{yz}b_{yz}^2 + 16a_{yz}a_{zz}b_{yz}^2 \\
&+ 16a_{yz}^2b_{zz}b_{yz} + 9a_{yy}a_{zz}b_{zz}b_{yz} + 9a_{yy}a_{yz}b_{zz}^2\Big)\ell^2 \\
&+ 1008\Big(a_{zz}b_{yz} + a_{yz}b_{zz}\Big)\Big(a_{zz}a_{zz} - a_{yz}^2\Big)\Big]\ell^6
\end{aligned}
$$

$$
\begin{aligned}
m_{u42} = 6\Big[&\Big(a_{yz}b_{yy} + a_{zz}b_{yz} + a_{yy}b_{yz} + a_{yz}b_{zz}\Big)\Big(b_{zz}b_{yy} - b_{yz}^2\Big)\ell^2 \\
&+ 12\Big(a_{yz}b_{yy} + a_{yy}b_{yz}\Big)\Big(a_{yz}b_{yz} + a_{zz}b_{yy}\Big) \\
&+ 12\Big(a_{yy}b_{zz} + a_{yz}b_{yz}\Big)\Big(a_{zz}b_{yz} + a_{yz}b_{zz}\Big)\Big]\ell^9
\end{aligned}
$$

$$
\begin{aligned}
m_{u62} = -3\Big[&\Big(b_{zz}b_{yy} - b_{yz}^2\Big)^2\ell^8 \\
&+ 4\Big(6a_{zz}b_{yy} + 13a_{yz}b_{yz} + 7a_{yy}b_{zz}\Big)\Big(b_{zz}b_{yy} - b_{yz}^2\Big)\ell^6 \\
&+ 24\Big(6a_{zz}^2b_{yy}^2 + 26a_{yz}a_{zz}b_{yz}b_{yy} + 28a_{yy}a_{zz}b_{zz}b_{yy} \\
&+ 7a_{yy}^2b_{zz}^2 - 14a_{yz}^2b_{zz}b_{yy} \\
&+ 41a_{yz}^2b_{yz}^2 - 14a_{yy}a_{zz}b_{yz}^2 + 28a_{yy}a_{yz}b_{zz}b_{yz}\Big)\ell^4 \\
&- 24\Big(a_{zz}b_{yz} + a_{yz}b_{zz}\Big)^2\ell^4 \\
&+ 4032\Big(a_{zz}b_{yy} + 2a_{yz}b_{yz} + a_{yy}b_{zz}\Big)\Big(a_{yy}a_{zz} - a_{yz}^2\Big)\ell^2 \\
&+ 24{,}192\Big(a_{yy}a_{zz} - a_{yz}^2\Big)^2\Big]\ell^5
\end{aligned}
$$

$$
\begin{aligned}
m_{u33} = 12\Big[&13\Big(b_{zz}b_{yy} - b_{yz}^2\Big)^2\ell^8 \\
&+ 6\Big(49a_{zz}b_{yy} + 101a_{yz}b_{yz} + 52a_{yy}b_{zz}\Big)\Big(b_{zz}b_{yy} - b_{yz}^2\Big)\ell^6 \\
&+ 24\Big(70a_{zz}^2b_{yy}^2 + 287a_{yz}a_{zz}b_{yz}b_{yy} + 294a_{yy}a_{zz}b_{zz}b_{yy} \\
&+ 78a_{yy}^2b_{zz}^2 - 147a_{yz}^2b_{zz}b_{yy} \\
&+ 442a_{yz}^2b_{yz}^2 - 147a_{yy}a_{zz}b_{yz}^2 + 287a_{yy}a_{yz}b_{zz}b_{yz}\Big)\ell^4 \\
&+ 24\Big(a_{yz}b_{yy} + a_{yy}b_{yz}\Big)^2\ell^4
\end{aligned}
$$

$$+ 2016\left(20a_{zz}b_{yy} + 41a_{yz}b_{yz} + 21a_{yy}b_{zz}\right)\left(a_{yy}a_{zz} - a_{yz}^2\right)\ell^2$$

$$- 241{,}920\left(a_{yy}a_{zz} - a_{yz}^2\right)^2\Big]\ell^3$$

$$m_{u43} = -2\Big[11\left(b_{zz}b_{yy} - b_{yz}^2\right)^2\ell^8$$

$$+ 33\left(7a_{zz}b_{yy} + 15a_{yz}b_{yz} + 8a_{yy}b_{zz}\right)\left(b_{zz}b_{yy} - b_{yz}^2\right)\ell^6$$

$$+ 36\Big(35a_{zz}^2b_{yy}^2 + 147a_{yz}a_{zz}b_{yz}b_{yy} + 154a_{yy}a_{zz}b_{zz}b_{yy}$$

$$+ 44a_{yy}^2b_{zz}^2 - 77a_{yz}^2b_{zz}b_{yy} + 233a_{yz}^2b_{yz}^2 - 77a_{yy}a_{zz}b_{yz}^2$$

$$+ 165a_{yy}a_{yz}b_{zz}b_{yz}\Big)\ell^4 + 72\left(a_{yz}b_{yy} + a_{yy}b_{yz}\right)^2\ell^4$$

$$+ 3024\left(10a_{zz}b_{yy} + 21a_{yz}b_{yz} + 11a_{yy}b_{zz}\right)\left(a_{yy}a_{zz} - a_{yz}^2\right)\ell^2$$

$$+ 181{,}440\left(a_{yy}a_{zz} - a_{yz}^2\right)^2\Big]\ell^4$$

$$m_{u73} = 18\Big[3\left(b_{zz}b_{yy} - b_{yz}^2\right)^2\ell^8$$

$$+ 4\left(21a_{zz}b_{yy} + 39a_{yz}b_{yz} + 18a_{yy}b_{zz}\right)\left(b_{zz}b_{yy} - b_{yz}^2\right)\ell^6$$

$$+ 16\Big(35a_{zz}^2b_{yy}^2 + 133a_{yz}a_{zz}b_{yz}b_{yy} + 126a_{yy}a_{zz}b_{zz}b_{yy}$$

$$+ 27a_{yy}^2b_{zz}^2 - 63a_{yz}^2b_{zz}b_{yy}$$

$$+ 188a_{yz}^2b_{yz}^2 - 63a_{yy}a_{zz}b_{yz}^2 + 117a_{yy}a_{yz}b_{zz}b_{yz}\ell\Big)\ell^4$$

$$- 16\left(a_{yz}b_{yy} + a_{yy}b_{yz}\right)^2\ell^4$$

$$+ 1344\left(10a_{zz}b_{yy} + 19a_{yz}b_{yz} + 9a_{yy}b_{zz}\right)\left(a_{yy}a_{zz} - a_{yz}^2\right)\ell^2$$

$$+ 80{,}640\left(a_{yy}a_{zz} - a_{yz}^2\right)^2\Big]\ell^3$$

$$m_{u83} = \Big[13\left(b_{zz}b_{yy} - b_{yz}^2\right)^2\ell^8$$

$$+ 6\left(63a_{zz}b_{yy} + 115a_{yz}b_{yz} + 52a_{yy}b_{zz}\right)\left(b_{zz}b_{yy} - b_{yz}^2\right)\ell^6$$

$$+ 72\Big(35a_{zz}^2b_{yy}^2 + 133a_{yz}a_{zz}b_{yz}b_{yy} + 126a_{yy}a_{zz}b_{zz}b_{yy}$$

$$
\begin{aligned}
& + 26a_{yy}^2 b_{zz}^2 - 63a_{yz}^2 b_{zz} b_{yy} \\
& + 187a_{yz}^2 b_{yz}^2 - 63a_{yy}a_{zz}b_{yz}^2 + 115a_{yy}a_{yz}b_{zz}b_{yz}\Big)\ell^4 \\
& - 144\Big(a_{yz}b_{yy} + a_{yy}b_{yz}\Big)^2 \ell^4 \\
& + 6048\Big(10a_{zz}b_{yy} + 19a_{yz}b_{yz} + 9a_{yy}b_{zz}\Big)\Big(a_{yy}a_{zz} - a_{yz}^2\Big)\ell^2 \\
& + 362{,}880\Big(a_{yy}a_{zz} - a_{yz}^2\Big)^2\Big]\ell^4
\end{aligned}
$$

$$
\begin{aligned}
m_{u44} = 4\Big[& \Big(b_{zz}b_{yy} - b_{yz}^2\Big)^2 \ell^8 \\
& + 3\Big(7a_{zz}b_{yy} + 15a_{yz}b_{yz} + 8a_{yy}b_{zz}\Big)\Big(b_{zz}b_{yy} - b_{yz}^2\Big)\ell^6 \\
& + 18\Big(7a_{zz}^2 b_{yy}^2 + 28a_{yz}a_{zz}b_{yz}b_{yy} \\
& + 28a_{yy}a_{zz}b_{zz}b_{yy} - 14a_{yz}^2 b_{zz}b_{yy} - 14a_{yy}a_{zz}b_{yz}^2 \\
& + 43a_{yz}^2 b_{yz}^2 + 30a_{yy}a_{yz}b_{zz}b_{yz} + 8a_{yy}^2 b_{zz}^2\Big)\ell^4 \\
& + 18\Big(a_{yz}b_{yy} + a_{yy}b_{yz}\Big)^2 \ell^4 \\
& + 3024\Big(a_{zz}b_{yy} + 2a_{yz}b_{yz} + a_{yy}b_{zz}\Big)\Big(a_{yy}a_{zz} - a_{yz}^2\Big)\ell^2 \\
& + 18{,}144\Big(a_{yy}a_{zz} - a_{yz}^2\Big)^2\Big]\ell^5
\end{aligned}
$$

$$
\begin{aligned}
m_{u84} = -3\Big[& \Big(b_{zz}b_{yy} - b_{yz}^2\Big)^2 \ell^8 \\
& + 4\Big(7a_{zz}b_{yy} + 13a_{yz}b_{yz} + 6a_{yy}b_{zz}\Big)\Big(b_{zz}b_{yy} \\
& - b_{yz}^2\Big)\ell^6 + 24\Big(7a_{zz}^2 b_{yy}^2 + 28a_{yz}a_{zz}b_{yz}b_{yy} \\
& + 28a_{yy}a_{zz}b_{zz}b_{yy} - 14a_{yz}^2 b_{zz}b_{yy} - 14a_{yy}a_{zz}b_{yz}^2 \\
& + 41a_{yz}^2 b_{yz}^2 + 26a_{yy}a_{yz}b_{zz}b_{yz} + 6a_{yy}^2 b_{zz}^2\Big)\ell^4 \\
& - 24\Big(a_{yz}b_{yy} + a_{yy}b_{yz}\Big)^2 \ell^4 \\
& + 4032\Big(a_{zz}b_{yy} + 2a_{yz}b_{yz} + a_{yy}b_{zz}\Big)\Big(a_{yy}a_{zz} - a_{yz}^2\Big)\ell^2 \\
& + 24{,}192\Big(a_{yy}a_{zz} - a_{yz}^2\Big)^2\Big]\ell^5
\end{aligned}
\tag{6.190}
$$

Rotary Inertia

$$\mathbf{m}_\theta^e = \frac{\rho^*}{30D^2}$$

$$\times \begin{bmatrix} m_{\theta 11} & & & & & & & \\ m_{\theta 21} & m_{\theta 22} & & & & & & \\ m_{\theta 31} & m_{\theta 32} & m_{\theta 33} & & & \text{symmetric} & & \\ m_{\theta 41} & m_{\theta 42} & m_{\theta 43} & m_{\theta 44} & & & & \\ -m_{\theta 11} & -m_{\theta 21} & -m_{\theta 31} & -m_{\theta 41} & m_{\theta 11} & & & \\ m_{\theta 21} & m_{\theta 62} & m_{\theta 32} & m_{\theta 42} & -m_{\theta 21} & m_{\theta 22} & & \\ -m_{\theta 31} & -m_{\theta 32} & -m_{\theta 33} & -m_{\theta 43} & m_{\theta 31} & -m_{\theta 32} & m_{\theta 33} & \\ m_{\theta 41} & m_{\theta 42} & m_{\theta 43} & m_{\theta 84} & -m_{\theta 41} & m_{\theta 42} & -m_{\theta 43} & m_{\theta 44} \end{bmatrix} \tag{6.191}$$

The entries of this matrix are

$$\begin{aligned} m_{\theta 11} = 36\Big[& I_z\left(b_{zz}b_{yy} - b_{yz}^2\right)^2 \ell^4 \\ & + 24I_z\left(a_{zz}b_{yy} + a_{yz}b_{yz}\right)\left(b_{zz}b_{yy} - b_{yz}^2\right)\ell^2 \\ & + 144I_y\left(a_{zz}b_{yz} + a_{yz}b_{zz}\right)^2 + 144I_z\left(a_{zz}b_{yy} + a_{yz}b_{yz}\right)^2\Big]\ell^5 \\ m_{\theta 21} = 3\Big[& I_z\left(b_{zz}b_{yy} - b_{yz}^2\right)^2 \ell^6 \\ & + 12I_z\left(2a_{zz}b_{yy} - 3a_{yz}b_{yz} - 5a_{yy}b_{zz}\right)\left(b_{zz}b_{yy} - b_{yz}^2\right)\ell^4 \\ & + 144I_z\left(a_{zz}^2 b_{yy}^2 - 3a_{yz}a_{zz}b_{yz}b_{yy} - 10a_{yy}a_{zz}b_{zz}b_{yy} + 5a_{yz}^2 b_{zz}b_{yy}\right. \\ & \left. + 5a_{yy}a_{zz}b_{yz}^2 - 9a_{yz}^2 b_{yz}^2 - 5a_{yy}a_{yz}b_{zz}b_{yz}\right)\ell^2 \\ & + 864I_y\left(a_{zz}b_{yz} + a_{yz}b_{zz}\right)\ell^2 \\ & -8640I_z\left(a_{zz}b_{yy} + a_{yz}b_{yz}\right)\left(a_{yy}a_{zz} - a_{yz}^2\right)\Big]\ell^4 \\ m_{\theta 31} = -432\Big[& I_y\left(a_{zz}b_{yz} + a_{yz}b_{zz}\right)\left(b_{zz}b_{yy} - b_{yz}^2\right)\ell^2 \\ & + I_z\left(a_{yz}b_{yy} + a_{yy}b_{yz}\right)\left(b_{zz}b_{yy} - b_{yz}^2\right)\ell^2 \\ & + 12I_y\left(a_{yz}b_{yz} + a_{yy}b_{zz}\right)\left(a_{zz}b_{yz} + a_{yz}b_{zz}\right) \\ & + 12I_z\left(a_{yz}b_{yy} + a_{yy}b_{yz}\right)\left(a_{zz}b_{yy} + a_{yz}b_{yz}\right)\Big]\ell^5 \end{aligned}$$

$$m_{\theta 41} = 36\Big[I_y\Big(a_{zz}b_{yz} + a_{yz}b_{zz}\Big)\Big(b_{zz}b_{yy} - b_{yz}^2\Big)\ell^4$$
$$+ 6I_z\Big(a_{yz}b_{yy} + a_{yy}b_{yz}\Big)\Big(b_{zz}b_{yy} - b_{yz}^2\Big)\ell^4$$
$$+ 12I_y\Big(a_{yy}b_{zz} - 4a_{yz}b_{yz} - 5a_{zz}b_{yy}\Big)\Big(a_{zz}b_{yz} + a_{yz}b_{zz}\Big)\ell^2$$
$$+ 72I_z\Big(a_{yz}b_{yy} + a_{yy}b_{yz}\Big)\Big(a_{zz}b_{yy} + a_{yz}b_{yz}\Big)\ell^2$$
$$- 720I_y\Big(a_{zz}b_{yz} + a_{yz}b_{zz}\Big)\Big(a_{yy}a_{zz} - a_{yz}^2\Big)\Big]\ell^4$$

$$m_{\theta 22} = 4\Big[I_z\Big(b_{zz}b_{yy} - b_{yz}^2\Big)^2\ell^8$$
$$+ 3I_z\Big(8a_{zz}b_{yy} + 13a_{yz}b_{yz} + 5a_{yy}b_{zz}\Big)\Big(b_{zz}b_{yy} - b_{yz}^2\Big)\ell^6$$
$$+ 36I_z\Big(4a_{zz}^2b_{yy}^2 + 13a_{yz}a_{zz}b_{yz}b_{yy} + 10a_{yy}a_{zz}b_{zz}b_{yy}$$
$$- 5a_{yz}^2b_{zz}b_{yy} - 5a_{yy}a_{zz}b_{yz}^2$$
$$+ 24a_{yz}^2b_{yz}^2 + 25a_{yy}a_{yz}b_{zz}b_{yz} + 10a_{yy}^2b_{zz}^2\Big)\ell^4$$
$$+ 324I_y\Big(a_{zz}b_{yz} + a_{yz}b_{zz}\Big)^2\ell^4$$
$$+ 2160I_z\Big(a_{zz}b_{yy} + 5a_{yz}b_{yz} + 4a_{yy}b_{zz}\Big)\Big(a_{yy}a_{zz} - a_{yz}^2\Big)\ell^2$$
$$+ 51{,}840I_z\Big(a_{yy}a_{zz} - a_{yz}^2\Big)^2\Big]\ell^3$$

$$m_{\theta 32} = -36\Big[6I_y\Big(a_{zz}b_{yz} + a_{yz}b_{zz}\Big)\Big(b_{zz}b_{yy} - b_{yz}^2\Big)\ell^4$$
$$+ I_z\Big(a_{yz}b_{yy} + a_{yy}b_{yz}\Big)\Big(b_{zz}b_{yy} - b_{yz}^2\Big)\ell^4$$
$$+ 72I_y\Big(a_{yz}b_{yz} + a_{yy}b_{zz}\Big)\Big(a_{zz}b_{yz} + a_{yz}b_{zz}\Big)\ell^2$$
$$+ 12I_z\Big(a_{zz}b_{yy} - 4a_{yz}b_{yz} - 5a_{yy}b_{zz}\Big)\Big(a_{yz}b_{yy} + a_{yy}b_{yz}\Big)\ell^2$$
$$- 720I_z\Big(a_{yz}b_{yy} + a_{yy}b_{yz}\Big)\Big(a_{yy}a_{zz} - a_{yz}^2\Big)\Big]\ell^4$$

$$m_{\theta 42} = 18\Big[I_y\Big(a_{zz}b_{yz} + a_{yz}b_{zz}\Big)\Big(b_{zz}b_{yy} - b_{yz}^2\Big)\ell^4$$
$$+ I_z\Big(a_{yz}b_{yy} + a_{yy}b_{yz}\Big)\Big(b_{zz}b_{yy} - b_{yz}^2\Big)\ell^4$$
$$- 12I_z\Big(5a_{yy}b_{zz} + 4a_{yz}b_{yz} - a_{zz}b_{yy}\Big)\Big(a_{yz}b_{yy} + a_{yy}b_{yz}\Big)\ell^2$$

$$- 12I_y\left(5a_{zz}b_{yy} + 4a_{yz}b_{yz} - a_{yy}b_{zz}\right)\left(a_{zz}b_{yz} + a_{yz}b_{zz}\right)\ell^2$$

$$- 720I_y\left(a_{zz}b_{yz} + a_{yz}b_{zz}\right)\left(a_{yy}a_{zz} - a_{yz}^2\right)$$

$$- 720I_z\left(a_{yz}b_{yy} + a_{yy}b_{yz}\right)\left(a_{yy}a_{zz} - a_{yz}^2\right)\Big]\ell^5$$

$$m_{\theta 62} = -\Big[I_z\left(b_{zz}b_{yy} - b_{yz}^2\right)^2\ell^8$$

$$+ 12I_z\left(2a_{zz}b_{yy} + 7a_{yz}b_{yz} + 5a_{yy}b_{zz}\right)\left(b_{zz}b_{yy} - b_{yz}^2\right)\ell^6$$

$$+ 144I_z\Big(a_{zz}^2b_{yy}^2 + 7a_{yz}a_{zz}b_{yz}b_{yy} + 10a_{yy}a_{zz}b_{zz}b_{yy} - 5a_{yz}^2b_{zz}b_{yy}$$

$$- 5a_{yy}a_{zz}b_{yz}^2 + 6a_{yz}^2b_{yz}^2 - 5a_{yy}a_{yz}b_{zz}b_{yz} - 5a_{yy}^2b_{zz}^2\Big)\ell^4$$

$$- 1296I_y\left(a_{zz}b_{yz} + a_{yz}b_{zz}\right)^2\ell^4$$

$$+ 8640I_z\left(a_{zz}b_{yy} - 2a_{yy}b_{zz} - a_{yz}b_{yz}\right)\left(a_{yy}a_{zz} - a_{yz}^2\right)\ell^2$$

$$- 103{,}680I_z\left(a_{yy}a_{zz} - a_{yz}^2\right)^2\Big]\ell^3$$

$$m_{\theta 33} = 36\Big[I_y\left(b_{zz}b_{yy} - b_{yz}^2\right)\ell^4$$

$$+ 24I_y\left(a_{yz}b_{yz} + a_{yy}b_{zz}\right)\left(b_{zz}b_{yy} - b_{yz}^2\right)\ell^2$$

$$+ 144I_y\left(a_{yz}b_{yz} + a_{yy}b_{zz}\right)^2 + 144I_z\left(a_{yz}b_{yy} + a_{yy}b_{yz}\right)^2\Big]\ell^5$$

$$m_{\theta 43} = -3\Big[I_y\left(b_{zz}b_{yy} - b_{yz}^2\right)^2\ell^6$$

$$+ 12I_y\left(2a_{yy}b_{zz} - 3a_{yz}b_{yz} - 5a_{zz}b_{yy}\right)\left(b_{zz}b_{yy} - b_{yz}^2\right)\ell^4$$

$$+ 144I_y\Big(5a_{yz}^2b_{zz}b_{yy} + 5a_{yy}a_{zz}b_{yz}^2 - 5a_{yz}a_{zz}b_{yz}b_{yy}$$

$$- 10a_{yy}a_{zz}b_{zz}b_{yy} + a_{yy}^2b_{zz}^2 - 3a_{yy}a_{yz}b_{zz}b_{yz} - 9a_{yz}^2b_{yz}^2\Big)\ell^2$$

$$+ 864I_z\left(a_{yz}b_{yy} + a_{yy}b_{yz}\right)^2\ell^2$$

$$- 8640I_y\left(a_{yz}b_{yz} + a_{yy}b_{zz}\right)\left(a_{yy}a_{zz} - a_{yz}^2\right)\Big]\ell^4$$

$$m_{\theta 44} = 4\Big[I_y\left(b_{zz}b_{yy} - b_{yz}^2\right)^2\ell^8$$

$$+ 3I_y\left(5a_{zz}b_{yy} + 13a_{yz}b_{yz} + 8a_{yy}b_{zz}\right)\left(b_{zz}b_{yy} - b_{yz}^2\right)\ell^6$$

$$
\begin{aligned}
&+ 36I_y\Big(10a_{zz}^2b_{yy}^2 + 25a_{yz}a_{zz}b_{yz}b_{yy} + 10a_{yy}a_{zz}b_{zz}b_{yy}\\
&- 5a_{yz}^2b_{zz}b_{yy} - 5a_{yy}a_{zz}b_{yz}^2\\
&+ 24a_{yz}^2b_{yz}^2 + 13a_{yy}a_{yz}b_{zz}b_{yz} + 4a_{yy}^2b_{zz}^2\Big)\ell^4\\
&+ 324I_z\Big(a_{yz}b_{yy} + a_{yy}b_{yz}\Big)^2\ell^4\\
&+ 2160I_y\Big(4a_{zz}b_{yy} + 5a_{yz}b_{yz} + a_{yy}b_{zz}\Big)\Big(a_{yy}a_{zz} - a_{yz}^2\Big)\ell^2\\
&+ 51{,}840I_y\Big(a_{yy}a_{zz} - a_{yz}^2\Big)^2\Big]\ell^3
\end{aligned}
$$

$$
\begin{aligned}
m_{\theta 84} = &-\Big[I_y\Big(b_{zz}b_{yy} - b_{yz}^2\Big)^2\ell^8\\
&+ 12I_y\Big(5a_{zz}b_{yy} + 7a_{yz}b_{yz} + 2a_{yy}b_{zz}\Big)\Big(b_{zz}b_{yy} - b_{yz}^2\Big)\ell^6\\
&- 720I_y\Big(a_{zz}^2b_{yy}^2 + a_{yz}a_{zz}b_{yz}b_{yy} - 2a_{yy}a_{zz}b_{zz}b_{yy}\\
&- a_{yz}^2b_{zz}b_{yy} + a_{yy}a_{zz}b_{yz}^2\Big)\ell^4\\
&+ 144I_y\Big(6a_{yz}^2b_{yz}^2 + 7a_{yy}a_{yz}b_{zz}b_{yz} + a_{yy}^2b_{zz}^2\Big)\ell^4\\
&- 1296I_z\Big(a_{yz}b_{yy} + a_{yy}b_{yz}\Big)^2\ell^4\\
&+ 8640I_y\Big(a_{yy}b_{zz} - a_{yz}b_{yz} - 2a_{zz}b_{yy}\Big)\Big(a_{yy}a_{zz} - a_{yz}^2\Big)\ell^2\\
&- 103{,}680I_y\Big(a_{yy}a_{zz} - a_{yz}^2\Big)^2\Big]\ell^3
\end{aligned} \tag{6.192}
$$

Example 6.12 Influence of Shear Deformation on the Natural Frequencies of a Beam. Consider again the cantilever beam with a trapezoidal cross section of Example 6.9 and Fig. 6.17 to ascertain the effect of shear deformation on the natural frequencies.

Express the natural frequencies (hertz) of this cantilever beam as

$$
f_{yi} = \frac{\lambda_{yi}^2}{2\pi L^2}\left(\frac{EI_z}{\rho}\right)^{1/2} \qquad f_{zi} = \frac{\lambda_{zi}^2}{2\pi L^2}\left(\frac{EI_y}{\rho}\right)^{1/2} \qquad i = 1, 2, 3, \ldots \tag{1}
$$

where f_{yi} and f_{zi} correspond to the y and z principal bending axes, respectively.

The coefficients λ_{yi} and λ_{zi} are listed in Table 6.12 as λ_{yi}^0 and λ_{zi}^0 for the Bernoulli–Euler beam theory, when no shear deformation effects are considered.

The ratios of $\lambda_{yi}/\lambda_{yi}^0$ and $\lambda_{zi}/\lambda_{zi}^0$ for various sets of shear deformation coefficients are shown in Table 6.13. Detailed explanations of the meaning of the columns of this Table are provided in Example 6.9. It can be observed in moving from the first to the

TABLE 6.12 Coefficients $\lambda_{yi} = \lambda_{yi}^0$, $\lambda_{zi} = \lambda_{zi}^0$ for Natural Frequencies of a Cantilever Beam Based on Bernoulli–Euler Beam Theory

i	$\lambda_{yi} = \lambda_{yi}^0$	$\lambda_{zi} = \lambda_{zi}^0$
1	1.873392	1.862955
2	4.664652	4.499925
3	7.740505	7.177770
4	10.711110	9.499564
5	13.581400	11.539000
6	16.343046	
7	18.997060	

fourth columns that there is a change (presumably, to a higher degree of accuracy) in the natural frequencies, especially for the higher modes.

A comparison of the third and fourth columns of Table 6.13 demonstrates the effect of coupling between the motions in the y and z directions. For this trapezoidal cross section, the introduction of the shear deformation cross coefficient α_{yz} does

TABLE 6.13 Ratios of Natural Frequencies of the Cantilever Beam with Various Sets of Shear Deformation Coefficients

	Bernoulli–Euler Beam Theory	Frequently Used Default Values of α_{yy} and α_{zz}	Utilize Proper Values for α_{yy} and α_{zz}, but Set $\alpha_{yz} = 0$	Use the Correct Values for α_{yy}, α_{zz}, and α_{yz}
α_{yy}	0	1	1.159977	1.159977
α_{zz}	0	1	1.476062	1.476062
α_{yz}	0	0	0	−0.050085
$\lambda_{y1}/\lambda_{y1}^0$	1	0.9953167	0.9931080	0.9931078
$\lambda_{y2}/\lambda_{y2}^0$	1	0.9696793	0.9561353	0.9561319
$\lambda_{y3}/\lambda_{y3}^0$	1	0.9370653	0.9109106	0.9109165
$\lambda_{y4}/\lambda_{y4}^0$	1	0.9023160	0.8644692	0.8642957
$\lambda_{y5}/\lambda_{y5}^0$	1	0.8744223	0.8275235	0.8274083
$\lambda_{y6}/\lambda_{y6}^0$	1	0.8493372	0.7959447	0.7958172
$\lambda_{y7}/\lambda_{y7}^0$	1	0.8333934	0.7748016	0.7746413
$\lambda_{z1}/\lambda_{z1}^0$	1	0.9689476	0.9642118	0.9642123
$\lambda_{z2}/\lambda_{z2}^0$	1	0.8485505	0.8298480	0.8298514
$\lambda_{z3}/\lambda_{z3}^0$	1	0.7729115	0.7488483	0.7490240
$\lambda_{z4}/\lambda_{z4}^0$	1	0.7224356	0.6949695	0.6951051
$\lambda_{z5}/\lambda_{z5}^0$	1	0.6961900	0.6662446	0.6663878

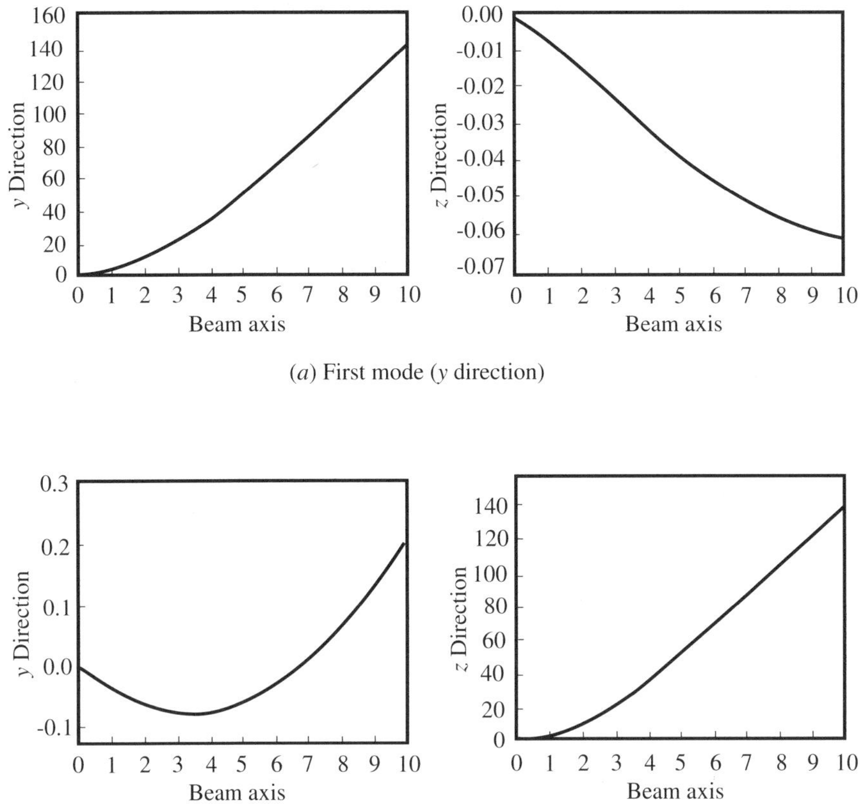

(a) First mode (y direction)

(b) Second mode (z direction)

Figure 6.19 Mode shapes of a cantilever beam with an asymmetrical cross section.

not significantly affect the natural frequencies, although the difference tends to increase with an increase in the order of the mode. Note, for example, that $\lambda_{yz}/\lambda_{yz}^0$ is 0.7748016 when α_{yz} is zero and 0.7746413 when α_{yz} is given its proper value. There is, however, a difference that is evident by studying the mode shapes. Figure 6.19 illustrates that the mode shapes are spatial curves, not just planar curves, when α_{yz} is not taken to be zero. The mode shapes for the Bernoulli–Euler theory, with no shear deformation effects, will lie completely in the y direction for the first mode and in the z direction for the second mode. These results demonstrate that the transverse motion cannot be uncoupled into two planes if shear deformation effects are included.

6.4.6 Loading Vectors

The stiffness matrix obtained by a rearrangement of the transfer matrix can be accompanied by a loading vector such as Eq. (2.65). The loading vector $\mathbf{p}^{eo}$ corresponding to the stiffness matrix of Eq. (6.158) is provided in Pilkey et al. (1995). Using this loading vector permits thermal loading and in-span distributed applied forces to be taken into account.

6.4.7 Elasticity-Based Beam Theory

A general beam theory stiffness matrix that considers bending, torsion, and shear deformation is derived from the theory of elasticity in Schramm et al. (1997). As discussed in Section 6.3.1, the shear deformation coefficients are supplemented with torsional deformation coefficients. The stiffness matrix can be applied to arbitrary cross-sectional shapes.

An exact theory for linearly elastic beams of arbitrary cross section and loading by Ladavèze (1983) and Ladevèze and Simmonds (1996) was mentioned in Section 6.3. In this theory a conjugate generalized displacement is defined, which comprises displacementlike and rotationlike quantities. The generalized displacements are not cross-sectional averages of three-dimensional displacements, but rather are defined as cross-sectional integrals involving three-dimensional displacements and three-dimensional stresses. The beam problem is decomposed into an interior portion, explicitly computable from the solutions of associated one-dimensional exact beam equations and a decaying, edge-effect portion.

6.5 CURVED BARS

Although this book deals primarily with the computational analysis and design of straight beams, curved beams can be handled with similar methods. Deformation analyses are available in numerous sources, including formula books such as Pilkey (1994). The same books typically tabulate analytical formulas for normal and shear stresses. Computational methods for the stress analysis of circularly curved bars are developed in Thasanatorn and Pilkey (1979, 1981).

REFERENCES

Cowper, G. R. (1966). The shear coefficient in Timoshenko's beam theory, *J. Appl. Mech.*, Vol. 33, pp. 335–340.

Fung, Y. C. (1969). *An Introduction to the Theory of Aeroelasticity*, Dover, New York.

Hutchinson, J. R. (2001). Shear coefficients for Timoshenko beam theory, *J. Appl. Mech.*, Vol. 68, pp. 87–92.

Kaneko, T. (1975). On Timoshenko's correction for shear in vibrating beams, *J. Phys.*, Vol. D8, pp. 1927–1936.

Kennedy, D., and Williams, F. W. (1991). More efficient use of determinants to solve transcendental structural problems reliably, *Comput. Struct.*, Vol. 41, pp. 973–979.

Ladevèze, P. (1983). Sur le principe de Saint-Venant en èlasticitè, *J. Mècanique Théorique Appliquée*, Vol. 1, pp. 161–184.

Ladevèze, P., and Simmonds, J. G. (1996). New concepts for linear beam theory with arbitrary geometry and loading, *Comptes Rendus Acad. Sci. Paris*, Vol. 332, Ser. IIb, pp. 455–462.

Mason, W. E., and Herrmann, L. R. (1968). Elastic shear analysis of general prismatic beams, *J. Eng. Mech.*, Vol. 94, pp. 965–983.

Oden, J. T., and Ripperger, E. A. (1981). *Mechanics of Elastic Structures*, 2nd ed., McGraw-Hill, New York

Pilkey, W. D. (1994). *Formulas for Stress, Strain, and Structural Matrices*, Wiley, New York.

Pilkey, W. D., and Kang, W. (1996). Exact stiffness matrix for a beam element with axial force and shear deformation, *Finite Elements Anal. Des.*, Vol. 22, pp. 1–13.

Pilkey, W. D., and Wunderlich, W. (1994). *Mechanics of Structures: Variational and Computational Methods*, CRC Press, Boca Raton, Fla.

Pilkey, W. D., Kang, W., and Schramm, U. (1995). New structural matrices for a beam element with shear deformation, *Finite Elements Anal. Des.*, Vol. 19, pp. 25–44.

Renton, J. D. (1991). Generalized beam theory applied to shear stiffness, *Int. J. Solids Struct.*, Vol. 27, pp. 1955–1967.

Renton, J. D. (1997). A note on the form of the shear coefficient, *Int. J. Solids Struct.*, Vol. 34, pp. 1681–1685.

Romano, G., Rasati, L., and Ferro, G. (1992). Shear deformability of thin-walled beams with arbitrary cross sections, *Int. J. Numer. Methods Eng.*, Vol. 35, pp. 283–306.

Rubinstein, I., and Rubinstein, L. (1993). *Partial Differential Equations in Classical Mathematical Physics*, Cambridge University Press, Cambridge.

Sanchez, P. (2001). Mise en oeuvre et illustrations de la théorie exacte des poutres, Doctor's Thesis, L'École Normale Supérieure de Cachan, Paris, France.

Schramm, U., Kitis, L., Kang, W., and Pilkey, W. D. (1994). On the shear deformation coefficient in beam theory, *Finite Elements Anal. Des.*, Vol. 16, pp. 141–162.

Schramm, U., Rubenchik, V., and Pilkey, W. D. (1997). Beam stiffness matrix based on the elasticity equations, *Int. J. Numer. Methods Eng.*, Vol. 40, pp. 211–232.

Sokolnikoff, I. S. (1956). *Mathematical Theory of Elasticity*, McGraw-Hill, New York.

Thasanatorn, C., and Pilkey, W. D., (1979). Torsional stresses in circularly curved bars, *J. Struct. Div., ASCE*, Vol. 105, pp. 2327–2342.

Thasanatorn, C., and Pilkey, W. D., (1981). Shear stresses in circularly curved bars, *J. Struct. Div., ASCE*, Vol. 107, pp. 1907–1919.

Timoshenko, S. P. (1921). On the correction for shear of the differential equation for transverse vibrations of prismatic bars, *Philos. Mag.*, Vol. 41, pp. 744–746.

Trefftz, E. (1936). Über den Schubmittelpunkt in einem durch eine Einzellast gebogenen Balken, *Z. Angew. Math. Mech.*, Vol. 15, pp. 220–225.

CHAPTER 7

RESTRAINED WARPING OF BEAMS

Most of this book deals with computational methods for the theory of elasticity based analysis of beams, especially for cross-sectional properties and stresses. The computational solutions here are, in general, applicable to thick (solid) and open and closed thin-walled cross sections. In this chapter, some of the formulations and solutions developed could be classified as belonging to the traditional analytical methods for studying thin-walled beams. In particular, this material is related to the Vlasov (1961) theory of thin-walled beams. The treatment here is somewhat restricted in that the reference coordinate system is always located at the centroid of the cross section. A systematic, thorough development of analytical thin-walled beam theory, including open and closed sections, is on the author's website. Also, rigorous accounts of the engineering theory of thin-walled structures can be found in many sources, including Murray (1984).

Saint-Venant torsion of Chapter 5 applies to bars for which there are no restraints on warping. If the warping of the bar is restrained, *warping stresses* are generated. In this chapter we define the cross-sectional properties for the restrained warping of beams and discuss warping stresses.

7.1 RESTRAINED WARPING

With the exception of bars with solid and hollow circular cross sections, bars subject to torsion tend to warp. The fundamental assumption in Saint-Venant's pure torsion analysis, described in Chapter 5, is that cross sections are free to warp without restraint. This assumption is not satisfied if the beam has external supports, or if the beam is not prismatic, or when the torsional moment varies along the length of the beam. Even in these cases, Saint-Venant's hypothesis of free warping gives good approximate results for beams with solid or closed thin-walled cross sections. For open thin-walled cross sections, however, this is no longer true. The twisting resistance of such sections is so small and the axial displacements are so large that the axial stress σ_x caused by restrained warping cannot be neglected.

In pure torsion, the location of the axis of twist is of no consequence when calculating torsional stresses. This means that in Fig. 7.1, the reference point O can be arbitrarily chosen. The torsional shear stresses are independent of this choice. In restrained warping, this is no longer the case, and the point about which the rotations are calculated are shown in this section to be uniquely determined. Define Cyz to be the coordinate system shown in Fig. 7.1 with its origin at the centroid C of the cross section. Let Oy^*z^* be the coordinate system obtained by translating the origin to another point O. Suppose that the position vector of O with respect to C is written as

$$\mathbf{r}_{OC} = b\mathbf{j} + c\mathbf{k} \tag{7.1}$$

where $\mathbf{j}$, $\mathbf{k}$ are the unit vectors parallel to the y, z axes. The coordinates y^*, z^* of a point in the coordinate system Oy^*z^* are related to the coordinates y, z of the point in Cyz by

$$y^* = y - b \qquad z^* = z - c \tag{7.2}$$

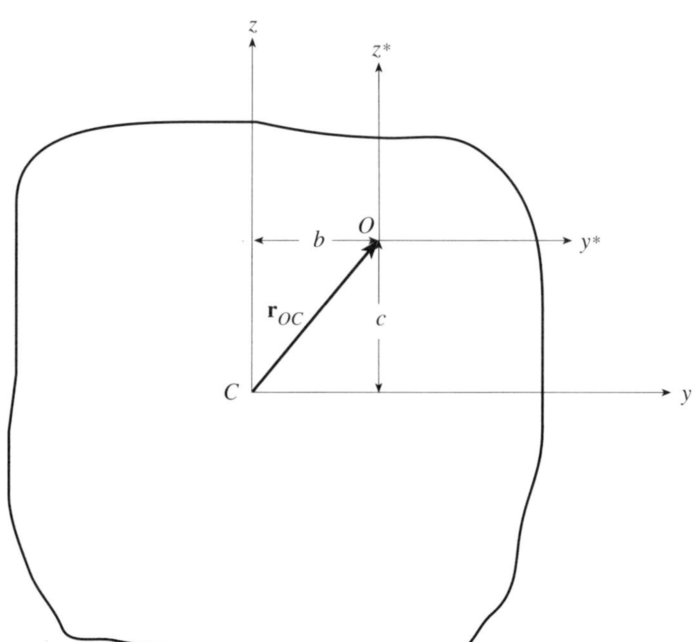

Figure 7.1 Translation of coordinates.

Let $u_z^*(y^*, z^*)$ be the warping function calculated in the coordinate system Oy^*z^*. Suppose that Saint-Venant's problem is solved in the coordinate systems Oy^*z^* and Cyz, assuming in the first case that the section rotates about O and in the second case that it rotates about C. The assumed displacement field in the latter case, as described in Chapter 5, Eqs. (5.1) and (5.2), is

$$u_y = -zx\theta \qquad u_z = yx\theta \qquad u_x = \theta\omega(y, z) \tag{7.3}$$

where θ is the angle of twist per unit length. In the former case, the assumed displacement fields are

$$\begin{aligned} u_y^* &= -z^*x\theta = -x\theta(z - c) \\ u_z^* &= y^*x\theta = x\theta(y - b) \\ u_x^* &= \theta\omega^*(y^*, z^*) \end{aligned} \tag{7.4}$$

The stresses corresponding to the displacements u_x^*, u_y^*, u_z^* are given by Hooke's law and the strain–displacement relations (Eq. 5.4)

$$\begin{aligned} \tau_{xy} &= G\theta\left(\frac{\partial\omega^*}{\partial y^*} - z^*\right) = G\theta\left(\frac{\partial\omega^*}{\partial y} - z + c\right) \\ \tau_{xz} &= G\theta\left(\frac{\partial\omega^*}{\partial z^*} + y^*\right) = G\theta\left(\frac{\partial\omega^*}{\partial z} + y - b\right) \end{aligned} \tag{7.5}$$

The equilibrium equation would again be Laplace's equation (Eq. 5.8) with the expressions above for the stresses

$$\nabla^2\omega^* = 0 \tag{7.6}$$

The boundary condition for the cylindrical surface of the beam to be free of traction, which requires that (Eq. 6.112)

$$\tau_{xy}n_y + \tau_{xz}n_z = 0 \tag{7.7}$$

becomes, with Eq. (7.5),

$$\left(\frac{\partial\omega^*}{\partial y} + c\right)n_y + \left(\frac{\partial\omega^*}{\partial z} - b\right)n_z = zn_y - yn_z \tag{7.8}$$

In vector notation this can be expressed as

$$\mathbf{n}\cdot\nabla\left(\omega^* + cy - bz\right) = \mathbf{n}\cdot\mathbf{g} \tag{7.9}$$

where

$$\mathbf{n} = n_y\mathbf{j} + n_z\mathbf{k}$$
$$\mathbf{g} = z\mathbf{j} - y\mathbf{k}$$
$$\nabla = \mathbf{j}\frac{\partial}{\partial y} + \mathbf{k}\frac{\partial}{\partial z}$$

Since the directional derivative of a point of a surface along a normal is defined as

$$\mathbf{n} \cdot \nabla = \frac{\partial}{\partial n} = n_y \frac{\partial}{\partial y} + n_z \frac{\partial}{\partial z} \tag{7.10}$$

An alternative form for Eq. (7.9) is

$$\frac{\partial}{\partial n}\left(\omega^* + cy - bz\right) = \mathbf{n} \cdot \mathbf{g} \tag{7.11}$$

Since ω^* satisfies Laplace's equation (Eq. 7.6), so does the function $\omega^* + cy - bz$, which also satisfies the same boundary condition as ω. The uniqueness of the Neumann problem therefore implies that ω and ω^* can differ at most by a constant. Hence

$$\omega^* + cy - bz = \omega(y, z) + C$$

or

$$\omega^* = \omega(y, z) - cy + bz + C \tag{7.12}$$

The stresses computed from ω^* are the same as those computed from ω:

$$\begin{aligned} \tau_{xy} &= G\theta\left(\frac{\partial \omega^*}{\partial y} - z + c\right) = G\theta\left(\frac{\partial \omega}{\partial y} - z\right) \\ \tau_{xz} &= G\theta\left(\frac{\partial \omega^*}{\partial z} + y - b\right) = G\theta\left(\frac{\partial \omega}{\partial z} + y\right) \end{aligned} \tag{7.13}$$

The displacements differ in the two cases (u_y, u_z, u_x and u_y^*, u_z^*, u_x^*) by at most a rigid-body displacement. Since the stresses for the two cases are the same, the choice of the axis of twist is immaterial in Saint-Venant's pure torsion problem.

Next, to account for restrained warping in an approximate way, the assumed axial displacement is taken as

$$u_x(x, y, z) = \theta(x)\omega^*(y, z) \tag{7.14}$$

This contrasts with the axial (warping) displacement of Eq. (5.2), which is not a function of x, the axial coordinate. Equation (7.14) allows the angle of twist θ per unit length to vary with x, and the use of ω^*, with the yet-undetermined reference point O, leaves undetermined the point through which the axis of twist passes. The effect of allowing the axial displacement to be a function of x is to allow axial stresses σ_x to develop:

$$\sigma_x = E\epsilon_x = E\frac{\partial u_x}{\partial x} = E\omega^*\frac{\partial \theta}{\partial x} \tag{7.15}$$

It is assumed that the only significant normal stress in the constitutive relations of Eq. (1.18) is σ_x. Since no axial forces or bending moments are applied at the section, this normal stress must be equivalent to a zero force–couple system, which means that the conditions (Eqs. 1.45 and 1.46)

$$\int \sigma_x \, dA = 0 \qquad M_z = -\int y\sigma_x \, dA = 0 \qquad M_y = \int z\sigma_x \, dA = 0 \tag{7.16}$$

must be satisfied. Substitute σ_x of Eq. (7.15) into Eq. (7.16) and the integrals of Eq. (7.16), in terms of the warping function, become

$$\int \omega^* \, dA = 0 \qquad \int y\omega^* \, dA = 0 \qquad \int z\omega^* \, dA = 0 \tag{7.17}$$

The first condition in Eq. (7.17), with ω^* of Eq. (7.12), is

$$\int \omega^* \, dA = \int \omega \, dA - c\int y \, dA + b\int z \, dA + CA = Q_\omega + CA = 0 \tag{7.18}$$

where the middle two terms are zero because y, z are centroidal axes and the *first moment of warping*, defined by

$$Q_\omega = \int \omega \, dA \tag{7.19}$$

is a geometric property of the cross section. From Eq. (7.18), the constant C is determined to be

$$C = -\frac{Q_\omega}{A} \tag{7.20}$$

The second and third conditions in Eq. (7.17) require that

$$\int y\omega^* \, dA = \int y\omega \, dA - c\int y^2 \, dA + b\int yz \, dA = I_{y\omega} - cI_z + bI_{yz} = 0$$

$$\int z\omega^* \, dA = \int z\omega \, dA - c\int yz \, dA + b\int z^2 \, dA = I_{z\omega} - cI_{yz} + bI_y = 0$$

where $I_{y\omega}$ and $I_{z\omega}$ are defined as

$$I_{y\omega} = \int y\omega \, dA$$

$$I_{z\omega} = \int z\omega \, dA \tag{7.21}$$

The solution of these equations for b and c yields

$$\begin{aligned} b &= \frac{I_{yz} I_{y\omega} - I_z I_{z\omega}}{I_y I_z - I_{yz}^2} \\ c &= \frac{I_y I_{y\omega} - I_{yz} I_{z\omega}}{I_y I_z - I_{yz}^2} \end{aligned} \tag{7.22}$$

These are the Trefftz shear center coordinates y_S and z_S given in Eqs. (6.118) and (6.119). Thus, the axis of twist, the choice of which is arbitrary in Saint-Venant torsion, passes through the shear center in a restrained torsion analysis.

7.2 THIN-WALLED BEAMS

Figure 7.2 shows an open thin-walled beam cross section. The arc length s is measured along the median line of the cross section in the counterclockwise direction starting from one end of the section, where s is assigned a zero value. The thickness $t(s)$ may vary along the median line of the cross section; it is assumed to be constant along the beam axis x. The cross section is subjected to transverse shear forces.

On two perpendicular planes, the shear stresses acting on the line of intersection in the direction perpendicular to that line are equal in magnitude and are either both directed away from or both directed toward the line of intersection (Fig. 7.3*a*). Since the cylindrical surface of the beam is free of traction, the shear stresses at the inside and outside edges of the section must act in the direction tangent to the edges (Fig. 7.3*b*). The variation of the shear stresses across the thickness is assumed to be negligible, because the wall thickness is small. At any s, the constant shear stress τ_{xs}

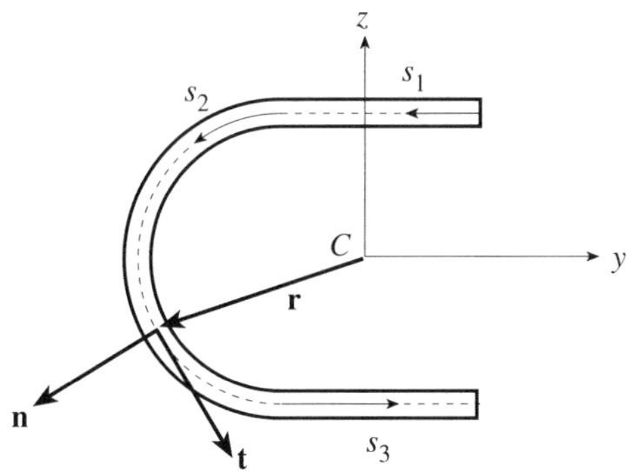

Figure 7.2 Open thin-walled cross section.

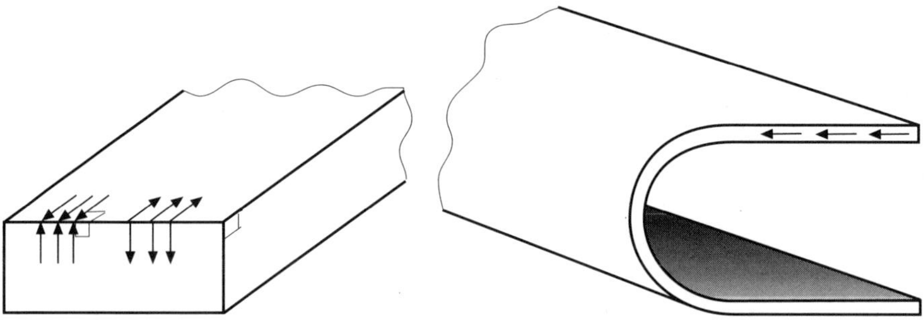

(a) (b)

Figure 7.3 Shear stress patterns for thin-walled cross section.

multiplied by the thickness $t(s)$ represents a shear force per unit length and is called *shear flow* (Fig. 5.14):

$$q(s) = \tau_{xs}(s)t(s) \tag{7.23}$$

Assume that the bending stress σ_x is the only stress acting on the section in addition to the shear flow at the median line of the section. Consider an element as shown in Fig 7.4*b* that is taken from the beam wall (Fig. 7.4*a*). The equilibrium of forces on this wall element gives

$$t(s)\frac{\partial \sigma_x}{\partial x} + \frac{\partial q}{\partial s} = 0 \tag{7.24}$$

For restrained warping in a thin-walled section, the axial stress is (Eq. 7.15)

$$\sigma_x(x, s) = E\omega^*(s)\theta'(x) \tag{7.25}$$

which is a simplification of the elasticity definition because the warping function ω^* is being considered as a function of s only. Substitution of the normal stress σ_x of Eq. (7.25) into Eq. (7.24) shows that the shear flow due to restrained warping satisfies the equation

$$t(s)E\omega^*(s)\theta''(x) + \frac{\partial q}{\partial s} = 0 \tag{7.26}$$

Hence

$$q(x, s) = -E\theta''(x)\int_0^s \omega^*(s)t(s)\,ds \tag{7.27}$$

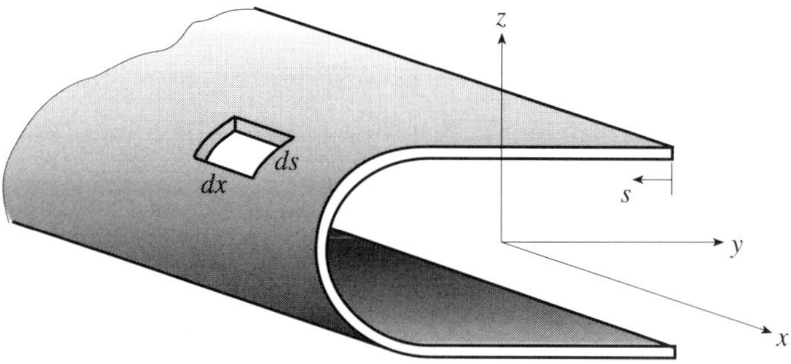

(a)

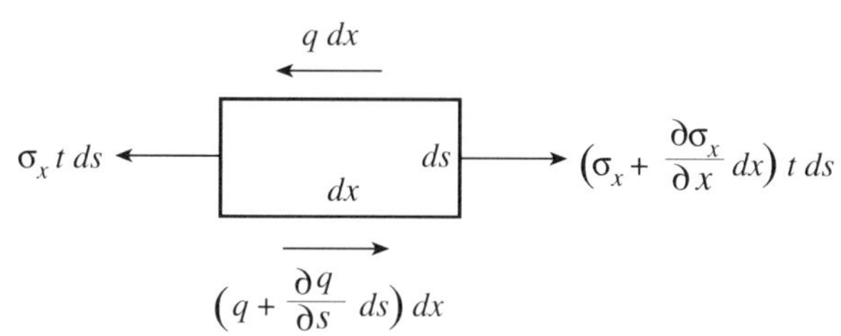

(b)

Figure 7.4 Forces on a wall element.

7.2.1 Saint-Venant Torsion

In Saint-Venant torsion, with unrestrained warping, a reasonable approximation to the torsional shear stresses in an open thin-walled section is obtained by assuming a linear distribution across the thickness, as shown for a narrow rectangular strip in Fig. 7.5 (Section 5.2.1 and Fig. 5.10). The shear stress is then zero at the median line, which in the case of the rectangle of Fig. 7.5 is the vertical dashed line in the center of the cross section. The shear strain along the median line is

$$\frac{\partial u_x}{\partial s} + \frac{\partial u_s}{\partial x} = \gamma_{xs} = \frac{\tau_{xs}}{G} = 0 \tag{7.28}$$

where u_s is the displacement along the tangent to the median line. For Saint-Venant torsion the displacement field assumptions are (Eqs. 5.1 and 5.2)

$$u_x = \theta\omega \qquad u_y = -zx\theta \qquad u_z = yx\theta \tag{7.29}$$

At the point whose position vector is **r** in Fig. 7.2, the unit tangent **t** and the unit outward normal **n** are related by (Eq. 1.34)

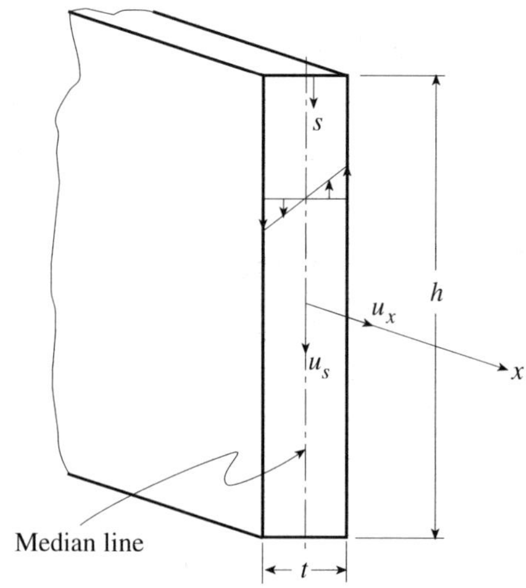

Figure 7.5 Shear stress distribution for a bar subjected to pure torsion with a narrow rectangular strip cross section.

$$\mathbf{n} = \mathbf{t} \times \mathbf{i} = t_z\mathbf{j} - t_y\mathbf{k} \tag{7.30}$$

where $\mathbf{t} = t_y\mathbf{j} + t_z\mathbf{k}$ and $\mathbf{n} = n_y\mathbf{j} + n_z\mathbf{k}$. From Eq. (7.30) it follows that (Eq. 1.33) $n_y = t_z$ and $n_z = -t_y$. The tangential displacement is calculated as follows:

$$u_s = (u_y\mathbf{j} + u_z\mathbf{k}) \cdot \mathbf{t} = (-zt_y + yt_z)x\theta = x\theta\mathbf{r} \cdot \mathbf{n} \tag{7.31}$$

where Eq. (7.29) has been introduced and $\mathbf{r} = y\mathbf{j} + z\mathbf{k}$. The shear strain at the median line of Eq. (7.28) is expressed in terms of the warping function as

$$\gamma_{xs} = \frac{\partial u_x}{\partial s} + \frac{\partial u_s}{\partial x} = \theta\frac{d\omega}{ds} + \theta\mathbf{r} \cdot \mathbf{n} = 0$$

so that

$$\omega(s) = \omega_0 - \int_0^s \mathbf{r} \cdot \mathbf{n}\, ds \tag{7.32}$$

where ω_0 is a constant. The integral in this expression is twice the area swept by the position vector $\mathbf{r}$ as its tip traces the median line from $s = 0$ to any value of s. Note that the warping function, which was introduced in Chapter 5 as the solution of Laplace's equation $\nabla^2\omega = 0$ along with certain boundary conditions, can be calculated for thin-walled sections in terms of purely geometric quantities, referred to as *sectorial characteristics*. A few sectorial properties were introduced briefly in Chapter 6.

Some Analytical Sectorial Properties of Thin-Walled Cross Sections

The warping function $\omega(s)$ defined by Eq. (7.32) can serve as the basis for finding warping-related properties and stresses, without the use of finite elements. This involves the use of sectorial properties. We begin with some basic definitions. The vector $\mathbf{r}$ in Eq. (7.32) is the position vector of a point of the median line of the thin-walled section measured from the centroid C (Fig. 7.2). Also, set the origin for the integration to be s_0 so that ω_0 is the value of $\omega(s)$ at $s = s_0$.

The sign of $\mathbf{r}(s) \cdot \mathbf{n}(s)$ is positive if the projection of $\mathbf{r}(s)$ onto the unit normal vector $\mathbf{n}(s)$ is positive. This sign is more conveniently identified from the sense of rotation of $\mathbf{r}(s)$ as it sweeps through the area in the direction of increasing s. If this rotation is clockwise, the contribution to the integral is negative, and if it is counterclockwise, the contribution is positive. This can be verified for any point of the median line by writing the projection of the positive vector $\mathbf{r}$ onto the unit normal in the form

$$\mathbf{r} \cdot \mathbf{n} = \mathbf{r} \cdot (\mathbf{t} \times \mathbf{i}) = (\mathbf{r} \times \mathbf{t}) \cdot \mathbf{i}$$

where $\mathbf{n}$, the unit vector normal to the median line; $\mathbf{t}$, the unit vector tangent to the median line in the direction of increasing s; and $\mathbf{i}$, the unit vector in the direction of the positive x axis (longitudinal axis of the bar), are defined so as to make a right-handed set of orthogonal vectors $\mathbf{i} = \mathbf{n} \times \mathbf{t}$ and $\mathbf{n} = \mathbf{t} \times \mathbf{i}$. It follows that when the rotation of the vector $\mathbf{r}$ is counterclockwise as its tip moves in the direction of $\mathbf{t}$, the cross product $\mathbf{r} \times \mathbf{t}$ is in the same direction as $\mathbf{i}$, making $\mathbf{r} \cdot \mathbf{n}$ positive.

The warping function, which is utilized in the first sectorial moment and in the sectorial products of area of Eq. (7.21), is defined for a particular pole and origin, both of which in our case are at the centroid. A pole for which the sectorial products of area are both zero is called a *principal pole*. If, for a given pole, the first sectorial moment of Eq. (7.19) is zero, there is a sectional origin such that the point is termed a *principal origin*. For a given cross section, it is possible to find a pole and an origin such that Q_ω, $I_{y\omega}$, and $I_{z\omega}$ are zero. A warping function satisfying these conditions is termed a *principal warping function*. The restrained warping function of the following section is a principal warping function. It is readily shown that the shear center and the principal pole are the same point.

For practical calculations it is useful to recognize that $\mathbf{r} \cdot \mathbf{n}$ is equal to the scalar r_n that is the perpendicular distance from a tangent of any point on the median line curve to point C (Fig. 7.6). With this definition

$$\omega = \omega_0 - \int r_n \, ds \tag{7.33}$$

The magnitude of this integral is equal to twice the shaded area of Fig. 7.7. The integration is taken over the segment of the curve corresponding to the shaded area beginning at an origin s_0. The integration is in the same direction as the direction of s.

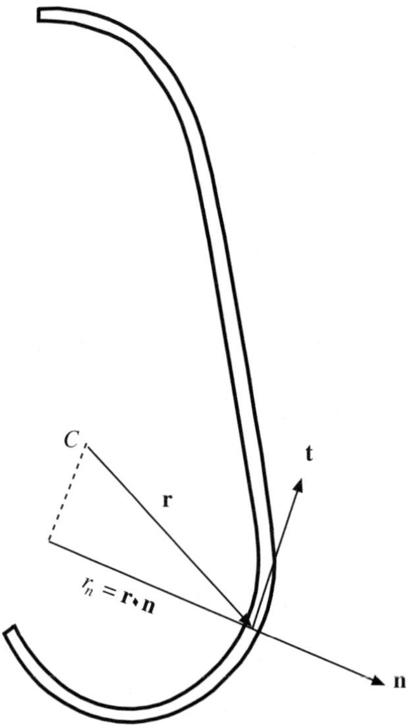

Figure 7.6 Definition of r_n.

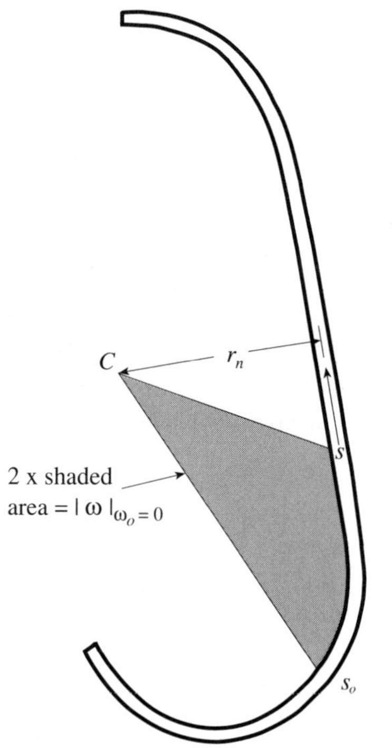

Figure 7.7 Notation for the sectorial area.

7.2.2 Restrained Warping

For restrained warping, the warping function is ω^* as given in Eqs. (7.12) and (7.20), with $b = y_S$ and $c = z_S$,

$$\omega^* = \omega - z_S\, y + y_S\, z - \frac{Q_\omega}{A} \tag{7.34}$$

Since z_S, y_S, and Q_ω/A are not functions of s, the derivative $d\omega^*/ds$ is simple to express. Introduce Eq. (1.36) and find that

$$\begin{aligned} \frac{d\omega^*}{ds} &= \frac{d\omega}{ds} - z_S\frac{dy}{ds} + y_S\frac{dz}{ds} = \frac{d\omega}{ds} - z_S t_y + y_S t_z \\ &= \frac{d\omega}{ds} + z_S n_z + y_S n_y = \frac{d\omega}{ds} + \mathbf{r}_S \cdot \mathbf{n} \end{aligned} \tag{7.35}$$

where $\mathbf{r}_S = y_S\mathbf{j} + z_S\mathbf{k}$ is the position vector of the shear center measured from the centroid. From Eq. (7.33), $d\omega/ds = -\mathbf{r}\cdot\mathbf{n}$. Then it is observed that the derivative of the restrained warping function with respect to the arc length may be rewritten as

$$\frac{d\omega^*}{ds} = -(\mathbf{r} - \mathbf{r}_S) \cdot \mathbf{n} \tag{7.36}$$

The vector in parentheses in Eq. (7.36) is the position vector of a point of the median line measured from the shear center. Integrate Eq. (7.36) to find that the warping function is

$$\omega^*(s) = \omega_0^* - \int_0^s (\mathbf{r} - \mathbf{r}_S) \cdot \mathbf{n}\, ds \tag{7.37}$$

The integral in this expression is twice the area swept by the position vector $\mathbf{r} - \mathbf{r}_S$ measured from the shear center as its tip moves along the median line of the thin-walled section from $s = 0$ to any value of s.

When shear forces act at any point of the cross section, it is normally necessary to replace them with their force–couple equivalent at the shear center. The moment of the couple is then added to any pure torques that may already be present at the section. The moment T_ω of the couple due to shear forces can be computed by

$$T_\omega = \int_0^S q(\mathbf{r} - \mathbf{r}_S) \cdot \mathbf{n}\, ds = -\int_0^S q\, d\omega^* \tag{7.38}$$

where S is the length of the median line and Eq. (7.36) has been introduced. The moment T_ω is referred to as the *warping torque* or the *warping shear*. Upon substitution of the shear flow q from Eq. (7.27),

$$T_\omega = E\theta''(x) \int_0^S \int_0^s \omega^* t\, ds\, d\omega^* \tag{7.39}$$

To simplify, apply integration by parts:

$$\int_0^S \int_0^s \omega^* t\, ds\, d\omega^* = \omega^* \int_0^s \omega^* t\, ds \Big|_0^S - \int (\omega^*)^2 t\, ds \tag{7.40}$$

The first integral on the right is zero because (Eq. 7.17)

$$\int_0^S \omega^* t\, ds = \int \omega^*\, dA = 0 \tag{7.41}$$

so that the torsional moment due to shear forces is

$$T_\omega = -E\theta'' \int (\omega^*)^2\, dA \tag{7.42}$$

The *warping constant* Γ with respect to the shear center is defined as the warping moment of inertia:

$$\Gamma = \int (\omega^*)^2\, dA \tag{7.43}$$

Then

$$T_\omega = -E\Gamma\theta'' \tag{7.44}$$

The total torque T at the section is expressed as the sum of the Saint-Venant torque $M_x = T_t$ and the torque T_ω due to the shear forces. From Eq. (5.18)

$$T_t = M_x = GJ\theta \tag{7.45}$$

Continue to use the notation of Chapter 5, where ϕ is the angle of twist. Since θ is the angle of twist per unit length

$$\theta = \frac{d\phi}{dx} \tag{7.46}$$

From Eqs. (7.44) and (7.45),

$$T = T_\omega + T_t = -E\Gamma\theta'' + GJ\theta \tag{7.47}$$

Introduce Eq. (7.46):

$$T = -E\Gamma\phi''' + GJ\phi' \tag{7.48}$$

If $\overline{m}_x$ is the applied torque per unit length of beam (Fig. 7.8), the condition of equilibrium for the moments about the x axis shows that

$$\frac{dT}{dx} = -\overline{m}_x \tag{7.49}$$

This is Eq. (5.22), except now the total torque T is involved, where $T = T_\omega + M_x$. It follows from Eqs. (7.48) and (7.49) that for constant E, Γ, G, and J the equation for the angle of twist is

$$E\Gamma\frac{d^4\phi}{dx^4} - GJ\frac{d^2\phi}{dx^2} = \overline{m}_x(x) \tag{7.50}$$

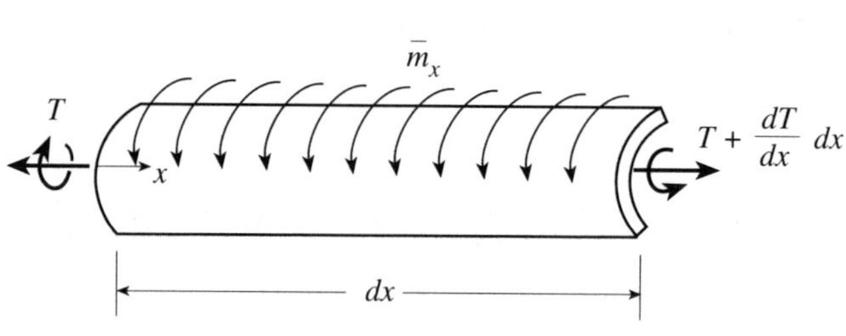

Figure 7.8 Element subject to distributed torque $\overline{m}_x$.

7.3 CALCULATION OF THE ANGLE OF TWIST

When restrained warping is to be taken into account, the angle of twist should be computed using the fourth-order differential equation of Eq. (7.50) rather than the second-order equation for Saint-Venant torsion. Equation (7.50) is relatively simple to solve, especially since the differential equations of motion for the bending of a beam are also fourth order, and with a change of constants the solutions are interchangeable.

7.3.1 Governing Equations

Define a *bimoment* M_ω in a fashion similar to the moment definitions of Eq. (1.45),

$$M_\omega = \int \omega^* \sigma_x \, dA \tag{7.51}$$

Substitute $\sigma_x = E\omega^* \, d^2\phi/dx^2$ from Eq. (7.15) into Eq. (7.51) and find that

$$M_\omega = E\phi''\Gamma \tag{7.52}$$

or

$$\frac{d^2\phi}{dx^2} = \frac{M_\omega}{E\Gamma} \tag{7.53}$$

The governing equations can now be assembled as:

Angle of twist:

$$\phi \tag{7.54a}$$

Rate of angle of twist:

$$\theta = \frac{d\phi}{dx} \tag{7.54b}$$

Bimoment:

$$M_\omega = E\Gamma \frac{d^2\phi}{dx^2} \tag{7.54c}$$

Warping torque:

$$T_\omega = -E\Gamma \frac{d^3\phi}{dx^3} \tag{7.54d}$$

Saint-Venant torque:

$$T_t = GJ\frac{d\phi}{dx} \tag{7.54e}$$

Total torque:

$$T = -E\Gamma\frac{d^3\phi}{dx^3} + GJ\frac{d\phi}{dx} \tag{7.54f}$$

Differential equation for the angle of twist:

$$E\Gamma\frac{d^4\phi}{dx^4} - GJ\frac{d^2\phi}{dx^2} = \overline{m}_x(x) \tag{7.54g}$$

7.3.2 Boundary Conditions

The most common boundary conditions are *fixed*, *simply supported*, and *free*. At a fixed support, no twisting or warping occurs. The second condition is satisfied by setting the warping component u_x of the displacement equal to zero, which as seen in Eq. (7.29) is proportional to $d\phi/dx$. Thus, the kinematical conditions at a fixed end are expressed by

$$\phi = 0 \qquad \frac{d\phi}{dx} = 0 \tag{7.55}$$

A simply supported end does not allow twisting and is free of normal stress. The normal stress due to restrained warping is (Eq. 7.15) $\sigma_x = E\omega^* d^2\phi/dx^2$, which is proportional to the bimoment M_ω. The boundary conditions at a simply supported end are then

$$\phi = 0 \qquad \frac{d^2\phi}{dx^2} = 0 \tag{7.56}$$

where the second condition expresses that the bimoment M_ω is zero.

At a free support there are two statical conditions, one expressing that there is no normal stress and the other that the total torque is zero. The first of these conditions corresponds to the bimoment M_ω being zero. The second of these conditions is

$$T = T_\omega + T_t = -E\Gamma\frac{d^3\phi}{dx^3} + GJ\frac{d\phi}{dx} = 0 \tag{7.57}$$

Thus, for a free support, the boundary conditions are

$$\frac{d^2\phi}{dx^2} = 0 \qquad -E\Gamma\frac{d^3\phi}{dx^3} + GJ\frac{d\phi}{dx} = 0 \tag{7.58}$$

7.3.3 Response Expressions

The general solution of Eq. (7.54g) is

$$\phi(x) = C_1 + C_2 x + C_3 \cosh cx + C_4 \sinh cx - \frac{1}{cGJ}\int_0^x [c(x-\xi) - \sinh c(x-\xi)]\,\overline{m}_x(\xi)\,d\xi \tag{7.59a}$$

where

$$c^2 = \frac{GJ}{E\Gamma}$$

and $C_k, k = 1, 2, 3, 4$, are the constants of integration, and one end of the beam is assumed to be at $x = 0$.

The bimoment of Eq. (7.54c) is taken from Eq. (7.59a) by two differentiations:

$$M_\omega(x) = JG(C_3 \cosh cx + C_4 \sinh cx) + \frac{1}{c}\int_0^x \overline{m}_x(\xi) \sinh c(x-\xi)\,d\xi \tag{7.59b}$$

The warping torque of Eq. (7.54d) requires a derivative of $M_\omega(x)$:

$$T_\omega(x) = -cGJ(C_3 \sinh cx + C_4 \cosh cx) - \int_0^x \overline{m}_x(\xi) \cosh c(x-\xi)\,d\xi \tag{7.59c}$$

The Saint-Venant (pure torsion) torque of Eq. (7.54e) is proportional to the first derivative of ϕ:

$$T_t(x) = GJ(C_2 + cC_3 \sinh cx + cC_4 \cosh cx) - \int_0^x \overline{m}_x(\xi)\,[1 - \cosh c(x-\xi)]\,d\xi \tag{7.59d}$$

and the total torque $T = T_t + T_\omega$ of Eq. (7.54f) is given by Eqs. (7.59c and d) as

$$T(x) = GJC_2 - \int_0^x \overline{m}_x(\xi)\,d\xi \tag{7.59e}$$

Example 7.1 Cantilever Beam with Concentrated Torques. A concentrated torque is applied at the unsupported end of the cantilever beam shown in Fig. 7.9a. For this loading condition, the distributed torque $\overline{m}_x(x)$ is zero, because there is no applied torque for the cross sections that lie between the two end sections. The applied torque is set equal to the total torque at $x = L$:

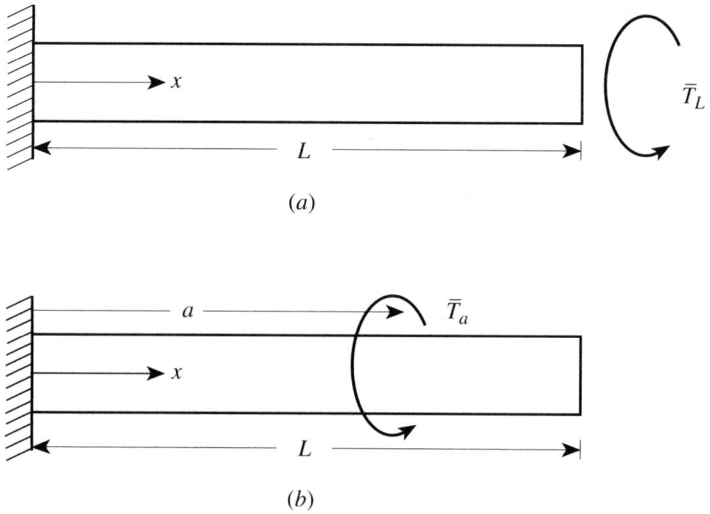

Figure 7.9 Cantilever beam with concentrated applied torque.

$$T(L) = GJC_2 = \overline{T}_L \tag{1}$$

where the total torque is given in Eq. (7.59e). The other boundary condition at $x = L$ is that the cross section is free of normal stress. From Eq. (7.59b):

$$M_\omega(L) = GJ(C_3 \cosh cL + C_4 \sinh cL) = 0 \tag{2}$$

From Eqs. (7.55) and (7.59a), the boundary conditions at the fixed end are

$$\begin{aligned} \phi(0) &= C_1 + C_3 = 0 \\ \frac{d\phi}{dx}(0) &= C_2 + cC_4 = 0 \end{aligned} \tag{3}$$

These four boundary conditions suffice to identify the four integration constants of the angle of twist of Eq. (7.59a):

$$\phi(x) = \frac{\overline{T}_L}{cGJ}[cx - \sinh cx - \tanh cL(1 - \cosh cx)] \tag{4}$$

Apply a torque at a point $x = a < L$ as shown in Fig. 7.9b. The concentrated torque of magnitude $\overline{T}_a$ can be expressed as a distributed torque in terms of the Dirac delta function:

$$\overline{m}_x(x) = \overline{T}_a \langle x - a \rangle^{-1} \tag{5}$$

The angle of twist is calculated from Eq. (7.59a) as

$$\phi(x) = C_1 + C_2 x + C_3 \cosh cx + C_4 \sinh cx - \frac{\overline{T}_a}{cGJ}\left[c(x-a) - \sinh c(x-a)\right]\langle x-a\rangle^0 \tag{6}$$

where $\langle x-a\rangle^0$ denotes the unit step function. The boundary conditions are

$$\begin{aligned}
\phi(0) &= C_1 + C_3 = 0\\
\phi'(0) &= C_2 + cC_4 = 0\\
T(L) &= GJC_2 - \overline{T}_a = 0\\
M_\omega(L) &= JG(C_3\cosh cL + C_4 \sinh cL) + \frac{\overline{T}_a}{c}\sinh c(L-a) = 0
\end{aligned} \tag{7}$$

The four integration constants can be evaluated by solving the four equations of (7). Then, from (6) the angle of twist becomes

$$\phi(x) = \frac{\overline{T}_a}{cGJ}\left[cx - \sinh cx + (1-\cosh cx)\left(\frac{\sinh c(L-a)}{\cosh cL} - \tanh cL\right)\right] - \frac{\overline{T}_a}{cGJ}\left[c(x-a) - \sinh c(x-a)\right]\langle x-a\rangle^0 \tag{8}$$

7.3.4 First-Order Governing Equations and General Solution

The governing relations of Eq. (7.54) can be expressed in first-order form as

$$\frac{d\phi}{dx} = \theta \tag{7.60a}$$

$$\frac{d\theta}{dx} = \frac{M_\omega}{E\Gamma} \tag{7.60b}$$

$$\frac{dT}{dx} = -\overline{m}_x \tag{7.60c}$$

$$\frac{dM_\omega}{dx} = GJ\theta - T \tag{7.60d}$$

where the warping torque is found using

$$T_\omega = -\frac{dM_\omega}{dx} \tag{7.60e}$$

With a change in constants the beam solution of Chapter 2 applies to this fourth-order restrained warping torsion problem. To use Table 2.2 define

Beam of Chapter 2	Restrained Warping of This Chapter
$\dfrac{1}{k_s GA}$	0
k	0
k^*	0
N	GJ
$\overline{p}_x$	$-\overline{m}_x$

(7.61a)

Then the transfer matrix $\mathbf{U}^e$ and the stiffness matrix $\mathbf{k}^e$ are obtained from Table 2.2 with

$$\zeta = -\frac{GJ}{E\Gamma} \qquad \eta = 0 \qquad \lambda = 0 \tag{7.61b}$$

This leads to the definitions of e_i,

$$\begin{gathered} e_0 = -\zeta B \qquad e_1 = A \qquad e_2 = B \qquad e_3 = \frac{1}{\zeta}(1 - A) \\ e_4 = \frac{1}{\zeta}(\ell - B) \quad e_5 = \frac{1}{\zeta}\left(\frac{\ell^2}{2} - e_3\right) \quad e_6 = \frac{1}{\zeta}\left(\frac{\ell^3}{6} - e_4\right) \end{gathered} \tag{7.61c}$$

where $A = \cosh \alpha\ell$ and $B = (\sinh \alpha\ell)/\alpha$ with $\alpha^2 = -\zeta$.

Finally, the results are interpreted using

Solution of Table 2.2	Restrained Warping of This Chapter
w	ϕ
θ	$-\theta$
V	$-T$
M	M_ω

(7.61d)

If dynamics, an axial force, and an elastic foundation are included, Eq. (7.60) becomes

$$\frac{\partial \phi}{\partial x} = \theta \tag{7.62a}$$

$$\frac{\partial \theta}{\partial x} = \frac{M_\omega}{E\Gamma} \tag{7.62b}$$

$$\frac{\partial T}{\partial x} = k_t \phi + \rho r_p^2 \frac{\partial^2 \phi}{\partial t^2} - \overline{m}_x(x, t) \tag{7.62c}$$

$$\frac{\partial M_\omega}{\partial x} = (GJ - C_p N)\theta - T \tag{7.62d}$$

where N is a compressive axial force, k_t is a torsional elastic foundation modulus, ρ is the mass per unit length, r_p is the polar radius of gyration, and C_p is a factor that depends on whether or not the shear center and centroid coincide (Pilkey, 1994).

The beam equations of Chapter 2 or Chapter 6 can be expanded to include torsion, either Saint-Venant or restrained warping. In the latter case, if the restrained warping torsion of Eq. (7.60) is to be introduced into the bending with shear deformation equations of Eq. (6.157), the static vector would be

$$\mathbf{z} = \begin{bmatrix} v & w & \theta_x & \theta & \theta_y & \theta_z & T & M_y & M_z & M_\omega & V_y & V_z \end{bmatrix}^{\mathrm{T}} \tag{7.63}$$

and the $\mathbf{A}$ and $\overline{\mathbf{P}}$ matrices of $d\mathbf{z}/dx = \mathbf{Az} + \overline{\mathbf{P}}$ are

$$\mathbf{A} = \begin{bmatrix}
0 & 0 & 0 & 0 & 0 & 1 & 0 & 0 & 0 & 0 & a_{yy} & a_{yz} \\
0 & 0 & 0 & 0 & -1 & 0 & 0 & 0 & 0 & 0 & a_{yz} & a_{zz} \\
0 & 0 & 0 & 1 & 0 & 0 & 0 & 0 & 0 & 0 & 0 & 0 \\
0 & 0 & 0 & 0 & 0 & 0 & 0 & 0 & 0 & \dfrac{1}{E\Gamma} & 0 & 0 \\
0 & 0 & 0 & 0 & 0 & 0 & 0 & b_{zz} & b_{yz} & 0 & 0 & 0 \\
0 & 0 & 0 & 0 & 0 & 0 & 0 & b_{yz} & b_{yy} & 0 & 0 & 0 \\
0 & 0 & 0 & 0 & 0 & 0 & 0 & 0 & 0 & 0 & 0 & 0 \\
0 & 0 & 0 & 0 & 0 & 0 & 0 & 0 & 0 & 0 & 0 & 1 \\
0 & 0 & 0 & 0 & 0 & 0 & 0 & 0 & 0 & 0 & -1 & 0 \\
0 & 0 & 0 & GJ & 0 & 0 & -1 & 0 & 0 & 0 & 0 & 0 \\
0 & 0 & 0 & 0 & 0 & 0 & 0 & 0 & 0 & 0 & 0 & 0 \\
0 & 0 & 0 & 0 & 0 & 0 & 0 & 0 & 0 & 0 & 0 & 0
\end{bmatrix} \tag{7.64a}$$

$$\overline{\mathbf{P}} = \begin{bmatrix}
0 \\ 0 \\ 0 \\ 0 \\ b_{yz} M_{Tz} + b_{zz} M_{Ty} \\ b_{yy} M_{Tz} + b_{yz} M_{Ty} \\ -\overline{m}_x \\ 0 \\ 0 \\ 0 \\ -\bar{p}_y(x) \\ -\bar{p}_z(x)
\end{bmatrix} \tag{7.64b}$$

where

$$a_{ij} = \frac{\alpha_{ij}}{GA} \qquad b_{ij} = \frac{I_{ij}}{E(I_{yy} I_{zz} - I_{yz}^2)} \qquad i, j = y, z$$

After these equations are solved for **z**, the warping torque T_ω is obtained from

$$T_\omega = -\frac{dM_\omega}{dx} \tag{7.65}$$

7.4 WARPING CONSTANT

The warping constant Γ is essential in calculations for restrained warping analyses. Formulas for Γ are derived in this section in terms of integrals involving the unrestrained warping function ω, because ω is the function calculated by solving Saint-Venant's boundary value problem for pure torsion. Substitute ω^* of Eq. (7.34) into the warping constant definition of Eq. (7.43):

$$\Gamma = \int (\omega^*)^2 \, dA = \int \left(\omega - z_S\, y + y_S\, z - \frac{Q_\omega}{A} \right)^2 dA \tag{7.66}$$

or

$$\Gamma = I_\omega - \frac{Q_\omega^2}{A} - 2z_S I_{y\omega} + 2y_S I_{z\omega} + z_S^2 I_z - 2y_S z_S I_{yz} + y_S^2 I_y \tag{7.67}$$

where

$$I_\omega = \int \omega^2 \, dA \qquad I_{y\omega} = \int y\omega \, dA \qquad I_{z\omega} = \int z\omega \, dA \tag{7.68}$$

are warping-dependent cross-sectional properties calculated in the centroidal coordinate system. The warping constant Γ is defined above with respect to the shear center. The quantities y_S and z_S are the coordinates of the shear center measured from the centroid. The first moment of warping Q_ω is computed using $Q_\omega = \int \omega(y, z) \, dA$, where y and z are measured from the centroid.

Since, from Eq. (7.21),

$$\begin{aligned} \int y\omega^* \, dA &= I_{y\omega} - z_S I_z + y_S I_{yz} = 0 \\ \int z\omega^* \, dA &= I_{z\omega} - z_S I_{yz} + y_S I_y = 0 \end{aligned} \tag{7.69}$$

the formula for Γ of Eq. (7.67) simplifies to

$$\Gamma = I_\omega - \frac{Q_\omega^2}{A} - z_S I_{y\omega} + y_S I_{z\omega} \tag{7.70}$$

which may also be written as

$$\Gamma = I_\omega - \frac{Q_\omega^2}{A} - \frac{I_z I_{z\omega}^2 + I_y I_{y\omega}^2 - 2 I_{yz} I_{y\omega} I_{z\omega}}{I_y I_z - I_{yz}^2} \tag{7.71}$$

or as

$$\Gamma = I_\omega - \frac{Q_\omega^2}{A} - y_S^2 I_y + 2y_S z_S I_{yz} - z_S^2 I_z \tag{7.72}$$

7.5 NORMAL STRESS DUE TO RESTRAINED WARPING

The normal stress due to restrained warping is (Eqs. 7.15 and 7.53)

$$\sigma_x = E\omega^* \frac{d^2\phi}{dx^2} = \frac{M_\omega \omega^*}{\Gamma} = \sigma_\omega \tag{7.73}$$

where Γ is the warping constant with respect to the shear center. In terms of the Saint-Venant warping function ω, the formula for the warping normal stress becomes

$$\sigma_\omega = \frac{M_\omega}{\Gamma}\left(\omega - z_S\, y + y_S\, z - \frac{Q_\omega}{A}\right) \tag{7.74}$$

where the coordinates y_S and z_S of the shear center S of the cross section are measured from the centroid, as are the coordinates y and z of the point where the stress is being determined.

For a nonhomogeneous cross section, if the area element of Eq. (7.51) is replaced with the modulus-weighted differential area $d\widetilde{A}$, and modulus-weighted section properties are used in the stress formula, the normal stress may be written as

$$\sigma_\omega = \frac{E}{E_r}\frac{M_\omega}{\widetilde{\Gamma}}\left(\omega - z_S\, y + y_S\, z - \frac{\widetilde{Q}_\omega}{\widetilde{A}}\right) \tag{7.75}$$

As in the bending normal stress formula of Eq. (1.71) for a nonhomogeneous beam, the reference elastic modulus E_r is chosen arbitrarily, and the value of the elastic modulus E is determined by the material of the element at a node of which σ_ω is being calculated.

The normal stress of Eq. (1.57) covers bending in two planes. A tensile axial force N adds $\sigma_x = N/A$ to Eq. (1.57). If the normal stress due to warping of Eq. (7.73) is included, Eq. (1.57) becomes

$$\sigma_x = \frac{N}{A} - \frac{I_{\overline{y}\,\overline{z}} M_y + I_{\overline{y}} M_z}{I_{\overline{y}} I_{\overline{z}} - I_{\overline{y}\,\overline{z}}^2}\overline{y} + \frac{I_{\overline{z}} M_y + I_{\overline{y}\,\overline{z}} M_z}{I_{\overline{y}} I_{\overline{z}} - I_{\overline{y}\,\overline{z}}^2}\overline{z} + \frac{M_\omega \omega^*}{\Gamma} \tag{7.76}$$

Thermal effects are readily incorporated in Eq. (7.76). If the material law $\sigma_x = E\epsilon_x$ of Eq. (1.49) is extended to cover a change of temperature ΔT, then

$$\sigma_x = E(\epsilon_x - \alpha\, \Delta T) \tag{7.77}$$

where α is the coefficient of linear thermal expansion. Define

$$\begin{aligned} \widehat{N} &= N + N_T & N_T &= \int_A E\alpha\, \Delta T\, dA \\ \widehat{M}_y &= M_y + M_{Ty} & M_{Ty} &= \int_A E\alpha\, \Delta T z\, dA \\ \widehat{M}_z &= M_z + M_{Tz} & M_{Tz} &= -\int_A E\alpha\, \Delta T y\, dA \\ \widehat{M}_\omega &= M_\omega + M_{T\omega} & M_{T\omega} &= \int_A E\alpha\, \Delta T \omega\, dA \end{aligned} \tag{7.78}$$

Then the normal stress of Eq. (7.36) is found to be

$$\sigma_x = \frac{\widehat{N}}{A} - \frac{I_{\bar{y}\,\bar{z}}\widehat{M}_y + I_{\bar{y}}\widehat{M}_z}{I_{\bar{y}}I_{\bar{z}} - I_{\bar{y}\bar{z}}^2}\bar{y} + \frac{I_{\bar{z}}\widehat{M}_y + I_{\bar{y}\,\bar{z}}\widehat{M}_z}{I_{\bar{y}}I_{\bar{z}} - I_{\bar{y}\,\bar{z}}^2}\bar{z} + \frac{\widehat{M}_\omega \omega^*}{\Gamma} - E\alpha\, \Delta T \tag{7.79}$$

7.6 SHEAR STRESS IN OPEN-SECTION BEAMS DUE TO RESTRAINED WARPING

Consider the forces on a small wall element of a thin-walled beam, as shown in Fig. 7.10a. The shear stress is determined from the x component of the force equilibrium equation for the wall element. If it is assumed that there is no longitudinal axial force on the surface

$$\frac{\partial q}{\partial s} + t(s)\frac{\partial \sigma_x}{\partial x} = 0 \tag{7.80}$$

where t is the thickness of the wall, and the shear flow q is defined by

$$q(x, s) = \tau_{xs}(x, s)t(s) \tag{7.81}$$

The shear flow is found by integrating the equilibrium equation with respect to s:

$$q(x, s) = q_0(x) - \int_0^s \frac{\partial \sigma_x}{\partial x} t(s)\, ds \tag{7.82}$$

where $q_0(x)$ is the shear flow at $s = 0$. If a free edge of a cross section is chosen as the origin of the coordinate s, $q_0(x)$ is zero. Substitution of σ_x of Eq. (7.76) with $N = 0$ gives the shear flow:

$$q(x, s) = \frac{I_{\bar{y}\,\bar{z}}\frac{\partial M_y}{\partial x} I_{\bar{y}}\frac{\partial M_z}{\partial x}}{I_{\bar{y}}I_{\bar{z}} - I_{\bar{y}\bar{z}}^2}Q_z'(s) - \frac{I_{\bar{z}}\frac{\partial M_y}{\partial x} + I_{\bar{y}\bar{z}}\frac{\partial M_z}{\partial x}}{I_{\bar{y}}I_{\bar{z}} - I_{\bar{y}\,\bar{z}}^2}Q_y'(s) - \frac{\frac{\partial M_\omega}{\partial x}Q_\omega'(s)}{\Gamma} \tag{7.83}$$

The quantities Q_y' and Q_z', are defined in Eq. (6.4), and Q_ω' is defined similarly:

$$Q_y'(s) = \int_{A'} \bar{z}\, dA \qquad Q_z'(s) = \int_{A'} \bar{y}\, dA \qquad Q_\omega'(s) = \int_{A'} \omega^*\, dA \tag{7.84}$$

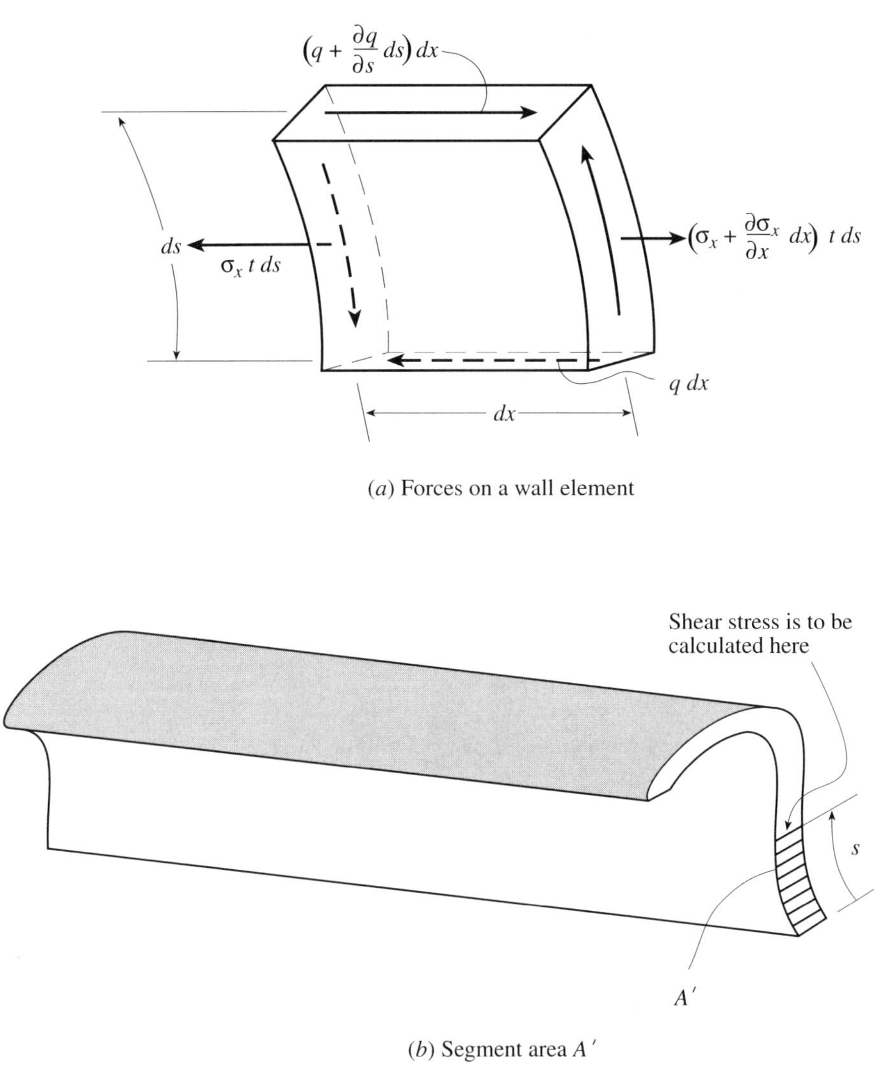

Figure 7.10 Shear flow and shear stress calculations.

with $A' = \int_{A'} dA$, where A' is the area of the segment of the cross section between the free edge and the coordinate s (Fig.7.10b). From Eqs. (6.157) and (7.65),

$$\frac{\partial M_z}{\partial x} = -V_y \qquad \frac{\partial M_y}{\partial x} = V_z \qquad \frac{\partial M_\omega}{\partial x} = -T_\omega \tag{7.85}$$

These expressions permit Eq. (7.83) to be written as

$$q(x, s) = \frac{I_{\overline{y}\,\overline{z}} V_z - I_{\overline{y}} V_y}{I_{\overline{y}} I_{\overline{z}} - I_{\overline{y}\,\overline{z}}^2} Q'_z(s) - \frac{I_{\overline{z}} V_z - I_{\overline{y}\,\overline{z}} V_y}{I_{\overline{y}} I_{\overline{z}} - I_{\overline{y}\,\overline{z}}^2} Q'_y(s) + \frac{T_\omega Q'_\omega(s)}{\Gamma} \tag{7.86}$$

This can be reorganized in the form

$$q(x,s) = -\frac{Q'_z I_{\overline{y}} - Q'_y I_{\overline{y}\,\overline{z}}}{I_{\overline{y}} I_{\overline{z}} - I^2_{\overline{y}\,\overline{z}}} V_y - \frac{Q'_y I_{\overline{z}} - Q'_z I_{\overline{y}\,\overline{z}}}{I_{\overline{y}} I_{\overline{z}} - I^2_{\overline{y}\,\overline{z}}} V_z + \frac{T_\omega Q'_\omega}{\Gamma} \tag{7.87}$$

Note that the resemblance of the shear flow of Eq. (7.87) for thin-walled cross sections of width (thickness) t to the shear stress of Eq. (6.6) for cross sections of any width b. Whereas the finite element solution provides the distribution of shear stress along the width (b), Eq. (6.6) can be used to calculate an average shear stress on a cut of width b. With width b replacing width t, Eq. (7.87) can be used to find an average shear stress τ for a "thick" section on a cut of width b. Then

$$\tau = -\frac{Q'_z I_{\overline{y}} - Q'_y I_{\overline{y}\,\overline{z}}}{b(I_{\overline{y}} I_{\overline{z}} - I^2_{\overline{y}\,\overline{z}})} V_y - \frac{Q'_y I_{\overline{z}} - Q'_z I_{\overline{y}\,\overline{z}}}{b(I_{\overline{y}} I_{\overline{z}} - I^2_{\overline{y}\overline{z}})} V_z + \frac{T_\omega Q'_\omega}{b\Gamma} \tag{7.88}$$

Equation (7.87) can be extended to include the axial force and the effect of a temperature change ΔT.

$$\begin{aligned} q(x,s) = {} & \frac{A'}{A}\frac{\partial \widehat{N}}{\partial x} - \frac{Q'_z I_{\overline{y}} - Q'_y I_{\overline{y}\overline{z}}}{I_{\overline{y}} I_{\overline{z}} - I^2_{\overline{y}\,\overline{z}}} \widehat{V}_y - \frac{Q'_y I_{\overline{z}} - Q'_z I_{\overline{y}\,\overline{z}}}{I_{\overline{y}} I_{\overline{z}} - I^2_{\overline{y}\overline{z}}} \widehat{V}_z \\ & + \frac{\widehat{T}_\omega Q'_\omega}{\Gamma} - \int_0^s E \frac{\partial}{\partial x}(\alpha \Delta T) t \, ds \end{aligned} \tag{7.89}$$

where

$$\widehat{V}_y = -\frac{\partial \widehat{M}_z}{\partial x} \qquad \widehat{V}_z = \frac{\partial \widehat{M}_y}{\partial x} \qquad \widehat{T}_\omega = -\frac{\partial \widehat{M}_\omega}{\partial x}$$

The total shear stress is obtained by adding the pure torsion (Saint-Venant) stresses of Chapter 5, which vary linearly through the thickness, to the expressions above.

Example 7.2 Stresses in a Cantilever Beam with I-Beam Cross Section. Stresses in a cantilever beam of length L, with its fixed end at $x = 0$ and its free end at $x = L$, will be analyzed. Place the y, z coordinates at the centroid. The cross section of the beam, shown in Fig. 7.11, is symmetric with respect to the y axis and is of constant thickness t. The load is a single vertical force of magnitude $\overline{P}$ applied at the free end of the beam. The point of application of $\overline{P}$ on the cross section is the lower end of the left flange.

The centroid is located by the dimension a from $\overline{P}$

$$a = \frac{h(2b_2 + h)}{2(h + b_1 + b_2)} \tag{1}$$

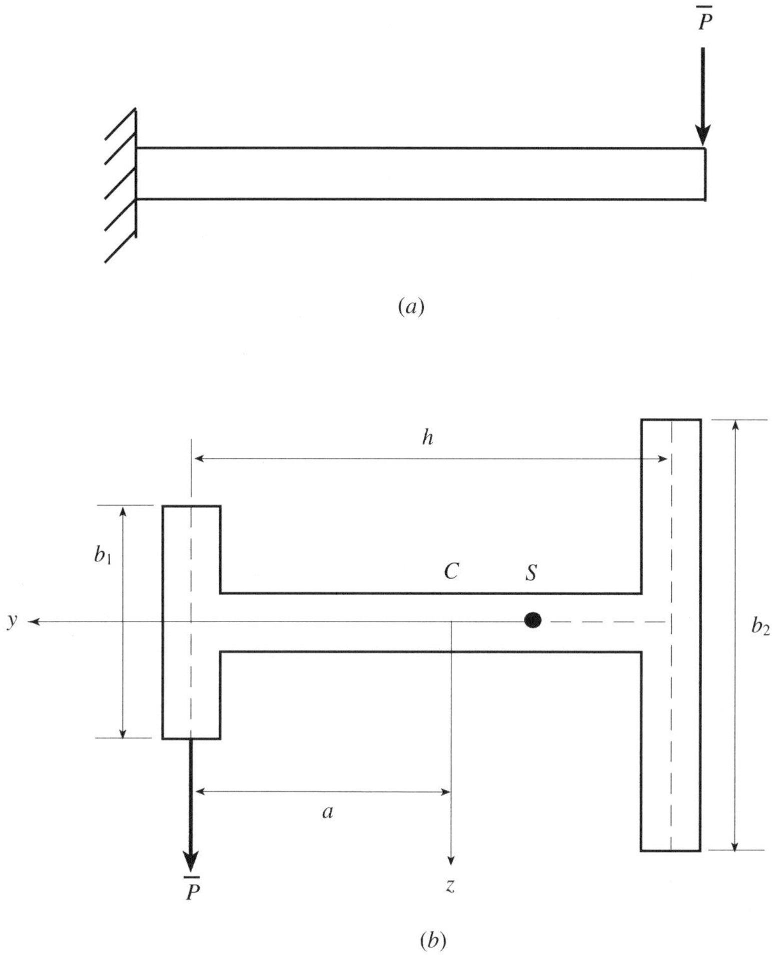

Figure 7.11 Cantilever beam with I-beam cross section.

The area moments of inertia are

$$
\begin{aligned}
I_y &\approx \frac{t}{12}\left(b_1^3 + b_2^3\right) \\
I_z &\approx tb_1a^2 + tb_2(h-a)^2 + \frac{t}{3}\left[a^3 + (h-a)^3\right] \qquad (2) \\
I_{yz} &= 0
\end{aligned}
$$

The warping function is calculated relative to the centroid C using Eq. (7.32) or, equivalently, Eq. (7.33).

$$
\begin{aligned}
\omega(s_1) &= -(h-a)s_1 \\
\omega(s_2) &= (h-a)s_2 \\
\omega(s_3) &= 0 \\
\omega(s_4) &= 0 \\
\omega(s_5) &= as_5 \\
\omega(s_6) &= -as_6
\end{aligned}
\tag{3}
$$

Note that in all cases ω_0 of Eq. (7.33) is zero. The quantity ω_0 is the value of $\omega(s)$ at the origin of integration for a segment of the thin-walled cross section. For the cross section of this example, the pole is at the centroid. The warping functions $\omega(s_3)$ and $\omega(s_4)$ are zero since r_n, the perpendicular distance from a tangent of a point on the median line curve to point C, is zero for both cases. The initial warping function ω_0 is zero at the origin (C) for $\omega(s_3)$ and $\omega(s_4)$. The initial warping functions ω_0 for the integration along s_1 or s_2 are both $\omega_0 = \omega(s_3)_{s_3=h-a} = 0$. Similarly, for the integration along s_5 and s_6, $\omega_0 = \omega(s_4)_{s_4=a} = 0$. The determination of the warping functions for poles other than the centroid are discussed in the treatment of the analysis of thin-walled beam theory on the author's website.

The shear center coordinates will be $z_S = 0$ and, since $I_{yz} = 0$, from Eq. (6.118)

$$
y_S = -\frac{I_{z\omega}}{I_y} \tag{4}
$$

The sectorial product of area $I_{z\omega}$ is calculated as Eq. (6.117)

$$
\begin{aligned}
I_{z\omega} &= \int z(s)\omega(s)\,dA = \int_0^{-b_2/2} s_1(-(h-a)s_1)t\,ds_1 + \int_0^{-b_2/2} s_2(h-a)s_2 t\,ds_2 \\
&\quad + \int_0^{-b_1/2} s_5(as_5)t\,ds_5 + \int_0^{b_1/2} s_6(-as_6)t\,ds_6 = \frac{t}{12}\left[(h-a)b_2^3 - ab_1^3\right]
\end{aligned}
\tag{5}
$$

From (4)

$$
y_S = -\frac{(h-a)b_2^3 - ab_1^3}{(b_1^3 + b_2^3)} = -\frac{h^2(b_2^3 - b_1^3) + 2hb_1b_2(b_2^2 - b_1^2)}{2(b_1 + b_2 + h)(b_1^3 + b_2^3)} \tag{6}
$$

which shows that S is to the right of the centroid for $b_2 > b_1$.

The warping constant Γ is found from Eq. (7.72):

$$
\begin{aligned}
\Gamma &= I_\omega - \frac{Q_\omega^2}{A} - y_S^2 I_y + 2y_S z_S I_{yz} - z_S^2 I_z \\
&= I_\omega - y_S^2 I_y
\end{aligned}
\tag{7}
$$

where it has been recognized that $I_{yz} = 0$ and $z_S = 0$. Also, Q_ω is zero since

$$Q_\omega = \int \omega t\, ds = \int_0^{-b_2/2} -(h-a)s_1 t\, ds_1 + \int_0^{b_2/2} (h-a)s_2 t\, ds_2 \\ + \int_0^{-b_1/2} a s_5 t\, ds_5 + \int_0^{b_1/2} -a s_6 t\, ds_6 = 0 \tag{8}$$

Then

$$\Gamma = \int \omega^2 dA - y_S^2 I_y = \int_0^{-b_2/2} (h-a)^2 s_1^2 t\, ds_1 \\ + \int_0^{b_2/2} (h-a)^2 s_2^2 t\, ds_2 + \int_0^{-b_1/2} a^2 s_5^2 t\, ds_5 \\ + \int_0^{b_1/2} a^2 s_6^2 t\, ds_6 - y_S^2 I_y = \frac{th^2}{12} \frac{b_1^3 b_2^3}{b_1^3 + b_2^3} \tag{9}$$

The restrained principal warping function is found by transforming ω according to Eq. (7.34)

$$\omega^*(s) = \omega(s) - z_S y + y_S z - \frac{Q_\omega}{A} = \omega(s) + y_S z \tag{10}$$

In terms of the branch coordinates $s_k, k = 1, 2, 3, 4, 5, 6$, defined in Fig. 7.12a, the restrained warping function is

$$\begin{aligned} \omega^*(s_1) &= -(h-a)s_1 - y_S s_1 = [-(h-a) - y_S]\, s_1 = -\frac{hb_1^3}{b_1^3 + b_2^3} s_1 \\ \omega^*(s_2) &= (h-a)s_2 - y_S s_2 = [(h-a) + y_S]\, s_2 = \frac{hb_1^3}{b_1^3 + b_2^3} s_2 \\ \omega^*(s_3) &= 0 \\ \omega^*(s_4) &= 0 \\ \omega^*(s_5) &= a s_5 - y_S s_5 = (a - y_S)s_5 = \frac{hb_2^3}{b_1^3 + b_2^3} s_5 \\ \omega^*(s_6) &= -a s_6 + y_S s_6 = (-a + y_S)s_6 = -\frac{hb_2^3}{b_1^3 + b_2^3} s_6 \end{aligned} \tag{11}$$

The applied force $\overline{P}$ at the free end of the beam does not pass through the shear center. The force–couple equivalent of $\overline{P}$ at the shear center S is the force $\overline{P}$ and the torsional moment T_0 of $\overline{P}$ about S

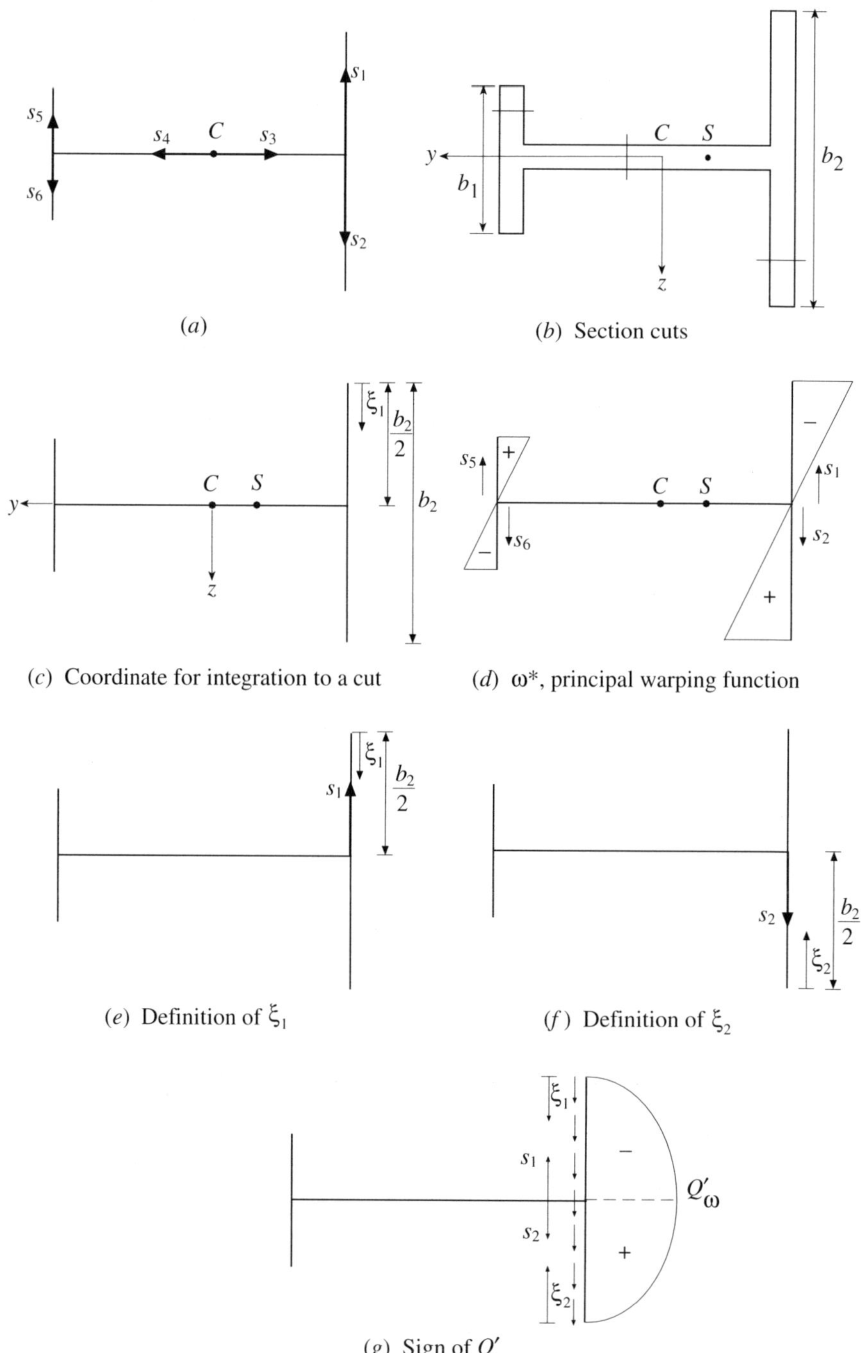

Figure 7.12 Configurations for I-beam solution.

$$T_0 = (a + \mid y_S \mid)\,\overline{P} = \frac{\overline{P} h b_2^3}{b_1^3 + b_2^3} \tag{12}$$

The angle of twist of the beam is, therefore (Eq. 4 of Example 7.1),

$$\phi(x) = \frac{T_0}{cGJ}\,[cx - \sinh cx - \tanh cL(1 - \cosh cx)] \tag{13}$$

For the torsional constant J, Saint-Venant's approximation can be used (Eq. 5.43):

$$J = \frac{t^3}{3}(h + b_1 + b_2) \tag{14}$$

The constant c depends on material constants and cross-sectional dimensions

$$c^2 = \frac{GJ}{EI_\omega} = \frac{Gt^3(h + b_1 + b_2)}{3Eh}\left(1 + \frac{b_2^3}{b_1^3}\right) \tag{15}$$

The internal forces at the clamped end are

$$V_z = \overline{P} \qquad M_y = -\overline{P}L \qquad T = T_0 \tag{16}$$

The torsional shear stress, which is proportional to $d\phi/dx$, is zero at the clamped end. The shear stress distribution over the cross section at the fixed end $x = 0$ is given by Eq. (7.87) as

$$\tau_{xs}(s) = -\frac{\overline{P}Q'_y(s)}{tI_y} + \frac{T_\omega Q'_\omega(s)}{t\Gamma} = \tau_V + \tau_\omega \tag{17}$$

In (17), the first moment area $Q'_y(s)$ is calculated using section cuts such as those indicated in Fig. 7.12*b*. For instance, with the section cut on the right flange, the integration for Q'_y is taken from the free edge at $z = -b_2/2$ to the cut. Then

$$Q'_y = \int_{A'} z\, dA \tag{18}$$

where the overbar on z of Eq. (7.84) has been ignored. If the integration is to be performed from the top of the flange ($z = -b_2/2$) to the cut, it is convenient to define a new coordinate $\xi_1, 0 \le \xi_1 \le b_2$, as shown in Fig. 7.12*c*. Then, with $z = -(b_2/2 - \xi_1)$

$$Q'_y = \int_0^{\xi_1} zt\, d\xi_1 = \int_0^{\xi_1} t\left(\xi_1 - \frac{b_2}{2}\right) d\xi_1 = \frac{t}{2}\xi_1(\xi_1 - b_2) \tag{19}$$

Upon substitution of $\xi_1 = z + b_2/2$ into (9),

$$Q'_y(z) = \frac{t}{2}\left(z^2 - \frac{b_2^2}{4}\right) \tag{20}$$

The corresponding shear stress distribution given by (17) is

$$\tau_V(z) = \frac{\overline{P}}{2I_y}\left(\frac{b_2^2}{4} - z^2\right) \tag{21}$$

Similarly, the shear stress in the left flange is given by

$$\tau_V(z) = \frac{\overline{P}}{2I_y}\left(\frac{b_1^2}{4} - z^2\right) \tag{22}$$

the first moment Q_y is zero for the web since the flanges have equal areas above and below the y axis and the shear stress τ_V in the web is zero.

The second contribution to the shear stress in (17) is the *warping shear stress*

$$\tau_\omega(s) = \frac{T_\omega Q'_\omega(s)}{t\Gamma} \tag{23}$$

where T_ω is equal to the entire torque T_0, since the pure torsion torque T_t is zero at the clamped end (Eq. 7.47). The first sectorial area moment $Q'_\omega(s) = \int_{A'} \omega^*\, dA$ (Eq. 7.84) of the principal warping function is calculated using (11) for the part of the cross section cut off at s, remembering that the integration starts at a free edge. The principal warping function is shown in Fig. 7.12*d*. Since the integration to find Q'_ω starts at a free edge, where $\tau_\omega = 0$, it is convenient to define another variable ξ_1 (Fig. 7.12*e*), $0 \le \xi_1 \le b_2/2$. Note that $\xi_1 = b_2/2 - s_1$ and $s_1 = b_2/2 - \xi_1$. This differs somewhat from the ξ_1 defined to calculate Q'_y where $0 \le \xi_1 \le b_2$. Then the integration is performed from the free edge with $\xi_1 = 0$ to ξ_1, where $\xi_1 \le b_2/2$.

$$\begin{aligned} Q'_\omega(s_1) &= \int_0^{\xi_1} \omega^*(\xi_1) t\, d\xi_1 = -\frac{hb_1^3 t}{b_1^3 + b_2^3}\int_0^{\xi_1}\left(\frac{b_2}{2} - \xi_1\right) d\xi_1 \\ &= -\frac{b_1^3 th}{2\left(b_1^3 + b_2^3\right)}(b_2 - \xi_1)\xi_1 = -\frac{b_1^3 th}{2\left(b_1^3 + b_2^3\right)}\left(\frac{b_2^2}{4} - s_1^2\right) \end{aligned} \tag{24}$$

For the bottom flange, define ξ_2 from the free edge as shown in Fig. 7.12*f*. Then, $s_2 = b_2/2 - \xi_2$ and

$$\begin{aligned} Q'_\omega(s_2) &= \int_0^{\xi_2} \omega^*(\xi_2) t\, d\xi_2 = \frac{b_1^3 th}{2\left(b_1^3 + b_2^3\right)}(b_2 - \xi_2)\,\xi_2 \\ &= \frac{b_1^3 th}{2\left(b_1^3 + b_2^3\right)}\left(\frac{b_2^2}{4} - s_2^2\right) \end{aligned} \tag{25}$$

Since the warping shear stress is given by $\tau_\omega(s) = T_\omega Q'_\omega(s)/(t\Gamma)$ (Eq. 23), the sign of the stress depends on the sign of $Q'_\omega(s)$ (Fig. 7.12*a*). For the right-hand flange

s	Sign of $Q'_\omega(s)$	Sign of τ_ω for $T_\omega > 0$
s_1	$-$	Same as direction of s_1, up
s_2	$+$	Opposite direction of s_2, up

The other first sectorial area moments are calculated in a similar fashion. In summary,

$$\begin{aligned}
Q'_\omega(s_1) &= -\frac{b_1^3 th}{2\left(b_1^3+b_2^3\right)}\left(\frac{b_2^2}{4}-s_1^2\right)\\
Q'_\omega(s_2) &= \frac{b_1^3 th}{2\left(b_1^3+b_2^3\right)}\left(\frac{b_2^2}{4}-s_1^2\right)\\
Q'_\omega(s_3) &= 0\\
Q'_\omega(s_4) &= 0\\
Q'_\omega(s_5) &= -\frac{b_2^3 th}{2\left(b_1^3+b_2^3\right)}\left(\frac{b_1^2}{4}-s_5^2\right)\\
Q'_\omega(s_6) &= \frac{b_2^3 th}{2\left(b_1^3+b_2^3\right)}\left(\frac{b_2^2}{4}-s_6^2\right)
\end{aligned} \tag{26}$$

The shear stresses are expressed by (23) as

$$\begin{aligned}
\tau_\omega(s_1) &= \frac{6T_0}{thb_2^3}\left(\frac{b_2^2}{4}-s_1^2\right)\\
\tau_\omega(s_2) &= -\frac{6T_0}{thb_2^3}\left(\frac{b_2^2}{4}-s_2^2\right)\\
\tau_\omega(s_3) &= 0\\
\tau_\omega(s_4) &= 0\\
\tau_\omega(s_5) &= -\frac{6T_0}{thb_1^3}\left(\frac{b_1^2}{4}-s_5^2\right)\\
\tau_\omega(s_6) &= \frac{6T_0}{thb_1^3}\left(\frac{b_1^2}{4}-s_6^2\right)
\end{aligned} \tag{27}$$

The sign of $\tau_\omega(s_1)$ is positive, which means that the shear flow is in the direction of increasing s_1, hence upward. Similarly, the sign of $\tau_\omega(s_2)$ is negative, so the shear flow is in the direction of decreasing s_2, hence upward. Thus, warping shear stresses on the right flange are directed upward, but the signs of $\tau_\omega(s_5)$ and $\tau_\omega(s_6)$ show that the warping shear stresses on the left flange are directed downward.

The normal stress at the fixed end due to bending is (Eq. 1.59)

$$\sigma_x(z) = \frac{M_y z}{I_y} = -\frac{\overline{P} L z}{I_y} \tag{28}$$

The normal warping stress is (Eq. 7.73)

$$\sigma_\omega(s) = \frac{M_\omega \omega^*(s)}{\Gamma} \tag{29}$$

where M_ω is the bimoment at the fixed end

$$M_\omega = -E\Gamma \frac{d^2\phi}{dx^2}(0) = -\frac{T_0}{c}\tanh cL \tag{30}$$

with $d^2\phi/dx^2$ taken from (13).

Sometimes the material constant of (30) is made more accurate. If the kinematic assumption that the median line of the cross section is inextensible is introduced, then the strain ϵ_s is zero, $\epsilon_s = (\sigma_s - \nu\sigma_x)/E = 0$ or $\sigma_s = \nu\sigma_x$. The longitudinal strain is then $\epsilon_x = (\sigma_x - \nu\sigma_s)/E = (1-\nu^2)\sigma_x/E$. It follows that an improved modulus of elasticity can be defined as

$$\overline{E} = \frac{E}{1-\nu^2} \tag{31}$$

Then (30) becomes

$$M_\omega = -\overline{E}\Gamma \frac{d^2\phi}{dx^2}(0) = -\frac{T_0}{c}\tanh cL \tag{32}$$

where $c^2 = GJ/\overline{E}\Gamma$. This modified constant, $\overline{E}$, can be used to replace E throughout this chapter.

To calculate the stresses numerically, the following dimensions will be assumed

$$b_1 = b \qquad h = b_2 = 2b \qquad L = 20b \qquad t = \frac{b}{10} \tag{33}$$

Poisson's ratio will be taken to be $\nu = 0.25$. The modulus of elasticity $\overline{E}$ and the shear modulus G are then

$$\overline{E} = \frac{E}{1-\nu^2} = \frac{16E}{15} \qquad G = \frac{E}{2(1+\nu)} = \frac{2E}{5} \tag{34}$$

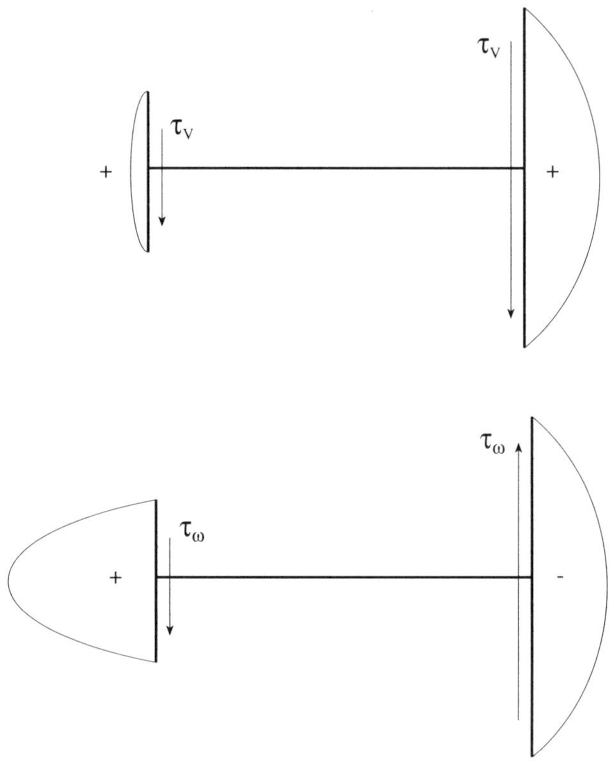

Figure 7.13 Transverse and warping shear stresses at the fixed end.

The transverse and warping shear stress distributions at the clamped end of the beam are sketched in Fig. 7.13. The force–couple equivalent of the transverse shear stress τ_V at the shear center S is a single force of magnitude $V_z = \overline{P}$. The warping shear stress τ_ω is statically equivalent to a couple. The total shear stress on the right flange is zero, so that, at the fixed end of the beam, all shear stresses are carried by the left flange.

The bending and warping normal stresses are shown in Fig. 7.14. The maximum normal warping stress exceeds the maximum bending stress. The bending stresses are statically equivalent to the bending moment $M_y = -\overline{P}L$. The warping stresses are statically equivalent to zero force and zero couple. When considered separately for the two flanges, these stresses are equivalent to two equal and opposite bending moments. The maximum stresses are shown in Table 7.1. The reference stress σ_0 is defined as

$$\sigma_0 = \frac{\overline{P}}{b^2} \tag{35}$$

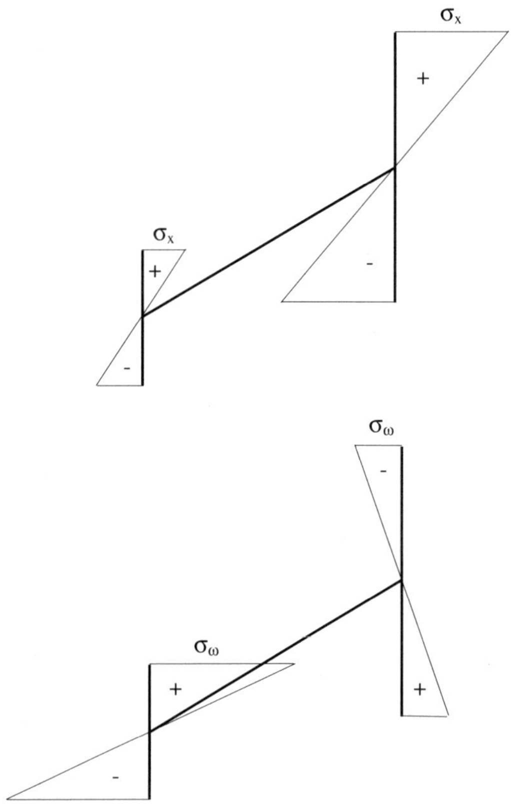

Figure 7.14 Bending and warping normal stresses at the fixed end.

TABLE 7.1 Maximum Stresses at the Clamped End

Maximum Stress	Right Flange	Left Flange
τ_V/σ_0 (Transverse)	6.7	1.7
τ_ω/σ_0 (Warping)	6.7	13.3
σ_x/σ_0 (Bending)	266.7	133.3
σ_ω/σ_0 (Warping)	91.3	365.0

Example 7.3 Stresses in a Simply Supported Beam. The stress distribution in the simply supported beam shown in Fig. 7.15*a* will be determined. The load is a vertical force of magnitude $\overline{P}$ at midspan. For the cross section of Fig. 7.15*b* it can be shown that the shear center coincides with the centroid. This can be accomplished analytically using Eq. (6.118) or reasoned physically (Gere and Timoshenko, 1997; Kollbrunner and Hajdin, 1972). The y, z axes are centroidal axes. The wall thickness t is the same for the flanges and the web.

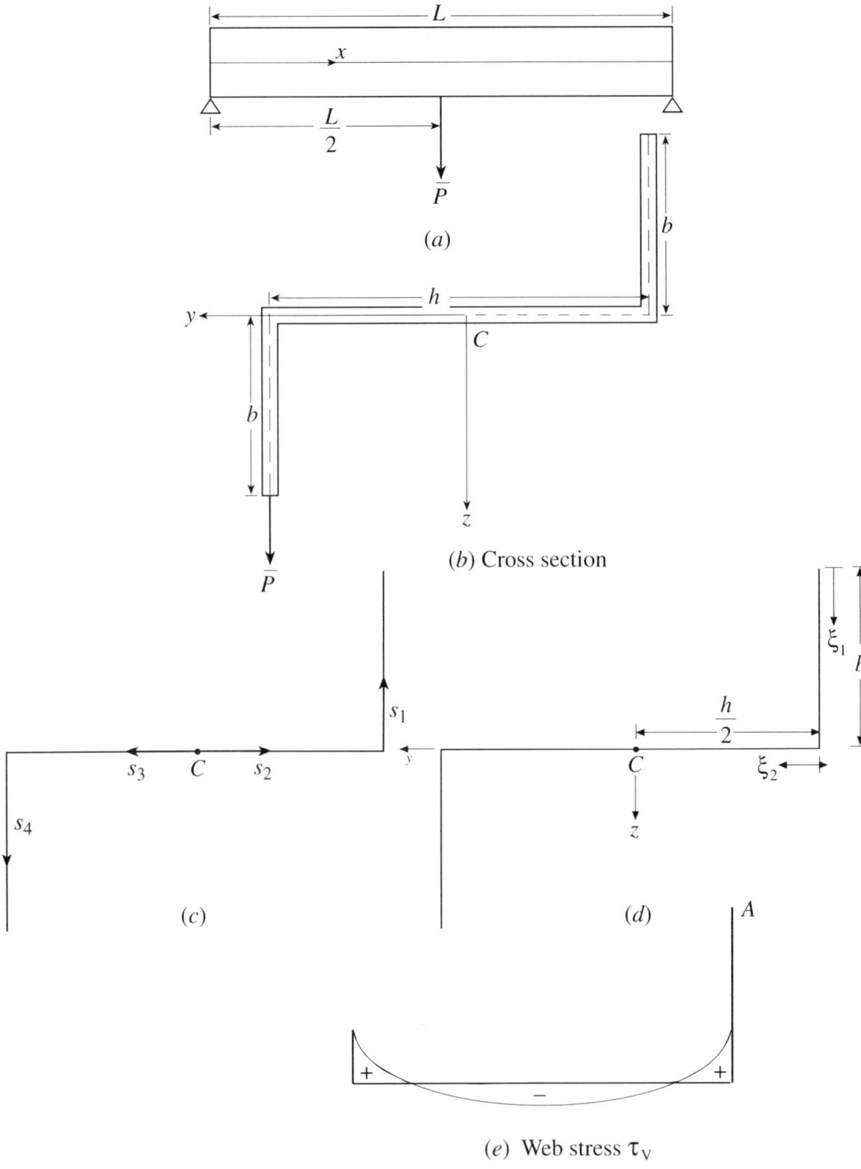

Figure 7.15 Simply supported beam with midspan load.

The area moments of inertia and the area product of inertia for this cross section are

$$I_y \approx \frac{2tb^3}{3} \qquad I_z \approx \frac{th^3}{12} + \frac{tbh^2}{2} \qquad I_{yz} \approx \frac{thb^2}{2} \tag{1}$$

The torsional constant, calculated from Saint-Venant's approximation, is (Eq. 5.43)

$$J = \frac{ht^3}{3} + \frac{2bt^3}{3} = \frac{t^3(h+2b)}{3} \tag{2}$$

The warping function with the pole and origin both at the centroid C is

$$\omega(s_1) = -\frac{h}{2}s_1 \qquad \omega(s_2) = 0 \qquad \omega(s_3) = 0 \qquad \omega(s_4) = -\frac{h}{2}s_4 \tag{3}$$

The principal warping function is given by Eq. (7.34) as

$$\omega^*(s) = \omega(s) - \overline{z}_S\, y + y_S\, z - \frac{Q_\omega}{A} = \omega(s) - \frac{Q_\omega}{A} \tag{4}$$

since $y_S = 0$ and $z_S = 0$. The first moment of warping is found to be

$$Q_\omega = \int_0^b \left(-\frac{h}{2}s_1\right) t\, ds_1 + \int_0^b \left(-\frac{h}{2}s_4\right) t\, ds_4 = -\frac{thb^2}{2} \tag{5}$$

and

$$A = t(h+2b) \tag{6}$$

so that the principal warping function of (4) for the right-hand flange is

$$\omega^*(s_1) = -\frac{hs_1}{2} + \frac{hb^2}{2(h+2b)} \tag{7}$$

Also

$$\omega^*(s_2) = \frac{hb^2}{2(h+2b)}$$

$$\omega^*(s_3) = \frac{hb^2}{2(h+2b)}$$

$$\omega^*(s_4) = \omega^*(s_1)$$

The warping constant is found from Eq. (7.70):

$$\Gamma = I_\omega - \frac{Q_\omega^2}{A} - z_S I_{y\omega} + y_S I_{z\omega} = \frac{th^2b^3(b+2h)}{12(h+2b)} \tag{8}$$

The applied load at $x = L/2$ is equivalent to a torsional couple T_0 and a transverse force $\overline{P}$ at the shear center:

$$T_0 = \frac{\overline{P}h}{2} \tag{9}$$

The applied torque per unit length can be written in terms of the Dirac delta function

$$\overline{m}(x) = T_0\, \delta\left(x - \frac{L}{2}\right) \tag{10}$$

The angle of twist is calculated from Eq. (7.59) as

$$\phi(x) = C_1 + C_2 x + C_3 \cosh cx + C_4 \sinh cx \tag{11}$$

for $0 \leqslant x \leqslant L/2$ and

$$\begin{aligned}\phi(x) = {} & C_1 + C_2 x + C_3 \cosh cx + C_4 \sinh cx \\ & - \frac{T_0}{cGJ}\left[c\left(x - \frac{L}{2}\right) - \sinh c\left(x - \frac{L}{2}\right)\right]\end{aligned} \tag{12}$$

for $L/2 \leqslant x \leqslant L$.

The boundary conditions at the two simple supports

$$\phi(0) = \phi(L) = 0 \qquad \frac{d^2\phi}{dx^2}(0) = \frac{d^2\phi}{dx^2}(L) = 0 \tag{13}$$

are solved for the integration constants

$$C_1 = 0 \qquad C_2 = \frac{T_0}{2GJ} \qquad C_3 = 0 \qquad C_4 = -\frac{T_0 \sinh cL/2}{cGJ \sinh cL} \tag{14}$$

For the left half of the beam

$$\begin{aligned}\phi(x) &= \frac{T_0}{2cGJ}\left(cx - 2\frac{\sinh cL/2}{\sinh cL}\sinh cx\right) \\ M_\omega(x) &= \frac{T_0}{c \sinh cL} \sinh \frac{cL}{2} \sinh cx \\ T_t(x) &= \frac{T_0}{2} - \frac{T_0 \sinh cL/2}{\sinh cL}\cosh cx \\ T_\omega(x) &= T_0 \frac{\sinh cL/2}{\sinh cL}\cosh cx\end{aligned} \tag{15}$$

The qualitative behavior of these functions over the entire span of the beam can be seen in Figs. 7.16, 7.17, and 7.18. In Fig. 7.18, the torques T_t and T_ω are shown as

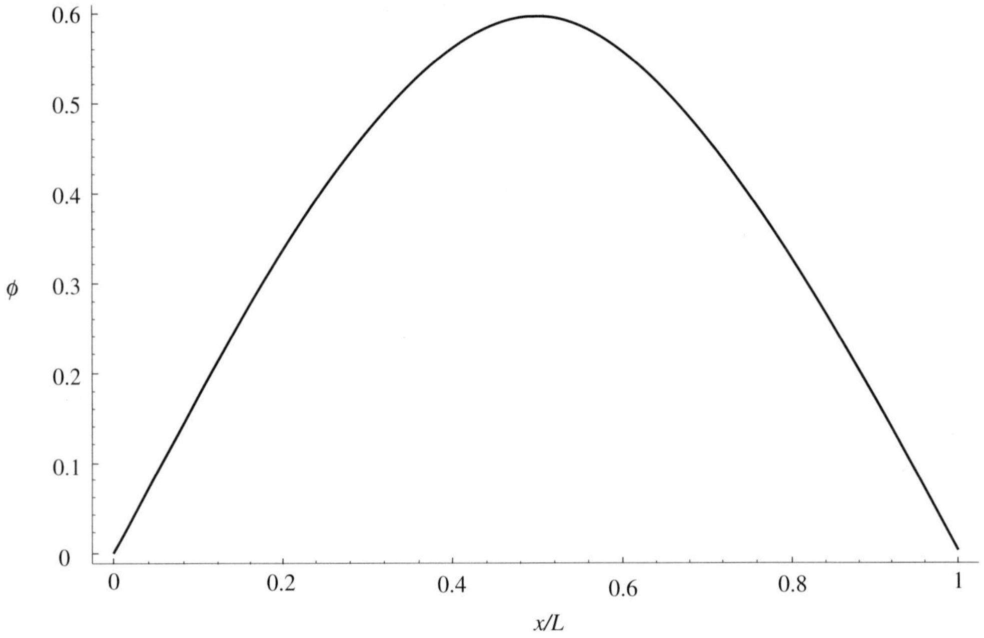

Figure 7.16 Angle of twist for the simply supported beam.

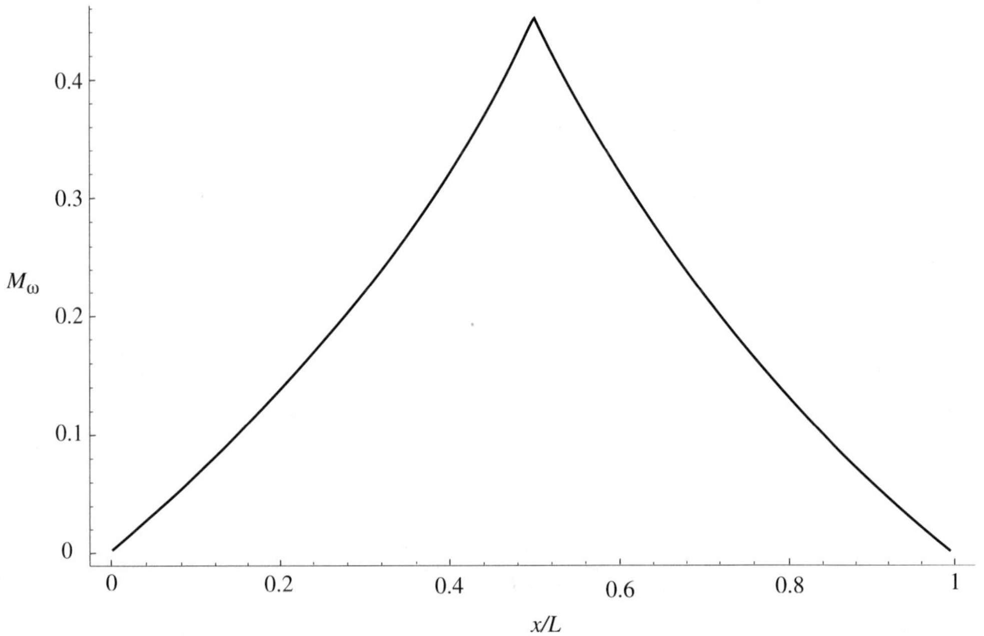

Figure 7.17 Bimoment for the simply supported beam.

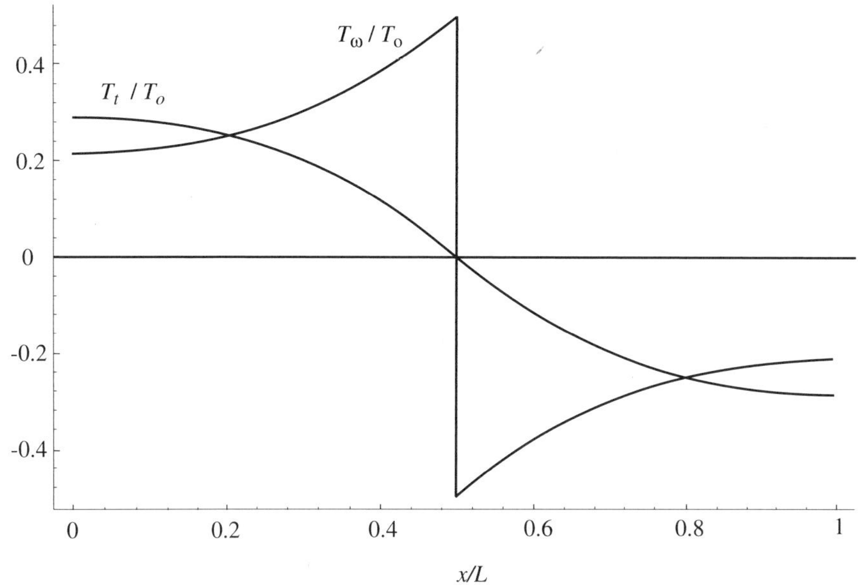

Figure 7.18 Pure torsion and warping torques for the simply supported beam.

fractions of the applied torque T_0. The total torque is the sum of T_t and T_ω

$$T(x) = \begin{cases} \dfrac{T_0}{2} & \text{if } x < \dfrac{L}{2} \\ -\dfrac{T_0}{2} & \text{if } x > \dfrac{L}{2} \end{cases} \tag{16}$$

The stresses at $x = L/2$ at the section just to the left of the applied torque will be calculated. The transverse shear stress τ_V at this section is (Eq. 7.87)

$$\tau_V(s) = -\frac{Q'_y(s)I_z - Q'_z(s)I_{yz}}{t\left(I_y I_z - I_{yz}^2\right)} V_z \tag{17}$$

where the reference y, z coordinate system remains at the centroid.

To aid in finding Q'_y and Q'_z for the right flange, define a new coordinate ξ_1 from the free edge as shown in Fig. 7.15*d*. Then, with $z(\xi_1) = -(b - \xi_1)$,

$$Q'_y(\xi_1) = \int_0^{\xi_1} z(\xi_1) t \, d\xi_1 = \int_0^{\xi_1} t(\xi_1 - b) \, d\xi_1 = t\xi_1 \left(\frac{\xi_1}{2} - b\right) \tag{18}$$

Since $s_1(\xi_1) = b - \xi_1$,

$$Q'_y = -\frac{t}{2}\left(b^2 - s_1^2\right) \tag{19}$$

Similarly,

$$Q_z' = -\frac{th}{2}(b - s_1) \tag{20}$$

The shear stress of (17) becomes

$$\tau_V(s_1) = \frac{3\overline{P}(b - s_1)(bh + hs_1 + 6bs_1)}{4tb^3(2h + 3b)} \tag{21}$$

The value of $\tau_V(s_1)$ for $s_1 < b$ is positive, which means that the stress is in the direction of increasing ξ_1, hence downward.

For the web, the first moment of area Q_y' is not zero, in spite of the thickness t being relatively small, because the integration begins at a free edge. Thus

$$Q_y'(s_2) = Q_y'(s_1 = 0) + \int_0^{s_2} zt\, ds_2 = Q_y'(s_1 = 0) = -\frac{tb^2}{2} \tag{22}$$

Also, where the coordinate ξ_2 of Fig. 7.15d is introduced with $\xi_2 = h/2 - s_2$

$$\begin{aligned} Q_z'(s_2) &= Q_z'(s_1 = 0) + \int_0^{\xi_2} \left(\xi_2 - \frac{h}{2}\right) t\, d\xi_2 = -\frac{thb}{2} + \frac{t}{2}\xi_2(\xi_2 - h) \\ &= \frac{t}{2}\left(\xi_2^2 - h\xi_2 - hb\right) = \frac{t}{2}\left(s_2^2 - hb - \frac{h^2}{4}\right) \end{aligned} \tag{23}$$

From (17)

$$\tau_V(s_2) = \frac{3\overline{P}}{4}\frac{\left(12s_2^2 - h^2\right)}{tbh(3b + 2h)} \tag{24}$$

A similar expression is obtained for the web to the left of C. This web stress is plotted in Fig. 7.15e. The integration to find τ_V for the web started at point A and ξ_2 may be considered as flowing from the free edge A so τ_V has the same direction as ξ_2 for $\tau_V > 0$ and opposes the flow direction of ξ_2 for $\tau_V < 0$.

On the left flange, it is found that

$$\tau_V(s_4) = -\frac{3\overline{P}(b - s_4)(bh + 6bs_4 + hs_4)}{4tb^3(2h + 3b)} \tag{25}$$

The warping shear stress τ_ω of Eq. (7.87) at $x = L/2$ is

$$\tau_\omega(s) = \frac{Q_\omega'(s)}{t\Gamma}T_\omega \tag{26}$$

where the warping torque T_ω is $T_0/2$, because the pure torsion torque T_t is zero at midspan. The integration to compute Q_ω' begins at a free end. With $\xi_1 = b - s_1$,

$$Q'_\omega = \int_0^{\xi_1} \left[-\frac{h}{2}(b-\xi_1) + \frac{hb^2}{2(h+2b)} \right] t \, d\xi_1 = -\frac{th(b-s_1)(hs_1+2bs_1+bh)}{4(2b+h)} \tag{27}$$

The Q'_ω for the other segments are computed in a similar fashion. The expressions for the warping shear stresses are found to be

$$\begin{aligned} \tau_\omega(s_1) &= \frac{T_0 h(b-s_1)(bh+hs_1+2bs_1)}{8\Gamma(h+2b)} \\ \tau_\omega(\xi_2) &= -\frac{T_0 hb^2(h-2\xi_2)}{8\Gamma(h+2b)} \\ \tau_\omega(s_4) &= \frac{T_0 h(b-s_4)(bh+hs_4+2bs_4)}{8\Gamma(h+2b)} \end{aligned} \tag{28}$$

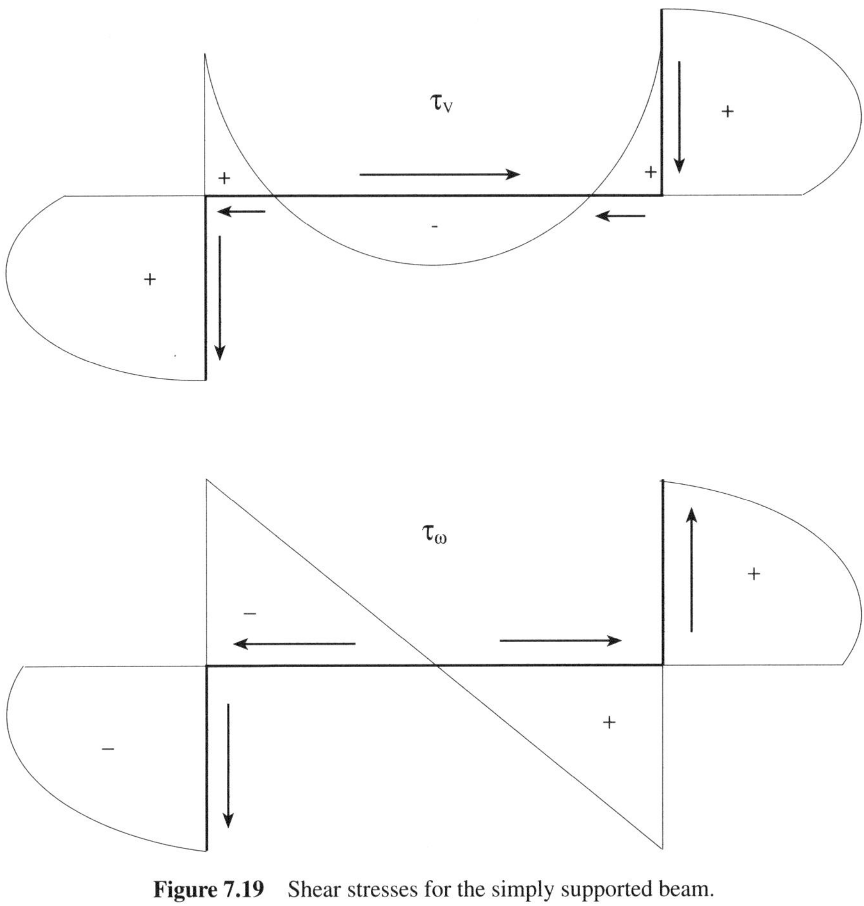

Figure 7.19 Shear stresses for the simply supported beam.

The normal stress distribution at $x = L/2$ due to bending is (Eq. 1.57)

$$\sigma_x = -\frac{I_{yz}M_y}{I_yI_z - I_{yz}^2}y + \frac{I_zM_y}{I_yI_z - I_{yz}^2}z \tag{29}$$

where $M_y = \overline{P}L/4$. The normal stress due to warping is (Eq. 7.73)

$$\sigma_\omega = \frac{M_\omega \omega^*}{\Gamma} \tag{30}$$

where Γ is given by (9) and

$$M_\omega = \frac{T_0}{2c}\tanh\frac{cL}{2} \tag{31}$$

The shear stress distribution at $x = L/2$ is sketched in Fig. 7.19. The force resultant of the transverse shear stress τ_V over the two flanges is equal to the total shear force $\overline{P}/2$. The transverse shear stresses over the web are statically equivalent to a

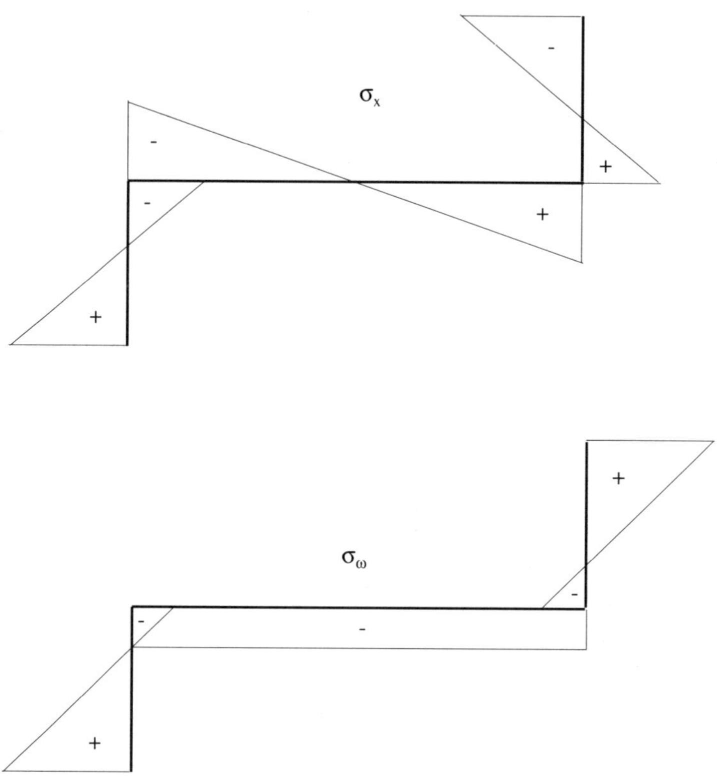

Figure 7.20 Normal stresses for the simply supported beam.

zero force couple. The warping shear stress τ_ω is equivalent to a torsional moment. The transverse shear stress adds to the warping shear stress over the left flange, but subtracts from it over the right flange.

The normal stress distribution at $x = L/2$ is sketched in Fig. 7.20. The normal stress σ_x due to bending is statically equivalent to a bending moment about the y axis. The warping stress σ_ω is statically equivalent to a zero force couple. The bending and warping stresses are additive over the left flange.

7.7 BEAMS FORMED OF MULTIPLE PARALLEL MEMBERS ATTACHED AT THE BOUNDARIES

Multimember cross-section problems are treated in this section. Of interest are beams formed of two or more members with unconnected (open) and attached (welded, closed) cross sections. Figure 7.21 shows an example with two members. The welded cross section, which occurs at particular locations along the beam, binds together the cross sections and restrains the warping of the unconnected sections. In Fig. 7.21, the welded sections are shown at $x = 0$ and $x = L$, so that the weld spacing in this beam is L. A substitute model for the beam, which is subjected to a constant torsional moment T, is to be found such that the Saint-Venant equation for the twisting of the beam can be used while the effects of restrained warping are approximately included. The formulation here leads to an approximate torsional constant J, which will be referred to as the *effective torsional constant.*

The modeling problem is solved by introducing this effective torsional constant J_{eff} as a cross-sectional property that includes both torsional and warping effects. The analysis starts with the differential equation for the angle of twist ϕ of a thin-walled beam (Eq. 7.54f):

$$GJ_0\phi' - E\Gamma_0\phi''' = T \tag{7.90}$$

where J_0 is the torsional constant and Γ_0 the warping constant for the open sections. The calculation of J_0 and Γ_0 is considered later in this section. It is assumed that the Saint-Venant equation for the rate of twist is applicable at the welded sections. This gives the boundary conditions

$$\phi'(0) = \frac{T(0)}{GJ_c} \qquad \phi'(L) = \frac{T(L)}{GJ_c} \tag{7.91}$$

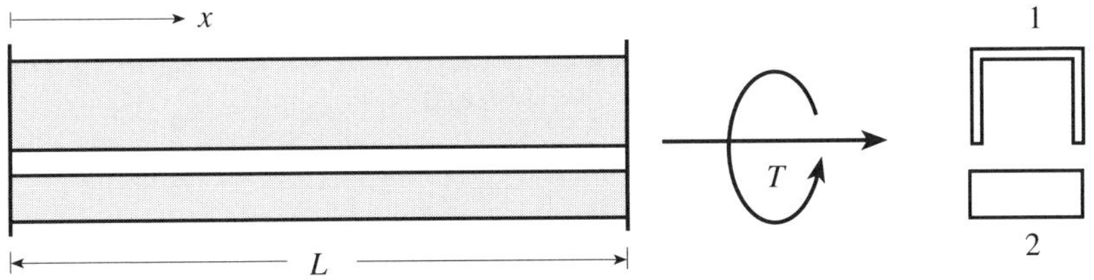

Figure 7.21 Beam with two cross-sectional members welded at $x = 0$ and $x = L$.

where J_c is the torsional constant of the closed end or welded sections. With the definition

$$\lambda = \sqrt{\frac{GJ_0}{E\Gamma_0}} \tag{7.92}$$

the governing equation for the rate of twist of Eq. (7.90) is rewritten as

$$\phi' - \frac{\phi'''}{\lambda^2} = \frac{T}{GJ_0} \tag{7.93}$$

The general solution of this equation is

$$\phi'(x) = \frac{T}{GJ_0}(C_1 \cosh \lambda x + C_2 \sinh \lambda x + 1) \tag{7.94}$$

where C_1, C_2 are constants to be determined from the boundary conditions.

The assumed boundary conditions of Eq. (7.91) give

$$\begin{aligned} \phi'(0) &= \frac{T}{GJ_0}(C_1 + 1) = \frac{T}{GJ_c} \\ \phi'(L) &= \frac{T}{GJ_0}(C_1 \cosh \lambda L + C_2 \sinh \lambda L + 1) = \frac{T}{GJ_c} \end{aligned} \tag{7.95}$$

from which the constants C_1 and C_2 are found to be

$$\begin{aligned} C_1 &= \frac{J_0}{J_c} - 1 \\ C_2 &= C_1 \frac{1 - \cosh \lambda L}{\sinh \lambda L} = -C_1 \tanh \frac{\lambda L}{2} = \left(1 - \frac{J_0}{J_c}\right) \tanh \frac{\lambda L}{2} \end{aligned} \tag{7.96}$$

If the rate of twist $\phi'(x)$ of Eq. (7.94) is integrated from $x = 0$ to $x = L$, the relative angle of twist between two welded sections is found:

$$\Delta\phi = \phi(L) - \phi(0) = \frac{TL}{GJ_0}\left[\frac{C_1}{\lambda L} \sinh \lambda L + \frac{C_2}{\lambda L}(\cosh \lambda L - 1) + 1\right] \tag{7.97}$$

Saint-Venant's formula expresses $\Delta\phi$ in terms of the effective torsional constant J_{eff}:

$$\Delta\phi = \frac{TL}{GJ_{\text{eff}}} \tag{7.98}$$

Equate the two expressions for $\Delta\phi$ of Eqs. (7.97) and (7.98):

$$\frac{J_0}{J_{\text{eff}}} = 1 + \frac{C_1}{\lambda L} \sinh \lambda L + \frac{C_2}{\lambda L}(\cosh \lambda L - 1)$$

$$
\begin{aligned}
&= 1 + \frac{C_1}{\lambda L} \tanh \frac{\lambda L}{2} (\cosh \lambda L + 1) + \frac{C_2}{\lambda L} (\cosh \lambda L - 1) \\
&= 1 + \frac{C_1}{\lambda L} \left[\tanh \frac{\lambda L}{2} (\cosh \lambda L + 1) - \tanh \frac{\lambda L}{2} (\cosh \lambda L - 1) \right] \\
&= 1 + \frac{2C_1}{\lambda L} \tanh \frac{\lambda L}{2} \\
&= 1 + \frac{2}{\lambda L} \left(\frac{J_0}{J_c} - 1 \right) \tanh \frac{\lambda L}{2}
\end{aligned}
\tag{7.99}
$$

This gives a formula for the effective torsional constant J_{eff}:

$$
J_{\text{eff}} = \frac{J_0}{1 - 2/\lambda L (1 - J_0/J_c) \tanh(\lambda L/2)} \tag{7.100}
$$

The limiting values of J_{eff} obtained from Eq. (7.100) as the weld spacing approaches zero or infinity are

$$
\lim_{\lambda L \to 0} J_{\text{eff}} = J_c \qquad \lim_{\lambda L \to \infty} J_{\text{eff}} = J_0 \tag{7.101}
$$

Figure 7.22 shows the variation of J_{eff} with λL for four values of the ratio J_c/J_0.

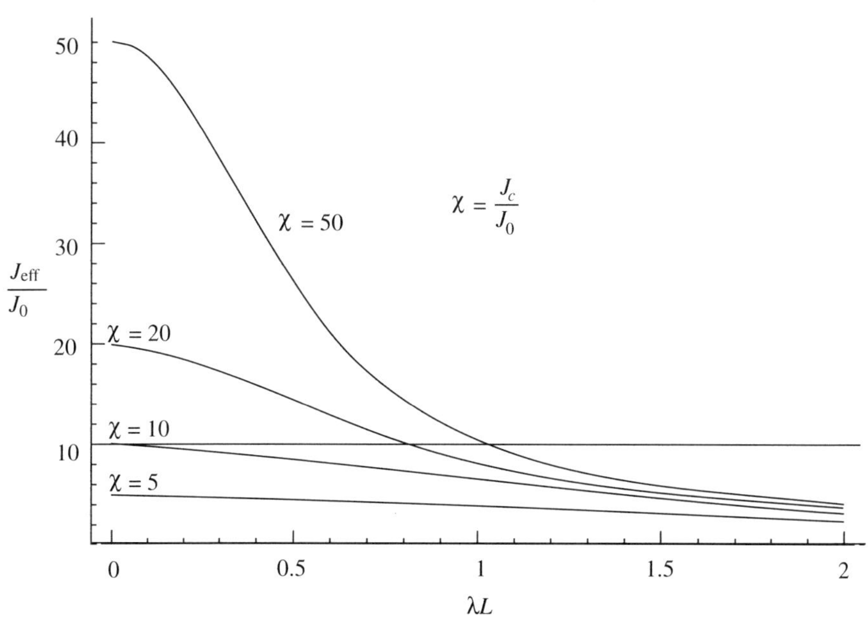

Figure 7.22 Effective torsional constant according to Eq. (7.100).

An energy method in which the strain energy of the substitute beam in Saint-Venant torsion and the strain energy of the actual multimember beam in nonuniform torsion are assumed to be equal may also be used to arrive at an effective torsional constant. The strain energy of the substitute beam in Saint-Venant torsion is

$$U_0 = \frac{GJ_0}{2} \int_0^L \phi'^2(x)\, dx = \frac{GJ_0}{2} \left(\frac{T}{GJ_{\text{eff}}} \right)^2 L = \frac{T^2 L}{2G} \frac{J_0}{J_{\text{eff}}^2} \tag{7.102}$$

For the actual beam, the rate of twist, which is calculated with the assumptions introduced in the derivation of Eq. (7.100), will be written as (Eq. 7.94)

$$\phi'(x) = \frac{T}{GJ_0} f(x) \tag{7.103}$$

where

$$f(x) = C_1 \cosh \lambda x + C_2 \sinh \lambda x + 1 \tag{7.104}$$

The second derivative of the angle of twist, which will be needed in the strain energy calculation, is

$$\phi''(x) = \frac{T}{GJ_0} \lambda g(x) \tag{7.105}$$

where

$$g(x) = \frac{f'(x)}{\lambda} = C_1 \sinh \lambda x + C_2 \cosh \lambda x \tag{7.106}$$

The strain energy of the beam in nonuniform torsion is

$$\begin{aligned} U_1 &= \frac{GJ_0}{2} \int_0^L \phi'^2(x)\, dx + \frac{E\Gamma_0}{2} \int_0^L \phi''^2(x)\, dx \\ &= \frac{GJ_0}{2} \left[\int_0^L \phi'^2(x)\, dx + \frac{1}{\lambda^2} \int_0^L \phi''^2(x)\, dx \right] \\ &= \frac{T^2}{2GJ_0} \left[\int_0^L f^2(x)\, dx + \int_0^L g^2(x)\, dx \right] \end{aligned} \tag{7.107}$$

The condition for the energy U_0 of Eq. (7.102) to be equal to U_1 of Eq. (7.107):

$$\frac{J_{\text{eff}}^2}{J_0^2} = \frac{L}{\int_0^L \left[f^2(x) + g^2(x) \right] dx} \tag{7.108}$$

gives a formula for the effective torsional constant J_{eff}

$$J_{\text{eff}} = \frac{J_0}{\sqrt{1 - (2/\lambda L)(1 - J_0^2/J_c^2) \tanh(\lambda L/2)}} \tag{7.109}$$

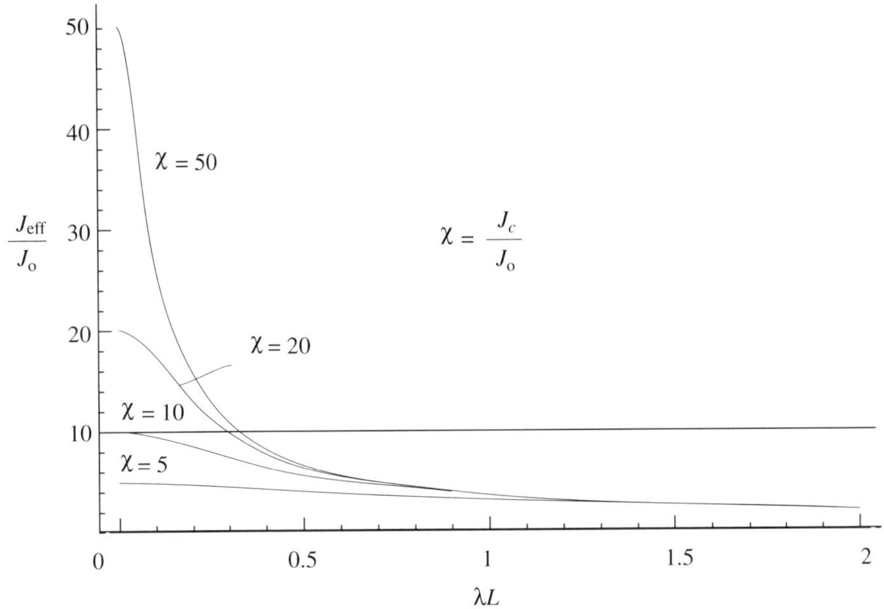

Figure 7.23 Effective torsional constant according to the strain energy criterion.

Figure 7.23 shows this torsional constant as a function of λL for four values of J_c/J_0. A comparison of Figs. 7.22 and 7.23 indicates that the strain energy criterion predicts smaller values for the effective torsional constant than those calculated from Eq. (7.100). This is also seen in Table 7.2, which lists the ratio of J_{eff} computed from Eq. (7.100) to that given by Eq. (7.109) at five values of λL for four values of J_c/J_0. It is apparent that the difference between the two effective torsional constants may become significant for large values of J_c/J_0.

TABLE 7.2 The Ratio of J_{eff} from Eq. (7.100) to J_{eff} from Eq. (7.109)

	$J_{eff}^{(7.100)}/J_{eff}^{(7.109)}$			
λL	$J_c/J_o = 5$	$J_c/J_o = 10$	$J_c/J_o = 20$	$J_c/J_o = 50$
0.2	1.03	1.12	1.43	2.62
0.4	1.09	1.36	2.00	3.54
0.6	1.17	1.56	2.29	3.54
0.8	1.24	1.68	2.35	3.25
1.0	1.29	1.73	2.29	2.93

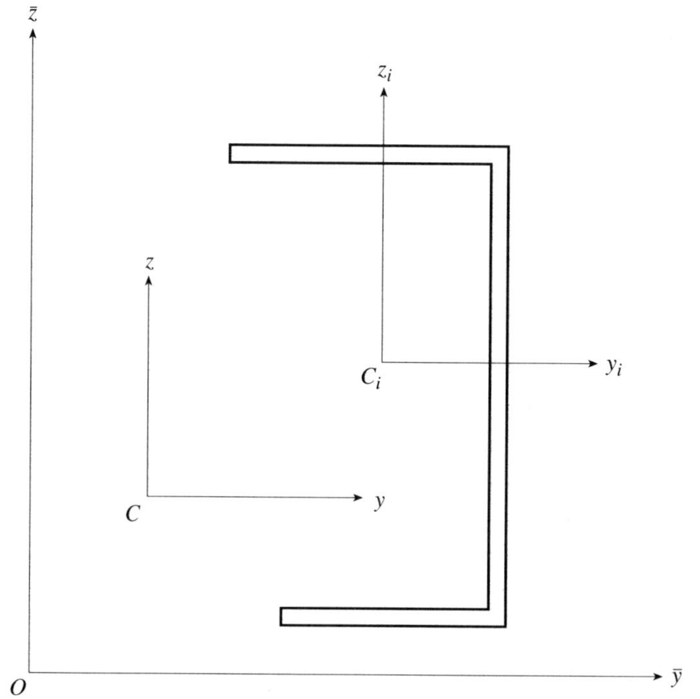

Figure 7.24 Coordinate systems for open-section analysis.

7.7.1 Calculation of Open-Section Properties

Figure 7.24 shows the coordinate systems to be used for open-section analysis. The user's coordinate system has origin O and axes $\bar{y}$, $\bar{z}$. The coordinate system having its origin at the centroid C of the entire cross section has its y axis parallel to $\bar{y}$ and its z axis parallel to $\bar{z}$. The origin C_i of the third coordinate system is the centroid of the ith cross-sectional member. The axes y_i and z_i are parallel to the y and z axes. The cross section is assumed to be made up of n unconnected members.

Suppose first that the following cross-sectional properties defined in Eq. (1.52) of the ith member have been calculated in the user's coordinate system:

$$
\begin{aligned}
Q_{\bar{y}}^i &= \int \bar{z}\, dA_i \\
Q_{\bar{z}}^i &= \int \bar{y}\, dA_i \\
I_{\bar{y}}^i &= \int \bar{z}^2\, dA_i
\end{aligned}
\tag{7.110}
$$

$$I^i_{\overline{z}} = \int \overline{y}^2 \, dA_i$$

$$I^i_{\overline{y}\,\overline{z}} = \int \overline{y}\,\overline{z} \, dA_i$$

The cross-sectional area of member i is denoted by A_i. The area integrals $Q^i_{\overline{y}}$, $Q^i_{\overline{z}}$ are the first moments of the area of member i about the $\overline{y}$, $\overline{z}$ axes. The area integrals $I^i_{\overline{y}}$, $I^i_{\overline{z}}$ are the area moments of inertia, and $I^i_{\overline{y}\,\overline{z}}$ is the area product of inertia of member i. The corresponding quantities for the entire cross section are given, in the user's coordinate system, by the sums

$$\begin{aligned} Q_{\overline{y}} &= \sum_{i=1}^{n} Q^i_{\overline{y}} \\ Q_{\overline{z}} &= \sum_{i=1}^{n} Q^i_{\overline{z}} \\ I_{\overline{y}} &= \sum_{i=1}^{n} I^i_{\overline{y}} \\ I_{\overline{z}} &= \sum_{i=1}^{n} I^i_{\overline{z}} \\ I_{\overline{y}\,\overline{z}} &= \sum_{i=1}^{n} I^i_{\overline{y}\,\overline{z}} \end{aligned} \tag{7.111}$$

where n is the number of individual members of the open section.

The coordinates of the centroid C are given, in the user's coordinate system, by (Eq. 1.56):

$$\overline{y}_C = \frac{Q_{\overline{z}}}{A} \qquad \overline{z}_C = \frac{Q_{\overline{y}}}{A} \tag{7.112}$$

where the total cross-sectional area A is found by summing the individual areas:

$$A = \sum_{i=1}^{n} A_i \tag{7.113}$$

If the coordinates of a point are y, z in the centroidal coordinate system Cyz of the section, the coordinates $\overline{y}$, $\overline{z}$ of the same point in the user's system are

$$\overline{y} = \overline{y}_C + y \qquad \overline{z} = \overline{z}_C + z \tag{7.114}$$

The area moment inertia I_z of the section about its centroidal axis z may be calculated as follows:

$$I_z = \int y^2\, dA = \int (\bar{y} - \bar{y}_C)^2\, dA = \int \left(\bar{y}^2 - 2\bar{y}\,\bar{y}_C + \bar{y}_C^2\right) dA = I_{\bar{z}} - A\bar{y}_C^2 \tag{7.115}$$

The product of inertia I_{yz} of the section is given by

$$I_{yz} = \int yz\, dA = \int (\bar{y} - \bar{y}_C)(\bar{z} - \bar{z}_C)\, dA = I_{\bar{y}\,\bar{z}} - A\bar{y}_C\bar{z}_C \tag{7.116}$$

The complete set of transformation equations from the user's coordinate system to the centroidal coordinate system are

$$\begin{aligned} I_y &= I_{\bar{y}} - A\bar{z}_C^2 \\ I_z &= I_{\bar{z}} - A\bar{y}_C^2 \\ I_{yz} &= I_{\bar{y}\,\bar{z}} - A\bar{y}_C\bar{z}_C \end{aligned} \tag{7.117}$$

These equations represent the parallel axis theorem of Eq. (1.60).

It is also possible to calculate the moments and products of inertia of the section in the centroidal coordinate system Cyz starting from the moments and product of inertia of member i in its own centroidal coordinate system $C_i y_i z_i$. The required transformation equations for member i are

$$\begin{aligned} I_y^i &= I_{y_i}^i + A_i z_{C_i}^2 \\ I_z^i &= I_{z_i}^i + A_i y_{C_i}^2 \\ I_{yz}^i &= I_{y_i z_i}^i + A_i y_{C_i} z_{C_i} \end{aligned} \tag{7.118}$$

The coordinates of the centroid C_i in these equations are measured in the coordinate system Cyz

$$\begin{aligned} y_{C_i} &= \bar{y}_{C_i} - \bar{y}_C = \frac{Q_{\bar{z}}^i}{A_i} - \bar{y}_C \\ z_{C_i} &= \bar{z}_{C_i} - \bar{z}_C = \frac{Q_{\bar{y}}^i}{A_i} - \bar{z}_C \end{aligned} \tag{7.119}$$

The section properties are given by

$$I_y = \sum_{i=1}^{n} I_y^i$$

$$I_z = \sum_{i=1}^{n} I_z^i \tag{7.120}$$

$$I_{yz} = \sum_{i=1}^{n} I_{yz}^i$$

7.7.2 Warping and Torsional Constants of an Open Section

The calculation of the warping constant of an open section relies on two assumptions. First, it is assumed that the shape of each member cross section remains unchanged during rotation, with each element undergoing the same twist as the entire cross section, so that

$$\phi_i = \phi \tag{7.121}$$

Second, it is assumed that the shear center of the welded section is the common shear center of each cross-sectional member of the unwelded section. This means that the warping constant Γ_0 of the unwelded section is calculated with respect to the shear center S of the welded section.

Let C_i be the centroid of the ith unconnected cross section, and let the coordinates of S measured in a centroidal coordinate system Cy_iz_i be denoted by y_{S_i} and z_{S_i} as shown in Fig. 7.25. Let ω_i be the warping function of the ith section, calculated in

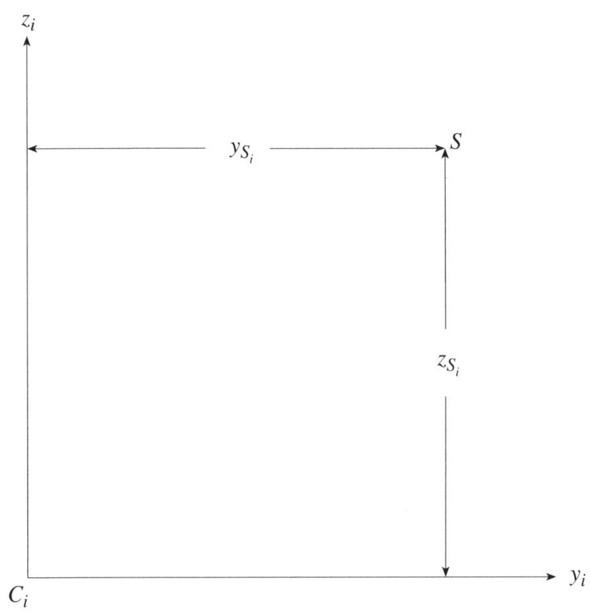

Figure 7.25 Centroidal coordinate system of cross-sectional member i and the shear center S of the closed section.

the coordinates of the centroid C_i. To transform from C_i to the shear center S of the closed section, use Eq. (7.34):

$$\omega_{S_i} = \omega_i - z_{S_i} y_i + y_{S_i} z_i - \frac{Q_{\omega_i}}{A_i} \tag{7.122}$$

where

$$Q_{\omega_i} = \int \omega_i \, dA_i \tag{7.123}$$

The warping constant Γ_{S_i} of the unwelded section member i with respect to the shear center S of the welded section is defined as

$$\Gamma_{S_i} = \int \omega_{S_i}^2 \, dA_i \tag{7.124}$$

Substitution of Eq. (7.122) into this integral leads to

$$\Gamma_{S_i} = I_{\omega_i} - \frac{Q_{\omega_i}^2}{A} - 2z_{S_i} I_{y_i\omega_i} + 2y_{S_i} I_{z_i\omega_i} + z_{S_i}^2 I_{z_i} + y_{S_i}^2 I_{y_i} - 2y_{S_i} z_{S_i} I_{y_i z_i} \tag{7.125}$$

In this equation, the cross-sectional properties that are independent of warping are the area moments and products of inertia, which are defined by

$$\begin{aligned} I_{z_i} &= \int y_i^2 \, dA_i \\ I_{y_i} &= \int z_i^2 \, dA_i \\ I_{y_i z_i} &= \int y_i z_i \, dA_i \end{aligned} \tag{7.126}$$

The cross-sectional properties that depend on the warping function are the warping constant and the sectorial products of area:

$$\begin{aligned} I_{\omega_i} &= \int \omega_i^2 \, dA_i \\ I_{y_i\omega_i} &= \int y_i \omega_i \, dA_i \\ I_{z_i\omega_i} &= \int z_i \omega_i \, dA_i \end{aligned} \tag{7.127}$$

The equation for the angle of twist of each cross-sectional member i may be written as

$$GJ_i \phi' - E\Gamma_{S_i} \phi''' = T_i \qquad 1 \le i \le n \tag{7.128}$$

In these equations, T_i is the torsional moment acting on the member i, whose torsional constant is J_i and whose warping constant with respect to the assumed common shear center S is Γ_{S_i}. The equilibrium condition for the torsional moments is

$$\sum_{i=1}^{n} T_i = T \tag{7.129}$$

where T is the torsional moment acting on the whole cross section.

The equation for the angle of twist of the whole section is

$$GJ_0\phi' - E\Gamma_0\phi''' = T \tag{7.130}$$

The sum of the equations of twist for the individual members gives

$$G\sum_{i=1}^{n} J_i\phi' - E\sum_{i=1}^{n} \Gamma_{S_i}\phi''' = \sum_{i=1}^{n} T_i = T \tag{7.131}$$

These two equations allow the identifications

$$J_0 = \sum_{i=1}^{n} J_i \tag{7.132}$$

$$\Gamma_0 = \sum_{i=1}^{n} \Gamma_{S_i} \tag{7.133}$$

Thus, for a multimember beam with n unconnected members, the torsional constant J_0 of the unwelded sections is the sum of the torsional constants of the individual members as given in Eq. (7.132). The torsional constant J_c of the welded section has to be computed separately. The warping constant Γ_0 of the unwelded section with respect to the shear center S of the welded section is found by summing the contributions of the unconnected section members as represented by Eq. (7.133).

7.7.3 Calculation of the Effective Torsional Constant

The computation of the effective torsional constant requires the following steps.

1. Determine the shear center S and the torsional constant J_c of the closed section.
2. For each unconnected cross-sectional member i, determine the centroid C_i, the torsional constant J_i, the area moments of inertia I_{y_i} and I_{z_i}, the product of inertia $I_{y_i z_i}$, the warping constant I_{ω_i} referred to the centroid C_i, and the sectorial products of area $I_{y_i\omega_i}$ and $I_{z_i\omega_i}$.
3. For each unconnected cross-sectional member i, calculate the warping constant with respect to S using Eq. (7.125).

4. Calculate the torsional constant J_0 of the open section by summing the contributions of the n unconnected cross-sectional members:

$$J_0 = \sum_{i=1}^{n} J_i \tag{7.134}$$

5. Calculate the warping constant Γ_0 of the open section by summing the contributions of the n unconnected cross-sectional members

$$\Gamma_0 = \sum_{i=1}^{n} \Gamma_{S_i} \tag{7.135}$$

6. Calculate λ:

$$\lambda = \sqrt{\frac{GJ_0}{E\Gamma_0}} \tag{7.136}$$

where E is the Young's modulus and G the shear modulus of the material.

7. Calculate the effective torsional constant J_{eff} using Eq. (7.100):

$$J_{\text{eff}} = \frac{J_0}{1 - 2/\lambda L(1 - J_0/J_c)\tanh(\lambda L/2)} \tag{7.137}$$

where L is the distance between the welded cross sections of the beam. Alternatively, use Eq. (7.109) to determine J_{eff}.

7.8 MORE PRECISE THEORIES

There is always concern about the relative accuracy of the various theories for finding beam responses. In the case of torsion, we have considered simple Saint-Venant torsion in Chapter 5 and restrained warping torsion here in Chapter 7. The careful use of a general-purpose structural analysis computer program (a finite element program) provides a means of assessing the accuracy of the theories for torsion. Other theories, which are more complex and precise than traditional theories and less complicated than the use of general-purpose three-dimensional structural analysis software, have been developed. An example is the torsion solution proposed by Reagan (2002), in which the bar deformations are represented by an exponentially decaying residual solution superimposed on the classical Saint-Venant solution. Figure 7.26 provides a comparison of the torsion solutions according to Saint-Venant, restrained warping, and a more precise theory for a fixed–free bar with a torque applied at the free end. The results are given for several cross-sectional shapes. In each case, the applied torque $\overline{M}_x$ is 100,000 in-lb, the length of the rod is 48 in., and the material is aluminum with $G = 384{,}615$ lb/in^2. In Fig. 7.26, θ^{SV} refers to the rate of twist for Saint-Venant torsion, θ^{RW} is the rate of twist according to the restrained warping

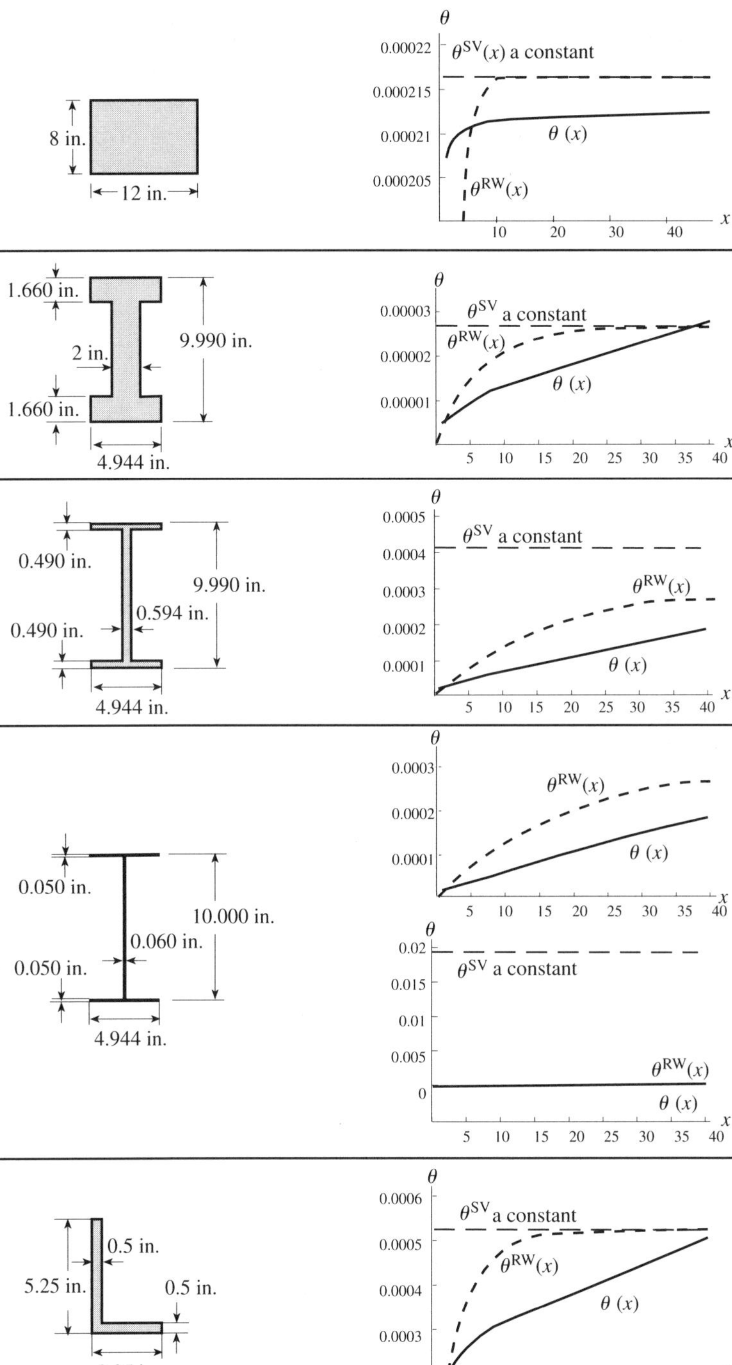

Figure 7.26 Comparison of rates of twist for Saint-Venant (SV) uniform torsion theory, the fourth-order restrained warping (RW) theory, and a more precise theory $\theta(x)$.

fourth-order theory of this chapter, and θ is determined from a more precise theory (Reagan, 2002). In general, it is seen that the more precise theory (θ) approaches the constant Saint-Venant value less rapidly than the restrained warping theory (θ^{RW}).

The exact beam theoery of Ladevèze (1983) and Ladevèze and Simmonds (1996), which is described in Section 6.4.7, applies to a broad range of beam problems, including those with restrained warping.

REFERENCES

Gere, J. M., and Timoshenko, S. P. (1997). *Mechanics of Materials*, 4th ed., PWS Publishing Co., Boston.

Gjelsvik, A. (1981). *The Theory of Thin Walled Bars*, Wiley, New York.

Göldner, H. (1979). *Lehrbuch Höhere Festigkeitslehre*, Veb Fachbuchverlag, Leipzig.

Hjelmstad, K. D. (1987). Warping effects in transverse bending of thin-walled beams, *J. Eng. Mech.*, Vol. 113, pp. 907–924.

Kazic, M., and Dong, S. B. (1990). Analysis of restrained torsion, *J. Eng. Mech.*, Vol. 116, pp. 870–891.

Kollbrunner, C., and Basler, K. (1969). *Torsion in Structures*, Springer-Verlag, Berlin.

Kollbrunner, C., and Hajdin, N. (1972). *Dünnwandige Stäbe*, Springer-Verlag, Berlin.

Krenk, S. and Jeppesen, B. (1989). Finite elements for beam cross-sections of moderate wall thickness, *Comput. Struct.*, Vol. 32, pp. 1035–1043.

Ladevèze, P. (1983). Sur le principe de Saint-Venant en élasticité, *J. Mécanique Théorique Appliquée*, Vol. 1, pp. 161–184.

Ladevèze, P., and Simmonds, J. G. (1996). New concepts for linear beam theory with arbitrary geometry and loading, *Comptes Rendus Acad. Sci. Paris*, Vol. 332, Ser. IIb, pp. 455–462.

Mason, W. E., and Herrmann, L. R. (1968). Elastic shear analysis of general prismatic beams, *J. Eng. Mech.*, Vol. 94, pp. 965–983.

Meek, J. L., and Ho, P. T. S. (1983). A simple finite element model for the warping torsion problem, *Comput. Methods Appl. Mech. Eng.*, Vol. 37, pp. 25–36.

Megson, T. H. G. (1974). *Linear Analysis of Thin-Walled Elastic Structures*, Wiley, New York.

Mottershead, J. E. (1988). Warping torsion in thin-walled open section beams using the semiloof beam element, *Int. J. Numer. Methods Eng.*, Vol. 26, pp. 231–243.

Murray, N. W. (1984). *Introduction to the Theory of Thin-Walled Structures*, Clarendon Press, Oxford.

Oden, J. T., and Ripperger, E. A. (1981). *Mechanics of Elastic Structures*, 2nd ed., McGraw-Hill, New York.

Pilkey, W. D. (1994). *Formulas for Stress, Strain, and Structural Matrices*, Wiley, New York.

Reagan, S. (2002). Constrained torsion of prismatic bars, Ph.D. dissertation, University of Virginia, Charlottesville, Va.

Romano, G., Rosati, L., and Ferro, G. (1992). Shear deformability of thin-walled beams with arbitrary cross sections, *Int. J. Numer. Methods Eng.*, Vol. 35, pp. 283–306.

Vlasov, V.Z. (1961). *Thin-Walled Elastic Beams*, translated from the Russian *Tonkostennye uprigie sterzhni* by the Israel Program for Scientific Translations for the N.S.F. and the U.S. Dept of Commerce, Office of Technical Services, Washington, D.C.

CHAPTER 8

ANALYSIS OF STRESS

Various normal and shear stresses have been defined in previous chapters. It is useful to review some of the techniques for combining these stresses. In this chapter, principal stresses, extreme shear stresses, and yield (failure) theories are discussed.

8.1 PRINCIPAL STRESSES AND EXTREME SHEAR STRESSES

8.1.1 State of Stress

Consider the three-dimensional state of stress defined in Section 1.1.2. Define three stress vectors

$$\begin{aligned} \boldsymbol{\sigma}_x &= \sigma_x \mathbf{i} + \tau_{xy}\mathbf{j} + \tau_{xz}\mathbf{k} \\ \boldsymbol{\sigma}_y &= \tau_{yx}\mathbf{i} + \sigma_y\mathbf{j} + \tau_{yz}\mathbf{k} \\ \boldsymbol{\sigma}_z &= \tau_{zx}\mathbf{i} + \tau_{zy}\mathbf{j} + \sigma_z\mathbf{k} \end{aligned} \tag{8.1}$$

on the x, y, z coordinate planes passing through a point. The vectors $\boldsymbol{\sigma}_x$, $\boldsymbol{\sigma}_y$, $\boldsymbol{\sigma}_z$ define the states of stress on the faces of a cube whose outward normals are **i**, **j**, **k**, respectively (see Fig. 1.1). The stress vector $\boldsymbol{\sigma}_n$ at a point on an arbitrarily oriented plane specified by the unit normal vector (Fig. 8.1)

$$\mathbf{n} = n_x\mathbf{i} + n_y\mathbf{j} + n_z\mathbf{k} \tag{8.2}$$

is given by (Pilkey and Wunderlich, 1994)

$$\boldsymbol{\sigma}_n = n_x\boldsymbol{\sigma}_x + n_y\boldsymbol{\sigma}_y + n_z\boldsymbol{\sigma}_z \tag{8.3}$$

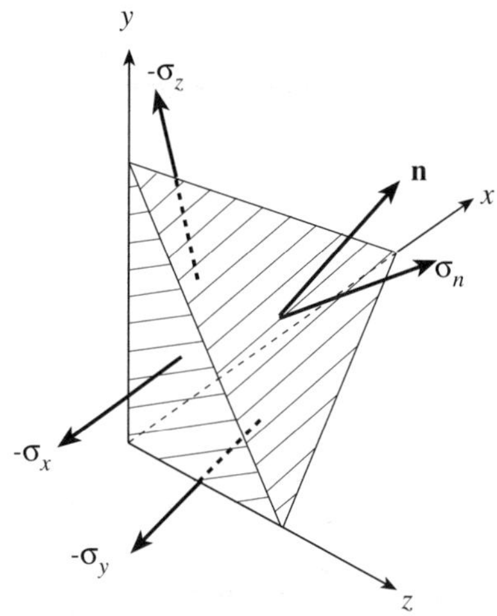

Figure 8.1 Stress vector on arbitrarily oriented plane with normal **n**.

The components n_x, n_y, n_z are the direction cosines of the unit normal **n**. This stress vector can be decomposed into normal and shear components. The normal component σ, which is parallel to the unit normal **n**, is found as

$$\sigma = \boldsymbol{\sigma}_n \cdot \mathbf{n} = \sigma_x n_x^2 + \sigma_y n_y^2 + \sigma_z n_z^2 + 2(\tau_{xy} n_x n_y + \tau_{yz} n_y n_z + \tau_{zx} n_z n_x) \tag{8.4}$$

If the orientation of this plane changes, the value of the normal stress σ varies.

8.1.2 Principal Stresses

Of special interest in an analysis of stresses are the extreme (minimum and maximum) values of σ and the corresponding orientations of the plane. The extreme values of σ need to be determined with the constraint that **n** is a unit vector

$$n_x^2 + n_y^2 + n_z^2 = 1 \tag{8.5}$$

To solve this extremum problem, introduce the Lagrange multiplier λ and find the orientation of the plane corresponding to extreme values of the function

$$F(n_x, n_y, n_z) = \sigma - \lambda(n_x^2 + n_y^2 + n_z^2 - 1) \tag{8.6}$$

The conditions for F to assume its extreme values, $\partial F/\partial n_x = 0$, $\partial F/\partial n_y = 0$, $\partial F/\partial n_z = 0$, lead to the three homogeneous linear equations for n_x, n_y, n_z

$$\begin{aligned}
(\sigma_x - \lambda)n_x + \tau_{xy}n_y + \tau_{xz}n_z &= 0 \\
\tau_{xy}n_x + (\sigma_y - \lambda)n_y + \tau_{yz}n_z &= 0 \\
\tau_{xz}n_x + \tau_{yz}n_y + (\sigma_z - \lambda)n_z &= 0
\end{aligned} \tag{8.7}$$

which is a set of three homogeneous algebraic equations for four unknowns, n_x, n_y, n_z, and λ. A fourth equation, $n_x^2 + n_y^2 + n_z^2 - 1 = 0$, which assures that $\mathbf{n}$ is a unit vector, is obtained from $\partial F/\partial\lambda = 0$. Any unit vector $\mathbf{n}$ (i.e., the components n_x, n_y, and n_z) satisfying the three relations of Eq. (8.7) is an eigenvector. From Cramer's rule the relations of Eq. (8.7) have a nontrivial solution if and only if the determinant of the coefficients is zero:

$$\begin{vmatrix} \sigma_x - \lambda & \tau_{xy} & \tau_{xz} \\ \tau_{xy} & \sigma_y - \lambda & \tau_{yz} \\ \tau_{xz} & \tau_{yz} & \sigma_z - \lambda \end{vmatrix} = 0 \tag{8.8}$$

Note that this problem of finding the extreme values of σ is algebraically the same as the eigenvalue problem for the 3×3 symmetric stress matrix (tensor)

$$\boldsymbol{\sigma} = \begin{bmatrix} \sigma_x & \tau_{xy} & \tau_{xz} \\ \tau_{xy} & \sigma_y & \tau_{yz} \\ \tau_{xz} & \tau_{yz} & \sigma_z \end{bmatrix} \tag{8.9}$$

The eigenvectors of this matrix are the normal vectors defining the planes on which σ assumes its extreme values, and eigenvalues of the matrix are the Lagrange multipliers λ.

A symmetric tensor with real elements has real eigenvalues and the eigenvectors corresponding to distinct eigenvalues are mutually orthogonal (Atanackovic and Guran, 1999). Thus, the three eigenvalues $\lambda_1, \lambda_2, \lambda_3$ of the stress matrix $\boldsymbol{\sigma}$ are real numbers, and the three eigenvectors corresponding to distinct eigenvalues are perpendicular. If precisely two of the eigenvalues are the same, there are two mutually perpendicular eigenvectors corresponding to these two eigenvalues. Moreover, any two mutually perpendicular vectors situated in the plane defined by these two eigenvectors are also eigenvectors corresponding to the two identical eigenvalues. For three identical eigenvalues, any three mutually perpendicular vectors are eigenvectors.

The columns of the stress matrix $\boldsymbol{\sigma}$ of Eq. (8.9) are the x, y, z stress vectors of Eq. (8.1). For a unit normal vector $\mathbf{n}$ (Eq. 8.2), the stress vector $\boldsymbol{\sigma}_n$ of Eq. (8.3) on an arbitrarily oriented plane is given by

$$\boldsymbol{\sigma}_n = \boldsymbol{\sigma}\mathbf{n} = [\boldsymbol{\sigma}_x \quad \boldsymbol{\sigma}_y \quad \boldsymbol{\sigma}_z]\mathbf{n} = n_x\boldsymbol{\sigma}_x + n_y\boldsymbol{\sigma}_y + n_z\boldsymbol{\sigma}_z \tag{8.10}$$

All vectors here have coordinates with respect to the unit vectors $\mathbf{i}, \mathbf{j}, \mathbf{k}$. From Eq. (8.10) it can be concluded that the x, y, z coordinates of a stress vector on the plane defined by the unit normal vector $\mathbf{n}$ can be calculated by multiplying the components n_x, n_y, n_z by the matrix $\boldsymbol{\sigma}$. The unit normal $\mathbf{n}$ multiplied by the stress vector $\boldsymbol{\sigma}_n$ gives the expression for the normal stress σ on the plane defined by $\mathbf{n}$:

$$\sigma = \mathbf{n} \cdot \boldsymbol{\sigma}_n = \mathbf{n} \cdot \boldsymbol{\sigma}\mathbf{n} \tag{8.11}$$

Suppose that the three eigenvalues of the stress tensor $\boldsymbol{\sigma}$ are $\lambda_1, \lambda_2, \lambda_3$ and the corresponding unit eigenvectors are $\mathbf{n}_1, \mathbf{n}_2, \mathbf{n}_3$. It follows from Eq. (8.11) that the normal stress σ_k, $k = 1, 2, 3$, on the planes defined by the eigenvectors $\mathbf{n}_k$, $k = 1, 2, 3$, are

$$\sigma_k = \mathbf{n}_k \cdot \boldsymbol{\sigma}\mathbf{n}_k = \mathbf{n}_k \cdot \lambda_k \mathbf{n}_k = \lambda_k \qquad k = 1, 2, 3 \tag{8.12}$$

where Eq. (8.5) has been utilized. It can be concluded, then, that the three values of the Lagrange multiplier λ, which are the eigenvalues of $\boldsymbol{\sigma}$, are the extreme values of the normal stress.

These eigenvalues of the stress tensor $\boldsymbol{\sigma}$ are called the *principal stresses*. As indicated in Eq. (8.12), these are $\sigma_1, \sigma_2, \sigma_3$. These are the minimum and maximum values assumed by normal stresses as a function of the orientation of the plane. The corresponding eigenvectors $\mathbf{n}_1, \mathbf{n}_2, \mathbf{n}_3$ are the *principal directions*. They can always be chosen to form a right-handed triad of unit vectors.

It is of interest to find the level of shear stress on the principal stress planes. Suppose that $\mathbf{n}$ is a unit eigenvector of the stress tensor $\boldsymbol{\sigma}$, with the corresponding eigenvalue λ, so that the stress vector $\boldsymbol{\sigma}_n$ is given by

$$\boldsymbol{\sigma}_n = \boldsymbol{\sigma}\mathbf{n} = \lambda\mathbf{n} \tag{8.13}$$

This relationship shows that the stress vector $\boldsymbol{\sigma}_n$ is formed only of a component along n. Thus, it is seen that the shear stress on the plane defined by $\mathbf{n}$ is zero. In other words, the principal stress planes are free of shear stress. Conversely, if $\mathbf{n}$ is the normal to a plane that is free of shear stress, the relationship for calculating $\mathbf{n}$ is the eigenvalue problem for the stress matrix $\boldsymbol{\sigma}$

$$\boldsymbol{\sigma}\mathbf{n} = \boldsymbol{\sigma}_n = \lambda\mathbf{n} \tag{8.14}$$

This establishes that a cutting plane through a point in a body is a principal stress plane if and only if it is free of shear stress.

8.1.3 Invariants of the Stress Matrix

The eigenvalues of $\boldsymbol{\sigma}$ are the three real roots of the cubic equation resulting from Eq. (8.8). That is, if $\mathbf{I}$ is the unit diagonal matrix,

$$|\ \boldsymbol{\sigma} - \lambda\mathbf{I}\ | = -\lambda^3 + I_1\lambda^2 - I_2\lambda + I_3 = 0 \tag{8.15}$$

where

$$\begin{aligned} I_1 &= \sigma_x + \sigma_y + \sigma_z \\ I_2 &= \sigma_x\sigma_y - \tau_{xy}^2 + \sigma_y\sigma_z - \tau_{yz}^2 + \sigma_x\sigma_z - \tau_{xz}^2 \\ I_3 &= \sigma_x\sigma_y\sigma_z - \sigma_x\tau_{yz}^2 - \sigma_y\tau_{xz}^2 - \sigma_z\tau_{xy}^2 + 2\tau_{xy}\tau_{xz}\tau_{yz} \end{aligned} \tag{8.16}$$

For a principal stress plane, the shear stresses are zero and the stress tensor is a diagonal matrix with the principal stresses on the main diagonal:

$$\begin{bmatrix} \sigma_1 & 0 & 0 \\ 0 & \sigma_2 & 0 \\ 0 & 0 & \sigma_3 \end{bmatrix}$$

In this case, Eq. (8.15) would appear as

$$(\sigma_1 - \lambda)(\sigma_2 - \lambda)(\sigma_3 - \lambda) = 0 \tag{8.17}$$

This relationship becomes

$$-\lambda^3 + (\sigma_1 + \sigma_2 + \sigma_3)\lambda^2 - (\sigma_1\sigma_2 + \sigma_2\sigma_3 + \sigma_1\sigma_3)\lambda + \sigma_1\sigma_2\sigma_3 = 0 \tag{8.18}$$

Comparison of Eqs. (8.15) and (8.18) shows that

$$\begin{aligned} I_1 &= \sigma_x + \sigma_y + \sigma_z = \sigma_1 + \sigma_2 + \sigma_3 \\ I_2 &= \sigma_x\sigma_y - \tau_{xy}^2 + \sigma_y\sigma_z - \tau_{yz}^2 + \sigma_x\sigma_z - \tau_{xz}^2 \\ &= \sigma_1\sigma_2 + \sigma_1\sigma_3 + \sigma_2\sigma_3 \\ I_3 &= \sigma_x\sigma_y\sigma_z - \sigma_x\tau_{yz}^2 - \sigma_y\tau_{xz}^2 - \sigma_z\tau_{xy}^2 + 2\tau_{xy}\tau_{xz}\tau_{yz} \\ &= \sigma_1\sigma_2\sigma_3 \end{aligned} \tag{8.19}$$

The quantities I_1, I_2, I_3 have the same values regardless of the choice of axes x, y, z in which the state of stress is given. Hence, I_1, I_2, I_3 are called *stress invariants*.

8.1.4 Extreme Values of Shear Stress

As the cutting plane orientation varies, the shear stresses will change. The extreme values of these stresses are calculated in this section. We choose to select the x, y, z axes as a set of principal axes at the point in question. Label these axes 1, 2, 3, with unit vectors $\mathbf{e}_1$, $\mathbf{e}_2$, $\mathbf{e}_3$. The stress vectors on the coordinate planes, on which the principal stresses σ_1, σ_2, σ_3 occur, are expressed as

$$\begin{aligned} \boldsymbol{\sigma}_1 &= \sigma_1\mathbf{e}_1 \\ \boldsymbol{\sigma}_2 &= \sigma_2\mathbf{e}_2 \\ \boldsymbol{\sigma}_3 &= \sigma_3\mathbf{e}_3 \end{aligned} \tag{8.20}$$

These are the stress vectors of Eq. (8.1) written along the principal axes for which the corresponding shear stresses are zero. Similar to Eq. (8.3), the stress vector on a plane whose unit normal is the vector $\mathbf{n}$ is

$$\boldsymbol{\sigma}_n = n_1\sigma_1\mathbf{e}_1 + n_2\sigma_2\mathbf{e}_2 + n_3\sigma_3\mathbf{e}_3 \tag{8.21}$$

Decompose the stress vector $\boldsymbol{\sigma}_n$ into two orthogonal components $\boldsymbol{\sigma}_n \cdot \mathbf{n}$ and a shear stress component of magnitude τ, expressed as

$$\tau^2 = \boldsymbol{\sigma}_n \cdot \boldsymbol{\sigma}_n - (\boldsymbol{\sigma}_n \cdot \mathbf{n})^2 \tag{8.22}$$

Introduce $\boldsymbol{\sigma}_n$ of Eq. (8.21) to find that

$$\begin{aligned} \tau^2 &= n_1^2\sigma_1^2 + n_2^2\sigma_2^2 + n_3^2\sigma_3^2 - (n_1^2\sigma_1 + n_2^2\sigma_2 + n_3^2\sigma_3)^2 \\ &= (\sigma_1 - \sigma_2)^2 n_1^2 n_2^2 + (\sigma_2 - \sigma_3)^2 n_2^2 n_3^2 + (\sigma_3 - \sigma_1)^2 n_3^2 n_1^2 \end{aligned} \tag{8.23}$$

As with Eq. (8.6) for the normal stresses, for τ to assume an extreme value, the function

$$F(n_1, n_2, n_3) = \tau^2 - \lambda(n_1^2 + n_2^2 + n_3^2 - 1) \tag{8.24}$$

must attain its extremum. Here λ is a Lagrange multiplier. The conditions for F to assume its extreme values,

$$\frac{\partial F}{\partial n_1} = 0 \qquad \frac{\partial F}{\partial n_2} = 0 \qquad \frac{\partial F}{\partial n_3} = 0 \qquad \frac{\partial F}{\partial \lambda} = 0 \tag{8.25}$$

lead to

$$\begin{aligned} (\sigma_1 - \sigma_2)^2 n_1 n_2^2 + (\sigma_1 - \sigma_3)^2 n_1 n_3^2 - \lambda n_1 &= 0 \\ (\sigma_2 - \sigma_3)^2 n_2 n_3^2 + (\sigma_2 - \sigma_1)^2 n_2 n_1^2 - \lambda n_2 &= 0 \\ (\sigma_3 - \sigma_1)^2 n_3 n_1^2 + (\sigma_3 - \sigma_2)^2 n_3 n_2^2 - \lambda n_3 &= 0 \\ n_1^2 + n_2^2 + n_3^2 - 1 &= 0 \end{aligned} \tag{8.26}$$

One solution of these equations corresponds to the trivial solution $\tau^2 = 0$. A second solution is

$$\begin{aligned} n_1 &= 0 \qquad n_2 = n_3 = \frac{1}{\sqrt{2}} \qquad \lambda = \frac{(\sigma_2 - \sigma_3)^2}{2} \\ n_2 &= 0 \qquad n_3 = n_1 = \frac{1}{\sqrt{2}} \qquad \lambda = \frac{(\sigma_3 - \sigma_1)^2}{2} \\ n_3 &= 0 \qquad n_1 = n_2 = \frac{1}{\sqrt{2}} \qquad \lambda = \frac{(\sigma_1 - \sigma_2)^2}{2} \end{aligned} \tag{8.27}$$

To find the extreme values of the shear stresses, substitute n_1, n_2, n_3 from Eq. (8.27) into Eq. (8.23). For example, for $n_1 = n_2 = 1/\sqrt{2}$ and $n_3 = 0$, Eq. (8.23) provides $\tau = \frac{1}{2} \mid \sigma_1 - \sigma_2 \mid$. In general, the extreme values of the shear stresses are

$$\tau = \frac{|\,\sigma_2 - \sigma_3\,|}{2} \qquad \tau = \frac{|\,\sigma_3 - \sigma_1\,|}{2} \qquad \tau = \frac{|\,\sigma_1 - \sigma_2\,|}{2} \tag{8.28}$$

It follows that the maximum shear stress can be expressed as

$$\tau_{\max} = \frac{\sigma_{\max} - \sigma_{\min}}{2} \tag{8.29}$$

where $\sigma_{\max}$ and $\sigma_{\min}$ are the maximum and minimum, respectively, of the principal stresses σ_1, σ_2, and σ_3. From the values of n_1, n_2, n_3 in Eq. (8.27), which identify the unit normal vector $\mathbf{n}$, it is apparent that the planes on which shear stresses reach their extreme values make a 45° angle with the principal directions. These planes where shear stresses have extreme values are, in general, not free of normal stress.

The primary conclusion of this section can be summarized as: The maximum shear stress is equal to one-half the difference between the maximum and minimum normal stresses and acts on a plane that bisects the angle between the directions of the maximum and minimum principal stresses.

8.1.5 Beam Stresses

For the beam theories presented here, including bending and torsion, the assumptions lead to three components of the stress tensor that are zero at any point of the beam cross section:

$$\sigma_y = 0 \qquad \sigma_z = 0 \qquad \tau_{yz} = 0 \tag{8.30}$$

The remaining three components define the state of stress, with the stress matrix given in the form

$$\boldsymbol{\sigma} = \begin{bmatrix} \sigma_x & \tau_{xy} & \tau_{xz} \\ \tau_{xy} & 0 & 0 \\ \tau_{xz} & 0 & 0 \end{bmatrix} \tag{8.31}$$

The normal stress σ_x is due to axial loads, bending moments, or cross-sectional warping. The shear stresses τ_{xy}, τ_{xz} are due to unrestrained torsion, transverse shear forces, or cross-sectional warping.

The eigenvalues of $\boldsymbol{\sigma}$ are the principal stresses. From Eq. (8.8) these eigenvalues are obtained by solving for λ from the equation

$$\begin{vmatrix} \sigma_x - \lambda & \tau_{xy} & \tau_{xz} \\ \tau_{xy} & -\lambda & 0 \\ \tau_{xz} & 0 & -\lambda \end{vmatrix} = 0 \tag{8.32}$$

The three values of λ obtained from Eq. (8.32) are the principal stresses $\sigma_1, \sigma_2, \sigma_3$.

These are

$$\begin{aligned}\sigma_1 &= \frac{\sigma_x}{2} + \sqrt{\left(\frac{\sigma_x}{2}\right)^2 + \tau_{xy}^2 + \tau_{xz}^2} \\ \sigma_2 &= 0 \\ \sigma_3 &= \frac{\sigma_x}{2} - \sqrt{\left(\frac{\sigma_x}{2}\right)^2 + \tau_{xy}^2 + \tau_{xz}^2}\end{aligned} \tag{8.33}$$

A unit vector

$$\mathbf{n} = n_x\mathbf{i} + n_y\mathbf{j} + n_z\mathbf{k} \tag{8.34}$$

which satisfies Eq. (8.7) is an eigenvector. There will be three eigenvectors $\mathbf{n}_1, \mathbf{n}_2, \mathbf{n}_3$, corresponding to the three eigenvalues $\lambda_1 = \sigma_1, \lambda_2 = \sigma_2, \lambda_3 = \sigma_3$. For beam theories, with $\sigma_y = 0, \sigma_z = 0$, and $\tau_{yz} = 0$, Eq. (8.7) reduces to

$$\begin{aligned}(\sigma_x - \lambda)\, n_x + \tau_{xy} n_y + \tau_{xz} n_z &= 0 \\ \tau_{xy} n_x - \lambda n_y \qquad\qquad &= 0 \\ \tau_{xz} n_x \qquad\qquad - \lambda n_z &= 0\end{aligned} \tag{8.35}$$

The eigenvectors are constructed by substituting one at a time $\lambda = \lambda_1, \lambda_2, \lambda_3$ into Eq. (8.35) and solving for n_x, n_y, n_z. Equations (8.35) are three linear, homogeneous equations for n_x, n_y, n_z. Hence, the components n_x, n_y, n_z can only be found in terms of each other. Typically, one of these components is arbitrarily set equal to a constant, usually with a value of 1. The resulting n_x, n_y, n_z are then substituted into the unit vector of Eq. (8.34), giving the eigenvectors $\mathbf{n}_1, \mathbf{n}_2, \mathbf{n}_3$.

As an example, consider the eigenvector for $\lambda_2 = \sigma_2 = 0$. Then Eq. (8.35) becomes

$$\begin{aligned}\sigma_x n_{2x} + \tau_{xy} n_{2y} + \tau_{xz} n_{2z} &= 0 \\ \tau_{xy} n_{2x} \qquad\qquad &= 0 \\ \tau_{xz} n_{2x} \qquad\qquad &= 0\end{aligned} \tag{8.36}$$

These give

$$n_{2x} = 0 \qquad n_{2y} = -\frac{n_{2z}\tau_{xz}}{\tau_{xy}} \tag{8.37}$$

With an arbitrary selection of $n_{2z} = 1$, the vector $n_{2x}\mathbf{i} + n_{2y}\mathbf{j} + n_{2z}\mathbf{k}$ corresponding to λ_2 becomes

$$-\frac{\tau_{xz}}{\tau_{xy}}\mathbf{j} + \mathbf{k} \tag{8.38}$$

This is converted to the unit eigenvector

$$\frac{1}{\sqrt{(\tau_{xz}/\tau_{xy})^2+1}}\left(-\frac{\tau_{xz}}{\tau_{xy}}\mathbf{j}+\mathbf{k}\right)=\frac{1}{\sqrt{\tau_{xz}^2+\tau_{xy}^2}}\left(-\tau_{xz}\mathbf{j}+\tau_{xy}\mathbf{k}\right) \tag{8.39}$$

The principal direction can be written simply as

$$\mathbf{n}_2=-\tau_{xz}\mathbf{j}+\tau_{xy}\mathbf{k} \tag{8.40}$$

This process leads to the three principal directions

$$\begin{aligned}\mathbf{n}_1 &= \sigma_1\mathbf{i}+\tau_{xy}\mathbf{j}+\tau_{xz}\mathbf{k}\\ \mathbf{n}_2 &= \qquad -\tau_{xz}\mathbf{j}+\tau_{xy}\mathbf{k}\\ \mathbf{n}_3 &= \sigma_3\mathbf{i}+\tau_{xy}\mathbf{j}+\tau_{xz}\mathbf{k}\end{aligned} \tag{8.41}$$

The resultant shear stress can be represented as

$$\boldsymbol{\tau}=\tau_{xy}\mathbf{j}+\tau_{xz}\mathbf{k} \tag{8.42}$$

Note that $\boldsymbol{\tau}\cdot\mathbf{n}_2=0$. It follows that direction $\mathbf{n}_2$ of the zero principal stress σ_2 is perpendicular to the resultant shear stress. The state of stress may therefore be considered biaxial, with σ_x as the only nonzero normal stress and the resultant shear stress:

$$\tau=\sqrt{\tau_{xy}^2+\tau_{xz}^2} \tag{8.43}$$

as the only shear stress.

The maximum shear stress acts on the plane whose normal bisects the angle between the principal planes of the maximum and minimum principal stresses and has a magnitude equal to one-half the difference between these two principal stresses. In our case, if $\sigma_3<0$, then

$$\tau_{\max}=\tfrac{1}{2}\mid\sigma_1-\sigma_3\mid=\sqrt{\left(\frac{\sigma_x}{2}\right)^2+\tau_{xy}^2+\tau_{xz}^2} \tag{8.44}$$

Details of Eigenvector Determination An in-depth study of the eigenvector problem leading to Eq. (8.40) is of interest. The goal is to find the eigenvectors corresponding to the eigenvalues $\sigma_1, \sigma_2, \sigma_3$ of Eq. (8.33).

To find the eigenvector $\mathbf{n}_2$ for the eigenvalue $\sigma_2=0$, we have $\boldsymbol{\sigma}\mathbf{n}_2=\sigma_2\mathbf{n}_2=0$, where

$$\mathbf{n}_2=n_{2x}\mathbf{i}+n_{2y}\mathbf{j}+n_{2z}\mathbf{k} \tag{8.45}$$

Suppose first that at least one of τ_{xy}, τ_{xz} is nonzero, say $\tau_{xy}\neq 0$. From

$$\begin{bmatrix} \sigma_x & \tau_{xy} & \tau_{xz} \\ \tau_{xy} & 0 & 0 \\ \tau_{xz} & 0 & 0 \end{bmatrix} \begin{bmatrix} n_{2x} \\ n_{2y} \\ n_{2z} \end{bmatrix} = \mathbf{0} \tag{8.46}$$

it follows that $\tau_{xy} n_{2x} = 0$. Since $\tau_{xy} \neq 0$, n_{2x} must be zero. The only remaining meaningful condition is

$$\tau_{xy} n_{2y} + \tau_{xz} n_{2z} = 0 \tag{8.47}$$

Arbitrarily choose n_{2z} to be τ_{xy} [rather than equal to 1 as for Eq. (8.38)] since $\tau_{xy} \neq 0$ and get

$$\tau_{xy} n_{2y} = -\tau_{xz} \tau_{xy} \qquad n_{2y} = -\tau_{xz} \tag{8.48}$$

Hence

$$\mathbf{n}_2 = -\tau_{xz} \mathbf{j} + \tau_{xy} \mathbf{k} \tag{8.49}$$

which corresponds to Eq. (8.41)

In case $\tau_{xz} \neq 0$, it follows again that $n_{2x} = 0$. The remaining condition is

$$\tau_{xy} n_{2y} + \tau_{xz} n_{2z} = 0 \tag{8.50}$$

so we may choose $n_{2y} = -\tau_{xz}$ and get $-\tau_{xy}\tau_{xz} = -\tau_{xz} n_{2z}$. Then $n_{2z} = \tau_{xy}$ and

$$\mathbf{n}_2 = -\tau_{xz} \mathbf{j} + \tau_{xy} \mathbf{k} \tag{8.51}$$

as before. It can be concluded that if at least one shear stress is nonzero, Eq. (8.40) holds.

Next, suppose that $\tau_{xy} = 0$ and $\tau_{xz} = 0$ and assume that $\sigma_x \neq 0$. The stress matrix of Eq. (8.31) reduces to

$$\boldsymbol{\sigma} = \begin{bmatrix} \sigma_x & 0 & 0 \\ 0 & 0 & 0 \\ 0 & 0 & 0 \end{bmatrix} \tag{8.52}$$

Then $\sigma_x n_{2x} = 0$ gives $n_{2x} = 0$. There are no other conditions, and $\mathbf{n}_2 = n_{2y}\mathbf{j} + n_{2z}\mathbf{k}$. The quantities n_{2y}, n_{2z} can be chosen arbitrarily as long as $\mathbf{n}_2$ is not the zero vector. For this case, eigenvectors can simply be chosen as

$$\begin{aligned} \mathbf{n}_1 &= \mathbf{i} && \text{for } \sigma_1 = \sigma_x \\ \mathbf{n}_2 &= \mathbf{j} && \text{for } \sigma_2 = 0 \\ \mathbf{n}_3 &= \mathbf{k} && \text{for } \sigma_3 = 0 \end{aligned} \tag{8.53}$$

This special case was not mentioned in the derivation of Eq. (8.40).

Consider now the determination of the eigenvector corresponding to the eigenvalue (Eq. 8.33)

$$\sigma_1 = \frac{\sigma_x}{2} + \sqrt{\left(\frac{\sigma_x}{2}\right)^2 + \tau_{xy}^2 + \tau_{xz}^2} \tag{8.54}$$

The eigenvalue problem is $(\boldsymbol{\sigma} - \sigma_1 \mathbf{I})\,\mathbf{n}_1 = \mathbf{0}$, where $\mathbf{I}$ is the unit diagonal vector, or

$$\begin{bmatrix} \sigma_x - \sigma_1 & \tau_{xy} & \tau_{xz} \\ \tau_{xy} & -\sigma_1 & 0 \\ \tau_{xz} & 0 & -\sigma_1 \end{bmatrix} \begin{bmatrix} n_{1x} \\ n_{1y} \\ n_{1z} \end{bmatrix} = \mathbf{0} \tag{8.55}$$

This matrix can have at most two independent rows. The condition $\sigma_1 \neq 0$ implies that the last two rows of this matrix are independent, so the only conditions are

$$\tau_{xy} n_{1x} - \sigma_1 n_{1y} = 0 \qquad \tau_{xz} n_{1x} - \sigma_1 n_{1z} = 0 \tag{8.56}$$

We are free to choose n_{1x}, so choose

$$n_{1x} = \sigma_1$$

where we know that n_{1x} is nonzero, as it has already been assumed that $\sigma_1 \neq 0$. Then

$$\sigma_1 \tau_{xy} - \sigma_1 n_{1y} = 0 \qquad \sigma_1 \tau_{xz} - \sigma_1 n_{1z} = 0 \tag{8.57}$$

gives $n_{1y} = \tau_{xy}$ and $n_{1z} = \tau_{xz}$. Then

$$\mathbf{n}_1 = \sigma_1 \quad \mathbf{i} + \tau_{xy} \quad \mathbf{j} + \tau_{xz} \quad \mathbf{k} \tag{8.58}$$

If $\sigma_1 = 0$, then

$$-\frac{\sigma_x}{2} = \sqrt{\left(\frac{\sigma_x}{2}\right)^2 + \tau_{xy}^2 + \tau_{xz}^2} \tag{8.59}$$

or

$$\left(\frac{\sigma_x}{2}\right)^2 = \left(\frac{\sigma_x}{2}\right)^2 + \tau_{xy}^2 + \tau_{xz}^2 \tag{8.60}$$

Thus $0 = \tau_{xy}^2 + \tau_{xz}^2$, so that $\tau_{xy} = 0$ and $\tau_{xz} = 0$. This becomes the problem already discussed with Eqs. (8.52) and (8.53). We see again that if the shear stresses τ_{xy} and τ_{xz} are zero, the eigenvectors of Eq. (8.41) are incorrect. Equation (8.41) corresponds to the case where at least one of the shear stresses τ_{xy}, τ_{xz} is nonzero.

The derivation of $\mathbf{n}_3$ for eigenvalue σ_3 is the same as that for σ_1. Note that

$$\mathbf{n}_1 \cdot \mathbf{n}_3 = 0 \qquad \mathbf{n}_1 \cdot \mathbf{n}_2 = 0 \qquad \mathbf{n}_2 \cdot \mathbf{n}_3 = 0. \tag{8.61}$$

8.2 YIELDING AND FAILURE CRITERIA

Failure for uniaxial states of stress is rather easy to define because simple tensile test material properties can be used directly. A structure is considered to undergo the transition to inelastic behavior at a certain level of stress, such as a yield stress σ_{ys}, an ultimate stress, or a fracture stress. Strains are treated similarly. For multiaxial states of stress, failure criteria have been developed to relate these states of stress to the behavior of a material in tension tests. A few of these *failure* or *strength theories* are discussed here.

8.2.1 Maximum Stress Theory

For the *maximum stress theory* or *Rankine theory*, failure is considered to occur when the structure is loaded such that the maximum principal stress at a point reaches the critical (yield, ultimate, or fracture) stress level. Arrange the principal stresses such that $\sigma_1 > \sigma_2 > \sigma_3$. Then, for example, yield occurs when

$$\sigma_1 = \sigma_{ys} \tag{8.62}$$

where σ_{ys} is the yield stress in tension.

8.2.2 Maximum Shear Theory

Maximum shear theory for failure is based on the hypothesis that yield or failure at a point in a structure occurs for a complex state of stress when the maximum shear stress as given by Eq. (8.28) reaches the value of the yield stress for shear of the material in a tensile test, $\sigma_{ys}/2$. It follows that yield or failure occurs if

$$\max(\mid \sigma_1 - \sigma_2 \mid, \mid \sigma_2 - \sigma_3 \mid, \mid \sigma_3 - \sigma_1 \mid) = \sigma_{ys} \tag{8.63}$$

Sometimes this is expressed as

$$\sigma_{\max} - \sigma_{\min} = \sigma_{ys} \tag{8.64}$$

where $\sigma_{\max}$ and $\sigma_{\min}$ are the maximum and minimum principal stresses. For instance, if $\sigma_1 > \sigma_2 > \sigma_3$, Eq. (8.63) or (8.64) would appear as $\sigma_1 - \sigma_3 = \sigma_{ys}$.

The maximum shear stress failure theory is known as the *Tresca* or *Guest theory*. Sometimes the left-hand side of either Eq. (8.63) or (8.64) is referred to as the *stress intensity*.

Beam Stresses It was shown in Section 8.1.5 that for the beam theories of this book, $\sigma_2 = 0$ (Eq. 8.33). If σ_1 is the maximum principal stress and $\sigma_3 < 0$, then $\tau_{\max} = \mid \sigma_1 - \sigma_3 \mid /2$ and the maximum shear stress yield criterion is

$$\mid \sigma_1 - \sigma_3 \mid = \sigma_{ys} \tag{8.65}$$

For $\sigma_3 \geq 0$, $\tau_{\max} = \mid \sigma_1 \mid /2$ and the yield criterion is

$$\mid \sigma_1 \mid = \sigma_{ys} \tag{8.66}$$

8.2.3 Von Mises Criterion

The von Mises theory of material failure is based on a state of stress on an octahedral plane, which is a plane whose normal has equal angles with the principal axes at a particular point in a solid. There are eight such planes, one is shown in Fig. 8.2. It is readily shown (Gould, 1994) that the normal stress on the octahedral plane is given

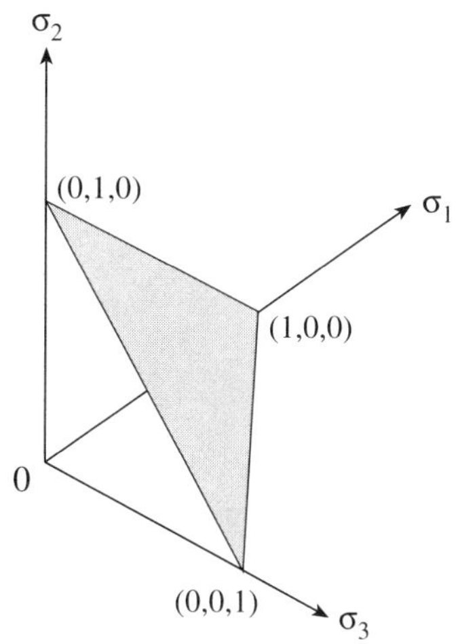

Figure 8.2 Octahedral plane.

by

$$\sigma_{\text{oct}} = \tfrac{1}{3}(\sigma_1 + \sigma_2 + \sigma_3) = \tfrac{1}{3} I_1 \tag{8.67}$$

which is referred to as the *hydrostatic* or *volumetric stress*. The shear stress is

$$\begin{aligned}\tau_{\text{oct}} &= \frac{1}{3}\sqrt{(\sigma_1 - \sigma_2)^2 + (\sigma_2 - \sigma_3)^2 + (\sigma_1 - \sigma_3)^2} \\ &= \frac{1}{3}\sqrt{\left(\sigma_x - \sigma_y\right)^2 + \left(\sigma_y - \sigma_z\right)^2 + (\sigma_x - \sigma_z)^2 + 6\left(\tau_{xy}^2 + \tau_{xz}^2 + \tau_{yz}^2\right)} \\ &= \frac{1}{3}\sqrt{2I_1^2 - 6I_2}\end{aligned} \tag{8.68}$$

In this theory, yielding of the complex state of stress occurs when the octahedral shear stress of Eq. (8.67) is equal to the octahedral shear stress at yield in tension ($\sigma_2 = \sigma_3 = 0$). This means that the criterion of failure by yielding is

$$\sqrt{\frac{(\sigma_1 - \sigma_2)^2 + (\sigma_2 - \sigma_3)^2 + (\sigma_1 - \sigma_3)^2}{2}} = \sigma_{ys} \tag{8.69}$$

This theory is called the *von Mises, Maxwell–Huber–Hencky, distortion energy,* and *octahedral shear stress theory.* Sometimes, the left-hand side of Eq. (8.69) is referred to as the *equivalent stress*.

Although Eqs. (8.63) and (8.69) are expressed in terms of the yield stress σ_{ys}, other failure modes, such as fatigue, ultimate stress, or fracture stress can be represented simply by replacing σ_{ys} by the appropriate tensile stress level.

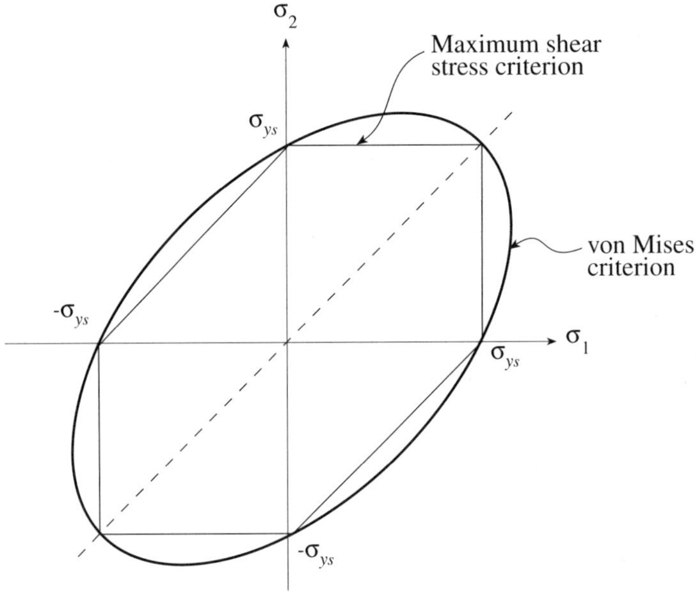

Figure 8.3 Maximum shear stress hexagon and von Mises ellipse.

Beam Stresses For beam theory with $\sigma_2 = 0$, Eq. (8.69) reduces to

$$\sqrt{\sigma_1^2 - \sigma_1\sigma_3 + \sigma_3^2} = \sigma_{ys} \tag{8.70}$$

In terms of the stress components,

$$\sqrt{\sigma_x^2 + 3(\tau_{xy}^2 + \tau_{xz}^2)} = \sqrt{\sigma_x^2 + 3\tau^2} = \sigma_{ys} \tag{8.71}$$

A plot of Eq. (8.70) for the von Mises criterion is shown in Fig. 8.3. Also shown is the maximum shear stress criterion of Eqs. (8.65) and (8.66). Both theories intersect the axes at the same points and also have in common the points of intersection with the bisectors of the first and third quadrants. Figure 8.3 shows that the maximum shear stress criterion is more conservative than the von Mises criterion.

REFERENCES

Atanackovic, T. M., and Guran, A. (1999). *Theory of Elasticity for Scientists and Engineers*, Birkhäuser, Boston.

Gould, P. L. (1994). *Introduction to Linear Elasticity*, 2nd ed., Springer-Verlag, New York.

Pilkey, W. D., and Wunderlich, W. (1994). *Mechanics of Structures: Variational and Computional Methods*, CRC Press, Boca Raton, Fla.

CHAPTER 9

RATIONAL B-SPLINE CURVES

It is desirable to be able to express the cross-sectional shape of a bar with functions of the coordinates y, z that contain only a few parameters. These parameters are helpful with the generation of the finite element mesh and can be the design parameters for an optimal cross-sectional shape design. B-spline curves are used to establish these functions. In particular, nonuniform rational B-splines, abbreviated NURBS, are introduced. An excellent reference on the subject of NURBS curves and surfaces is the book by Piegl and Tiller (1997).

9.1 CONCEPT OF A NURBS CURVE

NURBS is an acronym for *Non-Uniform Rational B-Spline*. NURBS were developed so that the shape of a curve could be controlled in a predictable manner by changing a few parameters. Previously, typically, curves were defined by points through which the curves passed. In the case of a NURBS curve the control parameters need not be on the curve. The fundamentals underlying NURBS were developed by P. Bézier in the 1960s and were utilized by the Renault company in the 1970s to assist in sculpting of surfaces for automobile bodies. A unified basis for representing standard and free-form shapes is provided by NURBS.

Some basic definitions are given in this section to provide a fundamental understanding of a NURBS curve. Subsequent sections contain more precise and complete definitions.

Degree, control point, knots, weights, and a mathematical formula define a NURBS curve. A *degree* p is a positive whole number. Degree 1 indicates that the curve is formed of linear lines; degree 2 corresponds to quadratic; degree 3 corresponds to cubic and is often used for free-form curves; degree five is quintic and is also used for free-form curves. The *order* of a NURBS curve is the degree plus 1 (i.e., $p + 1$). The degree of a NURBS curve can be increased without changing the shape of the curve, although a reduction in NURBS curve's degree will change the shape of the curve.

The *control points* $\mathbf{P}_i$ number at least $p + 1$. Control points are the vertices of a characteristic polygon (see Fig. 9.1). The geometry of a NURBS curve is readily changed by moving the control points. Associated with each control point is a *weight* w_i, which is usually a positive number. A rational function is a function expressed as the quotient of two polynomials. A curve is *non-rational* if all control points have the same weight; otherwise, the curve is said to be *rational*. Most NURBS curves are non-rational, although some curves such as circles and ellipses are always rational. The R in NURBS is included to indicate that there is the possibility that the curves are rational.

Knots are integer values that describe the connections between the vertices and order of the curve. Knots are a sequence of $m + 1$ numbers, where there are $n + 1$ control points, with $n = m - p - 1$. This sequence of numbers is referred to as a *knot vector*, designated as U. In this use, the word *vector* does not have the familiar meaning of a directed line segment. To satisfy certain conditions, it is necessary that the numbers in the knot vector do not decrease as you move along the

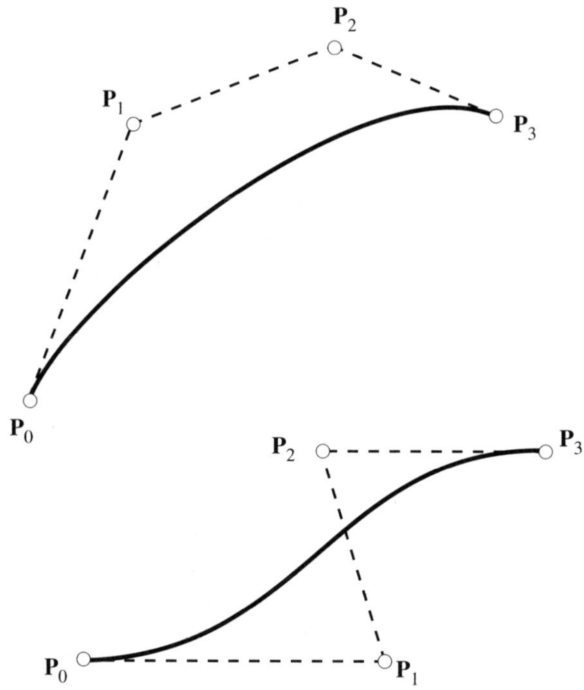

Figure 9.1 Cubic curves.

vector and that the number of duplicate values be limited to the order. As an example, consider a NURBS curve of degree 3 with 12 control points. The numbers 0, 0, 0, 0, 1, 2, 2, 3, 5, 5, 7, 7, 7, 7 form a satisfactory knot vector. However, the numbers 0, 0, 0, 0, 1, 3, 3, 3, 3, 3, 7, 7, 7, 7 are not acceptable because the number of 3's is larger than the order.

Knots multiplicity is the number of times a knot number is duplicated. In the satisfactory knot vector example, the multiplicity of the knot number 0 is four, 1 is one, 2 is two, 3 is one, 5 is two, and 7 is four. A *nonperiodic* knot vector occurs if the first and last knot values are repeated a number of times equal to the order. In the example, the knot vector is nonperiodic since the numbers 0, and 7 are repeated four times. A knot value that occurs only once is referred to as a *distinct knot*. The knot values 1 and 3 are distinct knots in the satisfactory example. A nonperiodic knot vector is *uniform* if the first $p + 1$ knots are followed by equally spaced, distinct interior knot values. An example of a curve with uniform knots would be a degree 3 NURBS curve with 10 control points and a knot vector 0, 0, 0, 0, 1, 2, 3, 4, 5, 5, 5, 5. The knot vector 0, 0, 0, 0, 1, 3, 4, 5, 6, 6, 6, 6 is not uniform or *nonuniform*. This explains the NU in NURBS. This NU indicates that the knots in a NURBS curve can be nonuniform.

Some characteristics of the NURBS curve depend on the knot values. For example, the NURBS curve becomes less smooth if duplicate knot values occur in the interior of a knot vector. A full-multiplicity knot in the interior of a knot vector means that a sharp kink can occur somewhere along the NURBS curve.

A particular knot is not usually associated with a single control point, although it is for degree 1 NURBS. In the case of higher-degree NURBS, a group of $2p$ knots corresponds to a group of $p + 1$ control points. For a degree 3 NURBS curve with seven control points and the knots 0, 0, 0, 1, 2, 4, 7, 7, 7, the first four control points are associated with first six knots. The second through fifth control points correspond to the knots 0, 0, 1, 2, 4, 7. The third through sixth control points are grouped with the six knot values 0, 1, 2, 4, 7, 7. The final four control points are associated with the final six knot values.

The mathematical formula that assigns a point on the NURBS curve is a function of B-spline basis functions, degree, knots, control points, and weights. The BS in NURBS refers to B-splines. The shape of a NURBS curve is modified by changing the knot vector, moving control points, and adjusting the magnitudes of the weights.

Since it tends to be difficult to determine how a curve will be modified in response to changes in the knot vector, most curve modifications are implemented by changing a control point or weight. If a weight w_i is increased or decreased in value, the curve moves toward or away from, respectively, the control point $\mathbf{P}_i$.

9.2 DEFINITION OF B-SPLINE BASIS FUNCTIONS

The concept of a NURBS curve was introduced in Section 9.1. A more thorough description of a B-spline curve is provided in the present section.

Let U be a nondecreasing sequence of $m+1$ real numbers:

$$U = \{u_0, u_1, \dots u_m\} \qquad u_i \le u_{i+1} \qquad i = 0, \dots, m \tag{9.1}$$

The elements u_i of the sequence U (the knot vector) are called *knots* or *knot values.* The ith B-spline basis function of *degree* p, or *order* $p+1$, denoted by $N_{i,p}(u)$, is defined recursively by

$$N_{i,0}(u) = \begin{cases} 1 & \text{if} \quad u_i \le u < u_{i+1} \\ 0 & \text{otherwise} \end{cases} \tag{9.2}$$

$$N_{i,p}(u) = \frac{u - u_i}{u_{i+p} - u_i} N_{i,p-1}(u) + \frac{u_{i+p+1} - u}{u_{i+p+1} - u_{i+1}} N_{i+1,p-1}(u) \tag{9.3}$$

A parametric representation of curves is chosen normally for most shapes in geometric modeling. Thus a two-dimensional curve is not represented by a single function such as $z = f(y)$, but by two functions $y = y(u)$ and $z = z(u)$ of the parameter u. The basis functions $N_{i,p}(u)$ are piecewise polynomials defined on the entire real line so that the *curve parameter* u lies in the open interval $(-\infty, +\infty)$. However, in practice, only the values of u that lie in the closed interval $[u_0, u_m]$ are actually used. The half-open interval $[u_i, u_{i+1})$ is referred to as the ith *knot span*; it can be empty since two consecutive knots may be identical. The B-spline basis function of degree 0, $N_{i,0}(u)$, is a unit step function, equal to zero except in the ith knot span. When the ith knot span is empty, $N_{i,0}(u)$ is identically zero everywhere:

$$N_{i,0}(u) = 0 \qquad \text{for } -\infty < u < \infty \text{ if } u_i = u_{i+1} \tag{9.4}$$

It will be noticed that Eq. (9.3) may generate an indeterminate quotient with zero numerator and zero denominator. For instance, if $p = 1$ and the knot span $i+1$ is empty, $N_{i+1,0}(u)$ and $u_{i+2} - u_{i+1}$ are both zero. It is part of the definition of $N_{i,p}(u)$ in Eq. (9.3) that such indeterminate forms are to be taken equal to zero.

The set of distinct knots in the knot vector U define the ranges of the curve parameter u for which the B-spline basis function is defined as a polynomial segment. For instance, if the distinct knots, arranged in increasing order, in U are a, b, and c, it is possible to write a polynomial in u for $a \le u \le b$ and another polynomial for $b \le u \le c$ to express any of the B-spline basis functions as explicit functions of u. Once the degree p, or the order $p+1$, is fixed the knots completely determine the B-spline basis functions. There are several kinds of knot vectors in use in the CAD literature. In this book, only *nonperiodic* knot vectors, which are also called *open* or *clamped* knot vectors, are used. The defining characteristic of a nonperiodic knot vector is that the first and last knots are repeated $p+1$ times:

$$U = \{\underbrace{a, \dots, a}_{p+1}, u_{p+1}, \dots, u_{m-p-1}, \underbrace{b, \dots, b}_{p+1}\} \tag{9.5}$$

Any knot other than the first knots a or the last knots b is referred to as an *interior knot*. When the number of knots is $m+1$, there are $n+1$ B-spline basis functions where

$$n = m - p - 1 \tag{9.6}$$

A nonperiodic knot vector is said to be *uniform* if all the interior knots are equally spaced, that is, if there is a real number d such that

$$u_{i+1} = u_i + d \qquad \text{for all } p \leq i \leq m - p - 1 \tag{9.7}$$

If this condition does not hold, the knot vector is *nonuniform*. In particular, the repeated occurrence of any interior knot makes the knot vector nonuniform.

In practice, the evaluation of the basis functions at a given value of the curve parameter u is done by a numerical algorithm designed for this purpose. It is generally not desirable or efficient to generate the piecewise polynomials defining the basis functions symbolically first. Example 9.1 shows the piecewise polynomials in symbolic form only to display the basis functions for understanding the definitions.

Example 9.1 Determination of Basis Functions. As an example knot vector U, let

$$\begin{aligned} U &= \{0, 0, 0, 1, 1, 2, 3, 3, 3\} = \{u_0, u_1, u_2, u_3, u_4, u_5, u_6, u_7, u_8\} \\ i &= 0, 1, 2, 3, 4, 5, 6, 7, 8 \end{aligned} \tag{1}$$

For this open knot vector the degree is $p = 2$, or the order is $p + 1 = 3$. The number of knots is $m + 1 = 9$. The number of basis functions is $n + 1$, where (Eq. 9.6)

$$n = m - p - 1 = 5 \tag{2}$$

The B-spline basis functions are

$$N_{i,2}(u) \qquad \text{for } 0 \leq i \leq 5 \tag{3}$$

The recursive nature of the definition of Eq. (9.3), however, does not allow these functions to be written out immediately. The functions

$$N_{i,1}(u) \qquad \text{for } 0 \leq i \leq 6 \tag{4}$$

and

$$N_{i,0}(u) \qquad \text{for } 0 \leq i \leq 7 \tag{5}$$

are also needed, the order of generation being $N_{i,0}$, $N_{i,1}$, $N_{i,2}$.

The knot vector U defines eight intervals, in this case

$$[0, 0) \quad [0, 0) \quad [0, 1) \quad [1, 1) \quad [1, 2) \quad [2, 3) \quad [3, 3) \quad [3, 3] \tag{6}$$

Among these intervals, the ones with nonzero length, where the basis functions $N_{i,0}$ are not identically zero, are

$$[0, 1) \quad [1, 2) \quad [2, 3) \tag{7}$$

The zeroth-degree basis functions are written from the definition of Eq. (9.2) to start the recursive evaluation.

$$
\begin{array}{lll}
\text{For } i = 0,\ N_{0,0}(u) = 0 & -\infty < u < \infty & \text{(first interval of (6))} \\
\text{For } i = 1,\ N_{1,0}(u) = 0 & -\infty < u < \infty & \text{(second interval of (6))} \\
\text{For } i = 2,\ N_{2,0}(u) = \begin{cases} 1 \\ 0 \end{cases} & \begin{array}{l} \text{if } \ 0 \leq u < 1 \\ \text{otherwise} \end{array} & \text{(third interval of (6))} \\
\text{For } i = 3,\ N_{3,0}(u) = 0 & -\infty < u < \infty & \text{(fourth interval of (6))} \\
\text{For } i = 4,\ N_{4,0}(u) = \begin{cases} 1 \\ 0 \end{cases} & \begin{array}{l} \text{if } \ 1 \leq u < 2 \\ \text{otherwise} \end{array} & \text{(fifth interval of (6))} \\
\text{For } i = 5,\ N_{5,0}(u) = \begin{cases} 1 \\ 0 \end{cases} & \begin{array}{l} \text{if } \ 2 \leq u < 3 \\ \text{otherwise} \end{array} & \text{(sixth interval of (6))} \\
\text{For } i = 6,\ N_{6,0}(u) = 0 & -\infty < u < \infty & \text{(seventh interval of (6))} \\
\text{For } i = 7,\ N_{7,0}(u) = 0 & -\infty < u < \infty & \text{(eighth interval of (6))}
\end{array}
\tag{8}
$$

The functions $N_{i,1}$, which are piecewise polynomials of degree 1, are computed from the recursive definition of Eq. (9.3). For instance, for $i = 0$ and $p = 1$,

$$
N_{0,1}(u) = \frac{u-0}{0-0} N_{0,0}(u) + \frac{0-u}{0-0} N_{1,0}(u) = 0 \tag{9}
$$

where the indeterminate quotient, zero over zero, arises and is set equal to zero as stipulated by the definition. The function $N_{1,1}$ ($i = 1$, $p = 1$) of Eq. (9.3) is computed as

$$
N_{1,1}(u) = \frac{u-0}{0-0} N_{1,0}(u) + \frac{1-u}{1-0} N_{2,0}(u) = \begin{cases} 1-u & \text{if } \ 0 \leq u < 1 \\ 0 & \text{otherwise} \end{cases} \tag{10}
$$

The remaining functions $N_{i,1}(u)$ are

$$
\begin{aligned}
N_{2,1}(u) &= \begin{cases} u & \text{if } \ 0 \leq u < 1 \\ 0 & \text{otherwise} \end{cases} \\
N_{3,1}(u) &= \begin{cases} 2-u & \text{if } \ 1 \leq u < 2 \\ 0 & \text{otherwise} \end{cases} \\
N_{4,1}(u) &= \begin{cases} u-1 & \text{if } \ 1 \leq u < 2 \\ 3-u & \text{if } \ 2 \leq u < 3 \end{cases} \\
N_{5,1}(u) &= \begin{cases} u-2 & \text{if } \ 2 \leq u < 3 \\ 0 & \text{otherwise} \end{cases} \\
N_{6,1}(u) &= 0 \qquad -\infty < u < \infty
\end{aligned}
\tag{11}
$$

The functions $N_{i,2}$ are piecewise polynomials of degree 2. In the definitions below, which are obtained from Eq. (9.3), the functions are understood to be zero everywhere except on the intervals specified

$$
\begin{aligned}
N_{0,2}(u) &= (u-1)^2 && \text{if} \quad 0 \le u < 1 \\
N_{1,2}(u) &= 2u(1-u) && \text{if} \quad 0 \le u < 1 \\
N_{2,2}(u) &= \begin{cases} u^2 \\ (u-2)^2 \end{cases} && \begin{matrix} \text{if} \quad 0 \le u < 1 \\ \text{if} \quad 1 \le u < 2 \end{matrix} \\
N_{3,2}(u) &= \begin{cases} \dfrac{-7+10u-3u^2}{2} & \text{if} \quad 1 \le u < 2 \\ \dfrac{(u-3)^2}{2} & \text{if} \quad 2 \le u < 3 \end{cases} \\
N_{4,2}(u) &= \begin{cases} \dfrac{(u-1)^2}{2} & \text{if} \quad 1 \le u < 2 \\ \dfrac{-15+14u-3u^2}{2} & \text{if} \quad 2 \le u < 3 \end{cases} \\
N_{5,2}(u) &= (u-2)^2 && \text{if} \quad 2 \le u < 3
\end{aligned}
\tag{12}
$$

These basis functions are shown in Fig. 9.2.

We now return to the general properties of the basis functions $N_{i,p}$. The *local support* property of the B-spline basis functions is

$$N_{i,p}(u) = 0 \qquad \text{if } u \notin [u_i, u_{i+p+1}) \tag{9.8}$$

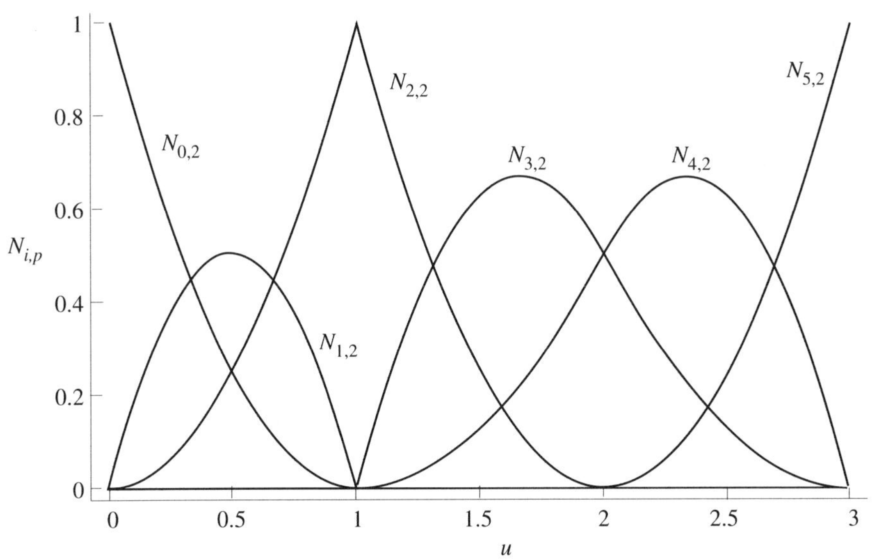

Figure 9.2 B-spline basis functions with knot vector $U = \{0, 0, 0, 1, 1, 2, 3, 3, 3\}$.

In Example 9.1, an example of the local support property is

$$N_{5,2}(u) = 0 \qquad \text{if } u \notin [u_5, u_8) = [2, 3) \tag{9.9}$$

which may be verified directly from the expression for $N_{5,2}(u)$ or from Fig. 9.2.

Another important property that is exploited when evaluating B-spline basis functions is that in any knot span $[u_j, u_{j+1})$, at most $p + 1$ of the basis functions can be nonzero. These basis functions are

$$N_{k,p} \qquad \text{for all } k \text{ such that } j - p \le k \le j \tag{9.10}$$

In Fig. 9.2 with $p = 2$, for instance, between any two distinct consecutive knots, there are at most three nonzero basis functions. Between $u_4 = 1$ and $u_5 = 2$, these functions are $N_{2,2}$, $N_{3,2}$, $N_{4,2}$. This is verified from Fig. 9.2 or the function definitions given in Eq. (12) of Example 9.1, in which the intervals on which the functions are zero have been explicitly specified.

The *nonnegativity* property of the B-spline basis functions states that

$$N_{i,p}(u) \ge 0 \qquad \text{for all } i,\ p,\ u \tag{9.11}$$

All B-spline basis functions $N_{i,p}(u)$ with nonzero degree p assume exactly one maximum value. All derivatives of $N_{i,p}(u)$ with respect to u exist in the interior of a knot span where $N_{i,p}(u)$ is a polynomial in u. At a knot, the number of times that $N_{i,p}(u)$ can be differentiated depends on the *multiplicity* of the knot. The term *multiplicity* refers here to the number of times a knot value is repeated within the knot sequence for a particular basis function. The knot sequence for a particular basis function is defined to be the knots over which the basis function is nonzero, as given by the local support property.

Example 9.2 Multiplicity and Derivatives of Basis Functions. In Example 9.1, with

$$U = \{0, 0, 0, 1, 1, 2, 3, 3, 3\} \tag{1}$$

the basis functions are computed over the following knot spans and are identically zero elsewhere:

$$\begin{array}{lll}
N_{0,2} & \text{from } u_0 \text{ to } u_3 & \{0, 0, 0, 1\} \\
N_{1,2} & \text{from } u_1 \text{ to } u_4 & \{0, 0, 1, 1\} \\
N_{2,2} & \text{from } u_2 \text{ to } u_5 & \{0, 1, 1, 2\} \\
N_{3,2} & \text{from } u_3 \text{ to } u_6 & \{1, 1, 2, 3\} \\
N_{4,2} & \text{from } u_4 \text{ to } u_7 & \{1, 2, 3, 3\} \\
N_{5,2} & \text{from } u_5 \text{ to } u_8 & \{2, 3, 3, 3\}
\end{array} \tag{2}$$

The multiplicity of a knot can be counted from the list of (2) for any given basis function. Thus, the multiplicity of the knot at $u = 1$ for the function $N_{2,2}$ which is nonzero over the knot span $\{0, 1, 1, 2\}$ is two, and the multiplicity of the knot at $u = 0$ for the function $N_{0,2}$ which is nonzero over the knot span $\{0, 0, 0, 1\}$ is three. At a knot, $N_{i,p}(u)$ is $p - k$ times continuously differentiable, where k is the multiplicity of the knot. For the example basis functions, this means that $N_{2,2}$ is $p - k = 0$ times continuously differentiable at $u = 1$, that is, it is continuous but not differentiable, as suggested by Fig. 9.2. For $N_{0,2}$ at the knot $u = 0$, $p - k = -1$, so this function is discontinuous at $u = 0$. This discontinuity is a jump discontinuity of magnitude 1 at $u = 0$, because $N_{0,2}$ is identically zero outside the interval from 0 to 1. The function $N_{1,2}$ has knot multiplicity 2 for its knot at $u = 0$, and it is continuous at $u = 0$ but not differentiable, as its derivative from the left and right sides do not have the same value.

The first derivative of $N_{i,p}(u)$ with respect to u is given by the formula

$$N'_{i,p} = \frac{p}{u_{i+p} - u_i} N_{i,p-1}(u) - \frac{p}{u_{i+p+1} - u_{i+1}} N_{i+1,p-1}(u) \tag{9.12}$$

This formula can be proved by induction on n (Piegl and Tiller, 1997). Let $N_{i,p}^{(k)}(u)$ denote the kth derivative of $N_{i,p}(u)$. By repeated differentiation of Eq. (9.12), the general differentiation formula is

$$N_{i,p}^{(k)}(u) = \frac{pN_{i,p-1}^{(k-1)}(u)}{u_{i+p} - u_i} - \frac{pN_{i+1,p-1}^{(k-1)}(u)}{u_{i+p+1} - u_{i+1}} \tag{9.13}$$

where k takes on integer values between 0 and p. For $k > p$, all derivatives are identically zero. In the derivative calculations, the denominators given by knot differences can become zero, in which case the quotient is defined to be zero.

9.3 B-SPLINE AND RATIONAL B-SPLINE CURVES

A curve defined by B splines, or simply a *B-spline curve*, of degree p is a parametric curve defined by

$$\mathbf{r}(u) = \sum_{i=0}^{n} N_{i,p}(u)\mathbf{P}_i \qquad a \le u \le b \tag{9.14}$$

where the $\mathbf{P}_i$ are the $n + 1$ *control points*, the $N_{i,p}$ are the B-spline basis functions of degree p defined on the nonperiodic, usually nonuniform, knot vector U:

$$U = \{\underbrace{a, \ldots, a}_{p+1}, u_{p+1}, \ldots, u_{m-p-1}, \underbrace{b, \ldots, b}_{p+1}\} \tag{9.15}$$

and $\mathbf{r}(u)$ is the position vector, or the point of the B-spline curve, at the curve parameter value u. There are $n+1$ control points and $m+1$ knots. If the number of control points is specified as well as the degree p of the B-spline curve, the number of required knots is determined since

$$m = n + p + 1 \tag{9.16}$$

Space curves are given by Eq. (9.14) when the $\mathbf{P}_i$ are specified as coordinates of points in three-dimensional space. For the applications at hand, B-spline curves are plane curves in the yz plane and $\mathbf{P}_i$ are points in this plane. The polygon formed by the control points is referred to as the *control polygon*. The control points are also sometimes referred to as *vertices*, meaning the vertices of the control polygon. The first and last points, $\mathbf{P}_0$ and $\mathbf{P}_n$, of the control polygon lie on the B-spline curve. This *endpoint interpolation* property is expressed by

$$\mathbf{r}(a) = \mathbf{P}_0 \qquad \mathbf{r}(b) = \mathbf{P}_n \tag{9.17}$$

Figure 9.3 shows an example B-spline curve. The knot vector U for this curve has been chosen to be uniform nonperiodic:

$$U = \{0, 0, 0, 0, 1, 2, 3, 4, 4, 4, 4\} \tag{9.18}$$

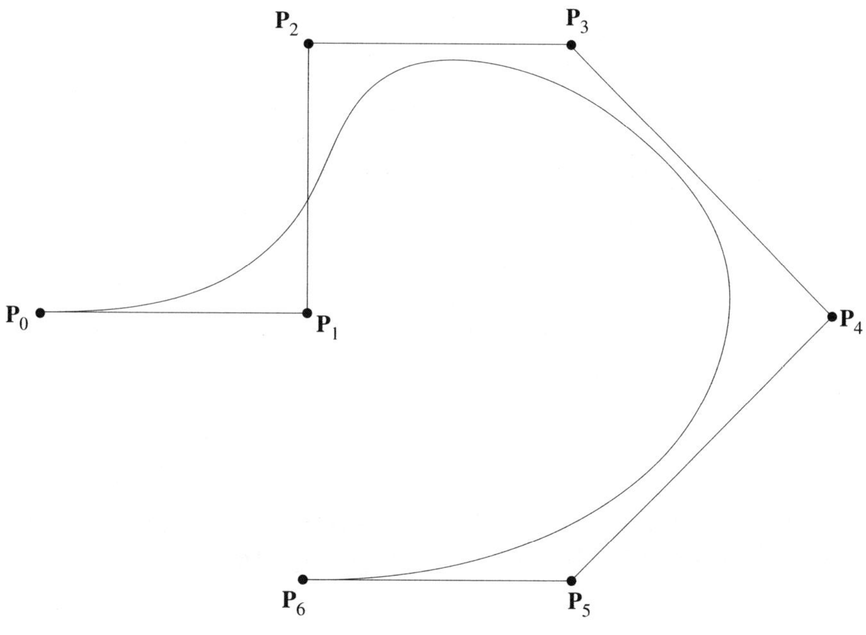

Figure 9.3 Example B-spline curve with knot vector $U = \{0, 0, 0, 0, 1, 2, 3, 4, 4, 4, 4\}$.

It follows from Eq. (9.15) that the B-spline basis function is of degree $p = 3$. The control point coordinates are

$$P = \{(0,0), (1,0), (1,1), (2,1), (3,0), (2,-1), (1,-1)\} \tag{9.19}$$

Since the basis functions $N_{i,p}(u)$ are piecewise polynomials, the curve given parametrically by $\mathbf{r}(u)$ in Eq. (9.14) is also a piecewise polynomial curve. Thus, if u_k and u_{k+1} are two distinct knots in the knot vector U, the part of the curve given by all points $\mathbf{r}(u)$ for u between u_k and u_{k+1} is a polynomial segment.

The control polygon is a piecewise linear approximation to the B-spline curve. Lowering the degree p in general causes the B-spline curve to follow the control polygon more closely. B-spline curves have the *local modification* property. If the control point $\mathbf{P}_k$ is changed, the curve $\mathbf{r}(u)$ itself changes only in the interval $[u_k, u_{k+p+1})$. This is a consequence of the local support property of the basis functions according to which $N_{k,p}(u)$, the coefficient of $\mathbf{P}_k$ in the formula for $\mathbf{r}(u)$, is identically zero outside the interval $[u_k, u_{k+p+1})$.

The B-spline curve $\mathbf{r}(u)$ has derivatives of all orders in the interior of knot spans. At a knot of multiplicity k, it is at least $p-k$ times continuously differentiable. These properties are a direct consequence of the continuity and differentiability properties of the B-spline basis functions discussed in the preceding section. The jth derivative of $\mathbf{r}(u)$ is obtained in terms of the jth derivatives of the basis functions as

$$\frac{d^j\mathbf{r}}{du^j} = \mathbf{r}^{(j)} = \sum_{i=0}^{n} N_{i,p}^{(j)}(u)\mathbf{P}_i \tag{9.20}$$

where a superscript in parentheses denotes the order of differentiation. At the first and last control points, the B-spline curve is tangent to the control polygon.

The ith *rational B-spline basis* function $R_{i,p}(u)$ of degree p is defined by

$$R_{i,p}(u) = \frac{N_{i,p}(u)w_i}{\sum_{j=0}^{n} N_{j,p}(u)w_j} \tag{9.21}$$

where the $n+1$ real numbers w_i are called *weights*. The weights will be assumed to be strictly positive:

$$w_j > 0 \qquad 0 \le j \le n \tag{9.22}$$

but this is not an essential requirement, and is occasionally relaxed to allow zero or negative weights. These basis functions are piecewise rational functions of u for $a \le u \le b$, a and b being the first and last knots as before. The parametric equation of a *nonuniform rational B-spline* (NURBS) curve of degree p, or order $p+1$, is

$$\mathbf{r}(u) = \sum_{i=0}^{n} R_{i,p}(u)\mathbf{P}_i \qquad a \le u \le b \tag{9.23}$$

Thus, a rational B-spline curve is specified by its degree p, the knot vector U, the control point coordinates $\mathbf{P}_i$, and the weights w_i. One function of the weights is to control the shape of the resulting curve. Thus, the introduction of the weights has provided a means of modifying the shape of the curve in addition to the effects of changing the knot vector and moving the control points. When w_j is increased, the curve is in general pulled toward the control point $\mathbf{P}_j$.

The rational basis functions $R_{i,p}$ inherit some of their most important characteristics from the B-spline basis functions $N_{i,p}$. With positive weights the *nonnegativity property* is preserved:

$$R_{i,p}(u) \geq 0 \qquad \text{for all } i,\ p,\ \text{and } a \leq u \leq b \tag{9.24}$$

The *local support* property is preserved:

$$R_{i,p}(u) = 0 \qquad \text{if } u \notin [u_i, u_{i+p+1}) \tag{9.25}$$

and in the knot span $[u_i, u_{i+1})$ at most $p + 1$ of the rational basis functions can be nonzero, namely $R_{i-p,p}, \ldots, R_{i,p}$. For $p > 0$, all rational basis functions attain exactly one maximum on the interval $a \leq u \leq b$. All derivatives of $R_{i,p}(u)$ exist in the interior of any knot span. At a knot, $R_{i,p}(u)$ can be differentiated $p - k$ times, where k is the multiplicity of the knot. If all weights are chosen to be identical, the rational basis functions become identical to the B-spline basis functions. That is, with identical weights $w_i = C$ for $0 \leq i \leq n$, the rational B-spline curve becomes a B-spline curve. This follows from Eq. (9.21) because $\sum_{j=0}^{n} N_{j,p}(u) = 1$.

Figure 9.4 shows the rational basis functions for which the knot vector was taken to be

$$U = \{0, 0, 0, 0, 1, 2, 3, 4, 4, 4, 4\} \tag{9.26}$$

The degree of the basis functions is $p = 3$ and the number of knots is $m + 1 = 11$, so that $m = 10$. There are (Eq. 9.6) $n + 1 = m - p = 7$ basis functions and therefore the number of weights is also 7. The weights were chosen to be

$$W = \{1, 2, 3, 4, 3, 2, 1\} \tag{9.27}$$

The derivatives of the rational basis functions $R_{i,p}(u)$ with respect to the curve parameter u can be computed in terms of the corresponding derivatives of the B-spline basis functions $N_{i,p}(u)$. Defining

$$D_p(u) = \sum_{j=0}^{n} w_j N_{j,p}(u) \tag{9.28}$$

we write the ith rational basis function as

$$R_{i,p}(u) = \frac{w_i N_{i,p}(u)}{D_p(u)} \tag{9.29}$$

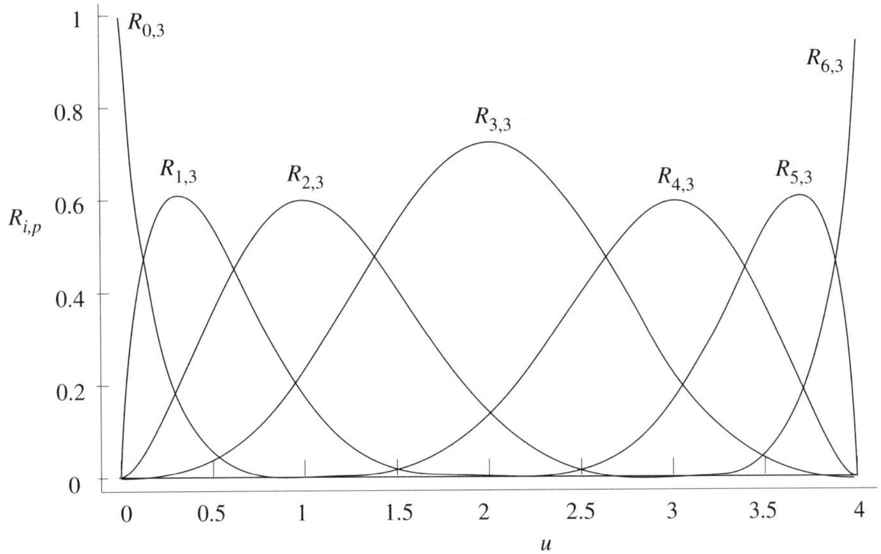

Figure 9.4 Cubic rational basis functions.

Apply Leibnitz's rule for differentiating a product to calculate the kth derivative:

$$\begin{aligned}\frac{d^k}{du^k}\left(D_p(u)R_{i,p}(u)\right) = (D_p(u)R_{i,p}(u))^{(k)} &= \sum_{j=0}^{k}\binom{k}{j} D_p^{(j)}(u)R_{i,p}^{(k-j)}(u) \\ &= D_p(u)R_{i,p}^{(k)}(u) + \sum_{j=1}^{k}\binom{k}{j} D_p^{(j)}(u)R_{i,p}^{(k-j)}(u)\end{aligned} \tag{9.30}$$

where the coefficients are the binomial coefficients

$$\binom{k}{j} = \frac{k!}{(k-j)!\,j!} \tag{9.31}$$

with $k! = 1 \cdot 2 \cdot 3 \cdots k$. Since, from Eq. (9.29),

$$(D_p(u)R_{i,p}(u))^{(k)} = w_i N_{i,p}^{(k)}(u) \tag{9.32}$$

we obtain from Eq. (9.30) the recursive formula for differentiating the rational basis functions:

$$R_{i,p}^{(k)}(u) = \frac{w_i N_{i,p}^{(k)}(u) - \sum_{j=1}^{k}\binom{k}{j} D_p^{(j)}(u)R_{i,p}^{(k-j)}(u)}{D_p(u)} \tag{9.33}$$

From the definition of Eq. (9.28), the derivatives of $D_p(u)$ are given in terms of the derivatives of the B-spline basis functions as

$$D_p^{(j)}(u) = \sum_{r=0}^{n} w_r N_{r,p}^{(j)}(u) \tag{9.34}$$

The derivatives of the position vector $\mathbf{r}(u)$ of Eq. (9.23) of a point on the rational B-spline curve are given in terms of the derivatives of the rational basis functions

$$\mathbf{r}^{(k)} = \sum_{i=0}^{n} R_{i,p}^{(k)}(u)\mathbf{P}_i \tag{9.35}$$

9.4 USE OF RATIONAL B-SPLINE CURVES IN THIN-WALLED BEAM ANALYSIS

The cross-sectional shape of a thin-walled beam can be specified by rational B-spline curves. The curves define the median line of the section, and a wall thickness given for each curve completes the specification of the cross-sectional geometry. When more than one curve is given, the curves should connect to each other to form either an open or a closed cross section. A rational B-spline curve is defined completely by its degree p, or equivalently, its order $p + 1$, its control points, knot vectors, and weights. If a knot vector is not given, a uniform knot vector can be introduced. If weights are not given, all weights are often assumed to be equal to 1. To describe complicated cross sections, it may sometimes be necessary to generate the rational B-splines by interpolation. Several interpolation algorithms are described in full detail in Piegl and Tiller (1997). Cross sections made of straight-line segments or simple curves can be described directly from a knowledge of how these components are represented as B-spline curves.

Example 9.3 B-Spline Curve for a Straight Line. A straight-line segment passing through the points $\mathbf{P}_0$ and $\mathbf{P}_1$ is a B-spline curve of degree $p = 1$. Since the number of control points is 2, $n + 1 = 2$ and the number of knots is (Eq. 9.6)

$$m + 1 = n + p + 2 = 4 \tag{1}$$

Choose the knot vector

$$U = \{0, 0, 1, 1\} \tag{2}$$

The B-spline basis functions for $0 \leq u \leq 1$ are

$$\begin{aligned} N_{0,0} &= 0 \\ N_{1,0} &= 1 \\ N_{2,0} &= 0 \end{aligned} \tag{3}$$

$$N_{0,1} = 1 - u$$

$$N_{1,1} = u$$

The B-spline curve is

$$\mathbf{r}(u) = N_{0,1}\mathbf{P}_0 + N_{1,1}\mathbf{P}_1 = (1 - u)\mathbf{P}_0 + u\mathbf{P}_1 \tag{4}$$

which is the parametric equation of a line segment joining point $\mathbf{P}_0$ to point $\mathbf{P}_1$. Thus, the specification of a straight-line segment requires only the specification of the coordinates of the two points $\mathbf{P}_0$, $\mathbf{P}_1$ and the order $p + 1 = 2$ of the B-spline curve. A uniform knot vector can be generated internally by a computer program, and the default weights would be appropriate since a B-spline curve is constructed, not a rational curve.

Example 9.4 ***Circles.*** A circle can be represented by rational quadratic B-splines. The order of the NURBS curve is 3, so that the degree p is 2. The control polygon has $n + 1 = 7$ control points. These points lie at the vertices and on the midpoints of the sides of an equilateral triangle, with the last point coinciding with the first (Fig. 9.5). The ordering of the control points is such that as the three sides of the triangle are traced in a counterclockwise sense, the control points are encountered in sequence. The knot vector is nonuniform:

$$U = \{0, 0, 0, 1, 1, 2, 2, 3, 3, 3\}$$

and the weights assigned to the control points are

$$W = \left\{1, \tfrac{1}{2}, 1, \tfrac{1}{2}, 1, \tfrac{1}{2}, 1\right\}$$

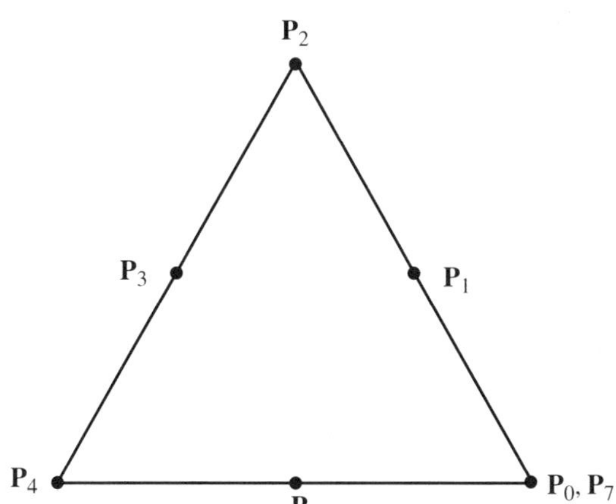

Figure 9.5 Control points and equilateral triangle for the representation of a circle using B-splines.

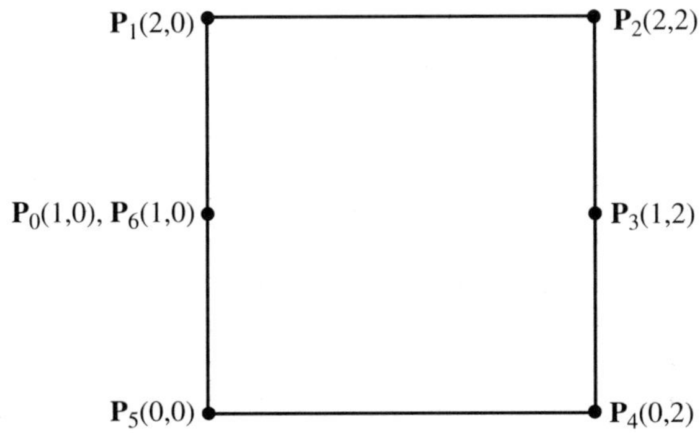

Figure 9.6 Control points and square for the representation of a circle using B-splines.

It is also possible to represent a circle by a seven-point square-based NURBS curve. In that case, the degree is $p = 2$ and the control points coordinates are

$$P = \{(1,0), (2,0), (2,2), (1,2), (0,2), (0,0), (1,0)\}$$

for a circle of radius 1 (Fig. 9.6). The knot vector is

$$U = \left\{0, 0, 0, \tfrac{1}{4}, \tfrac{1}{2}, \tfrac{1}{2}, \tfrac{3}{4}, 1, 1, 1\right\}$$

and the weights are

$$W = \left\{1, \tfrac{1}{2}, \tfrac{1}{2}, 1, \tfrac{1}{2}, \tfrac{1}{2}, 1\right\}$$

The representation of conics and circular arcs is described in detail in Chapter 7 of Piegl and Tiller (1997).

REFERENCES

Mortenson, M. (1997). *Geometric Modeling*, 2nd ed., Wiley, New York.

Piegl, L., and Tiller, W. (1997). *The NURBS Book*, Springer-Verlag, New York.

CHAPTER 10

SHAPE OPTIMIZATION OF THIN-WALLED SECTIONS

In this chapter we describe a shape design method for a thin-walled beam cross section. The cross-sectional shape of the beam is assumed to be defined by nonuniform rational B-spline curves as described in Chapter 9. The optimization problem is couched in terms of cross-sectional properties and stresses. The design variables are the wall thicknesses and the control points and weights of the B-spline curves that define the median line of the section. The optimized variable can be any of the cross-sectional properties or any of the stresses calculated by the methods of previous chapters.

A property of matrices that is used frequently in this chapter is that the transpose of the product of two matrices is equal to the product of the individual transposes taken in reverse order. For the two matrices **A** and **B**, this property is expressed as

$$(\mathbf{AB})^{\mathrm{T}} = \mathbf{B}^{\mathrm{T}}\mathbf{A}^{\mathrm{T}}$$

Typically in this chapter, $\mathbf{AB} = \alpha$, where α is a scalar. Then $\alpha = \alpha^{\mathrm{T}} = \mathbf{AB} = \mathbf{B}^{\mathrm{T}}\mathbf{A}^{\mathrm{T}}$.

10.1 DESIGN VELOCITY FIELD

In the shape design of a thin-walled beam cross section to be described in this chapter, the *design variables* of interest are the y, z coordinates of the control points of the NURBS curves defining the median line of the section, the weights assigned to these control points, and the wall thicknesses. The *sensitivity* of a dependent variable F with respect to a design variable β is defined as the partial derivative of F with respect to β. When any one of the design variables undergoes a change, the finite element mesh of the cross section changes. Thus, the y, z coordinates of all the nodes in the mesh have sensitivities with respect to the design variables. The set of all sensitivities of the nodal point coordinates is called the *design velocity field.*

The design velocity field plays a fundamental role in sensitivity analysis and optimization. If an analytical formula were available to compute all the nodal coordinates in the finite element mesh of the cross section, the sensitivities of the nodal coordinates would be calculable by differentiating that formula. This is normally not true of real-world mesh generators, which usually make use of complicated algorithms and heuristics to generate a mesh. The mesh generator of the thin-walled beam analysis program described in Appendix A, for example, uses special algorithms to generate the elements in the neighborhood of two intersecting NURBS curves. To calculate the design velocity field suitable for use with any mesh generator, it is necessary to know exactly how the mesh generator determines the nodal coordinates. In the present case it is also important to develop special algorithms to be able to differentiate the nodal coordinates that are shared between two NURBS curves.

Sensitivities of the Tangent and Normal Vectors In this section, formulas for the sensitivities of the unit tangent vector and the unit normal vector to a single NURBS curve are presented. The derivatives of these two vectors are fundamental to the determination of the design velocity field, because most nodal coordinates of the finite element mesh are obtained from formulas containing the normal and the tangent. For those nodal coordinates that are obtained from special considerations, special case-by-case formulas would be required, depending on the procedures built into the mesh generator.

For a given nonuniform rational B-spline among the curves defining the median line of the thin-walled section in the yz plane, let $\mathbf{P}_k$, $0 \leq k \leq n$, denote the position vectors of the $n + 1$ control points and let w_k be the weight assigned to the kth control point. As in Chapter 9, let u denote the curve parameter, and let $N_{i,p}(u)$ be the ith B-spline basis function of degree p. The kth rational basis function $R_{k,p}(u)$, abbreviated by suppressing the degree p to $R_k(u)$, is defined in Eq. (9.21) as

$$R_k(u) = \frac{w_k N_k(u)}{\sum_{i=0}^{n} w_i N_i(u)} \tag{10.1}$$

so that the position vector $\mathbf{r}(u)$ (Eq. 9.23) of the point at the curve parameter u may be written as

$$\mathbf{r}(u) = \sum_{k=0}^{n} R_k(u)\mathbf{P}_k \tag{10.2}$$

Together these expressions appear as

$$\mathbf{r}(u) = \frac{\sum_{i=0}^{n} w_i N_i(u)\mathbf{P}_i}{\sum_{i=0}^{n} w_i N_i(u)} \tag{10.3}$$

Let $\mathbf{t}(u)$ be the unit tangent vector to the NURBS curve at the curve parameter u, and let $\mathbf{n}(u)$ be the unit normal vector. The tangent vector is given by (Eq. 1.37)

$$\mathbf{t} = \frac{\mathbf{r}'}{|\,\mathbf{r}'\,|} \tag{10.4}$$

where the prime denotes partial differentiation with respect to u

$$\mathbf{r}' = \frac{\partial \mathbf{r}}{\partial u} \tag{10.5}$$

The derivative of the unit tangent vector with respect to any scalar variable β is perpendicular to $\mathbf{t}$. Thus,

$$\mathbf{t} \cdot \frac{\partial \mathbf{t}}{\partial \beta} = \frac{1}{2}\frac{\partial(\mathbf{t} \cdot \mathbf{t})}{\partial \beta} = 0 \tag{10.6}$$

since $\mathbf{t} \cdot \mathbf{t} = 1$. It follows that the derivative of the unit tangent vector with respect to any design variable can be written as

$$\frac{\partial \mathbf{t}}{\partial \beta} = \alpha \mathbf{n} \tag{10.7}$$

for some scalar α. The scalar factor α can be determined by differentiating Eq. (10.4), which defines $\mathbf{t}$, $\mathbf{r}' = |\,\mathbf{r}'\,|\mathbf{t}$, with respect to β:

$$\frac{\partial \mathbf{r}'}{\partial \beta} = \frac{\partial |\,\mathbf{r}'\,|}{\partial \beta}\mathbf{t} + |\,\mathbf{r}'\,|\frac{\partial \mathbf{t}}{\partial \beta} \tag{10.8}$$

The component along $\mathbf{n}$ of the vector on the left side of this equation is

$$\frac{\partial \mathbf{r}'}{\partial \beta} \cdot \mathbf{n} = \frac{\partial |\,\mathbf{r}'\,|}{\partial \beta}\mathbf{t} \cdot \mathbf{n} + |\,\mathbf{r}'\,|\frac{\partial \mathbf{t}}{\partial \beta} \cdot \mathbf{n} = |\,\mathbf{r}'\,|\alpha \mathbf{n} \cdot \mathbf{n} = |\,\mathbf{r}'\,|\alpha \tag{10.9}$$

from which it is seen that the factor α is given by

$$\alpha = \frac{\partial \mathbf{t}}{\partial \beta} \cdot \mathbf{n} = \frac{1}{|\,\mathbf{r}'\,|}\left(\frac{\partial \mathbf{r}'}{\partial \beta} \cdot \mathbf{n}\right) \tag{10.10}$$

where Eq. (10.4) has been introduced. Thus, from Eq. (10.7),

$$\frac{\partial \mathbf{t}}{\partial \beta} = \frac{1}{|\,\mathbf{r}'\,|}\left(\frac{\partial \mathbf{r}'}{\partial \beta} \cdot \mathbf{n}\right)\mathbf{n} \tag{10.11}$$

The derivative of the tangent vector with respect to the weight w_j is considered next. The derivative of the position vector $\mathbf{r}$ with respect to w_j is found by differentiating Eq. (10.3) in the form

$$\mathbf{r}(u)\sum_{i=0}^{n} w_i N_i(u) = \sum_{i=0}^{n} w_i N_i(u)\mathbf{P}_i \tag{10.12}$$

with respect to w_j, which gives

$$N_j(u)\mathbf{r}(u) + \frac{\partial \mathbf{r}}{\partial w_j} \sum_{i=0}^{n} w_i N_i(u) = N_j(u)\mathbf{P}_j \tag{10.13}$$

Hence, using Eq. (10.1) yields

$$\frac{\partial \mathbf{r}}{\partial w_j} = \frac{N_j(u)[\mathbf{P}_j - \mathbf{r}(u)]}{\sum_{i=0}^{n} w_i N_i(u)} = \frac{R_j(u)}{w_j}[\mathbf{P}_j - \mathbf{r}(u)] \tag{10.14}$$

The first derivative of $\mathbf{r}$ with respect to w_j is seen from this equation to be a vector pointing from the point $\mathbf{r}$ of the NURBS curve to the control point at $\mathbf{P}_j$. An increase in w_j has the effect of pulling curve points toward the control point $\mathbf{P}_j$ along straight lines joining the points to the point $\mathbf{P}_j$. Because the NURBS curves considered here pass through the first and last control points, as explained in Chapter 9, the sensitivity of the curve points $\mathbf{P}_0$ and $\mathbf{P}_n$ to any weight w_j is zero.

The derivative of $\mathbf{r}'(u)$ with respect to the weight w_j is found using Eq. (10.14):

$$\frac{\partial \mathbf{r}'}{\partial w_j} = \frac{\partial^2 \mathbf{r}}{\partial w_j \partial u} = \frac{\partial^2 \mathbf{r}}{\partial u \partial w_j} = \frac{R_j'}{w_j}(\mathbf{P}_j - \mathbf{r}) - \frac{R_j}{w_j}\mathbf{r}' \tag{10.15}$$

Equation (10.15) together with Eq. (10.11) yields the derivative of the unit tangent vector with respect to w_j:

$$\frac{\partial \mathbf{t}}{\partial w_j} = \frac{1}{|\,\mathbf{r}'\,|}\left(\frac{\partial \mathbf{r}'}{\partial w_j} \cdot \mathbf{n}\right)\mathbf{n} = \frac{R_j'}{w_j|\,\mathbf{r}'\,|}\left[(\mathbf{P}_j - \mathbf{r}) \cdot \mathbf{n}\right]\mathbf{n} \tag{10.16}$$

since $\mathbf{r}' \cdot \mathbf{n} = |\,\mathbf{r}'\,|\mathbf{t} \cdot \mathbf{n} = 0$. The derivative of the rational basis function R_j appearing in Eq. (10.16) is evaluated using Eq. (9.33) derived for the derivatives of the B-spline basis functions:

$$R_j' = \frac{w_j N_j' - R_j \sum_{i=0}^{n} w_i N_i'}{\sum_{i=0}^{n} w_i N_i} \tag{10.17}$$

The derivatives of $\mathbf{t}$ with respect to the coordinates y_j, z_j of the control point $\mathbf{P}_j$ are found by first writing the derivative of $\mathbf{r}$ of Eq. (9.23) in the form

$$\mathbf{r}' = \sum_{i=0}^{n} R_i' \mathbf{P}_i = \sum_{i=0}^{n} R_i'(y_i \mathbf{j} + z_i \mathbf{k}) \tag{10.18}$$

where $\mathbf{j}$, $\mathbf{k}$ are the unit vectors along the coordinate axes y and z. Since R_i' is independent of control point coordinates:

$$\begin{aligned} \frac{\partial \mathbf{r}'}{\partial y_j} &= R_j' \mathbf{j} \\ \frac{\partial \mathbf{r}'}{\partial z_j} &= R_j' \mathbf{k} \end{aligned} \tag{10.19}$$

and an application of Eq. (10.11) yields

$$\frac{\partial \mathbf{t}}{\partial y_j} = \frac{R_j'}{|\mathbf{r}'|}(\mathbf{j} \cdot \mathbf{n})\mathbf{n} \tag{10.20}$$

$$\frac{\partial \mathbf{t}}{\partial z_j} = \frac{R_j'}{|\mathbf{r}'|}(\mathbf{k} \cdot \mathbf{n})\mathbf{n} \tag{10.21}$$

The unit normal vector **n** can be obtained from (Eq. 1.34)

$$\mathbf{n} = \mathbf{t} \times \mathbf{i} \tag{10.22}$$

where **i** is the unit vector along the x axis. Thus, the derivatives of the unit normal **n** are determined from the derivatives of the unit tangent **t** by the formula

$$\frac{\partial \mathbf{n}}{\partial \beta} = \frac{\partial \mathbf{t}}{\partial \beta} \times \mathbf{i} \tag{10.23}$$

Since the derivative of **t** always has the form of Eq. (10.7), the derivative of the normal vector is

$$\frac{\partial \mathbf{n}}{\partial \beta} = \frac{\partial \mathbf{t}}{\partial \beta} \times \mathbf{i} = \alpha \mathbf{n} \times \mathbf{i} = -\alpha \mathbf{t} = -\frac{1}{|\mathbf{r}'|}\left(\frac{\partial \mathbf{r}'}{\partial \beta} \cdot \mathbf{n}\right)\mathbf{t} \tag{10.24}$$

Consequently, the sensitivity formulas for **n** are

$$\frac{\partial \mathbf{n}}{\partial w_j} = -\frac{R_j'}{w_j|\mathbf{r}'|}\left[(\mathbf{P}_j - \mathbf{r}) \cdot \mathbf{n}\right]\mathbf{t} \tag{10.25}$$

$$\frac{\partial \mathbf{n}}{\partial y_j} = -\frac{R_j'}{|\mathbf{r}'|}(\mathbf{j} \cdot \mathbf{n})\mathbf{t} \tag{10.26}$$

$$\frac{\partial \mathbf{n}}{\partial z_j} = -\frac{R_j'}{|\mathbf{r}'|}(\mathbf{k} \cdot \mathbf{n})\mathbf{t} \tag{10.27}$$

where Eqs. (10.15) and (10.19) have been introduced.

10.2 DESIGN SENSITIVITY ANALYSIS

Let β denote a parameter that influences the geometry of the cross section. The partial derivative with respect to β of any property found by integration is given by

$$\frac{\partial}{\partial \beta}\int f(y, z)\, dA = \frac{\partial}{\partial \beta}\int f(\eta, \zeta)\,|\mathbf{J}_e|\, d\eta\, d\zeta$$

$$
\begin{aligned}
&= \frac{\partial}{\partial \beta} \sum_{e=1}^{M} \int_{-1}^{1} \int_{-1}^{1} f(\eta, \zeta) \,|\, \mathbf{J}_e(\eta, \zeta) \,|\, d\eta \, d\zeta \\
&= \sum_{e=1}^{M} \int_{-1}^{1} \int_{-1}^{1} \left[\frac{\partial f}{\partial \beta} |\, \mathbf{J}_e \,| + f(\eta, \zeta) \frac{\partial |\, \mathbf{J}_e \,|}{\partial \beta} \right] d\eta \, d\zeta
\end{aligned}
\tag{10.28}
$$

Equations (4.21) and (4.27) were employed in setting up Eq. (10.28). In Eq. (10.28), e is the element number, $\mathbf{J}_e$ is the Jacobian matrix of element e, M is the number of elements, and η, ζ are the reference element coordinates, as shown in Figure 4.1.

The derivative of the Jacobian matrix of Eq. (4.15) is

$$
\frac{\partial \mathbf{J}_e}{\partial \beta} = \begin{bmatrix} \dfrac{\partial \mathbf{N}}{\partial \eta} \\ \dfrac{\partial \mathbf{N}}{\partial \zeta} \end{bmatrix} \begin{bmatrix} \dfrac{\partial \mathbf{y}_e}{\partial \beta} & \dfrac{\partial \mathbf{z}_e}{\partial \beta} \end{bmatrix} \tag{10.29}
$$

where $\mathbf{N}$ is the row vector of shape functions as in Chapter 4. Equation (10.29) shows that the derivative of the Jacobian matrix is obtained by substituting derivatives of the nodal coordinate vectors instead of the nodal vectors themselves in the formula for the Jacobian matrix. The derivative of $|\, \mathbf{J}_e \,| = J_{11} J_{22} - J_{12} J_{21}$ (Eq. 4.18) can therefore be written as

$$
\frac{\partial |\, \mathbf{J}_e \,|}{\partial \beta} = J'_{11} J_{22} + J_{11} J'_{22} - J'_{12} J_{21} - J_{12} J'_{21} \tag{10.30}
$$

where the prime denotes partial differentiation with respect to β.

As the formula of Eq. (10.29) for the derivative of $\mathbf{J}_e$ shows, the derivatives of the nodal coordinates play a central role in the application of Eq. (10.28) to calculate

TABLE 10.1 Formulas Needed in the Integrand of Eq. (10.28)

Property $\int f \, dA$	$f(\eta, \zeta)$	$\dfrac{\partial f}{\partial \beta}(\eta, \zeta)$
A	1 (Eq. 4.23)	0
Q_y	$\mathbf{N}\mathbf{z}_e$ (Eq. 4.36)	$\mathbf{N}\dfrac{\partial \mathbf{z}_e}{\partial \beta}$
Q_z	$\mathbf{N}\mathbf{y}_e$ (Eq. 4.36)	$\mathbf{N}\dfrac{\partial \mathbf{y}_e}{\partial \beta}$
I_y	$(\mathbf{N}\mathbf{z}_e)^2$ (Eq. 4.39)	$2\mathbf{N}\mathbf{z}_e\mathbf{N}\dfrac{\partial \mathbf{z}_e}{\partial \beta}$
I_z	$(\mathbf{N}\mathbf{y}_e)^2$ (Eq. 4.39)	$2\mathbf{N}\mathbf{y}_e\mathbf{N}\dfrac{\partial \mathbf{y}_e}{\partial \beta}$
I_{yz}	$\mathbf{N}\mathbf{y}_e\mathbf{N}\mathbf{z}_e$ (Eq. 4.41)	$\mathbf{N}\dfrac{\partial \mathbf{y}_e}{\partial \beta}\mathbf{N}\mathbf{z}_e + \mathbf{N}\mathbf{y}_e\mathbf{N}\dfrac{\partial \mathbf{z}_e}{\partial \beta}$

derivatives of cross-sectional properties found by quadrature. Table 10.1 shows that the derivatives of the integrands in these calculations are also dependent on the nodal coordinate derivatives. Thus, one essential step in design sensitivity analysis is the computation of the design velocity field

$$\frac{\partial \mathbf{y}_e}{\partial \beta} \qquad \frac{\partial \mathbf{z}_e}{\partial \beta} \tag{10.31}$$

for all elements e in the finite element mesh and all parameters β of interest.

10.2.1 Derivatives of Geometric Quantities

The coordinates of the centroid are given by Eq. (1.56) as $y_C = Q_z/A$ and $z_C = Q_y/A$. The partial derivatives of these coordinates are

$$\begin{aligned}
\frac{\partial y_C}{\partial \beta} &= \frac{1}{A}\frac{\partial Q_z}{\partial \beta} + Q_z \frac{\partial (1/A)}{\partial \beta} = \frac{1}{A}\frac{\partial Q_z}{\partial \beta} - \frac{Q_z}{A^2}\frac{\partial A}{\partial \beta} = \frac{1}{A}\left(\frac{\partial Q_z}{\partial \beta} - y_C \frac{\partial A}{\partial \beta}\right) \\
\frac{\partial z_C}{\partial \beta} &= \frac{1}{A}\frac{\partial Q_y}{\partial \beta} + Q_y \frac{\partial (1/A)}{\partial \beta} = \frac{1}{A}\frac{\partial Q_y}{\partial \beta} - \frac{Q_y}{A^2}\frac{\partial A}{\partial \beta} = \frac{1}{A}\left(\frac{\partial Q_y}{\partial \beta} - z_C \frac{\partial A}{\partial \beta}\right)
\end{aligned} \tag{10.32}$$

The derivatives on the right side of these equations are found using Eq. (10.28) and Table 10.1:

$$\begin{aligned}
\frac{\partial A}{\partial \beta} &= \sum_e \int_{-1}^{1}\int_{-1}^{1} \frac{\partial |\mathbf{J}_e|}{\partial \beta}\, d\eta\, d\zeta \\
\frac{\partial Q_y}{\partial \beta} &= \sum_e \int_{-1}^{1}\int_{-1}^{1} \left(\mathbf{N}\frac{\partial \mathbf{z}_e}{\partial \beta}|\mathbf{J}_e| + \mathbf{N}\mathbf{z}_e \frac{\partial |\mathbf{J}_e|}{\partial \beta}\right) d\eta\, d\zeta \\
\frac{\partial Q_z}{\partial \beta} &= \sum_e \int_{-1}^{1}\int_{-1}^{1} \left(\mathbf{N}\frac{\partial \mathbf{y}_e}{\partial \beta}|\mathbf{J}_e| + \mathbf{N}\mathbf{y}_e \frac{\partial |\mathbf{J}_e|}{\partial \beta}\right) d\eta\, d\zeta
\end{aligned} \tag{10.33}$$

The moments of inertia referred to the centroidal coordinate system are given by Eq. (1.60):

$$\begin{aligned}
I_{\bar{y}} &= I_y - z_C^2 A \\
I_{\bar{z}} &= I_z - y_C^2 A \\
I_{\bar{y}\bar{z}} &= I_{yz} - y_C z_C A
\end{aligned} \tag{10.34}$$

The derivatives of these quantities are given by

$$\frac{\partial I_{\bar{y}}}{\partial \beta} = \frac{\partial I_y}{\partial \beta} - 2Q_y \frac{\partial z_C}{\partial \beta} - z_C^2 \frac{\partial A}{\partial \beta}$$

$$\frac{\partial I_{\bar{z}}}{\partial \beta} = \frac{\partial I_z}{\partial \beta} - 2Q_z \frac{\partial y_C}{\partial \beta} - y_C^2 \frac{\partial A}{\partial \beta} \tag{10.35}$$
$$\frac{\partial I_{\bar{y}\bar{z}}}{\partial \beta} = \frac{\partial I_{yz}}{\partial \beta} - Q_y \frac{\partial y_C}{\partial \beta} - Q_z \frac{\partial z_C}{\partial \beta} - y_C z_C \frac{\partial A}{\partial \beta}$$

where the definitions of Eq. (1.56) have been introduced. These expressions are readily placed in the discrete form of Eq. (10.33).

10.2.2 Derivative of the Normal Stress

The normal stress of Eq. (1.101) can be expressed as

$$\sigma_x = \frac{E}{E_r} \left[\frac{N_x}{\widetilde{A}} + \frac{F}{D} \right] \tag{10.36}$$

where the abbreviations

$$F = (\widetilde{I}_{\bar{z}} M_y + \widetilde{I}_{\bar{y}\bar{z}} M_z)\bar{z} - (\widetilde{I}_{\bar{y}\bar{z}} M_y + \widetilde{I}_{\bar{y}} M_z)\bar{y}$$
$$D = \widetilde{I}_{\bar{y}} \widetilde{I}_{\bar{z}} - \widetilde{I}_{\bar{y}\bar{z}}^2$$

have been introduced. The derivative is found to be

$$\frac{\partial \sigma_x}{\partial \beta} = \frac{E}{E_r} \left(-\frac{N_x}{\widetilde{A}^2} \frac{\partial \widetilde{A}}{\partial \beta} + \frac{1}{D} \frac{\partial F}{\partial \beta} - \frac{F}{D^2} \frac{\partial D}{\partial \beta} \right) \tag{10.37}$$

The derivatives of D and F are

$$\frac{\partial D}{\partial \beta} = \widetilde{I}_{\bar{y}} \frac{\partial \widetilde{I}_{\bar{z}}}{\partial \beta} + \widetilde{I}_{\bar{z}} \frac{\partial \widetilde{I}_{\bar{y}}}{\partial \beta} - 2\widetilde{I}_{\bar{y}\bar{z}} \frac{\partial \widetilde{I}_{\bar{y}\bar{z}}}{\partial \beta}$$
$$\frac{\partial F}{\partial \beta} = \left(\frac{\partial \widetilde{I}_{\bar{z}}}{\partial \beta} M_y + \frac{\partial \widetilde{I}_{\bar{y}\bar{z}}}{\partial \beta} M_z \right) \bar{z} + (\widetilde{I}_{\bar{z}} M_y + \widetilde{I}_{\bar{y}\bar{z}} M_z) \frac{\partial \bar{z}}{\partial \beta} \tag{10.38}$$
$$- \left(\frac{\partial \widetilde{I}_{\bar{y}\bar{z}}}{\partial \beta} M_y + \frac{\partial \widetilde{I}_{\bar{y}}}{\partial \beta} M_z \right) \bar{y} - (\widetilde{I}_{\bar{y}\bar{z}} M_y + \widetilde{I}_{\bar{y}} M_z) \frac{\partial \bar{y}}{\partial \beta}$$

10.2.3 Derivatives of the Torsional Constant and the Shear Stresses

Next we describe a method for calculating the derivatives of the torsional constant J and the torsional shear stresses τ_{xy}, τ_{xz} with respect to a design parameter β. The description is given in terms of the nine-node element shown in Fig. 4.1. The definitions of the various matrices are given in Chapter 5.

From Eq. (5.119),

$$\mathbf{B}_e = \mathbf{J}_e^{-1} \begin{bmatrix} \dfrac{\partial \mathbf{N}}{\partial \eta} \\ \dfrac{\partial \mathbf{N}}{\partial \zeta} \end{bmatrix} \tag{10.39}$$

or

$$\mathbf{J}_e \mathbf{B}_e = \begin{bmatrix} \dfrac{\partial \mathbf{N}}{\partial \eta} \\ \dfrac{\partial \mathbf{N}}{\partial \zeta} \end{bmatrix} \tag{10.40}$$

Since the right side of the expression above is independent of β, the derivative of Eq. (10.40) is

$$\frac{\partial \mathbf{J}_e}{\partial \beta} \mathbf{B}_e + \mathbf{J}_e \frac{\partial \mathbf{B}_e}{\partial \beta} = 0 \tag{10.41}$$

so that the derivative of $\mathbf{B}_e$ with respect to β becomes

$$\frac{\partial \mathbf{B}_e}{\partial \beta} = -\mathbf{J}_e^{-1} \frac{\partial \mathbf{J}_e}{\partial \beta} \mathbf{B}_e \tag{10.42}$$

The derivative of the Jacobian of Eq. (4.15) is

$$\frac{\partial \mathbf{J}_e}{\partial \beta} = \begin{bmatrix} \dfrac{\partial \mathbf{N}}{\partial \eta} \\ \dfrac{\partial \mathbf{N}}{\partial \zeta} \end{bmatrix} \begin{bmatrix} \dfrac{\partial \mathbf{y}_e}{\partial \beta} & \dfrac{\partial \mathbf{z}_e}{\partial \beta} \end{bmatrix} = \mathbf{J}_e \mathbf{B}_e \begin{bmatrix} \dfrac{\partial \mathbf{y}_e}{\partial \beta} & \dfrac{\partial \mathbf{z}_e}{\partial \beta} \end{bmatrix} = \mathbf{J}_e \mathbf{A}_e \tag{10.43}$$

where the 2×2 matrix $\mathbf{A}_e$ is defined by

$$\mathbf{A}_e = \mathbf{B}_e \begin{bmatrix} \dfrac{\partial \mathbf{y}_e}{\partial \beta} & \dfrac{\partial \mathbf{z}_e}{\partial \beta} \end{bmatrix} \tag{10.44}$$

Substitution of Eq. (10.43) into Eq. (10.42) shows that the derivative of $\mathbf{B}_e$ becomes

$$\frac{\partial \mathbf{B}_e}{\partial \beta} = -\mathbf{A}_e \mathbf{B}_e \tag{10.45}$$

The element stiffness matrix of Eq. (5.122) is given as

$$\mathbf{k}^e = \int_{-1}^{1} \int_{-1}^{1} \mathbf{B}_e^{\mathrm{T}} \mathbf{B}_e |\, \mathbf{J}_e \,|\, d\eta \, d\zeta \tag{10.46}$$

The derivative of this matrix is

$$\frac{\partial \mathbf{k}^e}{\partial \beta} = \int_{-1}^{1}\int_{-1}^{1}\left(\frac{\partial \mathbf{B}_e^{\mathrm{T}}}{\partial \beta}\mathbf{B}_e|\,\mathbf{J}_e\,| + \mathbf{B}_e^{\mathrm{T}}\frac{\partial \mathbf{B}_e}{\partial \beta}|\,\mathbf{J}_e\,| + \mathbf{B}_e^{\mathrm{T}}\mathbf{B}_e\frac{\partial|\,\mathbf{J}_e\,|}{\partial \beta}\right) d\eta\, d\zeta \tag{10.47}$$

which, with the introduction of Eq. (10.45), leads to an expression for the derivative of the element stiffness matrix:

$$\frac{\partial \mathbf{k}^e}{\partial \beta} = \int_{-1}^{1}\int_{-1}^{1}\mathbf{B}_e^{\mathrm{T}}\left(-\mathbf{A}_e^{\mathrm{T}} - \mathbf{A}_e + \frac{1}{|\,\mathbf{J}_e\,|}\frac{\partial|\,\mathbf{J}_e\,|}{\partial \beta}\mathbf{I}\right)\mathbf{B}_e|\,\mathbf{J}_e\,|\, d\eta\, d\zeta \tag{10.48}$$

where $\mathbf{I}$ is the 2×2 identity matrix. The derivative of the system stiffness matrix, $\partial\mathbf{K}/\partial\beta$, is obtained by assembling the derivatives of element stiffness matrices of Eq. (10.48).

A manipulation similar to that just employed for the derivative of the element stiffness matrix leads to the derivative of the element load vector. From Eq. (5.123),

$$\mathbf{p}^e = \int_{-1}^{1}\int_{-1}^{1}\mathbf{B}_e^{\mathrm{T}}\mathbf{g}^e|\,\mathbf{J}_e\,|\, d\eta\, d\zeta \tag{10.49}$$

with

$$\mathbf{g}^e = \begin{bmatrix} \mathbf{N}\mathbf{z}_e \\ -\mathbf{N}\mathbf{y}_e \end{bmatrix}$$

The derivative of $\mathbf{p}^e$ of Eq. (10.49) is

$$\frac{\partial \mathbf{p}^e}{\partial \beta} = \int_{-1}^{1}\int_{-1}^{1}\left(\frac{\partial \mathbf{B}_e^{\mathrm{T}}}{\partial \beta}\mathbf{g}^e|\,\mathbf{J}_e\,| + \mathbf{B}_e^{\mathrm{T}}\frac{\partial \mathbf{g}^e}{\partial \beta}|\,\mathbf{J}_e\,| + \mathbf{B}_e^{\mathrm{T}}\mathbf{g}^e\frac{\partial|\,\mathbf{J}_e\,|}{\partial \beta}\right) d\eta\, d\zeta \tag{10.50}$$

which, with Eq. (10.45), can be written as

$$\frac{\partial \mathbf{p}^e}{\partial \beta} = \int_{-1}^{1}\int_{-1}^{1}\mathbf{B}_e^{\mathrm{T}}\left(-\mathbf{A}_e^{\mathrm{T}}\mathbf{g}^e + \frac{\partial \mathbf{g}^e}{\partial \beta} + \frac{1}{|\,\mathbf{J}_e\,|}\frac{\partial|\,\mathbf{J}_e\,|}{\partial \beta}\mathbf{g}^e\right)|\,\mathbf{J}_e\,|\, d\eta\, d\zeta \tag{10.51}$$

A necessary condition for the derivatives with respect to β of the element stiffness matrix and the load vector to be nonzero is that the nodal coordinate derivatives

$$\frac{\partial \mathbf{y}_e}{\partial \beta} \qquad \frac{\partial \mathbf{z}_e}{\partial \beta} \tag{10.52}$$

have some nonzero components. In other words, if the parameter β does not influence the nodal coordinates of a particular element, the derivatives of the stiffness matrix and load vector of that element with respect to β are zero.

The torsional constant J of Eq. (5.139) is

$$J = I_y + I_z - \boldsymbol{\omega}^{\mathrm{T}}\mathbf{P} \tag{10.53}$$

Note that the final term on the right-hand side is a scalar. The derivative of this expression is

$$\frac{\partial J}{\partial \beta} = \frac{\partial I_y}{\partial \beta} + \frac{\partial I_z}{\partial \beta} - \boldsymbol{\omega}^{\mathrm{T}} \frac{\partial \mathbf{P}}{\partial \beta} - \frac{\partial \boldsymbol{\omega}^{\mathrm{T}}}{\partial \beta} \mathbf{P} \tag{10.54}$$

Since the final term is a scalar,

$$\frac{\partial \boldsymbol{\omega}^{\mathrm{T}}}{\partial \beta} \mathbf{P} = \mathbf{P}^{\mathrm{T}} \frac{\partial \boldsymbol{\omega}}{\partial \beta}$$

and

$$\frac{\partial J}{\partial \beta} = \frac{\partial I_y}{\partial \beta} + \frac{\partial I_z}{\partial \beta} - \boldsymbol{\omega}^{\mathrm{T}} \frac{\partial \mathbf{P}}{\partial \beta} - \mathbf{P}^{\mathrm{T}} \frac{\partial \boldsymbol{\omega}}{\partial \beta} \tag{10.55}$$

The terms $\partial I_y/\partial \beta$ and $\partial I_z/\partial \beta$ are computed as described in Chapter 4. The derivative of $\mathbf{P}$ is obtained by assembling the derivatives of $\mathbf{p}^e$ of Eq. (5.123). The derivative of $\boldsymbol{\omega}$ is found by solving the differentiated system equation of Eq. (5.125):

$$\mathbf{K} \frac{\partial \boldsymbol{\omega}}{\partial \beta} + \frac{\partial \mathbf{K}}{\partial \beta} \boldsymbol{\omega} = \frac{\partial \mathbf{P}}{\partial \beta}$$

or

$$\mathbf{K} \frac{\partial \boldsymbol{\omega}}{\partial \beta} = \frac{\partial \mathbf{P}}{\partial \beta} - \frac{\partial \mathbf{K}}{\partial \beta} \boldsymbol{\omega} \tag{10.56}$$

The second vector on the right side of Eq. (10.56) is formed by assembling the element vectors of Eq. (5.124):

$$\frac{\partial \mathbf{k}^e}{\partial \beta} \boldsymbol{\omega}^e \qquad 1 \leq e \leq M \tag{10.57}$$

into a system vector.

It is possible to calculate the derivative of J without solving Eq. (10.55). The torsional constant J can be expressed as in Eq. (10.53) or, alternatively, from Eq. (5.139), as

$$J = I_y + I_z - \boldsymbol{\omega}^{\mathrm{T}} \mathbf{K} \boldsymbol{\omega} \tag{10.58}$$

The derivative of this expression is

$$\frac{\partial J}{\partial \beta} = \frac{\partial I_y}{\partial \beta} + \frac{\partial I_z}{\partial \beta} - \boldsymbol{\omega}^{\mathrm{T}} \mathbf{K} \frac{\partial \boldsymbol{\omega}}{\partial \beta} - \boldsymbol{\omega}^{\mathrm{T}} \frac{\partial \mathbf{K}}{\partial \beta} \boldsymbol{\omega} - \frac{\partial \boldsymbol{\omega}^{\mathrm{T}}}{\partial \beta} \mathbf{K} \boldsymbol{\omega}$$

which, since $\mathbf{K}$ is symmetric ($\mathbf{K} = \mathbf{K}^{\mathrm{T}}$), can be expressed as

$$\frac{\partial J}{\partial \beta} = \frac{\partial I_y}{\partial \beta} + \frac{\partial I_z}{\partial \beta} - \boldsymbol{\omega}^{\mathrm{T}} \left(2\mathbf{K} \frac{\partial \boldsymbol{\omega}}{\partial \beta} + \frac{\partial \mathbf{K}}{\partial \beta} \boldsymbol{\omega} \right) \tag{10.59}$$

The two vectors in the final parenthesized term of Eq. (10.59) can be found by assembling them from the corresponding element vectors. Alternatively, the assembly procedure can be bypassed in favor of a summation involving element vectors:

$$\frac{\partial J}{\partial \beta} = \frac{\partial I_y}{\partial \beta} + \frac{\partial I_z}{\partial \beta} - \sum_e \boldsymbol{\omega}^{e\mathrm{T}} \left(2\frac{\partial \mathbf{p}^e}{\partial \beta} - \frac{\partial \mathbf{k}^e}{\partial \beta} \boldsymbol{\omega}^e \right) \tag{10.60}$$

The shear stress vector $\boldsymbol{\tau}^e$ is given in Eq. (5.142) as

$$\boldsymbol{\tau}^e = \begin{bmatrix} \tau_{xy} \\ \tau_{xz} \end{bmatrix}^e = \frac{M_x}{J} \left(\mathbf{B}_e \boldsymbol{\omega}^e - \mathbf{h}^e \right) \tag{10.61}$$

with $\mathbf{h}^e = [\mathbf{Nz}_e \; -\mathbf{Ny}_e]^\mathrm{T}$. The design derivative is

$$\frac{\partial \boldsymbol{\tau}^e}{\partial \beta} = \frac{M_x}{J} \left(\mathbf{B}_e \frac{\partial \boldsymbol{\omega}^e}{\partial \beta} - \mathbf{A}_e \mathbf{B}_e \boldsymbol{\omega}^e - \frac{\partial \mathbf{h}^e}{\partial \beta} \right) - \frac{1}{J} \frac{\partial J}{\partial \beta} \boldsymbol{\tau}^e \tag{10.62}$$

It is necessary to solve the differentiated system equation (10.56) for $\partial \boldsymbol{\omega}/\partial \beta$ in order to evaluate the stress derivatives $\partial \boldsymbol{\tau}^e/\partial \beta$. The element connectivity matrix $\mathbf{C}$ (Section 5.5) is used to identify the entries of the element derivative $\partial \boldsymbol{\omega}^e/\partial \beta$ from the system derivative vector $\partial \boldsymbol{\omega}/\partial \beta$ by

$$\frac{\partial \omega_i^e}{\partial \beta} = \frac{\partial \omega_n}{\partial \beta} \qquad n = C_{ei} \quad 1 \le i \le 9 \tag{10.63}$$

10.3 DESIGN SENSITIVITY OF THE SHEAR DEFORMATION COEFFICIENTS

Formulas for differentiating the shear deformation coefficients of Chapter 6, α_y, α_z, α_{yz}, with respect to a design parameter β are derived in this section.

The shear coefficient α_y was defined in terms of a parameter κ_y (Eq. 6.131):

$$\alpha_y = \frac{\kappa_y A}{\Delta^2} \tag{10.64}$$

with $\Delta = 2(1+\nu)(I_y I_z - I_{yz}^2)$ and A equal to the cross-sectional area. From Eq. (6.134)

$$\kappa_y = \sum_e \int_{-1}^{1} \int_{-1}^{1} \left(\boldsymbol{\Psi}^{e\mathrm{T}} \mathbf{B}_e^\mathrm{T} \mathbf{B}_e \boldsymbol{\Psi}^e - 2\boldsymbol{\Psi}^{e\mathrm{T}} \mathbf{B}_e^\mathrm{T} \mathbf{d}^e + \mathbf{d}^{e\mathrm{T}} \mathbf{d}^e \right) |\,\mathbf{J}_e\,| \, d\eta \, d\zeta \tag{10.65}$$

with (Eqs. 6.69 and 6.70)

$$\mathbf{d}^e = \begin{bmatrix} I_y r^e - I_{yz} q^e \\ I_{yz} r^e + I_y q^e \end{bmatrix} \qquad r^e = (\mathbf{Ny}_e)^2 - (\mathbf{Nz}_e)^2 \qquad q^e = 2\mathbf{Ny}_e \mathbf{Nz}_e$$

The derivative of the shear coefficient α_y is determined as

$$\frac{\partial \alpha_y}{\partial \beta} = \frac{\kappa_y}{\Delta^2}\frac{\partial A}{\partial \beta} + \frac{A}{\Delta^2}\frac{\partial \kappa_y}{\partial \beta} - 2\frac{\kappa_y A}{\Delta^3}\frac{\partial \Delta}{\partial \beta}$$
$$= \frac{\alpha_y}{A}\frac{\partial A}{\partial \beta} + \frac{A}{\Delta^2}\frac{\partial \kappa_y}{\partial \beta} - 2\frac{\alpha_y}{\Delta}\frac{\partial \Delta}{\partial \beta} \tag{10.66}$$

Write κ_y of Eq. (10.65) as

$$\kappa_y = \sum_e \left(\mathbf{\Psi}^{e\mathrm{T}} \mathbf{k}^e \mathbf{\Psi}^e - 2\mathbf{\Psi}^{e\mathrm{T}} \mathbf{p}^e + \int_{-1}^{1}\int_{-1}^{1} \mathbf{d}^{e\mathrm{T}} \mathbf{d}^e |\, \mathbf{J}_e \,|\, d\eta \, d\zeta \right) \tag{10.67}$$

where (Eqs. 6.134 and 6.135)

$$\mathbf{k}^e = \int_{-1}^{1}\int_{-1}^{1} \mathbf{B}_e^{\mathrm{T}} \mathbf{B}_e |\, \mathbf{J}_e \,|\, d\eta \, d\zeta \qquad \mathbf{p}^e = \int_{-1}^{1}\int_{-1}^{1} \mathbf{B}_e^{\mathrm{T}} \mathbf{d}^e |\, \mathbf{J}_e \,|\, d\eta \, d\zeta$$

Assemble the matrices $\mathbf{k}^e$ and $\mathbf{p}^e$, giving

$$\kappa_y = \mathbf{\Psi}^{\mathrm{T}} \mathbf{K} \mathbf{\Psi} - 2\mathbf{\Psi}^{\mathrm{T}} \mathbf{P}_{y2} + \sum_e \int_{-1}^{1}\int_{-1}^{1} \mathbf{d}^{e\mathrm{T}} \mathbf{d}^e |\, \mathbf{J}_e \,|\, d\eta \, d\zeta$$
$$= \mathbf{\Psi}^{\mathrm{T}} \left(\mathbf{P}_y - 2\mathbf{P}_{y2}\right) + \sum_e \int_{-1}^{1}\int_{-1}^{1} \mathbf{d}^{e\mathrm{T}} \mathbf{d}^e |\, \mathbf{J}_e \,|\, d\eta \, d\zeta \tag{10.68}$$

with $\mathbf{K\Psi} = \mathbf{P}_y$ (Eq. 6.72) and $\sum_e \mathbf{\Psi}^{e\mathrm{T}} \mathbf{p}^e = \mathbf{\Psi}^{\mathrm{T}} \mathbf{P}_{y2}$. The derivative of κ_y becomes

$$\frac{\partial \kappa_y}{\partial \beta} = \frac{\partial}{\partial \beta}\left[\mathbf{\Psi}^{\mathrm{T}} \left(\mathbf{P}_y - 2\mathbf{P}_{y2}\right)\right]$$
$$+ \sum_e \int_{-1}^{1}\int_{-1}^{1} \left(\frac{\partial \mathbf{d}^{e\mathrm{T}}}{\partial \beta} \mathbf{d}^e |\, \mathbf{J}_e \,| + \mathbf{d}^{e\mathrm{T}} \frac{\partial \mathbf{d}^e}{\partial \beta} |\, \mathbf{J}_e \,| + \frac{\partial |\, \mathbf{J}_e \,|}{\partial \beta} \mathbf{d}^{e\mathrm{T}} \mathbf{d}^e \right) d\eta \, d\zeta$$
$$= \frac{\partial}{\partial \beta}\left[\mathbf{\Psi}^{\mathrm{T}} \left(\mathbf{P}_y - 2\mathbf{P}_{y2}\right)\right]$$
$$+ \sum_e \int_{-1}^{1}\int_{-1}^{1} \left(2 |\, \mathbf{J}_e \,| \mathbf{d}^{e\mathrm{T}} \frac{\partial \mathbf{d}^e}{\partial \beta} + \frac{\partial |\, \mathbf{J}_e \,|}{\partial \beta} \mathbf{d}^{e\mathrm{T}} \mathbf{d}^e \right) d\eta \, d\zeta \tag{10.69}$$

With Eq. (4.11) ($y(\eta, \zeta) = \mathbf{N}(\eta, \zeta)\mathbf{y}$ and $z(\eta, \zeta) = \mathbf{N}(\eta, \zeta)\mathbf{z}$), the load vector of Eq. (6.71) appears as

$$\mathbf{p}^e = \int_{-1}^{1}\int_{-1}^{1} \left[\frac{\nu}{2} \mathbf{B}_e^{\mathrm{T}} \mathbf{d}^e + 2(1+\nu)\mathbf{N}^{\mathrm{T}} \left(y I_y - z I_{yz} \right) \right] |\, \mathbf{J}_e \,|\, d\eta \, d\zeta \tag{10.70}$$

The element load vector $\mathbf{p}^e$ for shear in the y direction is split into two parts by defining

$$\begin{aligned}\mathbf{p}_1^e &= 2(1+\nu)\int_{-1}^{1}\int_{-1}^{1}\mathbf{N}^{\mathrm{T}}(yI_y - zI_{yz})|\,\mathbf{J}_e\,|\,d\eta\,d\zeta \\ \mathbf{p}_2^e &= \int_{-1}^{1}\int_{-1}^{1}\frac{\nu}{2}\mathbf{B}_e^{\mathrm{T}}\mathbf{d}^e|\,\mathbf{J}_e\,|\,d\eta\,d\zeta\end{aligned} \tag{10.71}$$

so that

$$\mathbf{p}^e = \mathbf{p}_1^e + \mathbf{p}_2^e \tag{10.72}$$

These element loading matrices can be assembled to form the global matrices

$$\mathbf{P}_y = \mathbf{P}_{y1} + \mathbf{P}_{y2} \tag{10.73}$$

Return to the derivative of κ_y of Eq. (10.69). The derivative of the first term of Eq. (10.69) is evaluated as follows:

$$\begin{aligned}\frac{\partial}{\partial\beta}\left[\boldsymbol{\Psi}^{\mathrm{T}}\left(\mathbf{P}_y - 2\mathbf{P}_{y2}\right)\right] &= \boldsymbol{\Psi}^{\mathrm{T}}\left(\frac{\partial\mathbf{P}_y}{\partial\beta} - 2\frac{\partial\mathbf{P}_{y2}}{\partial\beta}\right) + \frac{\partial\boldsymbol{\Psi}^{\mathrm{T}}}{\partial\beta}\left(\mathbf{P}_y - 2\mathbf{P}_{y2}\right) \\ &= \boldsymbol{\Psi}^{\mathrm{T}}\left(\frac{\partial\mathbf{P}_y}{\partial\beta} - 2\frac{\partial\mathbf{P}_{y2}}{\partial\beta}\right) + \left(\mathbf{P}_y^{\mathrm{T}} - 2\mathbf{P}_{y2}^{\mathrm{T}}\right)\frac{\partial\boldsymbol{\Psi}}{\partial\beta} \\ &= \boldsymbol{\Psi}^{\mathrm{T}}\left(\frac{\partial\mathbf{P}_{y1}}{\partial\beta} - \frac{\partial\mathbf{P}_{y2}}{\partial\beta}\right) + \left(\boldsymbol{\Psi}^{\mathrm{T}}\mathbf{K} - 2\boldsymbol{\Psi}_2^{\mathrm{T}}\mathbf{K}\right)\frac{\partial\boldsymbol{\Psi}}{\partial\beta} \\ &= \boldsymbol{\Psi}^{\mathrm{T}}\left(\frac{\partial\mathbf{P}_{y1}}{\partial\beta} - \frac{\partial\mathbf{P}_{y2}}{\partial\beta}\right) + \left(\boldsymbol{\Psi}^{\mathrm{T}} - 2\boldsymbol{\Psi}_2^{\mathrm{T}}\right)\left(\frac{\partial\mathbf{P}_y}{\partial\beta} - \frac{\partial\mathbf{K}}{\partial\beta}\boldsymbol{\Psi}\right)\end{aligned} \tag{10.74}$$

where $\mathbf{P}_y = \mathbf{P}_{y1} + \mathbf{P}_{y2}$, and $\boldsymbol{\Psi}$ and $\boldsymbol{\Psi}_2$, the solutions of $\mathbf{K}\boldsymbol{\Psi} = \mathbf{P}_y$ and $\mathbf{K}\boldsymbol{\Psi}_2 = \mathbf{P}_{y2}$, and

$$\frac{\partial\mathbf{P}_y}{\partial\beta} = \frac{\partial(\mathbf{K}\boldsymbol{\Psi})}{\partial\beta} = \mathbf{K}\frac{\partial\boldsymbol{\Psi}}{\partial\beta} + \frac{\partial\mathbf{K}}{\partial\beta}\boldsymbol{\Psi}$$

have been introduced. The derivatives of $\mathbf{P}_y$, $\mathbf{P}_{y1}$, and $\mathbf{P}_{y2}$ of Eq. (10.74) are found by assembling the derivatives of the element vectors $\mathbf{p}_1^e$ and $\mathbf{p}_2^e$. From Eq. (10.71), the derivative of $\mathbf{p}_1^e$ is given by

$$\begin{aligned}\frac{1}{2(1+\nu)}\frac{\partial\mathbf{p}_1^e}{\partial\beta} = \int_{-1}^{1}\int_{-1}^{1}\Bigg[\mathbf{N}^{\mathrm{T}}\left(I_y\frac{\partial y}{\partial\beta} + y\frac{\partial I_y}{\partial\beta} - I_{yz}\frac{\partial z}{\partial\beta} - z\frac{\partial I_{yz}}{\partial\beta}\right)|\,\mathbf{J}_e\,| \\ + \mathbf{N}^{\mathrm{T}}(yI_y - zI_{yz})\frac{\partial|\,\mathbf{J}_e\,|}{\partial\beta}\Bigg]\,d\eta\,d\zeta\end{aligned} \tag{10.75}$$

and the derivative of $\mathbf{p}_2^e$ is

$$\frac{2}{\nu}\frac{\partial \mathbf{p}_2^e}{\partial \beta} = \int_{-1}^{1}\int_{-1}^{1} \mathbf{B}_e^{\mathrm{T}} \left(-\mathbf{A}_e^{\mathrm{T}} |\, \mathbf{J}_e \,| \mathbf{d}^e + |\, \mathbf{J}_e \,| \frac{\partial \mathbf{d}^e}{\partial \beta} + \frac{\partial |\, \mathbf{J}_e \,|}{\partial \beta} \mathbf{d}^e \right) d\eta \, d\zeta \tag{10.76}$$

The final derivative needed to calculate $\partial \kappa_y / \partial \beta$ is $\partial \mathbf{d}^e / \partial \beta$. From Eq. (6.69),

$$\mathbf{d}^e = \nu \left(I_y \frac{y^2 - z^2}{2} - I_{yz} yz \right) \mathbf{j} + \nu \left(I_y yz + I_{yz} \frac{y^2 - z^2}{2} \right) \mathbf{k} \tag{10.77}$$

The derivative of $\mathbf{d}^e$ is then

$$\begin{aligned} \frac{\partial \mathbf{d}^e}{\partial \beta} &= \nu \left(z \frac{\partial y}{\partial \beta} + y \frac{\partial z}{\partial \beta} \right) (I_y \mathbf{k} - I_{yz} \mathbf{j}) + \nu yz \left(\frac{\partial I_y}{\partial \beta} \mathbf{k} - \frac{\partial I_{yz}}{\partial \beta} \mathbf{j} \right) \\ &\quad + \nu \left(y \frac{\partial y}{\partial \beta} - z \frac{\partial z}{\partial \beta} \right) (I_y \mathbf{j} + I_{yz} \mathbf{k}) + \nu \frac{y^2 - z^2}{2} \left(\frac{\partial I_y}{\partial \beta} \mathbf{j} + \frac{\partial I_{yz}}{\partial \beta} \mathbf{k} \right) \end{aligned} \tag{10.78}$$

The shear deformation coefficient α_z was defined in Chapter 6 as (Eq. 6.136)

$$\alpha_z = \frac{\kappa_z A}{\Delta^2} \tag{10.79}$$

From Eq. (6.137), κ_z can be expressed as

$$\kappa_z = \sum_e \int_{-1}^{1}\int_{-1}^{1} \left(\mathbf{\Phi}^{e\mathrm{T}} \mathbf{B}_e^{\mathrm{T}} \mathbf{B}_e \mathbf{\Phi}^e - 2 \mathbf{\Phi}^{e\mathrm{T}} \mathbf{B}_e^{\mathrm{T}} \mathbf{h} + \mathbf{h}^{\mathrm{T}} \mathbf{h} \right) |\, \mathbf{J}_e \,| \, d\eta \, d\zeta \tag{10.80}$$

The derivative of the shear coefficient α_z is

$$\frac{\partial \alpha_z}{\partial \beta} = \frac{\alpha_z}{A} \frac{\partial A}{\partial \beta} + \frac{A}{\Delta^2} \frac{\partial \kappa_z}{\partial \beta} - \frac{2\alpha_z}{\Delta} \frac{\partial \Delta}{\partial \beta} \tag{10.81}$$

Express κ_z of Eq. (10.80) as

$$\kappa_z = \sum_e \left(\mathbf{\Phi}^e \mathbf{k}^e \mathbf{\Phi}^e - 2 \mathbf{\Phi}^{e\mathrm{T}} \mathbf{p}^e + \int_{-1}^{1}\int_{-1}^{1} \mathbf{h}^{e\mathrm{T}} \mathbf{h}^e |\, \mathbf{J}_e \,| \, d\eta \, d\zeta \right) \tag{10.82}$$

Assemble the first two terms on the right-hand side:

$$\kappa_z = \mathbf{\Phi}^{\mathrm{T}} (\mathbf{P}_z - 2\mathbf{P}_{z2}) + \sum_e \int_{-1}^{1}\int_{-1}^{1} \mathbf{h}^{e\mathrm{T}} \mathbf{h}^e |\, \mathbf{J}_e \,| \, d\eta \, d\zeta \tag{10.83}$$

where $\mathbf{K\Phi} = \mathbf{P}_z$ and $\sum_e \mathbf{\Phi}^{e\mathrm{T}} \mathbf{p}^e = \mathbf{\Phi}^{\mathrm{T}} \mathbf{P}_{z2}$.

The formula for $\partial\alpha_z/\partial\beta$ of Eq. (10.81) requires an expression for $\partial\kappa_z/\partial\beta$. From Eq. (10.83),

$$\frac{\partial k_z}{\partial\beta} = \frac{\partial}{\partial\beta}\left[\mathbf{\Phi}^{\mathrm{T}}(\mathbf{P}_z - 2\mathbf{P}_{z2})\right] + \sum_e \int_{-1}^{1}\int_{-1}^{1}\left(2|\,\mathbf{J}_e\,|\mathbf{h}^{e\mathrm{T}}\frac{\partial\mathbf{h}^e}{\partial\beta} + \frac{\partial|\,\mathbf{J}_e\,|}{\partial\beta}\mathbf{h}^{e\mathrm{T}}\mathbf{h}^e\right)d\eta\, d\zeta \tag{10.84}$$

From Eq. (6.136), the element load vector of $\mathbf{\Phi}$ appears as

$$\mathbf{p}^e = \int_{-1}^{1}\int_{-1}^{1}\left[\frac{\nu}{2}\mathbf{B}_e^{\mathrm{T}}\mathbf{h}^e + 2(1+\nu)\mathbf{N}^{\mathrm{T}}\left(zI_z - yI_{yz}\right)\right]|\,\mathbf{J}_e\,|\,d\eta\, d\zeta \tag{10.85}$$

where y and z can be represented by Eq. (4.11). The element load vector $\mathbf{p}^e$ for shear in the z direction is split into two parts by defining

$$\begin{aligned}\mathbf{p}_1^e &= 2(1+\nu)\int_{-1}^{1}\int_{-1}^{1}\mathbf{N}^{\mathrm{T}}(zI_z - yI_{yz})|\,\mathbf{J}_e\,|\,d\eta\, d\zeta \\ \mathbf{p}_2^e &= \int_{-1}^{1}\int_{-1}^{1}\frac{\nu}{2}\mathbf{B}_e^{\mathrm{T}}\mathbf{h}^e|\,\mathbf{J}_e\,|\,d\eta\, d\zeta\end{aligned} \tag{10.86}$$

so that

$$\mathbf{p}^e = \mathbf{p}_1^e + \mathbf{p}_2^e \tag{10.87}$$

Assemble these element loading matrices giving the system matrices

$$\mathbf{P}_z = \mathbf{P}_{z1} + \mathbf{P}_{z2} \tag{10.88}$$

The derivative of κ_z of Eq. (10.84) can be evaluated in the same manner as the derivative of κ_y was determined. The derivative of the first term of Eq. (10.84) is

$$\frac{\partial}{\partial\beta}\left[\mathbf{\Phi}^{\mathrm{T}}\left(\mathbf{P}_z - 2\mathbf{P}_{z2}\right)\right] = \mathbf{\Phi}^{\mathrm{T}}\left(\frac{\partial\mathbf{P}_{z1}}{\partial\beta} - \frac{\partial\mathbf{P}_{z2}}{\partial\beta}\right) + \left(\mathbf{\Phi}^{\mathrm{T}} - 2\mathbf{\Phi}_2^{\mathrm{T}}\right)\left(\frac{\partial\mathbf{P}_z}{\partial\beta} - \frac{\partial\mathbf{K}}{\partial\beta}\mathbf{\Phi}\right) \tag{10.89}$$

where $\mathbf{\Phi}$ and $\mathbf{\Phi}_2$ are solutions of $\mathbf{K\Phi} = \mathbf{P}_z$ and $\mathbf{K\Phi}_2 = \mathbf{P}_{z2}$, respectively.

The derivatives of $\mathbf{P}_z$, $\mathbf{P}_{z1}$, and $\mathbf{P}_{z2}$ of Eq. (10.89) are obtained by assembling the derivatives of the element load vectors $\mathbf{p}_1^e$ and $\mathbf{p}_2^e$. From Eq. (10.86), the derivative of $\mathbf{p}_1^e$ is given by

$$\begin{aligned}\frac{1}{2(1+\nu)}\frac{\partial\mathbf{p}_1^e}{\partial\beta} = \int_{-1}^{1}\int_{-1}^{1}&\left[\mathbf{N}^{\mathrm{T}}\left(I_z\frac{\partial z}{\partial\beta} + z\frac{\partial I_z}{\partial\beta} - I_{yz}\frac{\partial y}{\partial\beta} - y\frac{\partial I_{yz}}{\partial\beta}\right)|\,\mathbf{J}_e\,|\right. \\ &\left. + \mathbf{N}^{\mathrm{T}}(zI_z - yI_{yz})\frac{\partial|\,\mathbf{J}_e\,|}{\partial\beta}\right] d\eta\, d\zeta\end{aligned} \tag{10.90}$$

and the derivative of $\mathbf{p}_2^e$ is

$$\frac{2}{\nu}\frac{\partial \mathbf{p}_2^e}{\partial \beta} = \int_{-1}^{1}\int_{-1}^{1} \mathbf{B}_e^{\mathrm{T}} \left[-\mathbf{A}_e^{\mathrm{T}} |\, \mathbf{J}_e \,| \mathbf{h}^e + |\, \mathbf{J}_e \,| \frac{\partial \mathbf{h}^e}{\partial \beta} + \frac{\partial |\, \mathbf{J}_e \,|}{\partial \beta} \mathbf{h}^e \right] d\eta \, d\zeta \tag{10.91}$$

Still needed to compute the derivative $\partial \kappa_z / \partial \beta$ of Eq. (10.84) is $\partial \mathbf{h}^e / \partial \beta$. From Eq. (6.69),

$$\mathbf{h}^e = \nu \left(I_z yz - I_{yz} \frac{y^2 - z^2}{2} \right) \mathbf{j} - \nu \left(I_z \frac{y^2 - z^2}{2} + I_{yz} yz \right) \mathbf{k} \tag{10.92}$$

Then

$$\begin{aligned} \frac{\partial \mathbf{h}^e}{\partial \beta} &= \nu \left(z \frac{\partial y}{\partial \beta} + y \frac{\partial z}{\partial \beta} \right) (I_z \mathbf{j} - I_{yz} \mathbf{k}) + \nu yz \left(\frac{\partial I_z}{\partial \beta} \mathbf{j} - \frac{\partial I_{yz}}{\partial \beta} \mathbf{k} \right) \\ &\quad - \nu \left(y \frac{\partial y}{\partial \beta} - z \frac{\partial z}{\partial \beta} \right) (I_{yz} \mathbf{j} + I_z \mathbf{k}) - \nu \frac{y^2 - z^2}{2} \left(\frac{\partial I_{yz}}{\partial \beta} \mathbf{j} + \frac{\partial I_z}{\partial \beta} \mathbf{k} \right) \end{aligned} \tag{10.93}$$

Equation (6.139) gives the shear deformation coefficient α_{yz} as

$$\alpha_{yz} = \frac{\kappa_{yz} A}{\Delta^2} \tag{10.94}$$

The derivative of α_{yz} is found from

$$\frac{\partial \alpha_{yz}}{\partial \beta} = \frac{\alpha_{yz}}{A}\frac{\partial A}{\partial \beta} + \frac{A}{\Delta^2}\frac{\partial \kappa_{yz}}{\partial \beta} - \frac{2\alpha_{yz}}{\Delta}\frac{\partial \Delta}{\partial \beta} \tag{10.95}$$

From Eq. (6.140),

$$\begin{aligned} \kappa_{yz} &= \sum_e \int_{-1}^{1}\int_{-1}^{1} \left(\boldsymbol{\Psi}^{e\mathrm{T}} \mathbf{B}_e^{\mathrm{T}} - \mathbf{d}^{e\mathrm{T}} \right) \left(\mathbf{B}_e \boldsymbol{\Phi}^e - \mathbf{h}^e \right) |\, \mathbf{J}_e \,| \, d\eta \, d\zeta \\ &= \sum_e \int_{-1}^{1}\int_{-1}^{1} \left(\boldsymbol{\Psi}^e \mathbf{B}_e^{\mathrm{T}} \mathbf{B}_e \boldsymbol{\Phi}^e - \mathbf{d}^{e\mathrm{T}} \mathbf{B}_e \boldsymbol{\Phi}^e - \boldsymbol{\Psi}^{e\mathrm{T}} \mathbf{B}_e^{\mathrm{T}} \mathbf{h}^e + \mathbf{d}^{e\mathrm{T}} \mathbf{h}^e \right) |\, \mathbf{J}_e \,| \, d\eta \, d\zeta \\ &= \mathbf{P}_y^{\mathrm{T}} \boldsymbol{\Phi} - \mathbf{P}_y^{\mathrm{T}} \boldsymbol{\Phi}_2 - \mathbf{P}_z^{\mathrm{T}} \boldsymbol{\Psi}_2 + \sum_e \int_{-1}^{1}\int_{-1}^{1} \mathbf{d}^{e\mathrm{T}} \mathbf{h}^e |\, \mathbf{J}_e \,| \, d\eta \, d\zeta \\ &= \mathbf{P}_y^{\mathrm{T}} \boldsymbol{\Phi}_1 - \mathbf{P}_z^{\mathrm{T}} \boldsymbol{\Psi}_2 + \sum_e \int_{-1}^{1}\int_{-1}^{1} \mathbf{d}^{e\mathrm{T}} \mathbf{h}^e |\, \mathbf{J}_e \,| \, d\eta \, d\zeta \end{aligned} \tag{10.96}$$

where $\boldsymbol{\Phi} = \boldsymbol{\Phi}_1 + \boldsymbol{\Phi}_2$.

To calculate $\partial \alpha_{yz} / \partial \beta$ from Eq. (10.95), an expression for $\partial \kappa_{yz} / \partial \beta$ is needed. From Eq. (10.96),

$$
\begin{aligned}
\frac{\partial \kappa_{yz}}{\partial \beta} &= \frac{\partial}{\partial \beta}\left(\mathbf{P}_y^{\mathrm{T}}\boldsymbol{\Phi}_1 - \mathbf{P}_z^{\mathrm{T}}\boldsymbol{\Psi}_2\right) \\
&\quad + \sum_e \int_{-1}^{1}\int_{-1}^{1}\left(|\mathbf{J}_e|\mathbf{h}^{e\mathrm{T}}\frac{\partial \mathbf{d}^e}{\partial \beta} + |\mathbf{J}_e|\mathbf{d}^{e\mathrm{T}}\frac{\partial \mathbf{h}^e}{\partial \beta} + \mathbf{d}^{e\mathrm{T}}\mathbf{h}^e\frac{\partial |\mathbf{J}_e|}{\partial \beta}\right) d\eta\, d\zeta
\end{aligned}
\tag{10.97}
$$

The derivative of the first two terms is

$$
\begin{aligned}
\frac{\partial}{\partial \beta}\left(\mathbf{P}_y^{\mathrm{T}}\boldsymbol{\Phi}_1 - \mathbf{P}_z^{\mathrm{T}}\boldsymbol{\Psi}_2\right) &= \frac{\partial}{\partial \beta}\left(\boldsymbol{\Psi}^{\mathrm{T}}\mathbf{P}_{z1} - \boldsymbol{\Phi}^{\mathrm{T}}\mathbf{P}_{z2}\right) \\
&= \boldsymbol{\Psi}^{\mathrm{T}}\frac{\partial \mathbf{P}_{z1}}{\partial \beta} + \frac{\partial \boldsymbol{\Psi}^{\mathrm{T}}}{\partial \beta}\mathbf{P}_{z1} - \boldsymbol{\Phi}^{\mathrm{T}}\frac{\partial \mathbf{P}_{y2}}{\partial \beta} - \frac{\partial \boldsymbol{\Phi}^{\mathrm{T}}}{\partial \beta}\mathbf{P}_{y2} \\
&= \boldsymbol{\Psi}^{\mathrm{T}}\frac{\partial \mathbf{P}_{z1}}{\partial \beta} + \boldsymbol{\Phi}_1^{\mathrm{T}}\left(\frac{\partial \mathbf{P}_y}{\partial \beta} - \frac{\partial \mathbf{K}}{\partial \beta}\boldsymbol{\Psi}\right) \\
&\quad - \boldsymbol{\Phi}^{\mathrm{T}}\frac{\partial \mathbf{P}_{y2}}{\partial \beta} - \boldsymbol{\Psi}_2^{\mathrm{T}}\left(\frac{\partial \mathbf{P}_z}{\partial \beta} - \frac{\partial \mathbf{K}}{\partial \beta}\boldsymbol{\Phi}\right)
\end{aligned}
\tag{10.98}
$$

The second and fourth terms are derived similarly. Consider, for example, the second term.

$$
\frac{\partial \boldsymbol{\Psi}^{\mathrm{T}}}{\partial \beta}\mathbf{P}_{z1} = \frac{\partial \boldsymbol{\Psi}^{\mathrm{T}}}{\partial \beta}\mathbf{K}\boldsymbol{\Phi}_1 = \boldsymbol{\Phi}_1^{\mathrm{T}}\mathbf{K}\frac{\partial \boldsymbol{\Psi}}{\partial \beta} = \boldsymbol{\Phi}_1^{\mathrm{T}}\left(\frac{\partial \mathbf{P}_y}{\partial \beta} - \frac{\partial \mathbf{K}}{\partial \beta}\boldsymbol{\Psi}\right)
\tag{10.99}
$$

The sensitivity analysis of shear coefficients is considerably simpler if Poisson's ratio ν may be assumed to be zero. In that case, $\mathbf{h} = 0$, $\mathbf{d} = 0$, and the shear coefficient relations of Section 6.3.3 reduce to

$$
\begin{aligned}
\kappa_y &= \boldsymbol{\Psi}^{\mathrm{T}}\mathbf{P}_y = \boldsymbol{\Psi}^{\mathrm{T}}\mathbf{P}_{y1} \\
\kappa_z &= \boldsymbol{\Phi}^{\mathrm{T}}\mathbf{P}_z = \boldsymbol{\Phi}^{\mathrm{T}}\mathbf{P}_{z1} \\
\kappa_{yz} &= \boldsymbol{\Phi}^{\mathrm{T}}\mathbf{P}_y = \boldsymbol{\Phi}^{\mathrm{T}}\mathbf{P}_{y1}
\end{aligned}
\tag{10.100}
$$

We need the derivatives of these variables with respect to β. In the first case,

$$
\begin{aligned}
\frac{\partial \kappa_y}{\partial \beta} &= \frac{\partial \boldsymbol{\Psi}^{\mathrm{T}}}{\partial \beta}\mathbf{P}_{y1} + \boldsymbol{\Psi}^{\mathrm{T}}\frac{\partial \mathbf{P}_{y1}}{\partial \beta} = \frac{\partial \boldsymbol{\Psi}^{\mathrm{T}}}{\partial \beta}\mathbf{K}\boldsymbol{\Psi} + \boldsymbol{\Psi}^{\mathrm{T}}\frac{\partial \mathbf{P}_{y1}}{\partial \beta} \\
&= \boldsymbol{\Psi}^{\mathrm{T}}\mathbf{K}\frac{\partial \boldsymbol{\Psi}}{\partial \beta} + \boldsymbol{\Psi}^{\mathrm{T}}\frac{\partial \mathbf{P}_{y1}}{\partial \beta} = \boldsymbol{\Psi}^{\mathrm{T}}\left(\frac{\partial \mathbf{P}_{y1}}{\partial \beta} - \frac{\partial \mathbf{K}}{\partial \beta}\boldsymbol{\Psi}\right) + \boldsymbol{\Psi}^{\mathrm{T}}\frac{\partial \mathbf{P}_{y1}}{\partial \beta}
\end{aligned}
\tag{10.101}
$$

where $\mathbf{P}_{y1} = \mathbf{K}\boldsymbol{\Psi}_1 + \mathbf{K}\boldsymbol{\Psi}$ and

$$
\frac{\partial \mathbf{P}_{y1}}{\partial \beta} = \frac{\partial \mathbf{K}}{\partial \beta}\boldsymbol{\Psi} + \mathbf{K}\frac{\partial \boldsymbol{\Psi}}{\partial \beta}
$$

have been introduced. Finally,

$$\frac{\partial \kappa_y}{\partial \beta} = \boldsymbol{\Psi}^{\mathrm{T}} \left(2\frac{\partial \mathbf{P}_{y1}}{\partial \beta} - \frac{\partial \mathbf{K}}{\partial \beta}\boldsymbol{\Psi} \right) \tag{10.102}$$

Similarly,

$$\begin{aligned} \frac{\partial \kappa_z}{\partial \beta} &= \boldsymbol{\Phi}^{\mathrm{T}} \left(2\frac{\partial \mathbf{P}_{z1}}{\partial \beta} - \frac{\partial \mathbf{K}}{\partial \beta}\boldsymbol{\Phi} \right) \\ \frac{\partial \kappa_{yz}}{\partial \beta} &= \boldsymbol{\Psi}^{\mathrm{T}} \left(\frac{\partial \mathbf{P}_{z1}}{\partial \beta} - \frac{\partial \mathbf{K}}{\partial \beta}\boldsymbol{\Phi} \right) + \boldsymbol{\Phi}^{\mathrm{T}} \frac{\partial \mathbf{P}_{y1}}{\partial \beta} \end{aligned} \tag{10.103}$$

10.4 DESIGN SENSITIVITY ANALYSIS FOR WARPING PROPERTIES

The warping properties appearing in the derivations in this section are defined in Chapter 7. It will be assumed that Saint-Venant's torsion problem has been solved in the form described in Chapter 5 (Eq. 5.125):

$$\mathbf{K}\boldsymbol{\omega} = \mathbf{P} \tag{10.104}$$

for the nodal vector $\boldsymbol{\omega}$ of warping function values. It will also be assumed that the derivative of $\boldsymbol{\omega}$ with respect to each design parameter β of interest has been found by solving the differentiated system relationship of Eq. (10.56):

$$\mathbf{K}\frac{\partial \boldsymbol{\omega}}{\partial \beta} = \frac{\partial \mathbf{P}}{\partial \beta} - \frac{\partial \mathbf{K}}{\partial \beta}\boldsymbol{\omega} \tag{10.105}$$

The element connectivity matrix $\mathbf{C}$ is used to identify the nodal values of the derivative of $\boldsymbol{\omega}^e$ for element e from the system derivative vector by

$$\frac{\partial \omega_i^e}{\partial \beta} = \frac{\partial \omega_n}{\partial \beta} \qquad n = C_{ei} \quad 1 \le i \le 9 \tag{10.106}$$

The warping-related properties Q_ω, I_ω, $I_{y\omega}$, and $I_{z\omega}$ are defined in Eqs. (7.19) and (7.48). Express the derivatives of

$$\begin{aligned} Q_\omega &= \int \omega\, dA = \sum_e \int_{-1}^{1}\int_{-1}^{1} \mathbf{N}\boldsymbol{\omega}^e |\, \mathbf{J}_e \,|\, d\eta\, d\zeta \\ I_\omega &= \int \omega^2\, dA = \sum_e \int_{-1}^{1}\int_{-1}^{1} (\mathbf{N}\boldsymbol{\omega}^e)^2 \,|\, \mathbf{J}_e \,|\, d\eta\, d\zeta \\ I_{y\omega} &= \int y\omega\, dA = \sum_e \int_{-1}^{1}\int_{-1}^{1} \mathbf{N}\mathbf{y}_e \mathbf{N}\boldsymbol{\omega}^e |\, \mathbf{J}_e \,|\, d\eta\, d\zeta \\ I_{z\omega} &= \int z\omega\, dA = \sum_e \int_{-1}^{1}\int_{-1}^{1} \mathbf{N}\mathbf{z}_e \mathbf{N}\boldsymbol{\omega}^e |\, \mathbf{J}_e \,|\, d\eta\, d\zeta \end{aligned} \tag{10.107}$$

TABLE 10.2 Functions for Calculating Derivatives of Warping-Related Properties

Derivative	Element Function, $F_e(\eta, \zeta)$
$\dfrac{\partial Q_\omega}{\partial \beta}$	$\mathbf{N}\left(\dfrac{\partial \boldsymbol{\omega}^e}{\partial \beta} + \dfrac{1}{\mid \mathbf{J}_e \mid}\dfrac{\partial \mathbf{J}_e}{\partial \beta}\boldsymbol{\omega}^e\right)$
$\dfrac{\partial I_\omega}{\partial \beta}$	$\mathbf{N}\boldsymbol{\omega}^e\left(2\mathbf{N}\dfrac{\partial \boldsymbol{\omega}^e}{\partial \beta} + \dfrac{1}{\mid \mathbf{J}_e \mid}\dfrac{\partial \mid \mathbf{J}_e \mid}{\partial \beta}\mathbf{N}\boldsymbol{\omega}^e\right)$
$\dfrac{\partial I_{y\omega}}{\partial \beta}$	$\mathbf{N}\boldsymbol{\omega}^e\mathbf{N}\dfrac{\partial \mathbf{y}_e}{\partial \beta} + \mathbf{N}\mathbf{y}_e\mathbf{N}\dfrac{\partial \boldsymbol{\omega}^e}{\partial \beta} + \dfrac{1}{\mid \mathbf{J}_e \mid}\dfrac{\partial \mid \mathbf{J}_e \mid}{\partial \beta}\mathbf{N}\mathbf{y}_e\mathbf{N}\boldsymbol{\omega}^e$
$\dfrac{\partial I_{z\omega}}{\partial \beta}$	$\mathbf{N}\boldsymbol{\omega}^e\mathbf{N}\dfrac{\partial \mathbf{z}_e}{\partial \beta} + \mathbf{N}\mathbf{z}_e\mathbf{N}\dfrac{\partial \boldsymbol{\omega}^e}{\partial \beta} + \dfrac{1}{\mid \mathbf{J}_e \mid}\dfrac{\partial \mid \mathbf{J}_e \mid}{\partial \beta}\mathbf{N}\mathbf{z}_e\mathbf{N}\boldsymbol{\omega}^e$

in the form

$$\sum_e \int_{-1}^{1}\int_{-1}^{1} F_e(\eta, \zeta) \mid \mathbf{J}_e \mid d\eta \, d\zeta \tag{10.108}$$

The functions $F_e(\eta, \zeta)$ are listed in Table 10.2.

The shear center coordinates y_S and z_S according to the Trefftz theory are (Eqs. 6.118 and 6.119)

$$\begin{aligned} y_S &= \frac{I_{yz}I_{y\omega} - I_z I_{z\omega}}{D} \\ z_S &= \frac{I_y I_{y\omega} - I_{yz}I_{z\omega}}{D} \end{aligned} \tag{10.109}$$

with $D = I_y I_z - I_{yz}^2$. The derivatives are given by

$$\begin{aligned} \frac{\partial y_S}{\partial \beta} &= \frac{1}{D}\left(I_{y\omega}\frac{\partial I_{yz}}{\partial \beta} + I_{yz}\frac{\partial I_{y\omega}}{\partial \beta} - I_{z\omega}\frac{\partial I_z}{\partial \beta} - I_z\frac{\partial I_{z\omega}}{\partial \beta}\right) - \frac{y_S}{D}\frac{\partial D}{\partial \beta} \\ \frac{\partial z_S}{\partial \beta} &= \frac{1}{D}\left(I_{y\omega}\frac{\partial I_y}{\partial \beta} + I_y\frac{\partial I_{y\omega}}{\partial \beta} - I_{z\omega}\frac{\partial I_{yz}}{\partial \beta} - I_{yz}\frac{\partial I_{z\omega}}{\partial \beta}\right) - \frac{z_S}{D}\frac{\partial D}{\partial \beta} \end{aligned} \tag{10.110}$$

where

$$\frac{\partial D}{\partial \beta} = I_y\frac{\partial I_z}{\partial \beta} + I_z\frac{\partial I_y}{\partial \beta} - 2I_{yz}\frac{\partial I_{yz}}{\partial \beta} \tag{10.111}$$

These derivatives determine the derivative of the warping constant Γ (Eq. 7.50) referred to the shear center

$$\frac{\partial \Gamma}{\partial \beta} = \frac{\partial I_\omega}{\partial \beta} - \frac{2Q_\omega}{A}\frac{\partial Q_\omega}{\partial \beta} + \frac{Q_\omega^2}{A^2}\frac{\partial A}{\partial \beta} - I_{y\omega}\frac{\partial z_S}{\partial \beta} - z_S\frac{\partial I_{y\omega}}{\partial \beta} + I_{z\omega}\frac{\partial y_S}{\partial \beta} + y_S\frac{\partial I_{z\omega}}{\partial \beta} \tag{10.112}$$

10.5 DESIGN SENSITIVITY ANALYSIS FOR EFFECTIVE TORSIONAL CONSTANT

The computation of the derivative of the effective torsional constant of Section 7.7 with respect to β requires several steps

1. Calculate the derivative of Γ_{S_i}:

$$
\begin{aligned}
\frac{\partial \Gamma_{S_i}}{\partial \beta} &= \frac{\partial I_{\omega_i}}{\partial \beta} - \frac{2Q_{\omega i}}{A_i}\frac{\partial Q_{\omega i}}{\partial \beta} + \frac{Q_{\omega i}^2}{A_i^2}\frac{\partial A_i}{\partial \beta} \\
&\quad - 2\frac{\partial z_{S_i}}{\partial \beta} I_{y_i \omega_i} - 2 z_{S_i}\frac{\partial I_{y_i \omega_i}}{\partial \beta} + 2\frac{\partial y_{S_i}}{\partial \beta} I_{z_i \omega_i} + 2 y_{S_i}\frac{\partial I_{z_i \omega_i}}{\partial \beta} \\
&\quad + 2 z_{S_i}\frac{\partial z_{S_i}}{\partial \beta} I_{z_i} + z_{S_i}^2 \frac{\partial I_{z_i}}{\partial \beta} + 2 y_{S_i}\frac{\partial y_{S_i}}{\partial \beta} I_{y_i} + y_{S_i}^2 \frac{\partial I_{y_i}}{\partial \beta} \\
&\quad - 2\frac{\partial y_{S_i}}{\partial \beta} z_{S_i} I_{y_i z_i} - 2 y_{S_i}\frac{\partial z_{S_i}}{\partial \beta} I_{y_i z_i} - 2 y_{S_i} z_{S_i}\frac{\partial I_{y_i z_i}}{\partial \beta}
\end{aligned}
\tag{10.113}
$$

2. Calculate the torsional constant J_0 of the open section by summing the contributions of the n unconnected cross-sectional members:

$$
J_0 = \sum_{i=1}^{n} J_i \tag{10.114}
$$

Find the derivative of J_0 by

$$
\frac{\partial J_0}{\partial \beta} = \sum_{i=1}^{n} \frac{\partial J_i}{\partial \beta} \tag{10.115}
$$

3. Calculate the warping constant Γ_0 of the open section by summing the contributions of the n unconnected cross-sectional members:

$$
\Gamma_0 = \sum_{i=1}^{n} \Gamma_{S_i} \tag{10.116}
$$

and find the derivative of Γ_0 by

$$
\frac{\partial \Gamma_0}{\partial \beta} = \sum_{i=1}^{n} \frac{\partial \Gamma_{S_i}}{\partial \beta} \tag{10.117}
$$

4. Calculate λ:

$$
\lambda = \sqrt{\frac{G J_0}{E \Gamma_0}} \tag{10.118}
$$

Calculate the derivative of λ:

$$\frac{\partial \lambda}{\partial \beta} = \frac{\lambda}{2}\left(\frac{1}{J_0}\frac{\partial J_0}{\partial \beta} - \frac{1}{\Gamma_0}\frac{\partial \Gamma_0}{\partial \beta}\right) \tag{10.119}$$

5. Calculate the effective torsional constant J_{eff} using Eq. (7.90):

$$J_{\text{eff}} = \frac{J_0}{1 - (2/\lambda L)\,(1 - (J_0/J_c))\tanh(\lambda L/2)} \tag{10.120}$$

Calculate the derivative of J_{eff}:

$$\begin{aligned}\frac{\partial J_0}{\partial \beta} = \frac{J_{\text{eff}}}{J_0}\Bigg[\frac{\partial J_0}{\partial \beta} &+ \frac{J_{\text{eff}}}{\lambda}\frac{\partial \lambda}{\partial \beta}\left(1 - \frac{J_0}{J_c}\right)\left(\operatorname{sech}^2\frac{\lambda L}{2} - \frac{2}{\lambda L}\tanh\frac{\lambda L}{2}\right)\\ &+ \frac{2J_{\text{eff}}}{\lambda L J_c}\left(\frac{J_0}{J_c}\frac{\partial J_c}{\partial \beta} - \frac{\partial J_0}{\partial \beta}\right)\tanh\frac{\lambda L}{2}\Bigg]\end{aligned} \tag{10.121}$$

10.6 OPTIMIZATION

The cross-sectional shape optimization of a thin-walled beam can be stated as a constrained optimization problem. The independent variables of this problem, denoted by β_i for $1 \le i \le n$, represent the parameters that define the cross-sectional shape, such as wall thickness, control point coordinates, and control point weights. The independent variables are also referred to as the design variables. Let W be the objective function, a function of the n variables β_i, $1 \le i \le n$, which is to be minimized. The constrained optimization problem is to determine the β_i for which W is minimized while the inequality and equality constraints of the problem are satisfied. Thus, the goal is to find the β_i, $1 \le i \le n$, that minimize

$$W(\beta_1, \ldots, \beta_n) \tag{10.122a}$$

subject to the constraints

$$g_j(\beta_1, \ldots, \beta_n) \le 0 \qquad 1 \le j \le m \tag{10.122b}$$

$$h_k(\beta_1, \ldots, \beta_n) = 0 \qquad 1 \le k \le l \tag{10.122c}$$

$$\beta_i^L \le \beta_i \le \beta_i^U \qquad 1 \le i \le n \tag{10.122d}$$

For convenience, the constraints on the design variables β_i (Eq. 10.122d) have been separated from the other "general" constraints of Eq. (10.122b and c). That is, the general constraints are written as inequalities or equalities for the functions g_j and h_k, respectively, while the side constraints, which force the independent variables β_i to be bounded from below by β_i^L and bounded from above by β_i^U, are stated separately.

The objective function W is usually a cross-sectional property such as the area, an area moment of inertia, or the torsional constant. The constraints g_j may state bounds on cross-sectional properties or stresses. When the independent variables of the problem are varied while searching for a minimizing set, the cross-sectional shape is modified, and the optimization problem is a search for the optimal cross-sectional shape. The equality constraints h_k are usually included to state the dependencies among the design variables.

The shape optimization problem is nonlinear in the design variables. Many optimization techniques have been developed to solve this type of problem. For example, it is usually possible to linearize the problem about an initial estimate of the solution and then find a solution to this linear approximation by a linear programming method. The problem can then be linearized about the approximate solution found, and the new linear programming problem can be solved to arrive at another approximate solution. These computations are repeated until a solution to the original problem is obtained. This approach is referred to as *sequential linear programming*.

Let $\mathbf{X}$ denote the vector of design variables

$$\mathbf{X} = [\beta_1, \ldots, \beta_n]^{\mathrm{T}} \tag{10.123}$$

and let $\mathbf{X}^0$ be the current approximate solution in the sequential linear programming approach. The next approximation $\mathbf{X}$, to be determined, is written as

$$\mathbf{X} = \mathbf{X}^0 + \delta\mathbf{X} \tag{10.124}$$

The vector $\delta\mathbf{X}$ will contain the independent variables of the linearized problem, which is obtained from the first-order Taylor series expansion of the optimization problem. Thus, the linearized objective function is written as

$$W(\mathbf{X}) = W(\mathbf{X}^0) + \nabla W(\mathbf{X}^0) \cdot \delta\mathbf{X} \tag{10.125}$$

and the linearized constraints are

$$\begin{aligned} g_j(\mathbf{X}) &= g_j(\mathbf{X}^0) + \nabla g_j(\mathbf{X}^0) \cdot \delta\mathbf{X} \le 0 \qquad & 1 \le j \le m \\ h_k(\mathbf{X}) &= h_k(\mathbf{X}^0) + \nabla h_k(\mathbf{X}^0) \cdot \delta\mathbf{X} = 0 \qquad & 1 \le k \le l \\ X_i^L &\le X_i^0 + \delta X_i \le X_i^U & 1 \le i \le n \end{aligned} \tag{10.126}$$

The linearized problem is seen to be a linear programming problem to be solved for $\delta\mathbf{X}$. This method of solution usually produces a sequence of improving but infeasible designs. If it is desirable to approach the optimum solution from inside the feasible region, a modification of this method, called the *method of centers*, can be used (Vanderplaats, 1984).

REFERENCE

Vanderplaats, G. N. (1984). *Numerical Optimization Techniques for Engineering Design*, McGraw-Hill, New York.

APPENDIX A

USING THE COMPUTER PROGRAMS

In this appendix we describe how to prepare input data files for some of the computer programs on the web site. The programs discussed here can be used to analyze beams with thin-walled open or closed cross sections of arbitrary shape. Most of the computational formulations in the book do not distinguish between solid, thin-walled, open, or closed cross sections.

A.1 OVERVIEW OF THE PROGRAMS

Two programs, `ThinWall` and `PlotStress`, are considered in this appendix. The first program calculates cross-sectional properties and stresses. Calculation of the stresses requires the specification of internal stress resultants at the cross section. These may originate from a full structural analysis or be hypothetical design loads. The program `ThinWall` has no built-in structural analysis capability. In the case of a full structural analysis, any of the standard general-purpose programs will provide the requisite stress resultants, such as moments, torques, and shear forces. The second program, `PlotStress`, reads the stress output files from `ThinWall` and makes a three-dimensional plot of the stress magnitude distribution over the cross section. As all the numerical information necessary to make these plots is contained in the output files from `ThinWall`, users may also use any postprocessor of their choice to view or print the plots.

The programs are in Fortran 90, and were written with portability in mind. The module `stdtypes.f`, which contains kind type and other parameter definitions, the platform-dependent lines of code will need adjustments. Aside from this, the only other platform-dependent code is in the two calls to the `getarg` intrinsic subroutine, one in ThinWall and the other in PlotStress, for getting the input file name from the command line.

A.2 INPUT DATA FILE FOR CROSS-SECTION ANALYSIS

The input data file is read as a sequence of blank-separated strings. Any of these strings that are to be interpreted as numerical values are later converted by the program to the correct type. The parser will recognize only the keywords described in this section. Upper- and lowercase letters are treated as distinct when a keyword is read. If the comment character # is encountered anywhere in the input file, any remaining characters on the same line are ignored by the parser.

The optional Title: keyword may be used to assign a title to the job run by the input file. The title specified is then printed as the first line of the output files generated by the run. The colon is expected to be followed by a character string of maximum length 128 on the same line as the keyword, as shown below:

```
Title: Test File for an Unsymmetric Channel Section
```

The Vertices keyword begins the definition of the control points for all the B-spline definitions to follow. The B splines define the median line of the cross section. An example of the Vertices keyword is

```
Vertices
    1     0     0
    2     8     0
    3     8    16
    4     0    16
    5    -8    16
    6    -8     0
    7     0     0
End Vertices
```

The vertices are assigned global integer identifiers, which are later used to associate them with the splines forming the median line of the section. The identifiers in the example above are the integers 1, 2, 3, 4, 5, 6, 7. Each of these identifiers is followed by the y and z coordinates of the vertex, or the control point. The coordinates are real numbers but may be input as integers as in the example above if they happen to have no fractional part. Vertex identifiers must be unique integers, but the same coordinates may be assigned more than one identifier, as in the case of vertex 1 and vertex 7. One use of this multiple assignment is in defining an open curve that is geometrically closed, as in the case of a circular tube with a slit. If the first and last control points of the circle are both given the same identifier, the program will generate a closed circular tube. If, on the other hand, the first and last control points have different identifiers, the program will assign different node numbers to the edge elements and will generate an open circular tube with a slit. In general, end-to-end connections of different curves in forming the median line of the section are determined by the identifiers of the vertices, not by the corresponding y, z coordinates.

The `Materials` keyword defines material identifiers, which must be integers, and the corresponding modulus of elasticity and Poisson's ratio, which are real numbers. For example, the following input lines define the two materials: aluminum with $E = 10.4 \times 10^6$ and $\nu = 0.3$ and copper with $E = 18.5 \times 10.6$ and $\nu = 0.3$.

```
Materials
  ID  1  Elastic  10.4e6  Poisson  0.3  #Aluminum
  ID  2  Elastic  18.5e6  Poisson  0.3  #Copper
End  Materials
```

If no `Materials` keyword is found in the input file, default values are assigned and all elements in the finite-element mesh are assumed to be of the default material. The default values are material identifier equal to 1, modulus of elasticity equal to 210×10^6, and a Poisson's ratio of $\frac{1}{3}$. This modulus of elasticity corresponds to steel in kPa. If section forces are specified in kN, section torque in kNm, lengths in m, etc., then the stress output would be in kPa.

The `Splines` keyword begins the input of the B-splines that define the median line of the section. An example with a single spline identified as `Branch` is shown below.

```
Splines
  Branch  1
    Thickness 1.25     Material 1     Order 3
    Nodes 1 2  3  4  5  6  7 End Nodes
    Weights 1 0.5 0.5 1 0.5 0.5 1 End Weights
    Knots 0 0 0 0.25 0.5 0.5 0.75 1 1 1 End Knots
  End Branch
End Splines
```

The `Branch` keyword under the `Splines` keyword identifies a single spline by an integer label, in this case 1. This single spline is then defined before the `End Branch` keywords are encountered. The `Thickness` keyword under each `Branch` defines the wall thickness for the spline branch being defined. The `Thickness` keyword is mandatory for each branch and is followed by a real number equal to the wall thickness in the units chosen by the user, in this case 1.25. The `Material` keyword is followed by a material identifier. The modulus of elasticity and Poisson's ratio for the material identifier are defined as described earlier under the `Materials` keyword. If `Material` is not included for the branch, the material identifier for the branch is by default equal to 1. The `Order` keyword is mandatory. The integer following it is the order of the NURBS, which is one larger than its degree. The mandatory `Nodes` keyword is followed by a sequence of control point identifiers. The coordinates corresponding to these identifiers are defined under the `Vertices` keyword as described earlier. The `Weights` keyword is optional. If `Weights` is not encountered under `Branch`, the weights associated with the control points of the branch are all assumed to be unity. The rational B-spline then becomes an ordinary B-spline. The `Knots` keyword is also optional. If `Knots` is not encountered under `Branch`, a uniform knot vector from 0 to 1 is assumed. In that case, the nonuniform rational B-spline becomes a uniform B-spline.

Two additional keywords are possible under `Branch`. `NormalElements` and `AspectRatio`, both of which are mesh density control parameters. `NormalElements` is followed by an integer that specifies the number of elements to place through the wall thickness. The default for this is 2, so that two layers of elements in the direction normal to the median line of the cross section are formed. This is usually sufficient for the computation of cross-sectional properties, but a larger number may sometimes be desirable or necessary in stress distribution calculations. `AspectRatio` is followed by a real number that defines the aspect ratio for the mesh. The mesh aspect ratio is such that the mesh generator attempts to make the length of an element along the direction parallel to the median line equal to L:

$$L = \frac{\rho t}{N}$$

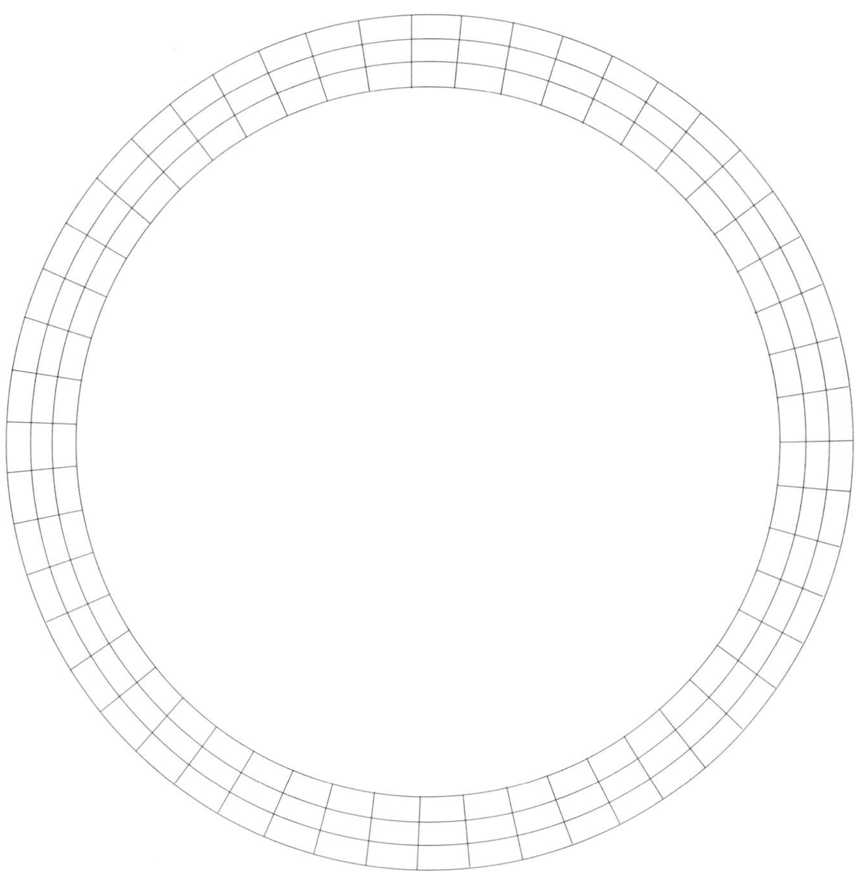

Figure A.1 Mesh with aspect ratio of 2. There are three elements through the thickness.

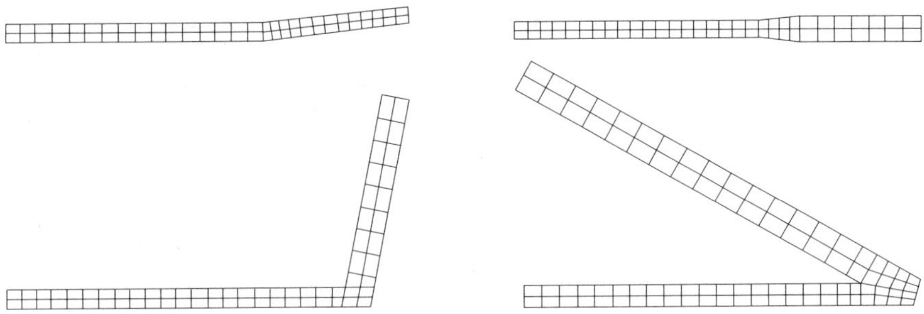

Figure A.2 Various types of patch meshes.

where ρ denotes the aspect ratio, t is the wall thickness, and N is the number of elements through the thickness as specified by `NormalElements`. The default value of ρ is set equal to the golden ratio 1.61803. Figure A.1 shows a mesh with aspect ratio $\rho = 2$ and `NormalElements` $N = 3$. This generates three layers of elements throughout the thickness, and the length L of each element is approximately twice its height.

When two branches are assigned the same control point identifier for their first or last control point, the branches are connected at that point and a patch mesh is made around it. Several types of patch meshes are defined, and the program determines which type to choose from the value of the intersection angle between the two connecting branches. Figure A.2 illustrates four types of patches that are generated at the intersection of two straight-line segments. Figure A.3 shows patch meshes at

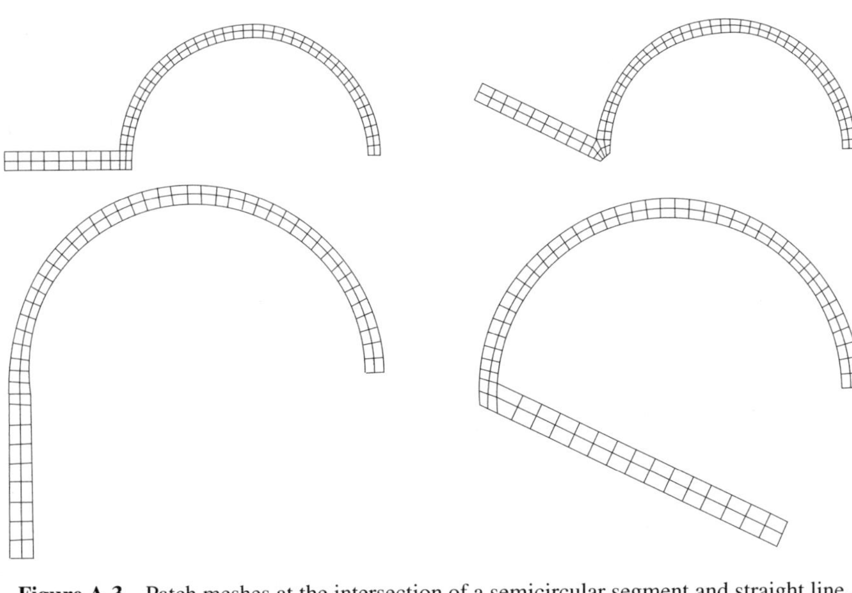

Figure A.3 Patch meshes at the intersection of a semicircular segment and straight line.

TABLE A.1 Data File for Intersecting Semicircular Segment and Straight Line

```
#Patch mesh example semicircle and intersecting horizontal line

Vertices
 1          10      0
 2          10     10
 3         -10     10
 4         -10      0
 5         -20      0
End vertices

Splines

Branch 1
Thickness 1 Order 3
Nodes 1 2 3 4 End Nodes
Knots 0 0 0 0.5 1 1 1 End Knots
Weights 1 0.5 0.5 1 End Weights
End Branch

Branch 2
Thickness 1.5 Order 2
Nodes  4 5 End Nodes
End Branch

End Splines
```

the intersection of a semicircular segment with a straight one at four different angles. The data file that produces the top left mesh in Fig. A.3 is listed in Table A.1.

This version of the program does not allow more than two splines to intersect, and the presence of three or more common end control points among the splines describing the cross-section walls may generate undetected errors or catastrophic crashes. Another way in which splines can connect to each other is provided, however, and this capability may sometimes be used to simulate splines intersecting at more than two points. The keyword `Weld` indicates that the branch identifiers following it are to be connected pairwise along their edges parallel to the spline. An example of this is shown in Fig. A.4. The middle vertical part of this cross section where there are four elements across the wall thickness contains two vertical splines that are joined to each other by the spline welding option. Because the welding is restricted to the parallel edges, it is necessary to put in short splines that are not welded near the T junctions to make the transition to the horizontal parts of the cross section-walls properly. For the welding of two splines to be successful, the nodes of the connected splines must have the same coordinates along the welded edge. If the aspect ratios of the welded splines are different, for example, a mismatch will occur. The program only checks the number of boundary points and the coordinates of one matching point. It is, therefore, possible to have an undetected meshing error, which may cause further undetected problems in the later parts of the computation. It goes

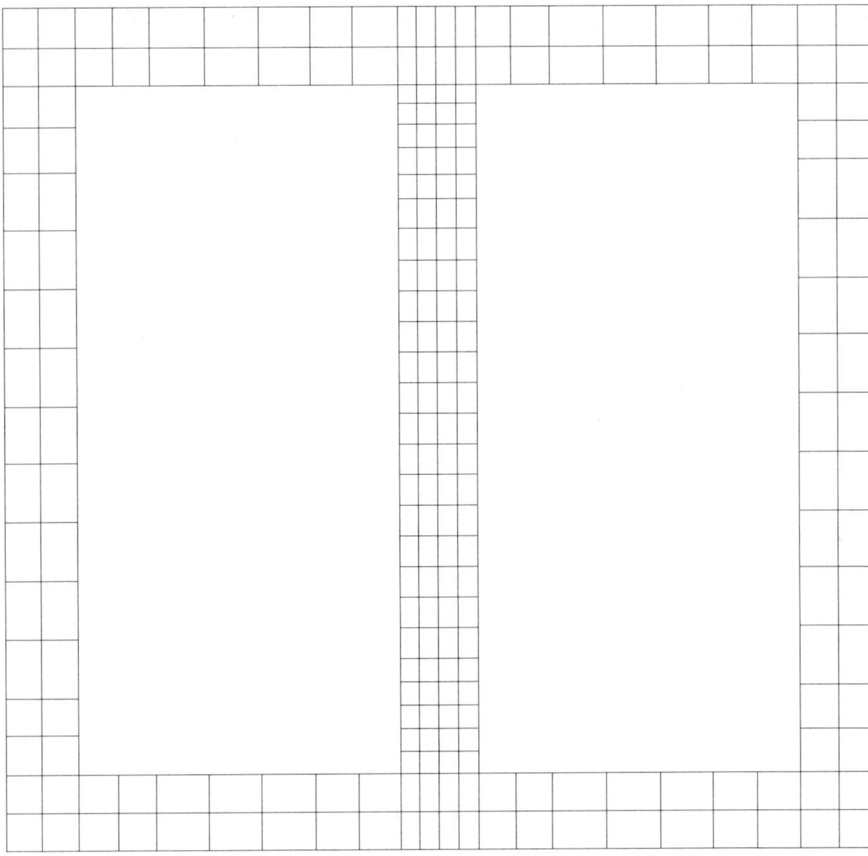

Figure A.4 The middle vertical lines in this mesh are connected along their edges.

without saying that users should always inspect the mesh visually before accepting any numerical results.

The keyword `Weld` is not intended to suggest that splines connected by it physically model a weld. The effect of welding is to assign the nodes along the matching edges the same global node numbers, and to treat these points as identical. The syntax of the `Weld` keyword is illustrated by this example:

```
Welds
  10  18  20  30  10  17
End Welds
```

This welds splines 10 and 18 together along a matching edge. The other edge of 10 is expected to match an edge of spline 17, so that 10 and 17 can also be connected along an edge. Splines 20, 30 are also attached to each other along a matching edge by the statement above. Table A.2 lists the data file that generates the mesh in Fig. A.4.

TABLE A.2 Data File for the Mesh in Fig. A.4

```
Vertices
1          0      0
2          0     10
3          5     10
4          5      8
5          5      2
6          5      0
7          5.5   10
8          5.5    8
9          5.5    2
10         5.5    0
11        10.5   10
12        10.5    0
End Vertices

Splines

Branch  1  Thickness 1    Order 2  Nodes  1   2  End Nodes  End Branch
Branch  2  Thickness 1    Order 2  Nodes  2   3  End Nodes  End Branch
Branch  3  Thickness 0.5  Order 2  Nodes  3   4  End Nodes  End Branch
Branch  4  Thickness 0.5  Order 2  Nodes  4   5  End Nodes  End Branch
Branch  5  Thickness 0.5  Order 2  Nodes  5   6  End Nodes  End Branch
Branch  6  Thickness 1    Order 2  Nodes  1   6  End Nodes  End Branch
Branch  7  Thickness 0.5  Order 2  Nodes  7   8  End Nodes  End Branch
Branch  8  Thickness 0.5  Order 2  Nodes  8   9  End Nodes  End Branch
Branch  9  Thickness 0.5  Order 2  Nodes  9  10  End Nodes  End Branch
Branch 10  Thickness 1    Order 2  Nodes 10  12  End Nodes  End Branch
Branch 11  Thickness 1    Order 2  Nodes  7  11  End Nodes  End Branch
Branch 12  Thickness 1    Order 2  Nodes 11  12  End Nodes  End Branch

End Splines

Welds  4  8  End  Welds
```

In addition to the possibility of setting the values of `NormalElements` and `AspectRatio` for each spline branch under the `Branch` keyword, it is frequently useful to be able to set these for the whole mesh. The keyword `Mesh` can be used for this purpose if the default settings are to be overridden. If, for example, four elements are wanted through the wall thickness with an aspect ratio of 2.25, the following `Mesh` statement is needed:

```
Mesh
  NormalElements 4
  AspectRatio    2.25
End Mesh
```

If some branches are to have different settings for these parameters, the `NormalElements` and `AspectRatio` keywords must be used with the desired values under the `Branch` keywords corresponding to those branches.

The appearance of the mesh can be controlled with the options provided under the `Graphics` keyword.

```
Graphics
  NodeNumbers
  ElementNumbers
  FullElements
  PageWidth  6
  LineThickness 0.25
EndGraphics
```

If the `NodeNumbers` option is included, the mesh picture will show the node numbers. This option should normally not be used, but it may occasionally help in debugging a data file. If the `ElementNumbers` option is included, the numbers assigned to the elements will be printed at the center of each element. The `FullElements` option shows each element with all nine nodes shown at the intersection of straight lines, as in Fig. 4.1. If the `FullElements` option is not included, each nine-node element is drawn as a quadrilateral. The keyword `PageWidth` may be used to set the page width in inches, and `LineThickness` to set the line width in points. The `Graphics` keyword is optional. If it is omitted, the page width is set equal to 6.5 in., the line width to 0.2, no node or element numbers are shown, and only the four edges of each element are drawn as straight lines.

The mesh generator does not check for geometric interference problems, such as one part of the wall overlapping another, which may arise because of thickness incompatibilities or overcrowding near a patch area. It may also happen that the length of a patch area determined by the meshing algorithm requires more than the length of the entire spline if the spline is very short. Since the algorithm expects to place at least one element on the main stem of the spline outside the patch, this will result in an untenable meshing requirement.

For stress analysis, the loads at the section are defined by the `Loads` keyword. Any loads that are zero may be omitted. An example is shown below.

```
Loads
  P  1000   Mx  8000  My -400.5  Mz 6000
  Vy 10000  Vz  30000.8
  Bimoment 875
  yP 0.5    zP   0.975
  yV 1.5    zV   1.85
End  Loads
```

The axial load of 1000 is specified by `P`, the section torque of 8000 by `Mx`, the bending moments of −400.5 and 6000 about the y and z axes, respectively, by `My` and `Mz`, the shear forces of 10000 and 30000.8 by `Vy` and `Vz`, and the bimoment of 875 by `Bimoment`. If there is an axial force, its point of application on the cross section is given by `yP`, `zP`. In this example, `yP` = 0.5 and `zP` = 0.975. If there are shear forces, their point of application is specified by `yV` and `zV`, here yV = 1.5 and zV = 1.85. The values assigned to these y, z coordinates are assumed to be measured in the user coordinate system. The inclusion of the keyword ShearAtShearCenter under Loads indicates that shear loads are to be applied at the shear center. Similarly, if the shear forces are to be applied at the centroid, the keyword ShearAtCentroid should appear under loads. The keyword AxialAtCentroid specifies that the point of application of the axial load is the centroid.

A.3 OUTPUT FILES

If the input data file described in the preceding section is named infile.dat and the executable for ThinWall is called twb, the program is run by the command

```
twb  infile.dat
```

This produces three output files if no stress analysis was requested in the input data file, and four files otherwise. The output file infile.mesh lists the connectivity matrix and the y, z coordinates of all the nodes. The PostScript file infile.ps may be viewed on the screen or printed to examine the mesh generated. The file infile.res lists the section properties calculated. The file infile.str lists the stress distributions calculated.

The mesh file infile.mesh may be read into a postprocessing program to draw the mesh if the PostScript file infile.ps cannot be used. The files infile.mesh and infile.str may be postprocessed for displaying the stress distributions. The PlotStress program writes PostScript files of three-dimensional stress plots by reading infile.mesh and infile.str.

The PlotStress program plots the normal stress due to axial and bending loads, the normal stress due to warping, the magnitude of the shear stress due to torsion, and the magnitude of the shear stress due to transverse shear forces. With the input data file named infile.dat, the stress plots are made by the command

```
twbplot  infile.grf
```

where twbplot is the name of the executable file for PlotStress, and the file infile.grf contains control commands for the plots.

The input file infile.grf for PlotStress has the keyword Plots for choosing what stresses to plot. If all four stresses are to be plotted, the following should appear in infile.grf:

```
Plots
  Normal
  Warping
  Torsional
  Shear
End Plots
```

The PostScript files for the four stresses are named `infilenrm.ps`, `infilewrp.ps`, `infiletor.ps`, and `infileshr.ps`. Stresses for which a plot is not needed, or those that are not caused by the loads specified, must be omitted from the list given under the `Plots` keywords.

The other keyword that can be included in `infile.grf` is the optional `View`, which may be used to control the appearance of the plots. The syntax is illustrated below.

```
View
  BoxRatios  1  1  0.4
  ViewCenter 0.5 0.5 0.2
  ViewPoint  5  5  5
  ViewVertical  0  0  1
  PageWidth  6.5
  LineThickness 0.20
End View
```

The values shown here are the defaults. The `BoxRatios` option places the three-dimensional plot into a rectangular box whose sides are in the ratios given following it. The `ViewCenter` is the central point of the perspective projection given in box coordinates. The default is to place this point at the center of the box. `ViewPoint`

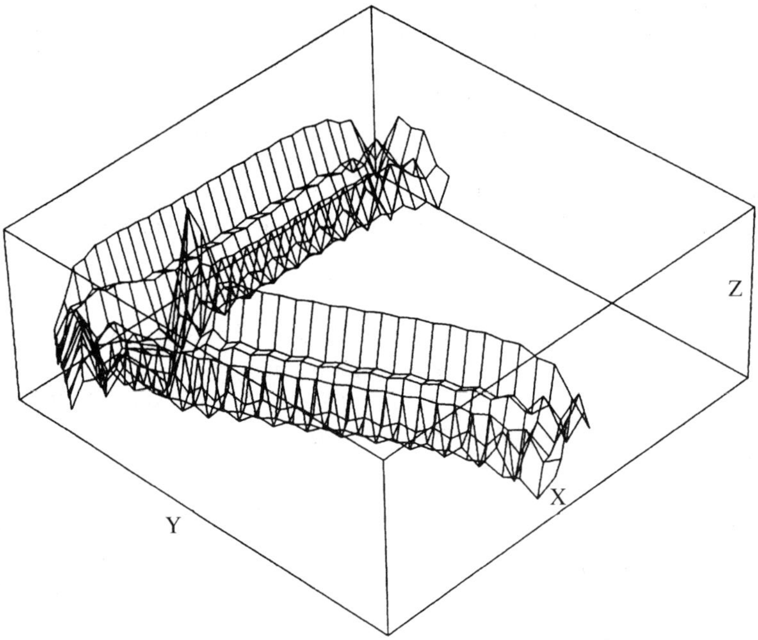

Figure A.5 Torsional stress plot.

is the projection reference point in box coordinates, which must lie outside the box. `ViewVertical` is the direction to be considered vertically upward. `PageWidth` is the width of the page in inches for the final two-dimensional projected points, and `LineThickness` is the line thickness in points. In the resulting plot, the direction marked `Z` is for stress and the other two box sides, labeled `X` and `Y`, define the plane on which the mesh nodes lie. An example of a stress plot of torsional stress magnitudes is shown in Fig. A.5.

APPENDIX B

NUMERICAL EXAMPLES

In this appendix we present numerical results for various cross sections obtained by the computer programs described in Appendix A. Some of the examples use approximate formulas for various cross-sectional shapes as a partial rough check of the computed results. A collection of such formulas is found in Pilkey (1994).

B.1 CLOSED ELLIPTICAL TUBE

The cross section of a closed elliptical tube is shown in Fig. B.1. The mesh of the cross section is shown in Fig. B.2. Set $t = 1$ in., $a = 5$ in., and $b = 8$ in. The input data file is given in Table B.1. An aspect ratio of unity was selected so that the elements are approximately square in shape. The user coordinate system has its origin at the center of the ellipse, with the z axis vertically upward and the y axis to the right. The results file is displayed in Table B.2. For this cross section, the torsional constant can be approximated by the formula

$$J = \frac{4\pi^2 a^2 b^2 t}{S} \tag{B.1}$$

where t is the wall thickness, a and b are the minor and major axes of the median line and S is the length of the median line. Since S is approximately

$$S = \pi \left[\frac{3(a+b)}{2} - \sqrt{ab} \right] \tag{B.2}$$

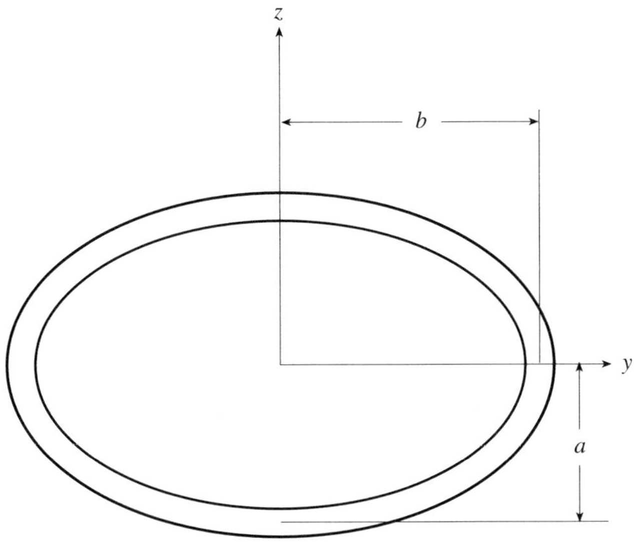

Figure B.1 Closed elliptical tube.

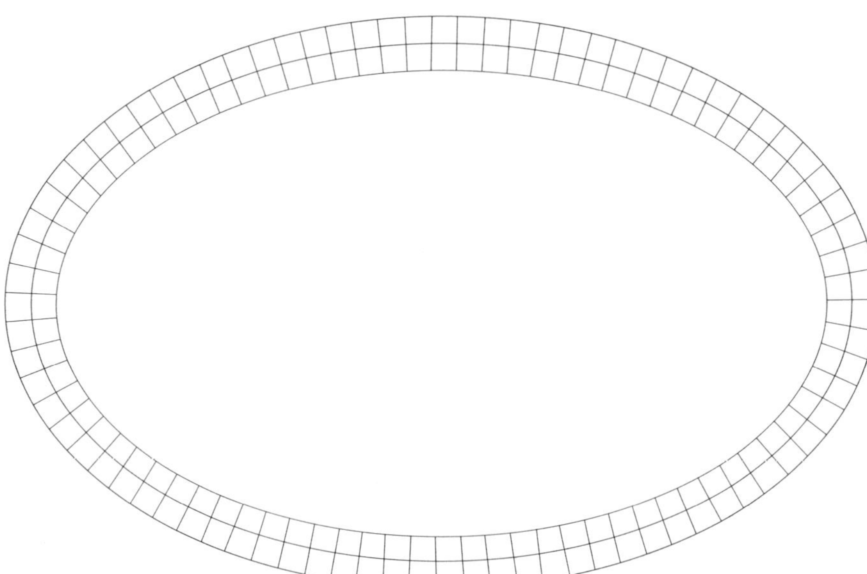

Figure B.2 Mesh for closed elliptical tube.

TABLE B.1 Input Data File for the Closed Elliptical Tube

```
Title: Closed Elliptical Tube Cross Section

Mesh AspectRatio 1 End Mesh
Graphics  PageWidth  5  End Graphics

Vertices
     1    8    0
     2    8    5
     3    0    5
     4   -8    5
     5   -8    0
     6   -8   -5
     7    0   -5
     8    8   -5
End Vertices

Splines

Branch  1
 Thickness  1    Order  3
 Nodes  1  2  3  4  5  6  7  8  1  End Nodes
 Knots  0  0  0  0.25  0.25  0.5  0.5  0.75  0.75  1  1  1  End knots

     Weights
  1  0.707107  1  0.707107  1  0.707107  1  0.707107  1
     End Weights

End Branch

End Spline
```

the formula for the torsional constant is

$$J = \frac{8\pi a^2 b^2 t}{3(a+b) - 2\sqrt{ab}} \tag{B.3}$$

For the numerical values in Table B.1, this gives $J = 1526 \text{ in}^4$, which is to be compared with the program result $J = 1537 \text{ in}^4$.

The default values for the modulus of elasticity and Poisson's ratio, $E = 210 \times 10^6 \text{ kN/m}^2$ (which corresponds to steel) and $\nu = \frac{1}{3}$, were employed by the program. These are listed as reference values in the results file. The reference elastic modulus plays no role in this case, because the cross section is homogeneous. The reference Poisson's ratio is used, however, in the shear boundary value problem solution. The shear coefficients and the shear center depend on Poisson's ratio, as discussed in Chapters 6 and 7. For this symmetric cross section, the shear center calculated by solving the shear boundary value problems is identical to the Trefftz shear center, which has no dependence on Poisson's ratio.

TABLE B.2 Results File for the Closed Elliptical Tube

Closed Elliptical Tube Cross Section	
Cross-Sectional Properties	
Cross-Sectional Area	41.38626
Y Moment of Area	-0.00002
Z Moment of Area	-0.00007
Y Centroid	0.00000
Z Centroid	0.00000
Y Shear Center	0.00000
Z Shear Center	0.00000
Y Shear Center wrt Centroid	0.00000
Z Shear Center wrt Centroid	0.00000
Y Shear Center wrt Centroid (Trefftz)	0.00000
Z Shear Center wrt Centroid (Trefftz)	0.00000
Moment of Inertia I_y	580.42697
Moment of Inertia I_z	1180.33120
Product of Inertia I_{yz}	-0.00015
Moment of Inertia I_{yC}	580.42697
Moment of Inertia I_{zC}	1180.33120
Product of Inertia I_{yzC}	-0.00015
Polar Moment of Inertia	1760.75817
Y Section Elastic Modulus	105.54076
Z Section Elastic Modulus	138.86247
Y Radius of Gyration	3.74495
Z Radius of Gyration	5.34040
Principal Bending Angle (rad)	1.57080
Principal Bending Angle (deg)	89.99998
Principal Moment of Inertia (max)	1180.33120
Principal Moment of Inertia (min)	580.42697
Reference Elastic Modulus	210000000.00000
Reference Poisson's Ratio	0.33333
Y Coordinate Extent	16.99997
Z Coordinate Extent	10.99899
Y Shear Coefficient	1.51457
Z Shear Coefficient	3.05985
YZ Shear Coefficient	0.00000
Torsional Constant	1537.38165
Warping Constant wrt Shear Center	451.90976
Warping Constant wrt Centroid	455.50816

B.2 SYMMETRIC CHANNEL SECTION

Figure B.3 shows a symmetric open- channel section, which is meshed as indicated in Fig. B.4. Choose $t = 1$ in., $h = 18$ in., $b = 8$ in., where b is the length of the flange and h is the height of the web. The input data file is listed in Table B.3. Since no knot vector is supplied in the input, a uniform knot vector from 0 to 1 is used in

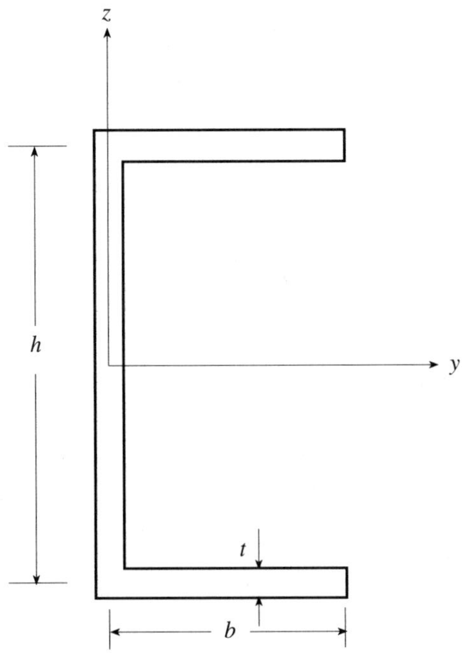

Figure B.3 Channel section symmetric about the y axis.

TABLE B.3 Input Data File for the Symmetric Channel Section

```
Title: Symmetric Channel Section

Graphics PageWidth  5  End Graphics

Vertices
1      8     -9
2      0     -9
3      0      9
4      8      9
End Vertices

Splines

Branch  1
   Thickness  1      Order  2     Nodes  1     2       End Nodes
End Branch

Branch  2
   Thickness  1      Order  2     Nodes  2     3       End Nodes
End Branch

Branch  3
   Thickness  1      Order  2     Nodes  3     4       End Nodes
End Branch

End Splines
```

Figure B.4 Mesh for symmetric channel section.

the program. Also, the default values of the weights equal to 1 are used. The origin of the user coordinate system is at the midpoint of the web, with the z axis vertically upward and the y axis to the right. The results file is listed in Table B.4.

An approximate formula for the torsional constant for open sections of constant wall thickness t is

$$J = \frac{St^3}{3} \tag{B.4}$$

where S is the length of the median line. For this channel section with the dimensions given, $S = 34$ in. and $J = 11.33$ in.^{4} The program result is $J = 11.29$ in.^{4}

TABLE B.4 Results File for the Open Symmetric Channel Section

Symmetric Channel Section Cross-Sectional Properties	
Cross-Sectional Area	34.00000
Y Moment of Area	0.00000
Z Moment of Area	63.75000
Y Centroid	1.87500
Z Centroid	0.00000
Y Shear Center	-2.86769
Z Shear Center	0.00000
Y Shear Center wrt Centroid	-4.74269
Z Shear Center wrt Centroid	0.00000
Y Shear Center wrt Centroid (Trefftz)	-4.74259
Z Shear Center wrt Centroid (Trefftz)	0.00000
Moment of Inertia I_y	1781.83333
Moment of Inertia I_z	342.83333
Product of Inertia I_{yz}	0.00000
Moment of Inertia I_{yC}	1787.83333
Moment of Inertia I_{zC}	223.30208
Product of Inertia I_{yzC}	0.00000
Polar Moment of Inertia	2011.13542
Y Section Elastic Modulus	188.19298
Z Section Elastic Modulus	36.45748
Y Radius of Gyration	7.25144
Z Radius of Gyration	2.56275
Principal Bending Angle (rad)	0.00000
Principal Bending Angle (deg)	0.00000
Principal Moment of Inertia (max)	1787.83333
Principal Moment of Inertia (min)	223.30208
Reference Elastic Modulus	210000000.00000
Reference Poisson's Ratio	0.33333
Y Coordinate Extent	8.50000
Z Coordinate Extent	19.00000
Y Shear Coefficient	3.40789
Z Shear Coefficient	2.15337
YZ Shear Coefficient	0.0000
Torsional Constant	11.28862
Warping Constant wrt Shear Center	12763.15184
Warping Constant wrt Centroid	283214.57041

The area moment of inertia I_y for this section is

$$I_y = \frac{(6b + h)h^2 t}{12} \tag{B.5}$$

The numerical value is $I_y = 1782$ in.^{4}, and the program calculates $I_y = 1788$ in.^{4} The y coordinate of the shear center may be computed as

$$y_S = \frac{3b^2}{6b + h} = 2.91 \text{ in.} \tag{B.6}$$

to be compared with the program result $y_S = 2.87$ in.

Finally, an approximate formula for the warping constant of the channel section gives the value

$$\Gamma = \frac{b^3 h^2 t(3b + 2h)}{12(6b + h)} = 12{,}567 \text{ in.}^6 \tag{B.7}$$

to be compared with the program value $\Gamma = 12,763$ in.^{6}, which should be considered the more accurate result, since the simplifying assumptions used in arriving at the approximate formulas are not made in the finite element analysis.

B.3 L SECTION WITHOUT SYMMETRY

A standard L section $\left(L8 \times 6 \times \frac{3}{4}\right)$ is shown in Fig. B.5, with a coordinate system origin on the median lines at the intersection of the horizontal and vertical legs, with

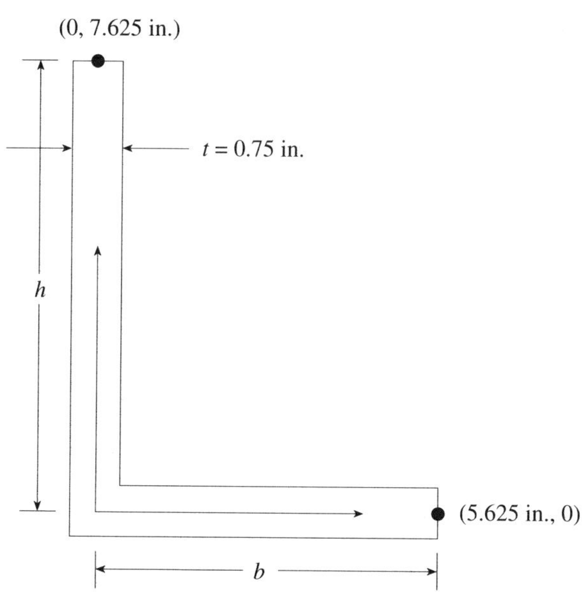

Figure B.5 Standard L section $\left(L8 \times 6 \times \frac{3}{4}\right)$.

TABLE B.5 Input Data File for the L Section

```
Title: A Standard L Section

Vertices
1      0      7.625
2      0        0
3    5.625      0
End Vertices

Splines

Branch    1
 Material    1  Thickness 0.75   Order   2
 Nodes   1   2  End Nodes
End Branch

Branch    2
 Material    1  Thickness 0.75   Order   2
 Nodes   2   3  End Nodes
End Branch

End Splines
```

the y, z axes in the horizontal and vertical directions, respectively. The input data for analyzing this L section are shown in Table B.5. The finite element mesh generated for this section is shown in Fig. B.6. This figure also shows the centroidal reference frame and the centroidal principal bending axes. The principal bending axes y', z' correspond to the maximum and minimum area moments of inertia, respectively.

TABLE B.6 Part of an Output File for the L Section Example

Cross-Sectional Properties	
Cross-Sectional Area	9.93750
Y Moment of Area	21.75000
Z Moment of Area	11.81250
Y Centroid	1.18868
Z Centroid	2.18868
Y Shear Center	0.00313
Z Shear Center	0.05735
Moment of Inertia I_{yC}	63.42455
Moment of Inertia I_{zC}	30.72142
Product of Inertia I_{yzC}	-25.85377
Principal Bending Angle (deg)	28.84407
Principal Moment of Inertia (max)	77.66369
Principal Moment of Inertia (min)	16.48228

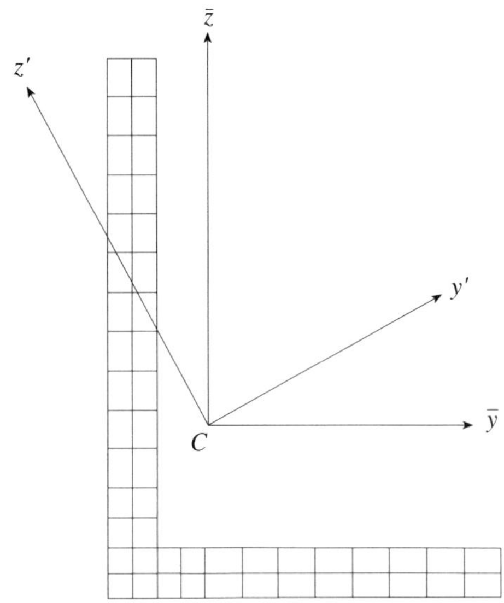

Figure B.6 Mesh for L section.

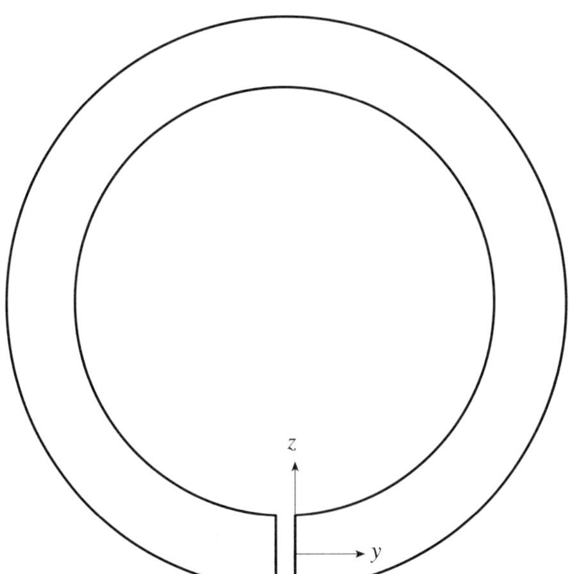

Figure B.7 Open circular cross section.

Table B.6 shows part of the output file that lists the cross-sectional properties of this L section.

B.4 OPEN CIRCULAR CROSS SECTION

An open cross section is shown in Figs. 6.8 and B.7. In the mesh shown in Fig. B.8, element 1 is not attached to element 99, and element 2 is not attached to element 100, although the left and right sides of these pairs of elements are geometrically coincident. In other words, the slit occurs along the left sides of elements 1 and 2. The input data file for the section is listed in Table B.7, and part of the output file for section properties is shown in Table B.8. The user coordinate system origin is at the upper left corner of element 1, with the z axis vertically upward and the y

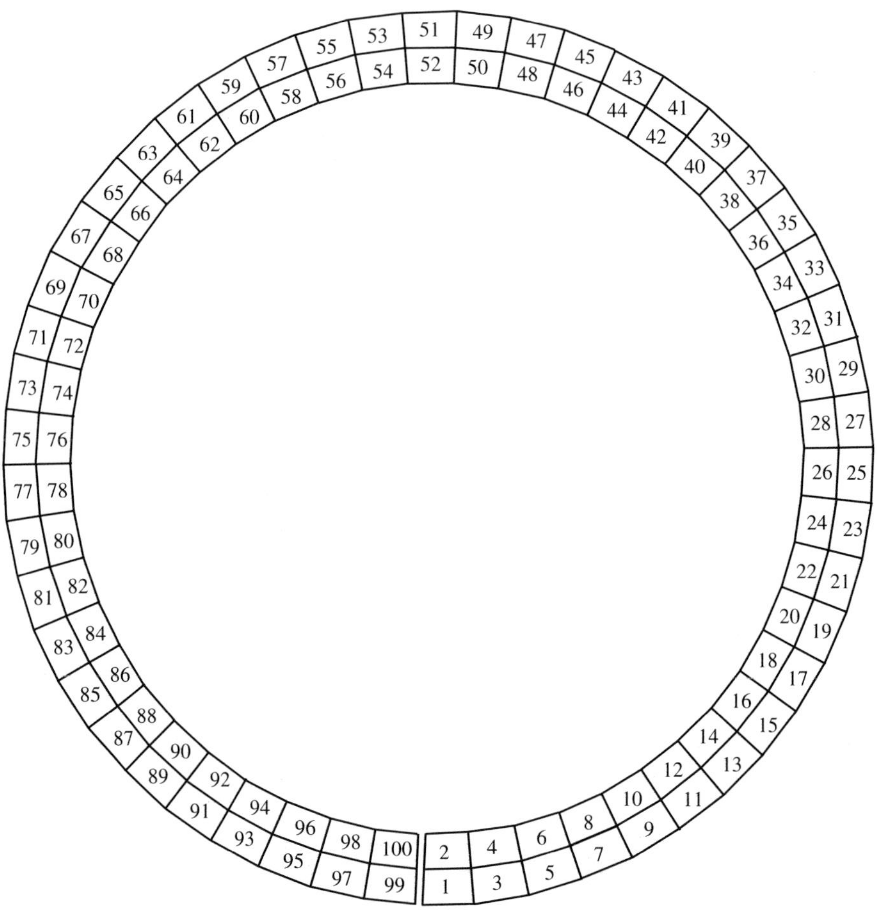

Figure B.8 Open circular cross section. Elements 1, 99, and elements 2, 100 are not connected.

TABLE B.7 Input Data File for Open Circular Cross Section

```
Title: Open Circle Cross Section
Vertices
1      0      0
2      8      0
3      8     16
4      0     16
5     -8     16
6     -8      0
7      0      0
End Vertices

Splines

Branch    1
Thickness   1.25
Nodes  1  2  3  4  5  6  7  End Nodes
Weights  1  0.5  0.5  1  0.5  0.5  1  End Weights
Knots  0  0  0  0.25  0.5  0.5  0.75  1  1  1 End Knots
Order  3
End Branch

End Splines
```

axis directed right. The shear center coordinates with respect to the centroid show that the shear center is approximately one diameter away from the centroid on the z axis, or one and a half diameters away from the slit. This result is derivable by hand calculation, if strength-of-materials assumptions are made.

B.5 WELDED HAT SECTION

A closed hat section is shown in Fig. B.9. Figure B.10 shows the mesh generated for this hat section with the input file listed in Table B.9. The default uniform knot vector from 0 to 1 is employed, so no knot vector is shown in the input. Since weights associated with the control points are not prescribed, the weights are set equal to unity by the program. Section properties calculated by the program are listed in Table B.10. The origin of the user coordinate system is at the midpoint of the left vertical edge of the mesh, with the z axis vertically upward and the y axis pointing right.

This example illustrates the welding of splines along their edges. The horizontal section wall at the bottom is generated by three straight spline segments, branches 2, 9, and 10 in the input data. The branches 2 and 10 are welded at their upper edges to the lower edges of branches 1 and 8. In specifying welds, it is important to keep in mind that there is no provision for welding of patch mesh areas, so that if a weld spline has a patch mesh area at either end, the program will attempt to connect only

TABLE B.8 Output File for the Open Circular Cross Section

Cross-Sectional Area	62.83182
Y Moment of Area	502.65457
Z Moment of Area	-0.00002
Y Centroid	0.00000
Z Centroid	8.00000
Y Shear Center	0.00000
Z Shear Center	23.90306
Y Shear Center wrt Centroid	0.00000
Z Shear Center wrt Centroid	15.90306
Y Shear Center wrt Centroid (Trefftz)	0.00000
Z Shear Center wrt Centroid (Trefftz)	15.90282
Moment of Inertia I_y	6044.12555
Moment of Inertia I_z	2022.88907
Moment of Inertia I_{yz}	0.00004
Moment of Inertia I_{yC}	2022.88890
Moment of Inertia I_{zC}	2022.88907
Moment of Inertia I_{yzC}	0.00018
Polar Moment of Inertia	4045.77796
Y Section Elastic Modulus	234.53784
Z Section Elastic Modulus	234.53789
Y Radius of Gyration	5.67409
Z Radius of Gyration	5.67409
Principal Bending Angle (rad)	-1.00958
Principal Bending Angle (deg)	-57.84458
Principal Moment of Inertia (max)	2022.88918
Principal Moment of Inertia (min)	2022.88878
Reference Elastic Modulus	210000000.00000
Reference Poisson's Ratio	0.33333
Y Coordinate Extent	17.24999
Z Coordinate Extent	17.25000
Y Shear Coefficient	5.93977
Z Shear Coefficient	1.98015
YZ Shear Coefficient	0.00000
Torsional Constant	32.23967
Warping Constant wrt Shear Center	331651.29223
Warping Constant wrt Centroid	3383583.01326

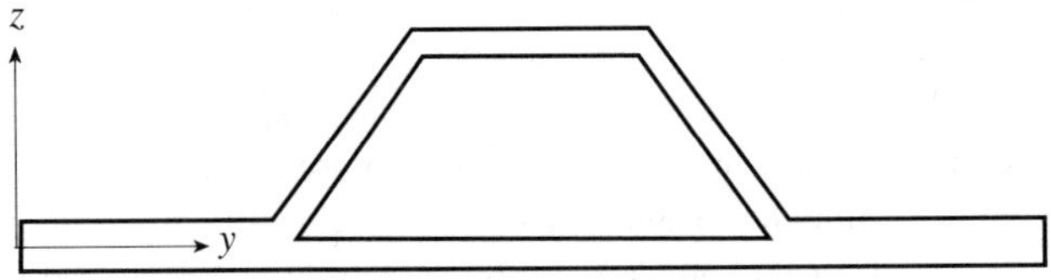

Figure B.9 Hat section.

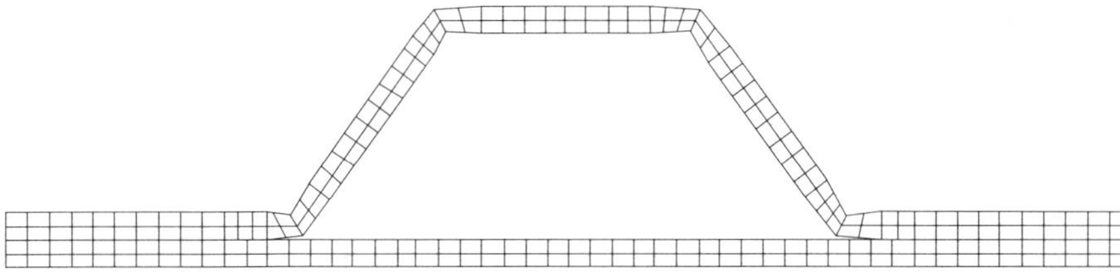

Figure B.10 Mesh for welded hat section.

TABLE B.9 Input Data File for the Hat Section

```
Title: Hat Section

Graphics PageWidth 5 End Graphics

Vertices
1       0     0.375
2       6     0.375
3       0    -0.375
4       6    -0.375
5       8     0.375
6      12     6
7      19     6
8      23     0.375
9      25     0.375
10     31     0.375
11     25    -0.375
12     31    -0.375
End Vertices

Splines

Branch   1  Thickness 0.75   Order  2  Nodes   1   2  End Nodes
End Branch
Branch   2  Thickness 0.75   Order  2  Nodes   3   4  End Nodes
End Branch
Branch   3  Thickness 0.75   Order  2  Nodes   2   5
End Branch
Branch   4  Thickness 0.75   Order  2  Nodes   5   6  End Nodes
End Branch
Branch   5  Thickness 0.75   Order  2  Nodes   6   7  End Nodes
End Branch
Branch   6  Thickness 0.75   Order  2  Nodes   7   8  End Nodes
End Branch
Branch   7  Thickness 0.75   Order  2  Nodes   8   9  End Nodes
End Branch
Branch   8  Thickness 0.75   Order  2  Nodes   9  10  End Nodes
End Branch
Branch   9  Thickness 0.75   Order  2  Nodes   4  11  End Nodes
End Branch
Branch  10  Thickness 0.75   Order  2  Nodes  11  12  End Nodes
End Branch
End Splines
Welds 1 2 8 10 End Welds
```

TABLE B.10 Output File for the Section Properties of the Hat Section

Title: Hat Section
Cross-Sectional Properties

Cross-Sectional Area	50.10830
Y Moment of Area	57.90772
Z Moment of Area	776.70074
Y Centroid	15.50044
Z Centroid	1.15565
Y Shear Center	15.50097
Z Shear Center	2.29918
Y Shear Center wrt Centroid	0.00053
Z Shear Center wrt Centroid	1.14353
Y Shear Center wrt Centroid (Trefftz)	0.00063
Z Shear Center wrt Centroid (Trefftz)	1.12956
Moment of Inertia I_y	316.10331
Moment of Inertia I_z	15874.79940
Product of Inertia I_{yz}	897.57735
Moment of Inertia I_{yC}	249.18219
Moment of Inertia I_{zC}	3835.59608
Product of Inertia I_{yzC}	-0.01777
Polar Moment of Inertia	4084.77827
Y Section Elastic Modulus	47.74201
Z Section Elastic Modulus	247.45078
Y Radius of Gyration	2.22999
Z Radius of Gyration	8.74906
Principal Bending Angle (rad)	1.57079
Principal Bending Angle (deg)	89.99971
Principal Moment of Inertia (max)	3835.59608
Principal Moment of Inertia (min)	249.18219
Reference Elastic Modulus	210000000.00000
Reference Poisson's Ratio	0.33333
Y Coordinate Extent	31.00000
Z Coordinate Extent	7.12500
Y Shear Coefficient	1.47731
Z Shear Coefficient	9.98545
YZ Shear Coefficient	0.00008
Torsional Constant	401.99583
Warping Constant wrt Shear Center	1856.54586
Warping Constant wrt Centroid	7269.40885

the main stem of the spline. In the hat section example, the weld splines connect to spline 9, which has the same thickness and the same slope at the connection, so that no patch mesh is made. The end of the welding is visible in this mesh, because at the point where elements are no longer attached, the side lengths of the upper layer of elements differ form those of the lower layer.

Visible inspection of the mesh is not sufficient to ensure that the welding has taken place. If the specifications under the `Welds` keyword are omitted entirely in the input data file, the resulting mesh will look identical to the mesh shown in Fig. B.10. The connection of the weld nodes will not have been made, however, although these nodes are geometrically coincident. The bandwidth minimizer will work its way through its algorithm, since it works for disconnected graphs. The Cholesky factorization, which is employed in the solution of system equations, will fail, because the system stiffness matrix will not be positive definite, and execution will stop at this point.

A key role is played by the bandwidth minimizer in improving the efficiency with which the finite element system equations are solved. The mesh is initially made in separate parts and these submeshes are later connected together to form the complete mesh. The node numbering in this process is, in general, far from optimal, and the bandwidth of the resulting system of equations can be quite large. For this hat section example, the initial bandwidth is 802, which is reduced to 53 by the bandwidth minimizer. The bandwidth algorithm used in the program is described in George and Liu (1981).

B.6 OPEN CURVED SECTION

Figure B.11 shows an open curved section, whose median line is a fourth-order spline. The input data file for this section is listed in Table B.11. The default uniform knot vector is used; hence, no knot vector is specified in the input. The default weight values of unity are employed. In meshing such sections, large or abrupt changes in curvature may cause the mesh generator to fail. Such failures are usually visible in the mesh drawing, because elements will overlap or become triangular. In some cases, the Jacobian determinant of elements become negative at the Gauss quadrature points, which indicates that the mesh cannot be used. This is a detected error, but will not stop execution. In some cases, the elements around the high-curvature area are distorted such that it is not possible to traverse the eight edge nodes in the clockwise direction. This is also a detected error, and the program will print an error message. Execution will continue so that, if possible, a mesh drawing is produced to make the meshing error visible.

For very complex median line curves, a data file of points along the curve may be available rather than the NURBS curve description. It is then necessary to construct a NURBS curve representation of the data. This can be done by a fitting method, using either interpolation or approximation. Several practical fitting algorithms are given in Piegl and Tiller (1997).

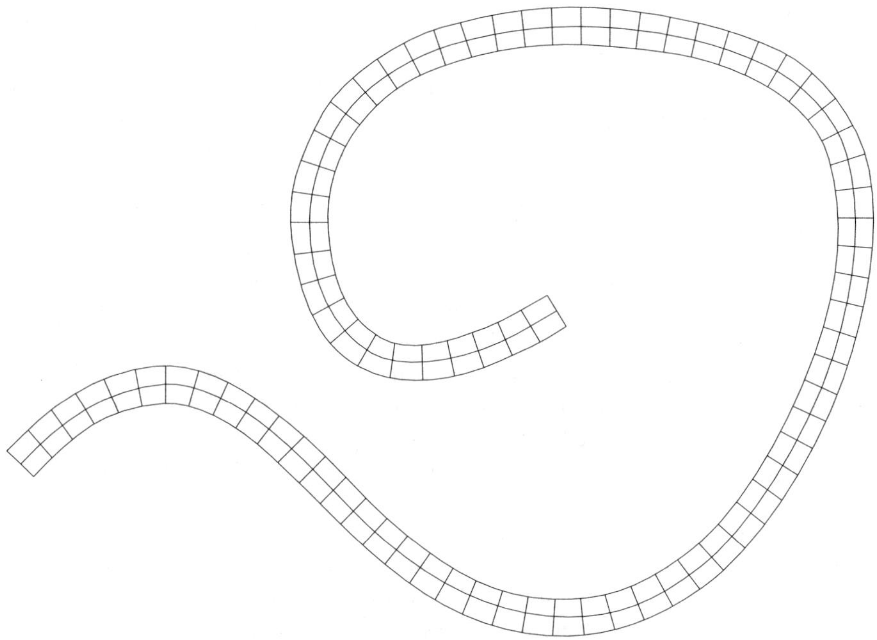

Figure B.11 Open curved section.

TABLE B.11 Input Data File for the Open Curved Section

```
Title: Open Curve with Fourth-Order Spline

Graphics PageWidth 5 End Graphics

Vertices
 1       0       0
 2       5       5
 3      10      -5
 4      20      -5
 5      25       8
 6      20      12
 7      10      12
 8       7       6
 9      10       1
10      15       4

End Vertices

Splines

Branch 1  Thickness 1   Order  4
Nodes 1  2  3  4  5  6  7  8  9  10  End Nodes End Branch

End Splines
```

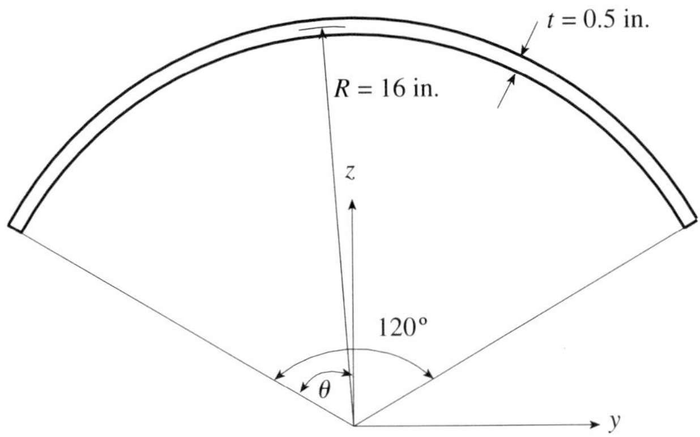

Figure B.12 Circular arc of 120° and 16-in. radius.

B.7 CIRCULAR ARC

Consider a circular arc of 120°, radius 16 in. (Fig. B.12). The user coordinate system has its origin at the center of the circle, with the z axis vertically upward and the y directed to the right. The section is symmetric with respect to the z axis. The input data file is listed in Table B.12. The mesh generated for this circular arc is shown in

TABLE B.12 Input Data File for the Circular Arc

```
Title: 120 Degree Circular Arc Radius=16

Graphics PageWidth 5 End Graphics

Vertices
1          13.8564064605510193      8
2           9.23760430703401525    16
3                             0    16
4          -9.23760430703401347    16
5         -13.8564064605510193      8

End Vertices

Splines

Branch 1
Thickness 0.50  Order 3
Nodes 1  2  3  4  5  End Nodes
Knots 0  0  0  0.5  0.5  1  1  1  End Knots
Weights 1 0.866025403784438597 1 0.866025403784438597 1  End Weights
End Branch

End Splines
```

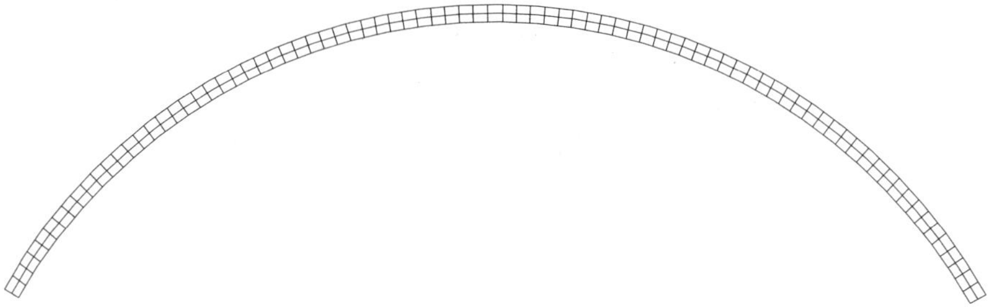

Figure B.13 Mesh for the circular arc.

Fig. B.13. An algorithm for generating a circular arc as a NURBS curve is given in Piegl and Tiller (1997).

The cross-sectional properties calculated are shown in Table B.13. The approximate engineering theory of thin-walled beams gives the shear center of this section at

$$z_S = 2R\frac{\sin\theta - \theta\cos\theta}{\theta - \sin\theta\cos\theta} = 17.84 \text{ in.} \tag{B.8}$$

where R is the radius and θ is half of the subtended angle, which in this example is 60°. In using Eq. (B.8), the angle should be in radians. The value of z_S is very close to the calculated value listed in Table B.13.

An approximate formula for the warping constant gives

$$\Gamma = \frac{2}{3}R^5 t\left[\theta^3 - \frac{6(\sin\theta - \theta\cos\theta)^2}{\theta - \sin\theta\cos\theta}\right] = 1014 \text{ in}^6$$

where the wall thickness $t = 0.5$ in. The computed value of $\Gamma = 1046$ in^6 is about 3% larger.

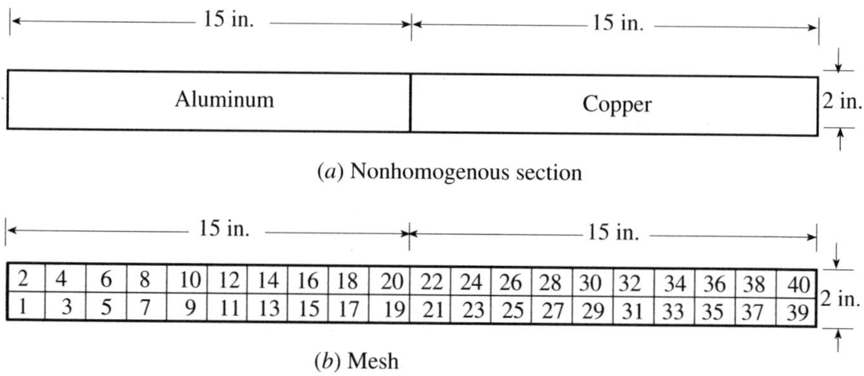

Figure B.14 Rectangular strip made of two materials.

TABLE B.13 Results File for the Circular Arc

120 Degree Circular Arc Radius =16
Cross-Sectional Properties

Cross-Sectional Area	16.75516
Y Moment of Area	221.72054
Z Moment of Area	0.00000
Y Centroid	0.00000
Z Centroid	13.23297
Y Shear Center	0.00000
Z Shear Center	17.83662
Y Shear Center wrt Centroid	0.00000
Z Shear Center wrt Centroid	4.60365
Y Shear Center wrt Centroid (Trefftz)	0.00000
Z Shear Center wrt Centroid (Trefftz)	4.60364
Moment of Inertia I_y	3032.21070
Moment of Inertia I_z	1258.15764
Product of Inertia I_{yz}	0.00001
Moment of Inertia I_{yC}	98.18931
Moment of Inertia I_{zC}	1258.15764
Product of Inertia I_{yzC}	0.00000
Polar Moment of Inertia	1356.34695
Y Section Elastic Modulus	18.32584
Z Section Elastic Modulus	89.40279
Y Radius of Gyration	2.42079
Z Radius of Gyration	8.66549
Principal Bending Angle (rad)	1.57080
Principal Bending Angle (deg)	90.00000
Principal Moment of Inertia (max)	1258.15764
Principal Moment of Inertia (min)	98.18931
Reference Elastic Modulus	210000000.00000
Reference Poisson's Ratio	0.33333
Y Coordinate Extent	8.14583
Z Coordinate Extent	8.37500
Y Shear Coefficient	1.50823
Z Shear Coefficient	4.60034
YZ Shear Coefficient	0.00000
Torsional Constant	1.38355
Warping Constant wrt Shear Center	1046.49221
Warping Constant wrt Centroid	140100.77369

TABLE B.14 Input Data File for the Example Section

```
Title: Two-material rectangular cross section

# Dimensions in inches

Materials
ID   1   Elastic   10.4e6   Poisson 0.3           #Aluminum
ID   2   Elastic   18.5e6   Poisson 0.3           #Copper
End Materials

Vertices
 1      0     0
 2     15     0
 3     30     0
End vertices

Splines

Branch   1
Thickness   2   Material  1
Nodes   1   2   End Nodes
Order    2
End Branch

Branch   2
Thickness  2    Order  2  Material  2
Nodes    2   3  End Nodes
End Branch

End Splines
```

B.8 COMPOSITE RECTANGULAR STRIP

A nonhomogeneous rectangular cross section is shown in Fig. B.14a. The left half of aluminum is meshed into elements 1 through 20 and the copper right half has elements 21 through 40 as indicated in Fig. B.14b. Table B.14 gives the input data file. The results are discussed in Example 5.13.

REFERENCES

George, A., and Liu, J. W. (1981). *Computer Solution of Large Sparce Positive Definite Systems*, Prentice Hall, Upper Saddle River, N.J.

Piegl, L., and Tiller, W. (1997). *The* NURBS *Book*, Springer-Verlag, New York.

Pilkey, W. D. (1994). *Formulas for Stress, Strain, and Structural Matrices*, Wiley, New York.

INDEX